D1624050

FOR THE RECORD

FOR THE RECORD

David Cameron

HARPER

An Imprint of HarperCollinsPublishers

FOR THE RECORD. Copyright © 2019 by David Cameron. All rights reserved. Printed in the United States of America. No part of this book may be used or reproduced in any manner whatsoever without written permission except in the case of brief quotations embodied in critical articles and reviews. For information, address HarperCollins Publishers, 195 Broadway, New York, NY 10007.

HarperCollins books may be purchased for educational, business, or sales promotional use. For information, please email the Special Markets Department at SPsales@harpercollins.com.

Published simultaneously in Great Britain in 2019 by William Collins, an imprint of HarperCollins Publishers.

FIRST U.S. EDITION

Library of Congress Cataloging-in-Publication Data has been applied for.

ISBN 978-0-06-268784-5

19 20 21 22 23   LSC   10 9 8 7 6 5 4 3 2 1

For Samantha

Contents

List of Illustrations

Drinking a beer with China's president Xi Jinping (Kirsty Wigglesworth/ WPA Pool/Getty Images)

Meeting Donald Tusk and Jean-Claude Juncker (Yves Herman/AFP/ Getty Images)

Inspecting the renegotiation documents with Tom Scholar and Ivan Rogers (Liz Sugg)

Addressing students and pro-EU 'Vote Remain' supporters (Christopher Furlong/Getty Images)

Watching the EU referendum results come in (Ramsay Jones)

Nancy, Elwen and Florence Cameron writing a letter for the incoming prime minister (Andrew Parsons)

Preparation for the final appearance at Prime Minister's Questions (Andrew Parsons)

The last official visit as prime minister (Chris J Ratcliffe/WPA Pool/Getty Images)

With family before leaving Downing Street (Andrew Parsons)

Visit to Alzheimer's UK (Edward Starr)

Every effort has been made to trace copyright holders and to obtain their permission for the use of copyright material. Where not explicitly referenced, the pictures are sourced from the author's personal archive. The publisher apologises for any errors or omissions in the above list and would be grateful if notified of any corrections that should be incorporated in future editions of this book.

Foreword

It is three years since the referendum on Britain's membership of the European Union. Not a day has passed that I haven't thought about my decision to hold that vote, and the consequences of doing so.

Yet during that time I have barely spoken publicly about it, or about any issues around my premiership. The reason is that I wanted to let my successor get on with the job. It is hard enough being prime minister – let alone one who has the momentous task of delivering Brexit – without your immediate predecessor giving a running commentary.

That silence has inevitably let certain narratives develop, for example about my motivations for holding the referendum and about my departure from Downing Street. There has been analysis of aspects of my premiership – from the campaign in Libya to the schemes that have helped so many people to buy their own home – with which I have disagreed deeply.

But my discomfort at not being able to respond to these things is nothing compared to the pain I have felt at seeing our politics paralysed and our people divided. It has been a bruising time for Britain, and I feel that keenly.

Yet just as I believe it is right for prime ministers to be allowed to get on with their job without interference, I also believe it's right for former prime ministers to set out what they did and why – and to correct the record where they think it is wrong.

Fortunately, I kept a record during my time in the job. Every month or so, my friend and adviser, the journalist Danny Finkelstein, would come to the flat above 11 Downing Street where I lived with my wife Samantha and our three young children. Danny and I would sit on the sofas in the bright sitting room that overlooked St James's Park, as he gently quizzed me about recent events.

Those recordings have helped me write this memoir, just as scribbles in a notebook or recordings on a dictaphone have assisted others. I sometimes quote directly from the recordings because they provide such an insight into how I felt at the time. Hearing them back – and writing this book – has helped me to understand how I feel about it all now.

A friend once asked Margaret Thatcher what, if she had her time again, she would do differently. There was a thoughtful pause, then she answered: 'I think I did pretty well the first time.' I don't feel quite the same. When I look back at my career in politics, I do have regrets. Lots. Not every choice we made during our economic programme was correct. There were many things that could have been handled better, like the health reforms. What happened after we prevented Gaddafi slaughtering his people in Benghazi was far from the outcome I'd have liked. The first parliamentary vote on intervention in Syria was a disaster.

And around the EU referendum I have many regrets. From the timing of the vote to the expectations I allowed to build about the renegotiation, there are many things I would do differently. I am very frank about all of that in the pages that follow. I did not fully anticipate the strength of feeling that would be unleashed both during the referendum and afterwards, and I am truly sorry to have seen the country I love so much suffer uncertainty and division in the years since then.

But on the central question of whether it was right to renegotiate Britain's relationship with the EU and give people the chance to have their say on it, my view remains that this was the right approach to take. I believe that, particularly with the Eurozone crisis, the organisation was changing before our very eyes, and our already precarious place in it was becoming harder to sustain. Renegotiating our position was my attempt to address that, and putting the outcome to a public vote was not just fair and not just overdue, but necessary and, I believe, ultimately inevitable. I know others may take a different view, but I couldn't see a future where Britain *didn't* hold a referendum. So many treaties had been agreed, so many powers transferred and so many promises about public votes made and then not fulfilled. With all that was happening in the EU, this could not be sustained.

Of course there are many who welcomed the referendum – and that was reflected in the overwhelming vote for it in Parliament and in the high public turnout when the time came. Far from being a flash in the pan, the referendum was announced more than a year before the 2014

European elections and more than two years before the general election. It was set out clearly in a manifesto that delivered an overall majority in the House of Commons.

But I know there are those who will never forgive me for holding it, or for failing to deliver the outcome – Britain staying in a reformed EU – that I sought. I deeply regret the outcome and accept that my approach failed. The decisions I took contributed to that failure. *I* failed. But, in my defence, I would make the case, as I think all prime ministers have, that especially when you are in the top job, *not* doing something, or putting something off, is also a decision.

And that is the thing that stands out for me when I look back over this time: decision-making. A prime minister these days is constantly in contact with their office by email, text and messenger services and is therefore making decisions, large and small, almost by the minute. They also, as I did over the EU referendum, consider the biggest decisions over months, even years. It's the most difficult, most stressful, yet most rewarding part of the job. In many ways, it *is* the job (and I almost called this book *Decisions* for that very reason).

Indeed, so many of Britain's problems we found when we came to power in 2010 were a result of decisions that had been put off. The government had spent and spent while the deficit and debt were left to grow and grow. Low pay and high taxes had been plugged by an ever increasing benefits system. Educational standards were sliding, but masked by increasingly generous grades. More and more people were going to university, but the system was becoming unsustainable. Big calls on infrastructure were avoided while new superpowers raced ahead. Immigration went up and up but without the control, integration and public consent that is needed to sustain such rapid changes to our society. Businesses were hamstrung by regulation and economic growth was excessively concentrated in the south-east. And yes, as I've said, Britain's unstable position in a changing EU was the biggest can kicked down the longest road.

In order to confront these issues we had to do several bold things. We had to modernise the party and make it electable once again – not with a modest change to our image but a full-blown overhaul of who we were, the issues we addressed, how we conducted ourselves and what we had to say to people in twenty-first-century Britain.

Then we had to do something just as bold: form the first coalition government since the Second World War (unpalatable for many in our party) *and* make it endure (impossible, according to many commentators). We then had to fix the country's finances after the worst crash in living memory. At the same time, we were bringing troops home from Afghanistan, while facing down security threats at home and around the globe.

None of these things was inevitable. They happened because we made them happen. Indeed, many things happened – as this book will show – simply because I got a bee in my bonnet about an issue and got the bit between my teeth (and like most modern politicians, mixed my metaphors along the way).

The youth volunteering programme National Citizen Service is something I am often stopped about in the street – and it was an idea I dreamt up many years ago. Technology is changing our world for the better in healthcare, finance, development, transport, the environment and much else besides, and – partly because of the support we gave in government – the UK is in the vanguard of all things 'tech'. The UK is also leading the world in dementia research and care – and putting it on the global agenda all started when, as an MP, I realised the extent and implications of diseases like Alzheimer's. Britain is one of the few countries to meet and keep its promise to the poorest in the world by spending 0.7 per cent of its national income on international aid and development. It's something I've always felt passionately about and wouldn't relent on in government. In fact, these four things – volunteering, tech, dementia and aid – have been my focus outside politics over the last few years.

For all the dissatisfaction with the futility of politics and the failures of politicians, the progress we made on them in government proves you can make a difference.

We were practising politics in the early twenty-first century, at a time when that dissatisfaction – with politics, with an entire global system – was on the rise. 9/11 and 7/7 led to a sense of physical and cultural insecurity. The 2008 financial crash led to economic insecurity. People looked to those in authority for answers. But all they saw were people in power failing – from MPs fiddling expenses to journalists hacking phones and bankers gambling on our global economy.

Much of what we did in government was focused on combatting economic insecurity. We helped create a record number of jobs, cut taxes for the lowest paid and substantially increased the minimum wage.

To address security concerns we established the National Security Council, backed our intelligence and security services and sharpened the focus on combatting all forms of Islamist extremism. Not just the appalling violence, but the poisonous narratives of exclusion and difference on which it feeds.

However, it was in the field of dealing with the sense of cultural insecurity that we failed most seriously. I support a world with global institutions and rules, and fundamentally believe that – on the whole – this is in Britain's interests. But an impression has grown that the interests of our country, our nation state, are on occasion secondary to some wider global or institutional goal. Most of the time this is nonsense. Occasionally – and the European Court of Human Rights is the most prolific offender here – it is correct. We should have done more to override this when true, and challenge it when not.

Most importantly, we failed to deliver effective control over levels of immigration in to our country and to convey a sense that the system we were putting in place was in the national interest.

Those who share my enthusiasm for free markets, open economies and diverse societies have got to recognise that none of these things will endure unless we deal with the insecurities – and demonstrate that doing so is absolutely vital to making our country more prosperous. The debate now seems to be 'pro globalisation' versus 'anti globalisation'. My point is that we have to listen to the genuine arguments of those who are 'anti' if we are to preserve what I believe we all ought to be 'pro'.

Readers might wonder why I have dedicated so much space to the early years of opposition and modernisation of the Conservative Party. It all seems rather distant, even irrelevant to today's troubles – hoodies, huskies, the Big Society are literally 'so 2008'. I disagree. It may be tempting to respond to these desperate times with desperate measures – to become louder and more extreme in our answers. But I believe the opposite is required. I look back at the approach we were taking in opposition, during the early years of this young century – moderate, rational, reasonable politics – and I realise those things are more important than ever.

In these difficult, disputatious times, as this young century reaches its twenties, I passionately believe the centre *can* hold. The centre is *still* the right place to be – a bold, radical, exciting place to be (which is why another working title for this book was *Right at the Centre*).

Winning the 2015 election after five years of coalition, difficult economic decisions and bold measures, like legislating for gay marriage, was proof that commanding the rational, centre ground can deliver good government and good politics too. It is the approach – in my view – that should be applied to Brexit. The most sensible, most rational (and the safest) approach would be to seek a very close partnership with the organisation that will remain our biggest source and destination for trade, as well as a vital partner for peace, security and development. Our aim in delivering the outcome of the referendum should be, as I put it in this book, to become contented neighbours of the EU rather than reluctant tenants.

I have tried throughout the book to mention as many people as I can who worked with me over the years, from those who mentored me when I was a young researcher starting out in politics, to my own special advisers when I was PM. I am sorry to anyone I've missed. I am so proud of you all – not only of what we achieved together but what so many have gone on to do, in finding centre-right answers to the biggest problems we face, from climate change to poverty, modern slavery, an ageing society and more.

I also want to thank those who helped me in writing this book. Danny Finkelstein, who listened to me download my thoughts over the years and helped me shape my arguments when the time came to write about it all. Jonathan Meakin, whose research and fact-checking capacity at times seemed equivalent to an entire government department. Arabella Pike at HarperCollins and the late Ed Victor, who enabled me to turn my proposal into a book and navigate what was for me a new world of publishing. Special thanks go to all those people who contributed, commented and reviewed various drafts – especially Nigel Casey, Peter Chadlington, Kate Fall, Andrew Feldman, Rupert Harrison, George Osborne, Hugh Powell, Oliver Letwin, Ed Llewellyn and Liz Sugg. The biggest thank you by far is to Jess Cunniffe, who first interviewed me on the campaign trail for a Milton Keynes newspaper, came to write my speeches in Downing Street and, eventually, helped me to write these memoirs.

I have been so lucky in so many ways in my life – I haven't tried to hide that in the pages that follow – but my greatest fortune has been to find a partner who has been the love of my life, my best friend and my rock. All these years on, I am *still* in awe of her. So I dedicate this book to Samantha. And I pay tribute at the same time to both our families and our friends. Being a spouse, friend, sibling, parent or child of someone in the public eye isn't always easy – particularly when they're prime minister, and even more so when they've held a controversial referendum. I want to recognise everyone, particularly Chris and Venetia Lockwood and Mary Wynne Finch, who have been so supportive during my time in politics, and since.

Sometimes Sam and I talk about how things would have been different if I had stayed on as prime minister for three months after the referendum – as I intended when I announced my departure. This is something that is not really discussed by commentators, but I think it is significant. Had I stayed on for that period, I would have had the chance to explain many of the things people wanted me to explain – the things I wanted to explain. I might have been able to help set the tone for what followed and for the early stages of our departure from the EU. But the 2016 leadership contest collapsed and I didn't get the chance to do so. Instead, it looked like I was beating a hasty retreat. Which I wasn't. As I set out later, having campaigned so passionately to remain in the EU, I would have had no authority or credibility to deliver the result of the referendum. The country needed a new prime minister. It would have been impossible for me to do the job.

So this book is my chance to say what I wanted to say then and what I want to say now. It is not a historical diary, or a political potboiler of who said what to whom and when. It is my take on my life and my political career done my way. It is to help us understand the past and give us some pause for the future. It is for us today, and – I hope – for posterity. It is *For the Record*.

1

Five Days in May

On Friday, 7 May 2010 I woke up in a dark, modern hotel room opposite the Houses of Parliament feeling deeply disappointed.

I had led the Conservative Party for half a decade, modernised it and steered it through a gruelling general election campaign. We had won more seats than any other party – more new seats than at any election for eighty years. We were the largest party in Parliament by far.

But it wasn't enough. For the first time in decades that glorious, golden building across the Thames was 'hung', because no single party had reached the absolute majority needed to form a government.

That wasn't just a blow to my party, it was – in my view – a blow to Britain. The country had just suffered the worst recession since the Second World War. Banks had been nationalised, businesses had folded and unemployment was climbing to a fifteen-year high. Just a few days earlier, Greece had been bailed out by the EU and the IMF. Athens was ablaze, our TV screens filled with images of protesters burning tyres and clashing with riot police in response to the austerity the bailout demanded.

Not only was our economy entwined with those on the continent. Our budget deficit was projected to be 11 per cent of GDP – the same as Greece's. We also needed dramatic reforms, and couldn't go on spending as we had. A stable, decisive government was more important than ever.

Yet we were far from that now. And while thirty million people had voted, what happened next would be largely down to just three of them: the serving Labour prime minister, Gordon Brown; the leader of the Liberal Democrats, Nick Clegg; and me.

So much has been written about the days that followed that election result. Documentaries, books and even films have catalogued every meeting and every moment, every twist and every turn. What can I add?

Well, the emotions I felt. The things that motivated me, and people who influenced me. An insight not just into the rooms in which events took place, but into my mind when the decisions were made. In short, what it was like to be right at the centre during that extraordinary time in British politics.

So, Friday started with disappointment. We had failed to win some of the seats we should have won – and failed to seal the deal with the British people. Thirteen long years of opposition still weren't over.

Of course, there was also a sense of relief. I had travelled 10,000 miles in the past month, trying to squeeze every last vote out of every marginal constituency, culminating in a twenty-four-hour length-and-breadth tour of Britain. I was exhausted.

The previous day, my team and I had met at the home of Steve Hilton, not far from my constituency home in the village of Dean, West Oxfordshire, and talked about the electoral outlook. Steve and I had worked together at the party's headquarters, Conservative Central Office, during our twenties. He had become renowned as a left-field thinker of the centre-right – passionate, bold, volatile, magnetic, and I'd made him my director of strategy. He was also a close friend to me and my wife, Samantha, and godfather to our first child, Ivan.

The magic number was 326: that was how many seats were needed for an absolute majority. But I knew all the marginal constituencies well, and I just didn't see us winning them all. I predicted we'd end up with between 300 and 310 seats.

One person who had come to the same conclusion – and we often reached the same conclusion – was George Osborne, shadow chancellor and chief of our general election campaign. Five years younger than me, he was my partner in politics: urban while I was more rural, realistic where I would sometimes let ideas run away with me, and more politically astute than anyone I'd ever met. He impressed me every single day.

The final tally of Conservative MPs was 306. While that was more or less what I had expected, what did surprise me was that the Lib Dems – in many ways the stars of the campaign, after Nick Clegg's initial success in Britain's first-ever TV election debates – had done worse than predicted, and lost seats. Labour – despite its unpopular leader, despite being obviously tired after thirteen years in power, despite having presided over the biggest financial crash in living memory, and despite many forecasts to the contrary – had done better than predicted.

I was surprised, too, by the ambiguity of the result. Whenever people had asked me beforehand what I would do in the event of a hung Parliament, I said I would do what democracy dictated. I thought that the result would point to an obvious outcome. If we were the largest party, we would form a minority government or – less likely – a coalition. If Labour was the largest party, it would do the same.

But that Friday morning I realised things hadn't turned out like that. Democracy hadn't been decisive, so I would have to be.

I was alone in that hotel room. Samantha, heavily pregnant with our fourth child, had gone home to get our children, Nancy and Elwen, ready for school. I ran through all the permutations. All I could think when I considered each was what my dad used to say to me: 'If you're not sure what to do, just do the right thing.'

A Conservative minority government was one clear option. With the most seats, we had a real claim to govern. But it would mean six months or more of playing politics day after day, trying to create the circumstances for a successful second general election. And at a time when the global economy was in peril, I knew instinctively that it would be the wrong option.

In any event, there was another real possibility: a 'rainbow coalition' of Labour, Lib Dems and other minor parties, which together constituted an anti-Tory majority. I knew that some in our party would say, let them get on with it. Wait while they forge a shaky alliance and then watch it collapse, forcing a new general election in months.

But as the instability of that morning stretched into the distance, I felt it would be wrong to help inflict such an outcome on a country that needed direction. At this time of national need, stability was paramount.

Another option was a Conservative minority government propped up by the Lib Dems through a 'confidence and supply' agreement. It would be less precarious than a minority government, but far from stable or effective. We would never be able to pass all the reforms that were so desperately needed.

They were needed not just to fix our broken economy, but to mend our broken society. Thirteen years of Labour had left us with a school system that, despite the beginnings of worthwhile reform, encouraged mediocrity. We had a welfare system that discouraged work, a health system that was struggling under the weight of new demands and bureaucracy, and a criminal justice system that undermined social responsibility. For all the money they had thrown at problems, Labour had neglected the family, patronised the elderly, and ignored some of our most ingrained ills, from addiction to abuse. In opposition we'd spent five years preparing to put these things right, but I didn't think a minority government with only a confidence and supply deal would be up to the task.

The final possibility was forming a full coalition between the Conservatives and the Lib Dems. Yet the Lib Dems were ideologically and historically closer to Labour than to us. Plus, minor parties never fared well in coalitions. What Lib Dem leader would be prepared to take such a risk?

Step forward Nick Clegg. His party, and its predecessor the Liberal Party, had been out of power for nearly a century, but his brand of sensible centrism and personal charisma gave it the biggest chance in decades to return to the forefront of British politics.

And what Conservative leader would want to join forces with a party that we had just been fighting ferociously for seats across much of the country, and that was seen by Conservative Party members and MPs as both left-wing and opportunistic?

Well, that would be me. I'd been MP for Witney in West Oxfordshire for nine years, and leader of my party for five. For most of my adult life I'd worked for the Conservative Party. I felt that my years navigating the British political system made me a match for this difficult task.

But more than that, I felt the courage of my convictions. I'd had about three hours' sleep over the last couple of nights, yet I saw with complete lucidity what needed to happen. It wasn't the obvious thing to do, but it was the right thing to do. I bounded out of bed and summoned my team – not to *ask* them what we should do, but to *tell* them.

The election result didn't feel like an accident, I said. Something different had happened, because people wanted something different. Parliament hadn't been hung for thirty-six years. I was advocating something that hadn't been done in peacetime for 150 years: forming a full coalition.

I called the 'big beasts' of the Conservative Party to inform them of my approach. John Major, the last Tory leader to have won an election, eighteen years previously. Former leaders like Michael Howard and Iain Duncan Smith. Party grandees, and my leadership rivals from five years earlier, Liam Fox and Ken Clarke. And the candidate who had made it into the final two with me, David Davis.

The feedback was overwhelmingly that it would be right to reach out to the Lib Dems, although there was the odd exception. 'Davis thinks it's a bad idea,' I reported to my team after I had hung up the phone. 'Which means I'm probably on the right track.'

Then Nick Clegg appeared briefly on the TV. He had led his party to new heights in the polls, and then, as I have said, lost seats. Still – and politics can be so strange like this – he found himself holding the balance of power. He stayed true to what he had said before the election: that if there was a hung Parliament he would talk first to the party with the largest number of seats. The door to power opened a crack.

Soon afterwards, the actual door to power – the big, black one with '10' on it – was flung open and Gordon Brown came out into Downing Street. He was ready, he said, to talk to the Lib Dems once they had spoken to us. I had thought that he would in some way concede that Labour had lost the election, and set the scene for his departure. George laughed at the suggestion: Brown, he said, would have to be prised out of No. 10 as he clung to the railings by his fingernails. He was right.

Fortunately, some of the spadework for a possible coalition with the Lib Dems had already been done. Before the election I had sanctioned George to compare our manifestos and prepare the ground for a deal with the potential kingmakers alongside my chief of staff, Ed Llewellyn. Diminutive and quietly spoken, Ed derived his authority from his intellect, decency and experience, having been chief of staff to Chris Patten in Hong Kong before the handover, and to Paddy Ashdown in Bosnia after the war.

They would work on this with Oliver Letwin, the West Dorset MP and the party's policy chief. Oliver was kind, endearing and clever. He may have looked like an old-fashioned Tory MP, with red corduroy trousers and matching complexion, but no one had been more influential in helping me develop my brand of 'modern, compassionate conservatism' over the past five years.

I hadn't taken part in any of the coalition preparation. I wanted to be single-minded about winning, and not to dissemble if people asked me what I had done to prepare for a coalition.

A huge amount would rest on the speech I would give, and we chose St Stephen's Club as the venue. Commentators made much of the fact that overlooking me was a portrait of Winston Churchill, the last prime minister to lead a coalition, in his case during the Second World War. But it was the ghost of another great PM, the club's first patron, Benjamin Disraeli, whose presence I really felt.

'England does not love coalitions,' Disraeli famously said. In many ways, I agreed. I had made endless speeches about supporting our electoral system because it produced decisive results and strong governments. In Europe it often took months to form a government – months of political instability that recession-battered Britain could not afford. But I felt that, given our circumstances, coalition really was the right choice – and I believed I could make it work.

I stepped up to the lectern to make my pitch. A strong, stable government that had the support of the public to take the difficult decisions was, I said, needed to put the country back on track. I didn't use the word 'coalition' – I didn't have to. It was clear that a coalition was on the table from the fact that I specifically talked about going beyond a confidence and supply deal.

I went through the key elements of the Lib Dem manifesto, and set out where we could 'give ground' and 'change priorities', giving prominence to cutting carbon emissions, raising the tax threshold for the lowest-paid and speeding up the introduction of a 'pupil premium', so schools with children from the poorest homes would receive more money. I indicated that we were also open to political and constitutional reform, which was hugely important to the Lib Dems, who had long campaigned for changes to the voting system.

The approach was generous and front-footed. We were making concessions before discussions had even started – and we were doing so in public.

I phoned Nick Clegg from our party's base, now known as Conservative Campaign Headquarters (CCHQ). He was keen to progress. He told me his four negotiators, and I named mine. William Hague had morphed from a much-caricatured party leader into a heavyweight shadow foreign secretary and an indispensable sage in my inner team. He would be joined by George Osborne, Ed Llewellyn and Oliver Letwin, making up the perfect quartet to secure a deal.

The first talks took place that evening in the Cabinet Office at 70 Whitehall – tantalisingly, tauntingly close to 10 Downing Street, where Gordon Brown was still holding firm. It was a key decision not to include civil servants in the discussions, as we thought that would enable us to get a deal without getting lost in problems and details.

Ed would call and update me with his usual cloak-and-dagger whispers that I could only half-hear. It turned out the Lib Dem team was pleasantly surprised by our concessions, and our team was pleasantly surprised at their willingness to go for a full coalition.

From then on there was a permanent pack outside 70 Whitehall: cameras, reporters, protesters, and the odd bemused tourist. The world was watching too. The pound had plummeted that day to a one-year low, and the markets wanted reassurance.

It was like waiting for a new Pope. When would the next signal come? What colour would the smoke be? Blue and yellow? Red and yellow? I had absolutely no idea.

The next morning, Saturday, I woke up at our home in North Kensington feeling positive. I weaved through the throng of cameras outside my house and went to buy the papers from the local shop. 'Squatter Holed Up in No. 10', said the *Sun*, depicting Brown as a fifty-nine-year-old man refusing to leave the central London property.

In an awkward twist, I came face-to-face with Brown and Clegg later that day as we marked the anniversary of VE Day. It was sixty-five years since the veterans lining Whitehall in their berets and bowler hats had liberated Europe and democracy had triumphed. And here the three of us were, the embodiment of democracy in its messiest form. As a testament to the confusion, some of the veterans even greeted me as 'Prime Minister'.

Before we were led to the Cenotaph to lay our wreaths, Brown started to engage Clegg about the discussions they had clearly already begun over the telephone. It felt inappropriate. 'He's still having a go at me,' Clegg whispered to me.

Our own conversation came later in the day. It wouldn't be our first interaction. Purely by accident, we had a good talk at the opening ceremony of the new Supreme Court in 2009. While Brown and the Queen undertook the formalities, Nick and I talked politics, families and life. He was only three months younger than me, and our lives were very similar. We shared a liberal outlook and an easy manner. I left thinking, what a reasonable, rational, decent guy.

As we sat down that Saturday night in a dingy room in Admiralty House, one of the government buildings on Whitehall, we discussed how we'd given the press the slip. Underground car park, I said. Switching cars outside the Home Office, he said.

We went through our two manifestos, and talked about compromises. But the detail was for the negotiators. For us, it was about the bigger picture – and it was about trust. We agreed that we could and should work together. There was a mutual recognition that we would both be judged forever on whether we could make something unprecedented work at a time when our nation needed it most.

We were both taking a big risk. For me, the risk would be angering those in my party who would not tolerate being in coalition, and might turn against me. But given the history of coalitions for minor parties, he was taking a greater risk.

'If we go for this I'll make it work,' I said to him. 'I'll make the deal a success, and I'll make it last.' I meant it, and I think he could see that.

Not only were the negotiations going well, but I felt confident in our position. If anyone had won the election, we had. We were the open ones, the democratic ones, the ones who were reading the national mood and responding to the public's wishes. A full coalition remained the lead option. A confidence and supply deal was just a fallback. That's why my mood the next day, Sunday, was calm.

Because I wasn't in the negotiating team, I tried to do some of the ordinary things I would do on a Sunday to get a sense of normality back into my life. I played tennis. I went shopping. I cooked for Sam and the kids while getting updates from that day's negotiations.

The updates were relatively reassuring. Crucially, it seemed the Lib Dems were willing to support a programme of spending cuts, including immediate ones. Without that, it would have been hard to form a stable government with clear purpose. The Budget affects every policy decision, and you have to see eye-to-eye on that. But given their voter base among public-sector workers, particularly in education, this willingness would damage the Lib Dems enormously.

In the early stages, the decisions for us weren't as difficult. We dropped our pledge to cut inheritance tax, something we could reluctantly but easily sacrifice. The hard stuff was still to come.

With things going well, I held a drop-in session for MPs. Some were less than keen on the idea of coalition. A group of backbench Tory MPs who tended to be on the anti-modernising end of the spectrum –

I referred to them as 'the usual suspects', because you were never surprised if they rejected any move to modernise, or rebelled on votes in Parliament – urged me to go into minority government and call a general election as soon as possible. Others simply said I should let the opposition parties form a rainbow coalition. I was undeterred: a full coalition was the right thing to do.

But when I met Clegg in my office that evening, something had changed. Though the negotiations were progressing, voting reform remained an obstacle. I had been offering an inquiry, but that wasn't enough for the Lib Dems, and the teams were now talking about the whole deal only in terms of confidence and supply. Perhaps that was the best we could do. I signed off on the wording of such a deal that Sunday night.

Then, at 11 p.m. I called Clegg from my Commons office. He'd had a meeting with the prime minister. Brown had made an offer on voting reform – to hold a referendum on implementing the Alternative Vote (AV) system, a sort of halfway house between the current first-past-the-post system and full proportional representation, where voters would rank candidates. AV did avoid the biggest problems with PR. Under it, every constituency would still have an MP, and every MP a constituency. But my party would find it extremely hard to stomach, and so would I. Most importantly, I didn't think the public wanted it either.

However, I realised that if we were asking the Lib Dems to make a political move they wouldn't have imagined possible, we would have to consider things we didn't imagine possible. Legislating directly for AV, of course not. But a referendum? That might be possible. After all, if one of my primary objections to AV was that the public didn't want it, a referendum would test that.

That late-night phone call had been set up by our aides to confirm that a full coalition was off the table, and we were now only looking at confidence and supply. But Clegg and I both went off script. 'Why are we doing this?' we asked each other. We agreed that we should try again to go the whole hog. I said I would have another look at an AV referendum, and push my party towards a full coalition.

By Monday, though, I was utterly dejected. The soaring hopes of the morning before had been trampled, as the Lib Dems signalled their annoyance at the lack of movement on voting reform.

Worse was to come. When I met Clegg again in Parliament he said that Brown had now offered him a deal that was *better* than a referendum on AV. This confirmed what I had been hearing from colleagues and press contacts: Labour was throwing everything at staying in power, even talking to the Lib Dems about changing the country's voting system to AV without asking the country.

I knew how hard it would be for Clegg to resist a full coalition and AV. But I also knew that Brown himself remained a huge barrier to a Lib–Lab deal. I appealed to Clegg as a democrat: 'You can't go with the guy who's just been voted out.' And I appealed to him as a rational human being: 'You know you can't work with him, but you know you *can* work with me.'

I gathered the shadow ministerial team in my Commons office for the second time that day. We hadn't met in the nearby Shadow Cabinet Room since before the election, because I said we'd never go in there again. I am not a superstitious person, but we needed all the luck we could get, even if it did force party grandees to perch on chair arms and tables.

I outlined the Lib Dem proposals for a referendum and a deal. 'We've *got* to offer something substantial on voting reform,' I said. 'And we've got to offer a full coalition.'

As we talked, Brown appeared on the television screen behind us. He said he would step down before the Labour conference in the autumn if that was what it would take for the Lib Dems to agree to a deal. It was a kamikaze mission. He was taking away one of the biggest obstacles to a Lib Dem deal with Labour.

Now it was clear what was at stake if we didn't move.

Still, Chris Grayling, the shadow home secretary, and Theresa Villiers, shadow transport, said that we shouldn't go ahead with the Lib Dems. But Andrew Lansley, the shadow health secretary, Theresa May, the shadow work and pensions secretary – even David Mundell, who said he would lose his seat under AV – spoke in favour. Eric Pickles, the Conservative Party chairman, said in his laconic Yorkshire voice, 'Go for it.' He was echoed by the education spokesman and former journalist Michael Gove, an intellectual force in my inner team and a close friend. George Osborne agreed, adding that an AV referendum was essential if we were to persuade the Lib Dems to support us.

The chief whip, Patrick McLoughlin, a former miner and a veteran of the Margaret Thatcher and John Major governments, put it bluntly: 'We have to live in the real world. Labour and the Lib Dems would be a legitimate government, and would command support in the country, especially with a new leader. We need to grasp this opportunity with both hands.'

I agreed. And I felt we had enough agreement round the room to proceed.

But I remained dejected. Brown's gambit had changed everything. By sacrificing himself, I felt a Lib–Lab coalition was becoming inevitable. And while I was winning round my shadow cabinet over an AV promise, I wasn't sure I could win over the party. 'Put the pictures back up on the wall,' I said as I walked out of my office, where everything had been packed up in bubble wrap, ready to be taken across the road to Downing Street. 'It's not going to happen.'

But even when things looked as hopeless as they did then, I knew I mustn't stop trying. I went for one final push by paying a visit to our backbenchers' forum, the 1922 Committee.

Along with the florist, the hair salon and the shooting gallery, the 22, as it is known, is one of many surprising features of Parliament: a trade-union-style meeting comprising, of all people, Tory MPs. Rather ominously for what we were about to embark on, it was named after the year Tory backbenchers decided to end the Lloyd George-led Liberal–Conservative alliance. It can often be a leader's toughest audience.

'Look,' I said. 'Brown's going. *And* they're offering a full coalition. *And* they'll go all the way on voting reform. The very least we can offer is a referendum on AV. It is the price of power. *Are you willing to pay the price?*'

I went home with the party's backing for what I was contemplating, but I still felt that it wasn't going to go our way.

'Would you mind if I went on leading the party in opposition?' I asked Sam. We had been talking about how a rainbow coalition would barely have a majority, and a shambolic government with a short shelf-life would need to be held to account.

'Of course,' she said. 'You must carry on.'

On Tuesday I woke to a text from Ed, telling me to call him. He informed me of the latest – maybe even the final – twist: the talks between Labour and the Lib Dems had broken down. We were back in the game. The negotiators got back to work, this time armed with our final big concession. My mood shifted once again, this time to anticipation. This might actually come off.

Things moved fast. That afternoon I was in my Commons office thrashing out the details with Clegg. We were still trying to establish how we'd reconcile our parties' very different approaches to Britain's nuclear deterrent, Trident. The word then came from the cabinet secretary that Brown wasn't leaving No. 10 tomorrow, he was going right now.

Before the sun went down – he hadn't wanted to leave in the dark – Brown resigned. I watched him addressing the cameras in Downing Street on the TV in my office, knowing that the time had come.

As I left Parliament for the final time as leader of the opposition, it wasn't my car waiting outside the Commons to take me to Buckingham Palace, but the prime ministerial Jaguar. 'You've worked so hard,' Sam said as she and I got in. We were both emotional. I was trying to savour the moment when my phone rang.

It was Gwen Hoare, my childhood nanny. Now eighty-nine, Gwen remained very much part of the family. 'How are you getting on, dear?' she asked. 'Well,' I said. 'I'm actually on my way to see the Queen.' Sam and I burst out laughing at the wonderful timing of it all.

I'd been to the palace in the past, but its splendour seemed brighter than ever as I arrived for this moment. I had met the Queen, too, and this time I was as awestruck as ever. However, she put me at ease immediately. Then came the formalities. I said I'd like to form a government, but I wasn't entirely sure what type of government it would be. I hoped, I added, that it would be a coalition.

She had seen it all during her fifty-eight-year reign – wars, crises, scandals, new dawns. But she had never seen the sort of five-day delay that had preceded her twelfth prime minister's entrance to this cere-mony of 'kissing hands' (no hands are actually kissed). I promised to report back on the true nature of the new government as soon as I could.

As our car pulled into Downing Street the sky was getting dark, but the street was lit up by camera flashes. A rainbow formed over us – welcoming not a rainbow coalition but the first Conservative-led government for thirteen years, and the first coalition government in seventy years.

There aren't many things that make me nervous, but this bank of cameras outside No. 10, the fact that this was No. 10, the fact I was now prime minister, was suddenly overwhelming. But Sam's presence calmed me. Also calming were the people who were on the street but out of shot: my team.

There was Ed Llewellyn and his deputy, Kate Fall, who had worked with me at the party in our twenties and joined me when I was an MP campaigning for the leadership. I valued her emotional intelligence and judgement more than anyone's.

There was Steve Hilton and his sparring partner Andy Coulson, the former *News of the World* editor who I'd appointed communications chief three years earlier. A question mark remained over whether he'd join us at No. 10 or move on. I very much hoped he would come.

There were Liz Sugg and Gabby Bertin, who had got me from A to B, fended off the press and made everything happen over the past five years. Laurence Mann, Kate Marley and Tim Chatwin had served me loyally for much of my leadership, and they were there too.

I made my way to the microphone stand in front of the famous black door. As on many previous occasions, I was going to deliver my words without notes.

'Compared with a decade ago, this country is more open at home and more compassionate abroad,' I began, wanting to strike a different, magnanimous tone by paying tribute to the good things Labour had achieved. 'I think the service our country needs right now is to face up to our really big challenges,' I went on, bracing people for the measures that were urgently needed to fix the economy. 'Real change is not what government can do on its own. Real change is when everyone pulls together.' This was the Big Society, the idea from which all our reforms would flow, being put front and centre of our programme. And I finished with a defining principle: 'Those who can, should, and those who can't, we will always help.' I had come up with this earlier, while talking with Steve. I would end up using it as a guide for much that I tried to do in that building, repeating it in my head like a mantra during those lonely moments when I was forced to make the most difficult decisions about people's lives.

Sam and I stepped through the big black door, passing between the civil servants lining the hallway and applauding – the traditional 'clapping in' – as we walked through to the prime minister's office. I felt exhausted, elated – but strangely at ease.

Not at ease in an entitled, born-to-rule sense. But because there is such a warmth from all the people in that building – and, for me, at least some familiarity. I'd been in No. 10 in my twenties as a young researcher and a special adviser. I had returned in my thirties for briefings on urgent issues as an MP and leader of the opposition. Now I was back, aged forty-three, as the youngest prime minister since Lord Liverpool in 1812.

But it was time to return to the 1922. As our backbench MPs clustered in the huge committee room, Samantha and I were led in by Patrick McLoughlin. 'Colleagues, the prime minister,' he said. There was an eruption of clapping, stamping and cheering.

Afterwards, I went back to my Commons office to thank the wider team.

I never did go back into the Shadow Cabinet Room.

In many ways, those five days in May were the most surreal and tense of my five years as opposition leader. But looking back, some of the things that looked as if they would hinder our path to power actually smoothed the way.

Take Nick Clegg. During the election campaign he had seemed like a big obstacle: the insurgent with a message of change. But the fact that we had similar temperaments and values, and were thinking the same way when crunch time came, meant that we were able to form this historic union when we had to.

And take Gordon Brown, with his determination to cling on in No. 10. While it seemed like another roadblock to us at the time, his stubbornness pushed us harder towards coalition, and bought us time to thrash things out with the Lib Dems. Had he gone straight away, we would have been forced into power in a minority arrangement that could well have failed.

People have since questioned whether I exaggerated the threat of AV being imposed without a referendum in order to get Tory MPs to agree to offer the Lib Dems something on voting reform. The truth is that I was absolutely convinced that Labour had put it on the table. Why wouldn't they? Brown was willing to sacrifice himself, so surely they were willing to do whatever it took.

Even if they hadn't offered AV without a referendum, they definitely were offering it *with* one. We would have had to match that anyway. Eventually it emerged that what had happened was somewhere between the two. Brown had said that, in the circumstances of an AV referendum, he would throw the full resources of Labour behind a 'yes' campaign. That was more than I was offering, and perhaps accounts for the confusing signals we were receiving at the time.

I am in no doubt that our flexibility and the concessions we were willing to make, combined with the tone we adopted from the outset, made a huge difference in bringing our two parties together.

In many ways, the boldest move wasn't the decision to form a coalition; it was the decision to make it work. There would be many difficult arguments and painful compromises to come. Sometimes there were full-on shouting matches and accusations of bad faith. Like all governments we made mistakes and missteps. But it was to prove one of the most stable – and, I would argue, most successful – governments anywhere in Europe. And I never once regretted the course we had taken.

2

A Berkshire Boy

So let's go back to the beginning.

I suppose every child grows up in his or her own world. You think that what you have is just, well, normal. I wasn't much different. Yet I think I did always know there was something special about it – that I was lucky.

My early years were ones of great privilege and comfort. My parents, Ian and Mary, inherited money and my dad worked hard to make us all comfortable. But the privilege wasn't solely material – it wasn't the wealth that determined the happy childhood, but the warmth. My parents and I shared an uncomplicated and unconditional love, and the simple values they taught me – to have respect for others, to understand the responsibility to contribute, or to 'put back in', as they would say – remain the cornerstone of my outlook on life.

I was born in London on 9 October 1966, and lived as a small child in Kensington's Phillimore Gardens. And then, in 1969, my father bought the Old Rectory, Peasemore, in Berkshire, which I've always thought of as my family home and still do. My older brother Alex lives there now with his family, and my mother lives in a cottage next door.

The schools I attended read like an English upper-middle-class cliché: Miss Emm's Nursery School, housed on a nearby country estate, Lockinge, outside Wantage. Greenwood private preparatory school near Newbury. Then Heatherdown – a classic boys' boarding school, where I went at the age of seven. Then, of course, Eton College. I was following my father, his father and his grandfather . . . as well as my mother's father, and his father . . . you get the picture.

My dad was an extraordinary man, and a huge influence on me. He was born with a pretty odd deformity. Legs that were far shorter than they should have been, no heels and three toes on one foot and four on the other. Sitting down, you would have thought he was well over six foot. Standing up, he was just over five.

Obviously, we children never knew any different, so it didn't seem odd at all. It was only as we got older that we started to understand what a stigma had been attached to disability when Dad was growing up. I remember the shock when he told me as a teenager that his father Donald was so ashamed about the disability that he had forbidden his wife, Dad's mother Enid, from having any more children. Much later, my father's aunt, a wonderfully eccentric woman we called 'Gav' – short for Great-Aunt Violet – told us that after Dad was born she had sat outside the hospital room night after night, worried that one of the other relatives would sneak in and 'snuff him out with a pillow over the head'.

As a result, Dad grew up an only child, with a father who struggled to love him and who would leave his mother for a beautiful Austrian aristocrat, who, just to make things complicated, was married to Great-Aunt Violet's brother-in-law. None of us children ever met our grandfather. Severely diabetic, possibly depressive and quite probably an alcoholic, he died in 1958.

Dad's stories of playing sport at school, determined not to be held back by his disability, were both inspiring and amusing. As hooker in a rugby scrum – or in the similar position, 'post', in the Eton Field Game – he would grab the ball between his short legs, heave himself up with his incredibly strong arms and shout at the rest of the pack to carry him over the line.

Looking back, you wouldn't have had to be a psychoanalyst to predict that his condition, his start in life and his subsequent success would make him the most wonderful 'can-do' optimist. And so they did. He was a glass-half-full man, normally with something pretty alcoholic in it. We all inherited his optimism – and his love of a good drink. But he taught us all more than optimism and a sunny outlook. He believed in hard work and responsibility. I recall him telling me that one of his proudest moments was looking after his mum and buying her a car after she was deserted by his father.

He worked for the same firm, the stockbrokers Panmure Gordon, for over forty years. While 'PG', as he called it, was a partnership, it was also something of a family firm: his father and grandfather had been senior partners before him. Dad himself became senior partner, built the business up and oversaw the company's successful takeover by the US giant

NationsBank during the 1980s 'big bang'. He never retired, and was still buying stocks and shares for a few remaining private clients just days before he died in 2010.

So, family first, hard work, do the right thing, take responsibility. These were all part of his make-up – and things he wanted us to take on too.

Us? When my parents were married they were told that they might not be able to have any children at all. The doctors didn't know if my father's condition was genetic, and Mum had been given warnings that she might not be able to conceive. But in the end there were four of us children. And that was a big part of the happiness: the large, argumentative but loving family. My brother Alex, three years older than me; then an eighteen-month gap to my sister Tania; then another eighteen-month gap to me; and a five-year break before my sister Clare. We were always a tight-knit set of siblings, sharing in each other's triumphs and disasters, and we remain so today.

Dad kept us entertained with his great sense of humour and his eccentricities. He really did believe in fairies at the end of the garden. In later life he commissioned small statues of Oberon and Titania. I have a clear picture in my mind's eye of him tottering off down the garden, even after he had lost both his legs, armed with a whisky and soda so he could spend quality time chatting to them and to any others that might turn up.

He also loved to impose obscure but apparently immovable rules, some based on his own experience, others seeming to come from nowhere. He forbade us, for instance, from becoming accountants, because he had found his own training so boring. Others were more obscure. 'Never sleep with a virgin.' 'Don't get married till you're twenty-six.' 'Never eat baked beans for breakfast.' 'Always travel in a suit.' And the perennial – and probably essential, in a large family – 'Nothing in life is fair.' They tripped off his tongue and made us all laugh, and most of us obeyed most of them, most of the time.

Politics? He followed it, and was an avid consumer of the news, but he was far from being politically active. I still remember being told to get down from the dinner table to go and 'warm up the television' for the 9 or 10 o'clock news. He was one of those who thought in the 1970s that Britain was so close to going to the dogs and collapsing that he started to stockpile emergency supplies in the cellar. It sounds mad now, but there were real fears of a military coup.

In the early 1980s, fears of military takeover were superseded by potential nuclear apocalypse, brought into sharper focus for us by the fact that home was pretty close to both Aldermaston, with its atomic weapons research establishment, and Greenham Common and its soon-to-arrive Cruise missiles. Dad had a theory that when the bomb went off, if you were drunk you would survive the blast and the radiation that followed, but would remain drunk in perpetuity. He loved this theory, and there were endless debates about how many people we could fit in the cellar, and what we would drink first.

I well remember watching films like *Threads*, a Barry Hines docudrama about the effects of a nuclear bomb being dropped on Sheffield, or *When the Wind Blows*, the animation of Raymond Briggs's book about the aftermath of nuclear war. But no one in our family – me included – was ever in much doubt: the Soviet Union were the bad guys; they had a bomb, so we needed one too.

My mother inherited her love of the countryside, and her belief in looking after others and putting back in, from her parents. She combined them with a great brain and a huge sense of fun. Very few women of her generation got the education they deserved, and had the chance to go to university and make the most of their intellectual talents. Mum wasn't one of them. Typically, she has never complained about this. After leaving school she worked at the Courtauld Institute under Anthony Blunt, whom she adored. When he was revealed as a communist spy in 1979, she was so shocked she couldn't sleep at night, and had to resort to sleeping pills. We teased Dad about 'reds' in his bed, not just underneath.

She served as a magistrate in Newbury for over thirty years, coping first with the Greenham Common women and then the Newbury Bypass protesters, including the briefly notorious 'Swampy'. On one occasion her younger sister Clare turned up in court for taking part in the anti-Cruise missile protests and Mum had to step down temporarily. The ethos of public service was something that mattered greatly to her, and I think it rubbed off on all of us. My older brother became a criminal barrister, and my younger sister has worked as a drug counsellor.

There was another key adult in our upbringing, the woman I spoke to on my way to Buckingham Palace that day in May 2010: Gwen Hoare. Yes, just to complete the picture of the old-fashioned, privileged set-up, I had a nanny. She was with our family for over seven decades.

Indeed, she was still living in a small cottage in the grounds of the Old Rectory, Peasemore, when sadly she passed away in June 2019, aged ninety-eight.

To say we loved Gwen as if she was part of the family would miss the point: she *was* part of the family. As well as the love and devotion she had always shown us – as children we would often bump into each other as we crawled into her bed at night – Gwen was a woman of strong values. In later years I used to wind her up by saying she could write *Daily Mail* editorials in her sleep, and that she made Queen Victoria look like a hippy.

Looking back over what I've written, it all sounds slightly old-fashioned and formal, even stiff. It wasn't like that. Unlike many fathers of his age, Dad was very physical – a hugger and kisser. He loved to talk and argue, always with a great sense of fun. The same with Mum. But they were both products of their age: born before the war, growing up during the austerity of the 1940s and 50s, and getting married at the start of the 1960s, before the sexual revolution was in full swing. Manners mattered, waste or excess were thoroughly frowned upon, and 'doing the right thing' was always important. These are values I still admire, and they undoubtedly shaped my politics.

When I tell my children today about the schools I went to, and some of the things that happened in them, it all seems incredibly old-fashioned. For starters, going away to boarding school aged just seven now seems brutal and bizarre. Of course I was homesick at first. I remember having one of those plastic cubes with pictures of my family on that I would look at in bed at night with tears welling up in my eyes.

Dad, as ever, was pretty phlegmatic, but Mum was torn, and later admitted that she only coped after waving me goodbye on the first day by taking a large dose of Valium. Dad would have approved – he was a famous self-medicator, and always had a squash bag full of various pills and potions. He even gave Samantha two Valium the night before our wedding, and advised her to 'Wash one down with a large gin and tonic – and if you don't pass out, have the other one tomorrow.' She happily followed his advice, and sailed serenely through the whole thing.

To say that Heatherdown was antiquated would be underplaying it. At bath time we had to line up naked in front of a row of Victorian metal baths and wait for the headmaster, James Edwards, to blow a whistle before we got in. Another whistle would indicate that it was time to get out. In between we would have to cope with clouds of smoke from the omnipresent foul-smelling pipe clenched between his teeth.

The school was tiny – fewer than a hundred boys – and the gene pool of those attending was even smaller. One contemporary of mine recalls that his 'dorm captains' (yes, we had those too) were the Duke of Bedford and Prince Edward.

The food was spartan. I lost a stone in weight during a single term. There was one meal that consisted of curry, rice – and maggots. In the school grounds were woods and a lake where we could play unsupervised in green boilersuits – it is something of a miracle that no one drowned.

Punishments were also old-fashioned. They included frequent beatings with the smooth side of an ebony clothes brush. If I shut my eyes I can see myself standing outside the headmaster's study, hearing the ticking of the grandfather clock and the thwack of the clothes brush on the backside of the boy in front of me, and feeling the dread of what was to follow.

Prince Edward was an exact contemporary of my brother, and I overlapped with both of them. Alex and Edward became friends, and Alex went to stay at Windsor Castle, even having breakfast once on the Queen's bed. I was madly jealous.

My own first brush with royalty was rather less successful. I was asked to read one of the lessons at our carol service – Isaiah, I think – and Her Majesty was in the front row. I did OK, but crucially forgot to say 'Thanks be to God' at the end. I remembered as I stepped away from the lectern, started to turn back, then realised it was too late to go back, panicked, and said, 'Oh shit.'

When I mentioned this to Her Majesty forty years later, she laughed, but fortunately said she had absolutely no recollection of the incident.

3

Eton, Oxford . . . and the Soviet Union

And then came Eton.

Eton and freedom. This may seem odd when you consider that you are away from home, dressed in a tailcoat, looking like a penguin, and punished severely for any wrongdoing. But when you arrive, the feeling – of having your own room, being allowed to walk around the small town from class to class, cooking your own tea and using your large amounts of free time as you choose – is enormously refreshing.

Another surprising thing about Eton is the extent to which you are able to find your own way. The teaching is first-class, and there is strong academic pressure to be a success in the classroom, and powerful social pressure to be a success on the playing field. But it is – or at least it was – a school that genuinely lets you, indeed encourages you to, forge your own path. The arts school, design studios, music facilities: they are all there for you. For someone like me – a jack of all trades – it suited me perfectly. I loved the place. I made friends. I was happy.

But it was far from all plain sailing. Trouble started brewing for me in my third year due to my growing sense of being slightly mediocre, a mild obsession about being trapped in my big brother's shadow, and a weakness for going with the crowd, even when the crowd was heading in the wrong direction. These things, combined with the temptations of drinking, smoking and thrill-seeking, nearly led to me being thrown out of school altogether.

In my political career I answered questions about drug use in my earlier life by saying 'Everyone is entitled to a private past,' and leaving it at that. But what happened did have a material effect on my career: not so much later, but when I was sixteen. A few friends had started getting hold of cannabis. In those days it was mostly in the form of hash, typically dark brown and crumbly, although occasionally some 'Red Leb', supposedly from the Bekaa Valley in Lebanon, would show up. Instead of popping behind the school theatre for a fag, we started going for a joint.

In my case – comically, as I now look back on it – three of us used to hire one of the school's double scull rowing boats and head off to a small island in the middle of the Thames called Queen's Eyot. Being quite small back then, I was the cox. Once there, we would roll up and spend a summer's afternoon gently off our heads.

This all came crashing down when the 'ringleaders' and so-called 'dealers' – the boys who had brought the drugs into the school – were caught and expelled. My two rowing friends were the first out of the door. I am not naming them now, not least because they've endured repeated approaches and entreaties from journalists to spill the beans on me. They never have.

I was one of the last to be rounded up. Boy after boy had been interrogated. It was getting close to half-term. As a minor offender, maybe I had got away with it? Not a bit of it.

I can still remember where I was sitting – in Jo Bradley's maths class – when the door opened and I was summoned to see my housemaster, John Faulkner, in the middle of the day. This was without doubt the worst moment of my life so far. The housemaster gave me no chance for weak excuses: 'It's no use denying it, David, we have signed confessions from others, and we know about at least one occasion when you took drugs.' The next stage was going to see the headmaster, Eric Anderson.

Eric is a wonderful man who has the probably unique distinction of having taught two prime ministers – Tony Blair at Fettes and me at Eton – and an heir to the throne – Prince Charles at Gordonstoun. He now lives in my old constituency, and we sometimes bump into each other in Chipping Norton or in his village of Kingham, where he lives opposite a pub I am particularly fond of.

The strange thing about that interview was that he seemed more nervous than me. I think he found the whole episode shocking, and he was clearly still coming to terms with the words for various drug paraphernalia. Because I was so keen not to implicate anyone else, I claimed – totally falsely – that I had only smoked cannabis once at Eton, and all the other times were 'at home in the village'. This involved me telling a more and more elaborate set of lies. I am not sure he believed a word I said, but my abiding memory is the moment he asked, 'Yes, Cameron, but who rolled the joint?'

The short-term consequences of my crime were tiresome, but I was so relieved at not being expelled that I would have been happy to accept any punishment. In the event I was 'gated' (restricted to within the school grounds), fined £20 for the smoking element, and made to write out one of Virgil's *Georgics* on the morning of the school's open day, 4 June. This involved copying out line after line of – as far as I was concerned – untranslatable Latin verse

The real punishment was telling my parents. During the course of the 4 June celebrations, which I joined late after having completed my *Georgic*, Mum could hardly look at me, while Dad simply said, in a rather British way, that it would not be mentioned that day, but he would have a serious talk to me in the morning. When morning came he was nursing a hangover, and made rather a mess of it all.

The long-term consequences of my drugs bust, however, were wholly beneficial. This was the shock I needed. First, I knew that one more misdemeanour would mean curtains for my time at Eton. Next, I realised that I needed to stop moping about lagging behind my brother and make my own way. Crucially, instead of drifting academically I needed to make a greater effort. It was time to pull my finger out.

My O-level results were, for Eton, distinctly mediocre. But as soon as I got going in the lower sixth year – 'B block' at Eton – I was a student transformed. I loved my subjects (history, economics and history of art), I adored my teachers, and my results started to improve rapidly.

Great teachers are the secret to any great school, and Eton is particularly blessed. The reason for singling a few out is that they so inspired me – including when it came to politics – that they really changed my life.

Michael Kidson, a wood-block-throwing eccentric, was a superb history teacher who rejected all forms of Marxist determinism and unashamedly taught the 'great men' version of history. He brought the nineteenth century alive. Brilliant but biased, he thought Disraeli was an utter charlatan and all politicians after the fall of Lloyd George, with the exception of Churchill, pygmies. His love for Gladstone was such that when he read the account of the grand old man's death in Philip Magnus's biography, tears streamed down his cheeks.

But while history was a subject I loved, and history of art the one from which I remember most, it was economics and politics that really set me alight. Here was something that was relevant, exciting, intellectually stimulating, and really seemed to matter. Instead of learning about past problems, you could learn the tools to solve new ones. And this was the era of mass unemployment, high inflation and persistent British economic underperformance. More than almost anything, studying what has wrongly been called the dismal science put me on the path to a life in politics.

To me at least, right from the start it was the radical monetarists and free marketeers who seemed to have the new and exciting ideas. There was a radical Institute of Economic Affairs pamphlet we were encouraged to read, 'What Price Unemployment?', which rejected all the old ideas about pumping more government spending into the economy and trying to control wages and prices. I think we were told to read it so that we could critique what was seen at the time as dangerous nonsense. I thought it made pretty good sense.

And so the mediocre sixteen-year-old became a good enough pupil to be awarded an exhibition to Brasenose College, Oxford, in the autumn of 1985 to study politics, philosophy and economics.

There are moments of life you never forget, such as your wedding day and the birth of your first child. To them I would add another: if you are ever in the fortunate position of having one, an Oxbridge interview belongs with those indelible moments. I still shiver at the memory. Three badly dressed and dishevelled dons sitting in front of you and trying to work out whether you are just the product of a good education, or genuinely bright. They were pretty convinced that I was the former, and had a really good go at me.

'Tell us which philosophers you have read.' I reeled off the very few I had read something *about*: Marx, Descartes and John Stuart Mill. 'Yes, and what others?' They waited until I had got to the end of my list, then started grilling me on the last name I had thought of, Immanuel Kant. It was agony.

Where did my fascination with politics come from? There was certainly plenty of politics on both sides of the family. My mother's forebears, 'the Mounts of Wasing', had served in Parliament for over a hundred years, starting off with a seat in the Isle of Wight before moving to Berkshire in the 1800s, where one of them became Newbury's first MP in 1885.

But I don't believe I inherited any political genes. I was influenced instead by what was happening as I came of age.

For anyone who was a teenager in the 1980s, the influence of 'Mrs

Thatcher', as we always referred to her, was massive. You couldn't be neutral about her. In simplistic terms, my generation was divided into those who hated her bourgeois capitalism, her slavish devotion to the US and the warmonger Ronald Reagan and her brutal suppression of the miners' strike, and those who saw her as a brave fighter for economic and political freedom who was determined to modernise Britain and free us from the grip of over-mighty trade unions.

I was securely in the second camp. I believed that what was being done by Thatcher was essential.

In 1984 I took a year off between school and university, during which I worked in the House of Commons for my godfather, the Conservative MP Tim Rathbone. Tim was what was then called 'a dripping wet', a Tory opponent of Margaret Thatcher. He was a great lover of the Conservative Party, Parliament and public service, with a passion for Europe and our membership of what was then the EEC. He had a deep interest in reforming British drug policy and the provision of nursery education. He asked me to carry out research for him on these two subjects, which I found very stimulating.

As well as doing research, Tim let me roam around the House of Commons and the House of Lords, attending select committees, watching what happened in the chambers, and absorbing the atmosphere of the place as it was in the middle of the Thatcher years. I remember the booming voice of Ian Paisley as I got into a lift with him. I recall going to watch Harold Macmillan's maiden speech in the House of Lords, in which he criticised Thatcher's handling of the miners' strike.

From the House of Commons I went to Hong Kong. I was determined to work abroad rather than just travel, and was fortunate in having connections through my father with the Keswick family and the age-old Far East trading company Jardine Matheson. Jardines had made its first fortune selling opium to the Chinese in the 1840s. That association would complicate some of my first visits to China as leader of the Conservative Party and as prime minister. In November 2010 the Chinese were incensed that my team and I were wearing poppies for Remembrance Sunday, suspecting that they symbolised the opium trade, and the two wars Britain fought to keep that trade open, with which they believed my family was personally associated. They insisted that we remove them before our meetings with the president and the premier in the Great Hall of the People. The official Foreign Office advice, delivered at the embassy in whispers accompanied by loud music, under the assumption that the Chinese were listening in to what was being said, was that we should acquiesce. We refused, and there was a brief stand-off during which it looked as if the entire trip would be cancelled. In the end they relented: the poppies stayed on, and the visit went ahead. The Chinese have very long memories.

Hong Kong in 1985 was a pretty nervous place. The joint declaration with China securing the end of British rule in 1997 had recently been signed, and Jardine Matheson had announced the transfer of its head-quarters from Hong Kong to Bermuda. I was sent to the Jardine Shipping Agency, where I found myself the only Westerner in an office of over a hundred Hong Kong Chinese. Our job was to book cargo onto container and other ships, and to look after the ships and their captains as they came into the harbour. My role was that of 'ship jumper', sent out in a small boat, or 'lighter', to board the big liners, meet the captain and ensure that everything went smoothly. It was an interesting job for some-one who was only eighteen, and I learned a lot.

From Hong Kong I travelled to Japan, and then to the Soviet Union. Ever since hearing the Russian Nobel Prize-winning author and dissi-dent Alexander Solzhenitsyn speak, I had wanted to visit the country. I had read about the Trans-Siberian Railway, and thought that would be a good way of doing so. The trip had a profound effect on me.

The train did not set off from its traditional starting point, the historic port of Vladivostok, which back then was a military base and closed to tourists. Instead we began our journey in a dreary grey town called Nakhodka, and joined the traditional route at Khabarovsk. I spent the next six days in 'hard class', sharing a compartment with two male students from East Germany and Russia, and a female student from Japan.

It's hard to convey now just how grim the Soviet Union looked and felt in the mid-1980s. I boarded the train full of excitement about this epic journey, but at the first stop most of the food disappeared as local people rushed on board and bought or bartered everything they possibly could. I had brought some oranges from Japan, as I'd been warned about the shortage of fruit and vegetables. I remember people on the train watching me with fascination as I peeled and ate them, as if they'd never seen these things before.

The well-worn clichés that young Russians and East Europeans were desperate for Western music and jeans turned out to be absolutely true, and these were the first subject of conversation with my two east European travelling companions. But the thing I will never forget is the stories they told me about just how grey and oppressive life was in East Germany and Russia, and how jealous they were of the West. How they knew that their leaders were lying to them, and that the propaganda about their countries' success was nonsense. My new East German friend flicked through my cassette collection and announced that, 'While you have great music records we just have tractor-production records, and they're all lies.'

I was a George Orwell junky, having read *Animal Farm*, *Nineteen Eighty-Four* and *Homage to Catalonia* several times over. I knew the history and the theory, and here was the living proof of communism's total failure. When, a couple of years earlier, Ronald Reagan had called the Soviet Union an 'evil empire', many people in the West had thought he was guilty of crass overstatement. I came firmly to the conclusion that he was totally right. From that train ride onwards I was never in any doubt that in the battle between the democratic, capitalist West and the communist, state-controlled East, we were on the right side. For my political generation the fall of the Berlin Wall in 1989 was a seminal moment, and for those of us who distrusted socialism, and hated what communism had done to eastern and central Europe, it was a moment of great ideological confirmation.

At the end of the epic railway journey – day upon day of mountains, rivers and never-ending silver birches – I met my friend Anthony Griffith in Moscow, and we travelled together around the Soviet Union. Most visitors at that time would be part of an official Intourist-organised group, and because we were travelling on our own we attracted quite a lot of interest from the authorities. We hadn't booked transfers from trains to hotels or anything like that, yet we tended to be met at every station or airport by a man in a long dark overcoat who already seemed to know where we were going.

Our itinerary caused me to make a diplomatic gaffe many years later. I was making small talk with Vladimir Putin at the G8, and told him about my extensive travels around his country. As I reeled off the list of cities I had visited – Moscow, Leningrad, Yalta, Kiev . . . – he stopped me to say, 'Yalta and Kiev are no longer part of my country.' In that moment I glimpsed the intense personal pain that the break-up of the Soviet Union had caused this old-fashioned Russian nationalist.

And it was in what is now Ukraine, on the beach at the Black Sea resort of Yalta, that Anthony and I were approached by two young men. One of them spoke perfect English, the other spoke French and some English. We never discovered what they were doing on a beach that was reserved for foreigners, but we didn't see any harm in accepting their invitation to have lunch and then dinner with them. They lavished vodka, sturgeon and caviar on us. We weren't naïve, and our suspicions increased when they started trying to goad us into criticising Britain and the British government. We made our excuses and left.

Later, when I arrived at university, I asked my politics tutor and mentor Vernon Bogdanor whether I had been right to be suspicious, and he was pretty convinced it was an attempt to recruit us.

As we crossed into a bleak and depressed Romania, most of my books about politics were confiscated by a bad-tempered border guard as 'inappropriate'. We then meandered our way through Transylvania to Hungary, and on to Vienna and Salzburg, where I was finally able to meet the Austrian woman, Marie Helene Schlumberger, who had run off with my now long-dead grandfather. She regaled us with stories of Austria before the war, the Russian occupation (which was only lifted in 1955) and my grandfather – 'my darling Donald' – while plying us with schnapps.

I was happy to be back in the West. It was time to go home, and then to university.

Although I went to Oxford frequently as a child, and although it is the capital of the county I represented in Parliament for fifteen years, I still feel a huge buzz every time I set foot back in the university part of the city.

I felt a great sense of privilege at being able to walk Oxford's streets, study in the university's great libraries and live in a magnificent and historic college. The college system brings people together in a way some other universities fail to do. The tutorial system means you have direct access, either on your own or in a very small group, to some of the finest minds in the world.

When people ask me what I most loved about being at Oxford, it wasn't the politics. I hardly took part. My fascination with politics was developing, but for some reason I didn't want to play at it. I visited the

Oxford Union a few times, and saw stars like Boris Johnson, already a very funny speaker, and masters of debate like Nick Robinson, who would later become political editor of the BBC.

It wasn't the sport that made Oxford special either. I briefly captained the Brasenose tennis team, and we reached the university finals. But the truth is that my teammates were so much better than me that I often had to drop myself from the squad.

My partner as third pair was a law student, Andrew Feldman, who became a lifelong friend. Andrew would raise the money for my 2005 leadership bid, and became chief fundraiser for the party, then its chief executive and finally party chairman. I would argue that he is the best chairman the Conservative Party has had in its entire history. The figures certainly back that up: we took over a party with £30 million of debt and handed it over eleven years later debt-free and with cash in the bank.

In Downing Street I kept reading that I was 'the essay-crisis prime minister', leaving vital work until the very last minute. I will come to how I made decisions as PM a bit later, but that certainly wasn't how I worked at Oxford. While most of my friends had late-night essay crises fuelled with black coffee and cigarettes, I hardly ever worked in the evening, and almost never at night. But I loved the life. I was fascinated by my studies. I made friends. I had fun. I argued. I gossiped. And I fell in love. Lots of times.

I can't, of course, write about Oxford without three dreaded words that haunted me for most of my political life: the Bullingdon Club.

When I look now at the much-reproduced photograph taken of our group of appallingly over-self-confident 'sons of privilege', I cringe. If I had known at the time the grief I would get for that picture, of course I would never have joined. But life isn't like that.

At the time I took the opposite view to Groucho Marx, and wanted to join pretty much any club that would have me. And this one was raffish and notorious. These were also the years after the ITV adaptation of *Brideshead Revisited*, when quite a few of us were carried away by the fantasy of an Evelyn Waugh-like Oxford existence.

The stories of excessive drunkenness, restaurant trashing . . . all these things are exaggerated. I was never arrested. I was never completely insensible from drink. However, it is true that the election ritual was being woken up in the middle of the night by a group of extremely rowdy men turning your rooms upside down. In my case this was made worse by the fact that I had had a party the night before, and there were dozens of empty wine bottles just outside my door. I have a pretty clear memory of walking from my bedroom into my sitting room to find a group of people making a terrible racket, with one of them standing on the legs of an upended table, using a golf club to smash bottles as they were thrown at him.

I can't swear that one of these people was Boris Johnson, but he was certainly a member at the time. Boris has claimed subsequently that he was unable to climb over the wall into my college. I'm not sure I believe his story. But I'm not totally certain of my own, either. So perhaps I should leave it there.

What did I love most about Oxford? I *did* love the work.

Vernon Bogdanor was, and still is, one of the leading experts on the UK constitution, electoral systems and – interestingly – referendums. The opposite of the fusty don in an ivory tower separated from the real world, he was always making us relate political history and constitutional theory to present-day politics.

I was taught economics by the brilliant Peter Sinclair, who could write simultaneous equations on a blackboard using both hands at the same time. His lectures were always packed, as he knew better than anyone how to bring the subject to life. Years later he surprised me by turning up unannounced to help me canvass when I first stood for Parliament, in Stafford in 1997. Peter bounded up to the first door, and told the unsuspecting inhabitant, 'I was your candidate's tutor at Oxford and he really is very clever.' Needless to say, the voter was both baffled and unmoved. I, on the other hand, was very touched.

One of the many things Oxford taught me was how to handle stress. Looking back, it seems unfair that we had just eight three-hour exams, squeezed into little more than a week, to justify our entire three years' work as an undergraduate. In my case there was no dissertation, no coursework, no pre-marking – nothing except for those exams. The stress was quite extraordinary.

Talking about getting a first at Oxford is probably almost as annoying as talking about going to the university in the first place. But psychologically it was an important moment for me.

I had absolutely no idea of what I wanted to do once I left. I certainly hadn't fixed on a political career. Like many others I did the so-called 'milk round', and was interviewed by management consultants, account-ancy practices and a few City firms, although as this was the year after the great stock-market crash of 1987, most of them had pretty much stopped recruiting. One interview was with a young management consultant working for McKinsey called William Hague. He didn't offer me a job – and neither did any of the other leading companies.

Jardines in Hong Kong were keen to have me back, but while I was considering this I saw an advertisement for the Conservative Research Department (CRD), and remembered coming across it when I was working with Tim Rathbone. There is no doubt that the interview that followed, and taking the job that was offered, changed my life even more than going to Oxford. It set me on the path of the political career that the rest of this book describes.

I only really knew I wanted to dedicate myself to politics and pursue a political career once I started working in it. But after that I was in absolutely no doubt. It was a vocation, the only thing I really wanted to do. I wanted to serve. I cared deeply about my country. I believed in public service. And I came to see – and to believe profoundly – that it is through political service that you can make the greatest difference.

4

Getting Started

It is one of the most famous moments in modern British politics. The chancellor of the exchequer is standing outside the Treasury, in front of the cameras, explaining that despite all his efforts and all his promises, Britain is suspending its membership of the Exchange Rate Mechanism of the European Monetary System.

It was a full-scale political disaster. And behind Norman Lamont's right shoulder, there I am.

So how did I come to be there?

If you want to learn about Conservative politics at the national level, there is no better place to be than the CRD. Neville Chamberlain founded it in 1929, and put it under the directorship of a most unusual man: Joseph Ball, half politician, half secret agent. Ever since, it has been a strange mixture of political intelligence service, policy workshop and finishing school for future politicians. Iain Macleod, Reginald Maudling, Enoch Powell, Douglas Hurd, then later Chris Patten and Michael Portillo, were all graduates of this academy. So were three members of my first cabinet: Andrew Lansley, George Osborne and me.

I was hired in the autumn of 1988 to cover trade, industry and energy, which meant following two government departments: the DTI, which was led by David Young, and the Department of Energy, where Cecil Parkinson was making a comeback after resigning over his affair with Sara Keays in 1983.

I liked Cecil enormously. He was a true believer in what Margaret Thatcher was doing, but he also believed politics should be engaging and fun. He asked me to help with his speeches, including the 1988 conference speech that pledged the privatisation of the coal industry.

David Young invited me to his weekly 'ministerial prayers' meeting, where his team gathered without civil servants present to talk about the challenges ahead. This was a great introduction to the many faces of the Tory party. The aggressively Thatcherite (Eric Forth), the ambitious and mainstream (Francis Maude), as well as the unassuming (Tony Newton), the affable (Robert Atkins) and the downright eccentric (Alan Clark). The meetings were a riot of argument and entertainment. When anyone reported back on bad news from the part of the country they represented, Alan Clark would tell them it was their fault for visiting their constituencies and listening to 'real people' in the first place.

It was a great pleasure – and a good decision, given his extraordinary dynamism – to welcome David to No. 10 as an adviser on business and enterprise twenty-two years later when I became prime minister. He was instrumental in delivering the 'Start-Up Loans' proposal that has created thousands of successful new businesses.

While a CRD desk officer learns a great deal about specialised areas of policy, one of the advantages, and challenges, of working there is that before long you have to be an expert on everything the government is doing. And in the process you become professional. Indeed, the things I learned in those years are, I think, part of the answer to the charge that we have too many 'professional politicians' in British politics.

Yes, we need people in Parliament from all walks of life, and with many different life experiences. And the Conservative parliamentary party is far broader in its make-up today than it was ten or twenty years ago. And yes, I gained hugely from the seven years I spent in business, outside the political world. But while politics is a vocation, it is also a profession. There are tools and skills that you need to master. Not just the speech-making, press handling or campaigning, but how you get things done in a political system, how you *make change happen*.

In my case, it wasn't long before I was briefing ministers for vital media appearances. It staggers me today to think of the access to senior cabinet ministers that I had when I was still in my twenties.

Most of the 'big beasts' lived up to their public images. Ken Clarke would spend most of the meeting telling you why the specific government policy you were pleading with him to defend was 'absolutely bonkers'. He would challenge the entire concept of a government 'line to take', and say pretty much what he liked. Twenty years later, when I asked him to join the shadow cabinet, not much had changed.

Michael Portillo – something of a hero to many of us in the CRD, as an alumnus with an apparently glittering future ahead of him – was both ferociously bright and warmly encouraging, as he had done the same briefing job himself. He once told me his key to success on BBC radio's *Any Questions*. Instead of being polite to your fellow guests at the dinner beforehand, and then fighting with them on the airwaves, do it the other way round. Be argumentative and objectionable in private, and then as soon as the microphones are switched on, be the voice of reason and consensus. Meanwhile, your fellow panellists are so steamed up and angry they come across on air as partisan and divisive.

I arrived at the CRD at the high-water mark of Thatcherism. The Conservative Party had won its third consecutive general election victory under her leadership, and she seemed to be at the height of her powers.

We viewed her with a mixture of admiration and terror. The first time I met her was at the CRD Christmas party, when she fixed me with a laser-like stare and asked what I did. After I had answered, she asked about the trade-deficit figures, which had come out that day. I hadn't seen them. At that moment, instant death – or even a lingering painful one – would have been a merciful release.

At around this time Robin Harris, CRD's staunchly Thatcherite director, told us that as there was so little effective opposition from Labour, we would have to provide it ourselves. He meant critique our own work, but it was a moment of supreme hubris. The party did provide opposition to itself, but not in quite the way Robin had envisaged. Within two years Margaret Thatcher was gone.

The history of this period has been written about extensively. An apparently cloudless sky in 1988 soon turned dark. It was the result of an overheated economy and the return of inflation, courtesy of shadowing the deutschmark, keeping interest rates too low for too long, and the encouragement of an unsustainable boom in house prices, partly through Nigel Lawson's 1988 Budget.

Then followed rows over Europe, with Thatcher's Bruges speech – which we in the CRD all applauded – the resignation of Lawson and the fateful decision to join the Exchange Rate Mechanism.

Then the dénouement. The resignation of Geoffrey Howe and the fall of Thatcher. In the middle of all this, there was the Poll Tax. And believe me, I was right in the middle of it all. Europe was the occasion of the Lady's fall, but the Poll Tax was the reason she couldn't get up again.

So many of the team that worked together at the CRD all those years ago ended up, twenty years later, in prominent positions in my government, including Ed Llewellyn, Kate Fall, Steve Hilton, Ed Vaizey and Jonathan Caine. All of us worked for Thatcher and then John Major. The late 1980s and early 1990s shaped us and our thinking. First we were labelled 'the brat pack', because of our age. Later 'the Notting Hill set', even though most of us didn't live there. Inasmuch as there was a clique – and I would argue that every successful politician needs a team – it was a CRD clique.

The fall of the Berlin Wall in 1989 confirmed us in the view that, with our beliefs in democratic politics and market economics, we were on the right side of history. The fall of Thatcher showed us that even the most successful authors of that history were mortal. It was an early experience of profound political trauma. She was the reason most of us were there in the first place.

My own view of her and the situation was nuanced. I was a supporter, but I did feel that 'late Thatcher' began to believe her own propaganda and somewhat lost touch with reality. I was a tremendous admirer of Nigel Lawson, and wanted those two titans to get over their differences.

Most supporters of Thatcher couldn't abide Michael Heseltine. Again, I took a different view. I didn't agree with his views on Europe, but I admired the muscular action he took to back British industry and to transform Liverpool and the inner cities. I liked the One Nation approach on poverty. And frankly, it looked as if he was being proved right on local government and the hated Poll Tax. If we were going to lose the Great One, wouldn't it be better if her replacement was someone with a plan, with passion and with election-winning charisma?

When I became prime minister twenty years later, few people were more helpful to me than Michael Heseltine. He backed the coalition. He gave strength to our regional policy, particularly through his unstinting support for elected mayors and the real devolution to our cities of both money and powers. He rolled up his sleeves, occupied an office in the Business Department, and, in his inimitable way, he got things done.

But back then, when Mrs Thatcher was on the brink, I felt that one of the reasons he didn't make it was a peculiar lack of charm. Not that he doesn't have any – he certainly does – but that he didn't always take the trouble to show it.

When the leadership challenge began, we at party headquarters were all supposed to be neutral. This order was not taken very seriously: most were passionate supporters of Mrs T. But by the end, while I was still loyal, I was unenthusiastic. I could see that her position was becoming untenable. When the new leadership campaign began after her fall, I was content to stay out of it altogether. What mattered, I thought a little piously, was continuing to implement Conservative ideas.

Despite my view that the end would come, when it did, the fall of Mrs Thatcher was still a political tragedy, one that affected all of us. More profound than personal feelings was the political impact: the leader of our country had been treated in a shabby and disloyal way by the very people she had helped to get elected in the first place. The resentments and divisions that this act of regicide created would affect Conservative politics for the next two decades. In fact, they still resonate.

Some of the lessons we learned from her fall were obvious: the importance of loyalty and teamwork; that leaders – particularly in our party – can never take their positions for granted. But there was something more subtle. We revered the *reality* of Thatcher, not the *mythology*.

The *reality* was a brilliantly effective prime minister who changed her country for the better, but who lost touch towards the end and was, in part, the author of her own downfall.

The *mythology* that grew and grew, particularly after her fall, was that she alone was ideologically pure; that she was always right and everyone else wrong; that she never compromised or backed down; and that she only ever did what was right, and never calculated what was politically deliverable.

This, of course, was nonsense. She backed down over many issues, like university tuition fees. She knew when to back off, as when giving in to the miners' demands in the early 1980s. She was a master of political calculation.

The subsequent problem for the Conservative Party in general, and for future Conservative leaders in particular, is simply put. Not only were we following a hugely successful, epoch-defining leader. Not only did we need to heal the divide between those who supported her to the end and those who brought about her fall. We were also being compared to the mythical Thatcher, rather than the real one.

At about this time, the ageing doyen of Fleet Street, Sir John Junor, asked me to supply him with political gossip for his *Mail on Sunday* column, and I duly obliged, seeing it as part of my efforts to expose splits in the Labour Party. He frequently took me to lunch, at which the exchange of information would all be in the other direction. I would sit back and listen to his stories of Beaverbrook, Churchill and Fleet Street before Murdoch, together with his personal obsessions with Princess Diana and Selina Scott.

The journalist Bruce Anderson would fill in for Sir John when he was away, and I continued the service for him, starting what would become a lifelong friendship. Bruce was close to John Major, and recommended to him that he bring me into No. 10 to help sharpen up his performances at Prime Minister's Questions.

This was the big call I had been waiting for, and I can still remember the thrill of walking through the famous black door to join the team that briefed the prime minister for what was then a twice-weekly encounter.

My partner in this endeavour was a rising star in the whips' office, the Boothferry MP David Davis. Fifteen years later we would become rivals for the leadership, but in 1990 he would come to my office very early on Tuesday and Thursday mornings, and we would discuss what bullets we could put into John Major's gun. We worked together well.

Some people look at Prime Minister's Questions, with all its noise, poor behaviour and often heavy-handed prepared jokes, and think it is somewhere between a national embarrassment and a complete waste of time. They miss the point. In our system, prime ministers have to be on top of their game and across every subject. PMQs exposes them if they are not. Weaknesses, failings, uncertainties, lack of knowledge – all these things and more are found out.

Not only does it help hold prime ministers to account; it gives them more power and control by enabling them to hold Whitehall to account. While serving in No. 10, I saw policy being either determined in double-quick time, or fundamentally changed, on many occasions because the spotlight was suddenly shining brightly on a particular area, and credible answers were urgently required. It is one of the mechanisms that makes our system so responsive. You can use it to change policy and override other ministers and departments. I did this a number of times when I became prime minister.

I got to know and like, and admire enormously, John Major. He was a passionate Conservative, but a practical one, not an ideologue. If he was unsure about how he should act on a particular issue, he seemed almost always to default to the decent thing. He had a temper, to be sure – and I was on the rough end of it once or twice – and parts of the job clearly weighed heavily on him. But he was a fundamentally good man.

He was also a very tactile one. I used to arrive early for the briefing sessions and sit at the bottom of the narrow flight of stairs that led to his No. 10 flat, just outside the door to the study where we held the briefing meeting. John had a habit of bounding down the stairs and, with a cheery hello, ruffling my hair.

My main job as leader of the political section of the CRD was taking apart the Labour opposition and preparing for the 1992 general election.

The tale of that election is extraordinary. The Conservative Party had ditched its most successful ever leader, caused inflation to rise, put up interest rates, seen the property market crash and the country tip head-long into recession. Meanwhile, the Labour Party under Neil Kinnock had scrapped some of its most unpopular policies, such as unilateral nuclear disarmament, and seemed hungry for, and perhaps even ready for, power.

Yet we won. And the scale of the victory should not be measured by the small parliamentary majority – twenty-one seats – that John Major achieved. The true scale of his victory was the fact that we were almost eight percentage points ahead of Labour, and he had attracted more votes than any other prime minister in British political history.

To be sure, we didn't expect it. I had a strong sense then that the only person who really thought we would win was John Major himself. He seemed to have an innate confidence that when given the choice, the British people would stick with him.

No one should underestimate the personal triumph for John Major. In the head-to-head with Neil Kinnock, people knew who they wanted as prime minister. But allied to this was the most systematic destruction of opposition policy that I have ever seen in a campaign. The mantra that 'Oppositions don't win elections, governments lose them' was turned on its head. The hubris of Labour's pre-polling day Sheffield rally, and the self-inflicted wound of its shadow Budget, in which Labour promised to raise taxes on people who saw themselves as middle-earners, are well documented. But those of us who were in the campaign team would argue that the costing of Labour's spending pledges, together with the blunderbuss of our advertising campaign, were what made the biggest difference. By polling day everyone knew that a Labour government meant higher taxes.

Norman Lamont's pre-election Budget was a political masterstroke: by stealing Labour's plan for a 20p starting rate of income tax he pushed them into making new tax proposals. So just at the moment they should have been talking about anything other than tax, they walked into our trap.

Election night, when predictions of Labour victory turned to the reality of a Conservative majority, was a moment of pure political joy. While I would experience the excitement of getting elected to Parliament in 2001, and the topsy-turvy night in 2010, the exhilaration of 1992 wouldn't really be matched until May 2015, twenty-three years later.

Victory in the general election, and my relationship with Norman Lamont, provided me with the chance to take the next step in my political career – becoming a special adviser, or 'spad', at the Treasury.

The Treasury today retains much of the power and aura it had back then, but the place I worked in nevertheless seems a world away. Women in white coats would wheel tea trolleys around the so-called 'magic circle' on the principal floor of the Treasury building in Whitehall where the key officials and spads sat. The office I had then – all to myself – was substantially bigger than the one I would have as prime minister.

And many of the rooms – particularly the chancellor's – were genuinely 'smoke-filled'. Norman smoked an endless succession of small cigars. His principal private secretary, later to become my cabinet secretary, Jeremy Heywood was rarely without a cigarette between his fingers. Chief economist Alan Budd and specialist economic adviser Bill Robinson were constantly puffing away. When the deputy governor of the Bank of England, Eddie George, came to see the chancellor he would light up too. Back then I was smoking twenty Marlboro Lights a day, and would happily join in. There were times when you couldn't see the other side of the room.

Going to the Treasury also introduced me properly to William Hague. Elected at a by-election in 1989, he was Lamont's parliamentary private secretary, the first rung on the ladder for a new MP.

William immediately struck me as one of the brightest and most talented Members of Parliament I'd ever come across. He had a huge understanding of the economic and other policy challenges we faced, while knowing his parliamentary colleagues and the complexity of Conservative politics back to front. We formed a friendship that has lasted ever since.

Seismic events were ahead for all of us. The decision to join the ERM – a fixed exchange rate between European currencies – was made by John Major and Margaret Thatcher in October 1990, before Norman Lamont arrived at the Treasury. Our task was to try to make the policy work. We failed.

The story of the end of Britain's membership of the ERM is simply told. Following reunification, the German economy required high interest rates. Following the Lawson boom, the United Kingdom economy was in recession and required low interest rates. It was Germany that drove the European economy, and the mighty Bundesbank had a critical say in the ERM. Naturally, they prioritised German domestic policy. In the end the ERM could not contain this fundamental structural imbalance. That, above all, is what lay behind Britain's exit.

But the ERM wasn't just a story about Britain's economic circumstances. It became an essential proxy in the Conservative war over the burgeoning European Union. Pro-Europeans made the argument then – and some still do now – that Britain joined at the wrong time and at the wrong rate, and if only different decisions had been made,the ERM might have worked. They also argue that leaving it so dramatically was unnecessary, that there was some middle way by which Britain could have been part of a realignment of Europe's currencies, thus avoiding the humiliation of either a very public devaluation of the pound, or the exit that eventually took place. Some anti-Europeans claimed then – and still claim now – that the ERM actually caused the British recession, and was therefore responsible for all the pain it would cause in terms of job losses and house repossessions.

Both these views are, in my view, wrong.

The pro-Europeans miss the real point. Of course it might have been better if Britain had joined the ERM at a different time or at a lower rate, but in the end what did so much damage to the British economy was not the precise exchange rate, but the high interest rates, and therefore high mortgage rates, required by the ERM because of what was happening in Germany. No country ever managed for any sustained period to have interest rates below those prevailing in Germany. So even if we had joined at a lower exchange rate, those high interest rates would still have been necessary.

The argument that a 'middle way', with a more general currency realignment, was possible simply doesn't stand up to scrutiny. All that was effectively on offer was a substantial British devaluation. Economists will continue to argue about whether that could have prevented our forced exit from the ERM. I would simply point out that several other countries had devalued more than once. The problem was interest rates.

The anti-European argument was equally bogus. The ERM did not cause the recession in Britain. It was caused by a rise in inflation at the end of Lawson's period as chancellor, and the need to raise interest rates to bring it under control. It is true that the ERM resulted in high interest rates for longer than was necessary, but most economists agree that that was for a matter of months, not years. The ERM made the recession longer and deeper; it didn't cause it in the first place.

The clear conclusion from all this was that fixed exchange rate systems can put huge pressures on the economies of different countries when their economies have different needs. The real lesson was that what was true for the ERM would be doubly true for a European single currency. It would be the ERM without an emergency exit route. My mantra became 'We cannot join the single currency, because it requires a single interest rate, and sometimes that will not suit us.'

The truth turned out to be even worse. During the Eurozone crisis, effective interest rates in the struggling economies like Spain, Greece and Portugal were far higher than in Germany because, in spite of the lack of a formal exit route from the euro, markets still thought departure was possible, and demanded a premium for funding governments, in the form of much higher rates in the most stricken countries.

William Hague's phrase describing the euro as a 'burning building with no exits' would prove doubly prescient.

By the time of the 1997 election I broke with the official policy that we would 'wait and see' on the euro, and joined the many who took a stronger position. My time in the Treasury had made me a Eurorealist, or a Eurosceptic. That did not, however, mean being anti-European. Norman Lamont and I drafted a pamphlet, 'Europe: A Community not a Superstate', to explain the consequences of what had happened, and the broader lessons for Britain's European policy. Membership of the EU was necessary for trade and cooperation, but Britain had never welcomed, and would never welcome, the political aspects of the Union. We wrote: 'No one would die for the European Union.' No. 10 asked us to take it out. We kept it in.

By this time Norman was in deep trouble. And politicians in trouble need everything to go right for them. They cannot afford any slip-ups, whether self-imposed or externally generated. Unfortunately, the campaign to save Norman's ministerial career got off to a bad start at the party conference in October.

We had spent too long crafting our pamphlet, and not enough time on his crucial conference speech. Getting the balance right, between a degree of contrition about the past and excitement about a future in which we could cut interest rates and generate growth, was a big challenge, which we failed. While the speech's reception in the hall seemed all right, the reaction of even quite friendly colleagues was that it was 'workmanlike'.

Whenever I've heard that word since to describe a performance, I know that what's really meant is 'bad'. And there's a rule with these things: if something is seen as quite bad on day one, it's a disaster on day two, and a career-shortening catastrophe by the end of the week.

The next task for Norman was to formulate an economic framework that would deliver the recovery the British economy so badly needed. Here he was in his element. Because he had seen that our ERM membership might well fail, he was ready to put a new policy in place. A credible domestic monetary policy to support the economy and deliver stable inflation. A tough, long-term fiscal policy to get the budget deficit under control. And supply-side reform to make our economy competitive.

This was pretty much the same medicine my government prescribed twenty years later. It worked well both times. But the right strategy needs the right implementation plan. And that is where we went wrong in 1992.

When you have to take lots of difficult and potentially unpopular decisions – including raising taxes – the trick is to separate those that are painful but deliverable from those that are potentially explosive. Step forward the proposal to put VAT on domestic heating bills. This was a mistake; and in many ways it was my mistake. We took the view then, just as we would in 2010, that we could not fairly and credibly reduce the budget deficit by cutting spending alone. Some tax increases would be needed. I looked carefully at all of the options, and came to the view that some of the zero rates on VAT were ripe for change. Energy prices were low, environmental concerns were growing, and we could protect the vulnerable from price increases through the benefits system.

Not for the last time in my political career, I had failed to spot the essential political equation: rational case versus emotional argument equals political disaster. And it *was* a disaster. We were defeated in the Commons, and had to revert to the much simpler (and less politically toxic) move of a small across-the-board increase in VAT. That taught me a lesson for the future – but it was another nail in Norman's coffin.

Meanwhile, the economic medicine was working. Cheap money, fiscal discipline and competitiveness ushered in a period of growth that would continue throughout the decade.

And so yet another lesson was learned: while, all things being equal, reductions in public spending can have an effect on the overall level of demand in an economy, in practice other things are not equal. Controlling public spending, in an open economy like the UK, helps to lower the exchange rate and support exports, and even more importantly it frees up monetary policy to support the economy.

The most powerful memory of my time in the Treasury is of course watching – and failing to prevent – the end of the chancellor's career.

I liked Norman Lamont immensely, and I still do. He was a thoughtful, intelligent, decent man. But he was also deeply sensitive, with a skin too thin for this sort of politics. We subsequently fell out over Brexit, of course. When I heard that he was coming out for the Leave campaign, I pleaded with him that while I had stayed true to the pamphlet we had written together all those years ago, knowing it was in the national interest to stay and fight, he – outside the responsibilities of office – was now arguing a more populist and easy case.

After the disasters of what became known as Black Wednesday and our departure from the ERM, Norman had travelled to America. When he returned he asked William Hague and me what we thought he should do. One of the reasons he was so against resigning was that he felt – rightly in some regards – that he had seen what was coming, and was warning others about it. And, more than anyone else, once we had left the ERM, he knew what needed to be done.

Could he have recovered his position without the other slips that took place: the reports of singing in the bath, the '*Je ne regrette rien*' remark at the Newbury by-election and the other controversies? Frankly, I doubt it. The truth, as we were all to learn, was that ERM membership may not have been a policy Norman invented, but he was responsible for it – and it failed. And above all, when the 'narrative' in the press changes so fundamentally, it is hard to fight against it.

I tended to be the bearer of the bad news. Indeed, I had to call Norman late one night to tell him about a call I had received from a deputy editor at the *Sun*: 'The good news is that your boss's picture is on the front page of tomorrow's newspaper. The bad news is that his head is in the middle of a cut-out-and-keep dartboard.'

Without going into all the horrors of the months that followed our ERM exit, one story stands out in my memory which demonstrates just how bad things had got. The suffix '-gate' is now appended to almost every so-called 'political crisis', no matter how minor or short-lived. My first serious 'gate' was the so-called 'Threshergate' affair, when the chancellor of the exchequer was basically accused of consorting with prostitutes and lying.

In November 1992 the *Sun* managed to get the details of Norman's credit card. It had – big deal – an outstanding balance. Along with whatever negative coverage could be squeezed out of such an unremarkable fact, one other thing caught the interest of the press: he had spent a small amount of money at a Threshers off-licence in Paddington. The hacks descended on what they assumed was the right shop in Praed Street, where an assistant, a Mr Onanugu, happily told them that the chancellor had popped in to buy some cheap champagne and Raffles cigarettes before heading out into the night in what was then, in part, a red-light district. The newspapers had a field day, with innuendos galore and cartoons featuring champagne bottles and ladies of the night.

After a day of stonewalling we decided we had to get to the truth. Norman told us he *had* been shopping in Paddington, but had only bought two bottles of wine for his family to drink at home. All Treasury business came to a complete halt as Norman hunted through his wallet in search of the receipt, while his wife Rosemary tried to find the bottles of wine in the No. 11 flat. All this time the shop was sticking rigidly to its story. And then came the moment of truth: Norman told us that the Threshers he went to wasn't in Praed Street. The only trouble was that he couldn't remember where it was.

By this stage he was at a European Council meeting in Edinburgh. I recall the absurdity of telephoning to pull him out of important discussions so he could describe the route he had taken that night, and where he had gone into the shop. I followed his directions with my finger on an A–Z, and we both concluded that it must have been in Connaught Street. I despatched an official in a taxi, and hallelujah – there *was* a Threshers in Connaught Street.

After the full pressure of the government was applied – I think it even took a call from the permanent secretary to the Treasury, Sir Terence Burns, to the head of the company that owned Threshers – finally the receipts were found and the puzzle solved. Norman was telling the truth. No cheap champagne. No cigarettes. And no prostitutes.

But even with all this evidence, the press didn't want to believe it. My final memory of the saga is wandering into the press gallery with a colleague and saying, 'For heaven's sake, who do you believe – the Chancellor of the Exchequer of the United Kingdom of Great Britain and Northern Ireland, or Mr Onanugu?' The unanimous cry came back: 'Mr Onanugu!' Today we would call it 'post-truth politics'. Back then it was the moment I should have known we were sunk.

Eventually the summons for Norman did come. In May 1993, John Major said he was having a reshuffle, and wanted to move Norman to be secretary of state for the environment. Norman was livid. I briefly tried to persuade him to stay on and rebuild, but it was no use. He would rather let it be known that he had been sacked by Major.

The end of Norman Lamont meant the end of my time as a special adviser at the Treasury. Terry Burns and Jeremy Heywood put in a good word for me with the new chancellor, Ken Clarke, and I joined his meetings on his first day, and even wrote part of his speech in the House of Commons debate that Labour had called following Norman's defenestration. But Ken's two special advisers, Tessa Keswick and David Ruffley, were keen to have complete control of the political side of the Treasury. So I was summoned to Ken's office and politely fired.

Ken was thoughtful about my future career, and had called up his old friend Michael Howard, now the new home secretary, and secured me a job as one of his advisers.

Michael was a man on a mission to reform the criminal justice system. His analysis, which I came to share, was pretty simple. More work was needed to prevent crime. The police needed to be freed from red tape to catch more criminals. The courts needed reform, so that there were more convictions. And sentences needed to be tougher, to send a clear message of deterrence. So he set us to work.

Patrick Rock was the senior special adviser, and we held meeting after meeting with experts and officials looking through every area where we could change the country's approach to crime. What came out of this was a package of measures that Michael announced in his party conference speech. It was a radical departure. Juvenile detention centres. Reforms to the right to silence. Proper use of DNA evidence. Restrictions on bail. Longer sentencing options. And far greater use of closed-circuit television cameras. These ideas led to an effective period of change which future home secretaries, Labour and Conservative, have generally stuck to. It ushered in a prolonged period of falling crime rates.

It's undoubtedly true that some of the motivation for this frenetic activity was the arrival of a new political figure as the Labour Party's shadow home secretary: Tony Blair.

I remember my first meeting with him. He had proposed an amendment to our criminal justice Bill on so-called 'video nasties'. He clearly cared about the issue, but also recognised that it was a brilliant 'wedge' issue: a Labour politician grabbing a small 'c' conservative theme and using it against a big 'C' Conservative government.

Meeting Tony Blair for the first time, I instantly realised that we were dealing with a different sort of politician. It wasn't just his mixture of charm, intelligence and a touch of star quality; he also struck me as a man with the common touch, full of common sense. This was to prove a lethal combination for the Conservative Party.

I remember exactly where I was on the evening of the day Blair's predecessor as Labour leader John Smith died. I was having an after-work pint outside the Two Chairmen pub in Westminster with Patrick Rock. The news had been shocking and tragic, but the political implications were clear. We looked at each other and said almost simultaneously, 'That's it. Tony Blair will become leader and we're stuffed.'

5

Samantha

Something else happened while I was a special adviser: properly meeting the love of my life – and my wife for the past twenty-three years – Samantha.

I say 'properly' because Samantha was a friend of my younger sister Clare, and we first met when she was just seventeen. I remember being struck by this laid-back, almost silent, waif-like thing lying on my parents' sofa, smoking rolled-up cigarettes and sniggering gently as my sister took the piss out of me.

We met properly on a holiday organised by my father four years later. Dad, who was always incredibly generous, decided to celebrate his and Mum's thirtieth wedding anniversary by inviting some of his best friends to a hotel in southern Italy, and he allowed each of us children to ask three friends along. Samantha was invited by Clare, who warned her in advance, 'Watch out – I think my brother fancies you.'

I did. And it was a blissful week.

I realise that what is meant to follow is a story about love at first sight. Neither of us being in any doubt. An instant recognition that we were partners for life. The truth is that neither of us felt like that. We had a lovely, romantic holiday amidst sunshine, friends, laughter and free-flowing cocktails. But when we got home neither of us was quite sure what would happen next.

Of course we were similar in some ways: brought up only twenty miles apart, with parents who, while of slightly different ages, moved in similar social circles. But our friends on both sides couldn't really understand what we were up to. I was the ambitious Tory apparatchik. She was the hippie-like art student. I was working in the Treasury for Norman Lamont. She was living in a Bristol flat with people who would have happily wrung his neck. I was trying to get invited to highbrow political dinner parties in Westminster. She was playing pool with the rapper Tricky in the trendiest part of Bristol.

Norman would frequently ring up early on a Saturday morning wanting to know what was in the papers. On more than one occasion Samantha, used to a student-style lie-in, would shout from under her pillow, 'If that's Norman asking about the newspapers, tell him to fuck off and buy them himself.' I would call him back, cramming 20p pieces into the student payphone to avoid being cut off.

Our courtship was a long one. Our first New Year was spent driving around Morocco in a battered Renault 5. The first night in Marrakesh was so cold and damp we slept with our clothes on. While there was a bit of an age gap, as well as the contrasts in our friends and our politics, there was something that kept bringing us together and helping us get to know and love each other more.

Part of that something was food. We are both greedy and somewhat obsessive. Restaurants, cooking, shopping, growing: all are part of that obsessiveness, as long as they end in eating.

It was in those early years that I first witnessed the 'Sam food panic' which has since become something of a family joke. When she is hungry she has an irrational fear that the restaurant, pub or shop we are heading to is about to close or run out of food. The panic won't end until I have called ahead to check that the kitchens are still open or the shelves aren't empty. Now at least I have the children on my side: as Sam shouts at us to check that the ice cream shop hasn't run out of ice cream, or the fish and chip shop hasn't run out of chips (both genuine recent 'food panic' examples) we all fall about laughing.

So we fell in love – and it was the deepest love I will ever know. But the falling took months and years, not days and nights; and I suspect it was longer for Sam than it was for me. But I believe the result has been something much stronger than either of us could ever have believed when we first got together under that powerful Italian sun. And it wouldn't just survive everything political life would throw at us, but also the worst fear of any parent, losing our beloved first-born child.

It wasn't until 1994 that I summoned the courage to propose. While Sam said yes, we decided to keep it secret for almost a year. I think she wanted some time to get used to the idea. She was still only twenty-one, so I thought it was only fair.

As our relationship developed, I decided to leave the Home Office and to go and work for the media company Carlton Communications plc. By this point I had applied to be on the Conservative Party's candidates' list, and after passing the assessment weekend – of interviews, practice speeches and written tests at a rather bleak hotel in Buckinghamshire – I was able to apply for parliamentary seats. For some time I had thought that getting more experience outside politics, specifically in business, would be good for me. I also needed to make some money, so my chosen career of politics would be more manageable.

Carlton fitted the bill perfectly. It was a FTSE 100 company, with all the corporate responsibilities that involved. Michael Green, its founder and chairman, was a swashbuckling entrepreneur from whom I would learn about business. The company was a part owner of ITV, and was involved in regulated industries, where my knowledge of government and Parliament would help.

I had always been clear to Samantha that my life would involve politics. I had found my calling – it was what I wanted to do. It wasn't just an option, or a possibility: as far as I was concerned, it was a near-certainty. She was very understanding.

Did we ever argue about politics? Yes, of course. My friends would say she helped to turn a pretty traditional Home Counties Tory boy into someone a bit more rounded, more questioning and more open-minded. But many of the arguments we had about politics were actually about logistics, rather than issues. Samantha worried hugely about how it would affect our life. Where would we live? How would we stay together? How much would we see of each other? She was right to ask all these questions: politics has been a destroyer of many strong marriages. For one person in the relationship it can become an obsession; for the other a duty, or even a burden. As much as you can try to choose your constituency, in the end it chooses you. And so much follows from that choice.

My first attempt at getting selected as a Conservative candidate was in Ashford in Kent, while I was still a special adviser at the Home Office. It was my first experience of a big selection audience of four hundred people or more. Today, nerves help me speak well. Back then, I think they didn't. In the event I was beaten by the far more experienced Damian Green. In third place – and it was the first time I had met her – was a shy yet assertive candidate called Theresa May.

I was torn between taking the traditional route of fighting a safe

Labour seat first, and trying to jump straight into a Conservative one. I pursued both strategies, even applying for Doncaster North, the rock-solid Labour seat that would later select Ed Miliband as its Member of Parliament.

In the end the choice was made for me, when in January 1996 I was called up as a reserve for the selection in Stafford, whose Conservative MP Bill Cash was moving to the newly created neighbouring constituency of Stone. It was a part of the world that I didn't really know, but something clicked. In the second round I would give my speech and take questions last. While we were waiting Samantha and I sat in the Castle Tavern opposite the constituency HQ. I fretted – and she drank. After two pints of cider we were summoned, and Samantha tripped on the way up the steps onto the stage. There was a gasp from the audience, but far from it being a disaster, it meant that everyone remembered at least something from my performance.

I was selected to be the candidate, and over the next year we worked as hard as we could, canvassing and campaigning, including Samantha's eccentric father, Sir Reginald Sheffield, who rather spoiled his hard work by loudly shouting into a mobile phone in a pub on polling day, 'We're about to lose.' But in the face of a nationwide Labour landslide, all our efforts weren't enough. By the end of the campaign, the result – a Labour majority of 4,314 – wasn't a surprise. While I did not have a previous election campaign to compare it with, I knew from the unanswered doors and the looks people gave me in the streets that the British people had had enough of the Conservatives.

6

Into Parliament

1997 was the start of the wilderness years for the Conservative Party. Just like Labour in 1979, we were set for a long period of opposition.

In the British system the blow of losing a general election is partly softened by becoming 'Her Majesty's Official Opposition', with a privileged status in Parliament and 'shadow' jobs to be handed out to ambitious MPs. Pretty soon you learn what a false comfort this is. You've lost. You're out of power. You don't make the decisions or achieve any of the things that made you want to get into politics in the first place. It is difficult to set the agenda, and hard to regain power after just one Parliament. What matters is not how well you oppose, but what you learn – and how fast.

After the 1997 election I was outside Parliament. Eight years later I was leading my party. And five more years after that we were back in power. It was a slow and painful ascent for the party, but an incredibly rapid rise for me.

I wasn't by any means the fastest to understand either the scale of our defeat or the profound nature of the change that was needed to our party. Michael Portillo and the group around him seemed to have thought more deeply than many others about this.

Nor was I the best parliamentary performer, speech-maker or even motivator of potential Conservative voters. William Hague was by far the most talented politician of my generation, and was superb at all of these vital tasks.

I wasn't the master tactician who could best plan the strategic thinking and the tactical moves that would help take the Conservative Party back to power. In those regards, I don't think George Osborne has an equal.

If a master strategist had sat down in 1997 to draw up the ideal sort of person to lead the Conservatives back to power, they would hardly have come up with a privileged Old Etonian who had worked for Norman Lamont when Britain was ejected from the ERM, and whose only experience outside politics was to work briefly for a London-based media conglomerate.

Good timing and good luck would combine to give me a chance after three heavy defeats. My principal opponent would change from the apparently unbeatable Tony Blair to the eminently beatable Gordon Brown. The long period of economic growth that started in 1992 would come to a juddering halt in 2008.

Added to that, I had Samantha, who humanised and rounded me. I had a constituency in Witney – both safe and close to London – that was an excellent springboard. And I had friends and supporters who were ready and able to back me. I learned a lot at Carlton about business, about management, and about people. And it may have been easier, out of Parliament, to see the big political picture.

Opinions differ about how effective Tony Blair's team was at driving through change between 1997 and 2001. However, as a political machine it was without equal. And the truth was that Tony Blair was the post-Thatcher leader the British people wanted. He combined pro-enterprise economics with a more compassionate approach to social policy and public services. He understood that in many ways Britain is a small 'c' conservative country. In opposition, and in government, he rarely gave his Conservative opponents room to breathe. He talked tough on crime, looked strong on defence, seemed concerned about school discipline, even posed as passionate about business.

At a supposedly 'off the record' dinner with journalists in October 2005 I said that just as Tony Blair had understood that he needed to be the 'heir to Thatcher', so we needed to understand the need to be 'heirs to Blair'. By this I didn't mean that we should imitate all of his political methods or adopt all of his policies and political positions. After all, that wasn't what Blair had done in relation to being the 'heir' to Thatcher. He had tossed aside many of her policies, and introduced some profound changes of his own. Out went subsidised private education for bright children from low-income homes, and tax relief for private healthcare. And in came devolution for Scotland and Wales, independence for the Bank of England, and the minimum wage.

But alongside that, he had carefully analysed and understood what had changed in the country since 1979, and therefore which elements of Thatcherism should be maintained and built on. He kept the trade-union reforms, the privatised industries, the low rates of income tax, the commitment to NATO, a largely pro-American foreign policy and a strong defence. These were natural Conservative policies, and by adopting them, Blair locked us out of Downing Street.

What were the equivalent moves for us? This was the question that Conservative Party 'modernisers' kept asking. Getting the answer right was an essential step in returning to power. And returning to power was what we needed to do. Despite his prowess at politics, and those reforms that had benefited Britain, Blair was, overall, taking the country in the wrong direction. It was more than just his policies – the unsustainable welfare system, the dumbing down of education standards, the neglect of some key overseas alliances, the increases in taxes, the overburdening of public finances and the failure to plan for the long-term future of everything from defence to the NHS. It was also about the culture those policies created. Something for nothing. Equality of outcome, not opportunity. Short-termism. I was absolutely not the heir to Blair in any policy or philosophical sense, and I was desperate to clear his government out.

But before I could do that I needed to find a safe parliamentary constituency. By the middle of 2000 I was heading towards the final selection rounds in Epsom, East Devon and Shaun Woodward's seat of Witney. It was Witney that I really wanted.

I knew Shaun from working with him at Conservative Central Office during the 1992 general election. I was well aware that he was wildly ambitious for high office, but I was just as surprised as everybody else when he chose to jump ship and join the Labour Party in December 1999. Surprise soon gave way to excitement: West Oxfordshire was an area I knew quite well, and the constituency would now need a new Conservative candidate for the next election. The town itself was just thirty miles north of where I was brought up, and West Oxfordshire was very similar to West Berkshire – a combination of market towns, attractive villages, rural enterprises, growing businesses and many talented people who commuted either to the university city of Oxford or to London.

The constituency party had concluded, unsurprisingly, that they'd made a dreadful mistake in selecting Shaun Woodward, who had turned out not to be the genuine article. So in an attempt to avoid making the same sort of mistake again, they employed an interesting ruse for the first round of interviews. The mayor of Carterton – the second town of West Oxfordshire after Witney – was a West Indian called Joe Walcott. He had come over from Jamaica during the war to serve in the RAF at Brize Norton, and had stayed on to become a prominent local figure and a passionate Conservative. As the prospective candidates arrived one by one to be interviewed at a pretty manor house in the village of Bampton, they found Joe standing on his own in the garden. Those who ignored him, or assumed he was a gardener or driver, were immediately struck off the list. Fortunately I gave him a warm handshake, and we would remain good friends until he died in November 2018.

Although I was competing against the former MP and highly effective minister Andrew Mitchell, and a talented young businesswoman called Sharon Buckle, I think my passion for the place shone through, and I was selected.

Within weeks I had new friends, a new home and the makings of a strong political base. Peter Gummer, Lord Chadlington, who I knew a little from when we had both worked for John Major, became something of a mentor, renting me a cottage in Dean, near Chipping Norton, the village where we have lived ever since. Together with Christopher Shale, who also became a firm friend and adviser, Peter helped me to rebuild the local Conservative association and its finances.

And I inherited from Shaun Woodward one of the best constituency agents in the country, Barry Norton, who was also the leader of the local district council. Witney born and bred, Barry is a workaholic with a thick Oxfordshire accent, and seemed to know about absolutely everything that happened anywhere in the 250 square miles of the Witney constituency.

While it is hard to characterise a whole local party, the West Oxfordshire Conservative Association, or WOCA as I came to know it, was, rather like its last two MPs, Shaun Woodward and Douglas Hurd, at the liberal, open-minded end of the party.

The general election, which took place in June 2001, was a fairly gentle affair. I spent the campaign travelling from village to village, having lunch in any number of extremely good West Oxfordshire pubs. It was a fun month, and at the end of it I had a majority of over 7,900. I was in.

In the run-up to the election, and for some time afterwards, when people asked me about my ambitions I would say that I wanted to be a

Member of Parliament because I believed it was an incredibly satisfying and worthwhile job. Serving the area, standing up for local people, getting things done, while taking part in debates about some of the big questions facing our country: that was what it was all about. Everything else beyond being a backbench MP – and I always hoped there would be more – would be a bonus.

There was soon the added stimulation and interest of sitting on the Home Affairs Select Committee. I was keen to take risks, and I fully supported the proposal that we look in depth at the issue of illegal drugs. I was later to disavow some of the most contentious conclusions we came to – downgrading ecstasy from class A to class B, for instance. It was, and remains, odd that ecstasy is in the same class as, for example, heroin. But I came to believe that the danger of signalling that certain drugs were more acceptable, or less dangerous, outweighed any benefit from being more scientifically accurate. But there is no doubt this report shifted the dial in terms of moving drugs policy away from criminalisation and towards treatment and education. This was something I would continue to promote as prime minister.

But while I loved the job, my joy at being a Member of Parliament was tempered by the hopelessness of our situation as a party. 1997 was the year of the Tory wipeout, and in 2001 we added precisely one to our historic low of 165 MPs. The Conservative parliamentary party looked very white, very rural, very male, and frankly rather irrelevant. Of the thirty-four new Tory MPs who had made it to Westminster, only one was a woman.

And during that Parliament a whole series of things happened that brought home to me just how wretched our situation was, and how simply waiting for something to happen was a useless strategy.

William Hague's resignation as leader after the 2001 election was sad, but not surprising. He had done his best in almost impossible circumstances. Throughout his leadership the Conservative Party had been divided and fractious, and was still trying, though often failing, to come to terms with defeat. Blair was always likely to be given a second chance by the electorate. But through the force of his performances, both inside and outside Parliament, William had kept the party together and the show on the road.

Because he changed tack partway through the Parliament, backing off from modernisation and returning to more traditional Tory themes such as law and order and Europe, it is easy to represent his leadership as a false start for the modernisation of the party and its policies. I don't think that's fair.

Timing is everything in politics. William did not have the support in the party for modernisation, and given that Blair was then at his peak, even a changed Conservative Party couldn't expect much reward from the electorate. Pressing on might have sacrificed core support without attracting new voters.

In any event, I can't claim any particular foresight: I backed him strongly in both phases. And the pressure on him was spectacular. William said to me some years later, when I was trying – successfully – to tempt him back into front-line politics, that the experience of leading the party after 1997 had nearly broken him.

In the leadership election of 2001 I was a committed supporter of Michael Portillo. I had seen how good he was in office, and the unexpected loss of his seat in 1997 had clearly made him think deeply about what needed to change. Re-elected to Parliament in a by-election in 1999, it seemed that he had a clear plan for change, and for a more liberal party with greater urban support. However, he had so fallen out of love with his own party that he couldn't really contemplate the hard work and compromises needed to reform it. Also, with his hard-man-of-the-right past, he struggled to convince all those who supported the modernising agenda.

And so we were left with Ken Clarke versus Iain Duncan Smith. It was a hopeless situation. One couldn't unite the party; the other couldn't win over the country.

I made what I thought was a rational choice, which was to support 'IDS', because I thought that if Ken won, the subsequent inevitable party split over Europe would be so bad as to make us both a laughing stock and wide open to a revolt from our right. Samantha said I was mad, and voted for Ken. Frankly, I was pretty happy that we cancelled each other out.

My only memory of the entire leadership campaign was of an event I organised in Witney for party supporters to hear John Bercow speaking for IDS and George Young speaking for Ken. Samantha asked Bercow whether he supported his candidate's views that abortion should be restricted and the death penalty restored. It was one of the many moments in our married life when I realised that she had seen the big picture rather quicker and clearer than I had. IDS was always going to be seen as an outdated old clunker.

But two days before he won the leadership contest on 13 September 2001, the world changed.

When the first plane struck the World Trade Center I was at home in Dean doing constituency work. Samantha was in New York starting the process of setting up a new Smythson store in Manhattan. For about four hours I was unable to get in touch with her because the telephone lines were down. I sat with the TV remote control in one hand and my mobile phone in the other, watching in shock and pressing redial over and over again. By the time I got through to her that evening I was staring out of the window on the train to London. Relief.

People now tend to jump straight from 9/11 to the war in Iraq, but that is unfair. Tony Blair's initial response to what had happened that day in New York was masterful. He moved fast, and set the agenda both at home and abroad. He correctly identified the problem of Islamist extremism, the inadequacy of our response both domestically and internationally, and supported – quite rightly in my view – the action to remove the Taliban regime from Afghanistan. Once it was clear that they would not stop al-Qaeda using the country as a safe haven, there was no realistic alternative.

Along with other relevant select committee members, I went to No. 10 for a briefing in late 2001. It was the first time I had been through the famous black door in years, and Blair impressed me then and in the many debates and statements that followed. Even as a relatively tribal Conservative, I felt strongly that at this moment Britain had the right prime minister. I even stopped Blair behind the speaker's chair after one statement in the Commons to say that in his clarity about the threat we faced, he was speaking for the whole country.

But what of Iraq? While anyone with an ounce of reason could see that the regime in Afghanistan was a legitimate target, it was impossible to be quite as certain when it came to Iraq. As I showed in the anguished *Guardian* columns I wrote at the time – I had a regular spot in the paper's online comment pages – I was a sceptic about the move to war.

Saddam Hussein's regime was brutal. He was in breach of countless resolutions passed by the UN, an organisation for which he showed only contempt. His people would unquestionably be better off without him. There was a risk that, left in place, he might start to work more closely with the extremist groups that threatened us. And, after all, he had employed 'weapons of mass destruction' against his own people when he used poison gas on the Kurds.

I bought all of these arguments, and still do, but as I put it at the time: 'We are being asked to swap deterrence with something new called pre-emptive war. I cannot be certain but I suspect that many of us will not support pre-emptive war unless Blair can produce either compelling evidence of the direct threat to the UK or a UN resolution giving it specific backing.'

As the evidence to satisfy the first condition was pretty unconvincing even at the time, and as Blair clearly failed on the second condition, why did a sceptic like me vote for military action?

The convenient answer would be to say I was 'duped' by the various dossiers and the claims about Britain being 'forty-five minutes from doom'. But that's not really the case. They only formed a small part of the reasons I gave publicly and to my many highly sceptical constituents.

I wrote at the time about the consequences of backing away. It would undermine the UK–US alliance. Saddam would win an invaluable propaganda victory. We would jeopardise any chance of a proper, multilateral approach. And, of course, while there was no 'second reso-lution' specifically mandating force, there were over a dozen resolu-tions dealing with Iraq, and the UN would look powerless if they weren't enforced.

Sitting in the Commons, it was also clear that a vote against military action wouldn't stop the war, it would just make it less of an international coalition. The Bush administration was going to have this war, the ques-tion was whether we would be involved.

And I listened to my closest colleagues and friends. Some, like George Osborne, who was a fairly enthusiastic 'neo-con', had no doubts. Others, like Oliver Letwin, who were wavering sceptics like me, decided on the balance of evidence to vote with Blair.

Samantha was totally opposed, and told me to stick with my initial scepticism. But this was a time in our marriage when we talked about politics very little.

Our first-born son Ivan was a year old, desperately ill and in hospital almost as much as he was at home. I would often leave his bedside in the morning after a night sleeping beside him, handing over to Samantha before heading off to the Commons for the next Iraq debate or statement. Less parenting by relay, and more time together, and she might have persuaded me.

But to be truthful, there was something else. I believed that the prime minister was entitled to something approaching the benefit of the doubt. I was all for Parliament voting on going to war – and I would subsequently help to entrench that convention as prime minister – but I don't start from the proposition that a prime minister asks for backing for a military conflict 'lightly or inadvisedly'. Indeed, I believe that if the prime minister comes to Parliament and says effectively, 'We are standing with our oldest allies, fighting a dictator who has brutalised his people, and we risk humiliation or worse if we falter,' then I would try to be supportive.

Assuming that other MPs shared this rational patriotism, or naïvety – take your pick – was to let me down several years later, in the vote on bombing Syria when I was prime minister. I regret what happened subsequently, and we will never know how things might have been if matters had been handled differently. But I take the view that if you vote for something, you should take your share of responsibility for the consequences rather than try to find some formulation to show that you were conned or misled. Without Saddam, Iraq at least has a chance of a better future; although even today it is probably still too soon to say whether that chance will be taken.

It wouldn't be fair to write off Iain's entire period leading the party. He understood that the Conservative Party needed fundamental change. But he wasn't capable of some of the basic requirements of leadership in British politics – building an effective team, performing at Prime Minister's Questions, and delivering big speeches and media interviews.

For PMQs, George Osborne and I were drafted in, together with a bright young staffer, James Cartlidge (now the MP for South Suffolk). From time to time we were joined by Boris Johnson, whose appearances grew less frequent the more obvious it became that we were marooned in the polls and heading for defeat.

They were pretty desperate sessions. Blair was at the height of his powers, and Iain was leaden and dull. Boris asked me after one particularly depressing prep session, 'Hey Dave, what's the plan?' He then grabbed me by the shoulders and said, 'Presumably it's like carrying an injured hooker in the scrum – we know he can't play but we just' – at this point he grunted and heaved me off the ground – 'pick him up and carry him over the line.'

George and Boris saw the writing on the wall much more clearly than I did. I didn't attend either of the IDS party conferences, as on both occasions I had to be at Ivan's bedside at St Mary's Hospital, Paddington. Despite this, I did catch his second conference speech – the one where he declared that 'the quiet man' was 'turning up the volume'. I watched it on an ancient hospital television, but even I could see that the multiple standing ovations were staged and looked ridiculous.

In the end, this right-winger with the potential to unite the party was overthrown by a combination of left and right after losing a fight he didn't need to pick. A government Bill on adoption and children was amended to enable unmarried couples to adopt children, opening the door for gay couples to adopt. It had already passed the House of Commons, but the Lords had rejected the amendment and reinstated the original 'married-only' rule.

Iain tried to whip the party against supporting unmarried couples' right to adopt. A small number of MPs rebelled completely and voted with the government. A larger number ignored the three-line whip – so-called because the whips underline the vote three times on the official notice, meaning that you must support the party line – and abstained. There were only three of us from the 2001 intake who did so: me, George and Boris.

Instead of ignoring the rebellion – as I frequently chose to do as party leader – Iain's lieutenants called an emergency press conference, telling the party to 'Unite or die.' IDS's personal authority was left mortally wounded, with more or less open discussion of plots to oust him. Only the Iraq War, which soon dominated the political discourse, diverted press and political attention from the travails of the Tory leadership.

But by October 2003 the party had had enough. A major donor announced on the radio that he and others were considering abandoning ship if IDS's leadership continued. Given the party's precarious financial situation, this new crisis stampeded the parliamentary party into action. Shortly afterwards a vote of confidence in Iain's leadership was triggered as the chairman of the 1922 Committee received the sufficient number of letters from Conservative MPs.

The day of the vote was also the day of PMQs. For once Boris turned up at our weekly prep meeting, and there was lots of gallows humour, including from Iain, about potential leadership bids. Afterwards, I asked to stay behind for a private word. I pleaded with Iain to resign, and not face the indignity of losing a vote of confidence. George was probably right, though, when he said the deed simply needed to be done.

The arrival of Michael Howard as leader provided yet more lessons in leading, and in losing. Michael handled the technical aspects of the job well. After two years of IDS, there was a sense that the grown-ups and the professionals were back in charge. PMQs was a fight once again. Conferences were well organised. There was a newly effective media operation.

Overall the Michael Howard leadership gave us a fighting chance. The critique of Blair was sharpened: over-regulation was holding back the economy, and over-centralisation was holding back public services. And the government was ignoring vital issues such as crime and immigration, on which Michael Howard could demonstrate both passion and expertise. And yet. Once again it didn't work.

Did I ever believe that we could win in 2005? While I thought we could take away Labour's majority, I was never confident that we could win outright. We simply hadn't won the right to be heard. Nor had we developed a clear enough description of what we needed to do.

Perhaps the biggest lesson of this whole period is something that is both hard to measure, and unfair. People make up their minds about the major party leaders pretty quickly. Iain couldn't escape his image of being old-fashioned, a hanger and flogger, and not quite up to the job. And Michael never shook off the 'something of the night about him' attack by Ann Widdecombe.

My view increasingly came to be blunt: a large share of the voting public had simply written off the Tories after 1997. They weren't going to listen to what they had come to believe was an arrogant bunch of politicians who they believed were more interested in looking after their own interests than anybody else's. And even when people did listen to something we said, they would mark it down, irrespective of whether they agreed with it or not, simply because it was 'the effing Tories' that were saying it.

What followed from this was that government failure, even if on an epic scale, wouldn't see us return to power. Simply put, as bad as Labour were, the electorate thought they were better than the alternative. We needed to prove that we had listened, learned and changed.

I am saying a lot about this period because it forms the backdrop to my later decision to stand for the leadership. Tony Blair and Gordon Brown had, respectively, eleven and twenty-four years in Parliament before leading their party. I had just four.

I had, however, joined the front bench, though the jobs I held before 2005 were not particularly significant. The first rung on the ladder was becoming one of several deputy chairmen of the Conservative Party. Being appointed deputy shadow leader of the House of Commons was only marginally less meaningless. After all, in opposition you have virtually no control of the parliamentary timetable, so there is little enough for the shadow leader to do, let alone their deputy.

But the non-job did give me an opportunity. My boss Eric Forth decided to take a week off one Thursday, and handed the task of Business Questions over to me. I made a reasonable fist of it, with a few funny jokes and a half-decent attack on the government. The parliamentary sketchwriters gave me the thumbs up. These things get noticed.

But the most important lessons from this period came from spending time with a very bright group of relatively young Conservative colleagues, commentators and former staffers who wanted to understand why we kept losing, and what needed to change.

Andrew Cooper, who had become the founder of the market research company Populus, and Daniel Finkelstein, now a *Times* columnist, had joined the CRD shortly before the 1997 debacle. After the 2001 election they had teamed up with George to begin pressing the party to change. In a series of papers, articles and polls they argued that the Tory Party would not win again unless it understood why people had turned away from it. I joined in with this group, and together with others like Michael Gove, Ed Vaizey and Nicholas Boles we began to meet, usually at Policy Exchange, the new modernising Conservative think-tank, and talk over pizzas and beer.

As a genuine, moderate and liberally minded One Nation Conservative, I was an enthusiast for change. At the time I wrote that there were three essential components for a successful modern conservatism: 'First, we need to reclaim the full set of values that makes conservatism whole. I joined up because the Conservative Party combined a message about aspiration – that everyone should be free to do what they could and be what they could – with compassion for the weak, the vulnerable and those left behind. Second, we must look outwards and forwards, not inwards and backwards. Parties should exist to identify and address the modern challenges that our country faces. Finally . . . conservatism is nothing if it is not practical. We need a relentless focus on the things that people care about in their daily lives: the public services they use, the taxes they pay and their hopes and fears about the future.'

In other words, pretty much everything needed to change. Instead of tax cuts, crime and Europe, we needed to shift our focus onto the issues the Conservative Party had ignored: health, education, and tackling entrenched poverty. It simply wasn't acceptable to have so few women MPs, so little representation from ethnic minorities, and such a poor geographical spread of Conservative seats. As I came to believe passionately, words alone do not work; you need positive action. It's no good simply telling talented British Asians or young businesswomen just how meritocratic you are when the first meeting they attend is a sea of white male faces.

And the Conservative Party had to stop putting people off with curtain-twitching moralising. Yes, there were genuine arguments about family breakdown and behaviour that needed to be made, but we were in no position to make them. We had to earn the right to be heard on these and other subjects.

Added to this, we all agreed that it was time for the Conservative Party to make a decisive step in favour of equal treatment for gay people. In 2003, Labour had repealed the law that banned councils from 'intentionally promoting homosexuality'. It was known as 'Section 28', after the clause in which it appeared in the Local Government Act 1988, passed by the Conservatives.

For me at the time, the reason this legislation had been passed was that councils were overstepping their role. What business had a local council promoting sexuality in any form? But by arguing this I was ignoring an even bigger question: what were *we* doing backing what looked like, and what was for many, an attack on homosexuality? As Nick Boles later put it to me, 'It's not about what councils should and shouldn't do. That's not the point. It makes gay people feel like they're worth less.'

In all of this, there was something we agreed shouldn't change: we were all convinced that the Conservative Party had become, and should remain, a Eurosceptic party.

While we were all at that time supporters of the UK staying in the

European Union, we certainly didn't see support for the EU, as it was currently constituted, as in any way 'modern'. But we did believe that 'banging on about Europe' (a phrase I was famously to use a year later) was damaging, because while it was just about in the top ten issues for the British public, it seemed to be the only thing that the Conservative Party really cared about.

The biggest influence on me in all these discussions was George Osborne. He was the most convinced, and the most convincing, moderniser. From the very start we built a genuine partnership of a kind that I believe is very rare in modern politics. We each wanted the other to succeed. There was no senior partner and no junior partner. Above all, what mattered most was trust: we came to know that we could tell each other anything, and it would not be passed on to others, and certainly not to the press.

This relationship, and our shared view of what needed to happen, would become stronger during the general election of 2005. Michael Howard gave us both key roles and ringside seats in the last of the contests that we would fight and lose together.

7

Our Darling Ivan

'You're the first, the last, my everything...' The lyrics of the Barry White song boomed across the operating theatre from a radio. I'd always been a fan of his music, but I was concerned that it was too loud, and the team of doctors and nurses hovering over Samantha wouldn't be able to concentrate.

I needn't have worried. Everything went smoothly. And within minutes I was holding our first-born son, Ivan.

It was 8 April 2002, and we were in Queen Charlotte's Hospital in Hammersmith. Samantha was having an emergency caesarean, because when her contractions started it turned out that Ivan was 'feet first'. In other words he was the wrong way round in the womb, or what they call an 'undiagnosed breech'.

Sam and I had been married for five years, and had built our life together in our house in North Kensington. Neither of us had any regrets about waiting before having children. Sam had the job she had worked so hard for, as the director of design at the Bond Street store Smythson. I had been elected to Parliament, representing a seat that suited me down to the ground. We had taken the risk of borrowing a lot of money to buy a small house in the constituency, in the hamlet of Dean, near Chipping Norton. There didn't seem to be a cloud on the horizon. But our life was about to change in a way we never expected.

When Ivan first arrived, there didn't seem to be anything wrong. With caesarean births, the dad is the first person to hold the baby. Bursting with pride, I squeezed him tight as we crossed the room to check his weight and carry out the initial tests. Ivan was a small baby, just over six pounds, but he passed all of them with high scores.

We were the typical proud parents. Grandmothers and grandfathers, sisters and brothers all came to visit the new arrival in a room that rather eerily overlooked the exercise yard of next-door Wormwood Scrubs prison. One of the first to come was my godfather Tim Rathbone, who was suffering from terminal cancer and was being treated at the next-door Hammersmith Hospital. I could see that he was dying, and it felt so poignant that he was there.

Once Samantha was well enough, we headed off to her mother and stepfather's house in Oxfordshire, where we were going to spend those supposedly idyllic first few days together. But then we noticed that something was wrong. Ivan was sleepy, like many premature babies. And, again like many others, he would sometimes wake with a start, hands outstretched. But we noticed that these sudden and jerky movements were happening more and more.

The worries mounted. He wasn't feeding properly. He was losing weight. And the movements got worse. He was tiny, but these looked like full-grown seizures. So, after a friendly but inconclusive visit from the local GP, we jumped in the car and headed for the John Radcliffe Hospital in Oxford.

And so the litany of specialists, children's wards, tests and treatments began. The staff at the hospital did all they could to reassure us. But when you watch your tiny baby undergoing multiple blood tests, your heart aches. When they bend him back into the foetal position to remove fluid from the base of his spine with a long, threatening-looking needle, it almost breaks.

The meeting with the consultant, Dr Mike Pike, for the initial verdict on all these tests is etched forever in my mind. As we sat down, a box of tissues was placed on the table by our side. 'Severely delayed development,' he said. These words were carefully chosen, and there is a whole industry of literature and thought behind them. But they don't mean much to the uninitiated new parent. I asked whether this meant he would struggle at sport, or spend his life in a wheelchair. 'I'm afraid it's more likely to be the latter,' was the reply.

It turned out that Ivan had 'Ohtahara Syndrome', named after the Japanese physician who first observed it. Like many of these diagnoses, it is more a description of a set of symptoms than an explanation of how it happened or what can be done about it. Put bluntly, the cause was unknown. The treatment options were uncertain. And there was no cure.

Ohtahara Syndrome is incredibly rare, but our Ivan was a typical case. What its sufferers tend to have in common is severe and often uncontrollable epilepsy, and very poor outcomes in terms of development. Most are quadriplegic (unable to use their limbs) and suffer severe developmental delay (unable to speak, or communicate properly).

The news hit us both very hard. Like all parents, we had worried about having a healthy baby. But, also like many others, it is something you don't think will actually happen to you. We were almost completely unprepared.

And when it does happen, the effect is sudden, deep and lasting. It takes a long time to understand what has taken place. You enter a period of mourning, trying to come to terms with the difference between the child you expected and longed for, and the reality that you now face.

But like so many things to do with the human spirit, there is a resilience that you didn't know you had. You feel such strong bonds of love, and such desire to protect this beautiful little creature, that something inside you helps you through.

We went home to Dean, and the tears flowed. How would we manage? What would it be like? Most of all, how could we cope with seeing our precious child suffer so much?

Today, when I think of Ivan, I think of how we *did* cope. I think of the smiles and the holidays. Covering his legs with warm sand on the beach in Devon. Or trying to get him to sit on a pony. Or lying with him for hours on my lap or on my tummy. Having a bath with him and the other children, with Nancy and Elwen gently washing his hair. Swinging in a hammock and listening to him gurgle with pleasure. The happy memories are now at the front of my mind.

But if I think for too long, I also remember the seizures. He could have twenty or thirty in a day, lasting for minutes, or sometimes hours, his small frame racked with spasms and what looked like searing pain. By the end his clothes would be drenched in sweat and his poor little body exhausted. And so often, there was nothing we could do. It was a torture that I can hardly bear to remember. For Samantha, the mother who bore him and who loved him so deeply, it was a torture that was tearing her apart.

In those early days after Ivan's birth we talked and talked together. On one car journey back from the John Radcliffe to Dean I remember saying, 'We *are* going to make it.' We had to. We hadn't wanted this. We weren't prepared for it. But we loved him, and we would find a way through. If we, with all our advantages, our security, our love for each other, couldn't manage, then who could? There would be many times in the subsequent months and years when we felt close to collapse, and would remind each other of this conversation.

Something had happened before Ivan's birth that did give me pause for thought – and at least some mental preparation. A constituent called Tussie Myerson who lived in a neighbouring village had asked me, as the new MP, to come and see her to talk about the care, or rather the lack of it, that her severely disabled daughter Emmy was receiving. When I arrived she sat me down at her kitchen table, wedged in with her nine-year-old daughter in a wheelchair next to me, so I couldn't move. She told me years later that she had done this on purpose: she wanted me to see just how difficult it was to cope with someone who couldn't feed themselves. Who couldn't communicate. Who was in permanent danger of choking. Who was frequently ill and prone to powerful seizures. Tussie never told me whether or not I passed the test. But as I look back and remember our discussion of care packages, respite breaks and special schools, and how little I knew then, my sense is that I only narrowly avoided outright failure.

After Ivan was born, Tussie got in touch and offered much sound advice, along with huge amounts of sympathy. She said, 'Always remember, you didn't volunteer for this. You're not angels, and you shouldn't pretend that you are. Do everything you can to keep your love for each other, and your marriage and family together.' I always remembered this, and have passed on similar advice to dozens of other parents with disabled children.

That said, we had no idea how difficult it was going to be. We soon moved from the John Radcliffe back to our home in London – and frequent visits to St Mary's Hospital, Paddington. More tests. More drugs. More attempts to stabilise Ivan's condition, with the aim of providing at least some limited quality of life.

From there he moved on to Great Ormond Street, which richly deserves its reputation as one of the best children's hospitals in the world. We tried different medications. Cocktails of anti-epileptic drugs, one added to another, with dosage levels changed to try to get control of the seizures. Too strong and he was crashed out, asleep for most of the day, with his chances of developing like other children set back even further. Too weak and the seizures would return, his little body convulsing and our hearts breaking all over again.

Most of the medicines tasted disgusting, and it was often impossible to get him to keep them down. He developed 'reflux', where everything – milk and medicines – would come shooting back up again, sometimes accompanied by a burp and a winning smile. It was almost as if he was telling us that nothing was going to work. Even when we could get the medicines down, the epilepsy always seemed smarter than the doctors. No matter what combinations of drugs and treatments we tried, it would emerge again, the seizures often stronger than before.

We tried steroid injections, which have helped other children. They made his weight balloon and his blood pressure rise, and his kidneys came close to failing. We ended up in the renal ward of Great Ormond Street, where Sam and I took turns to sleep on the floor by his bed. Most of the other children on the ward had kidney problems, and when Ivan was asleep I would read them stories to pass the long hours they were stuck in bed waiting for the next operation or dialysis session.

We certainly saw the best of the NHS, with consultants like Mike Pike at the John Radcliffe, Diane Smythe and Mando Watson at St Mary's and Helen Cross at Great Ormond Street. They have changed and improved the lives of so many children, and they did a lot to help Ivan. But I think they would all agree that he was one of the toughest cases they'd ever had to deal with.

We also saw at first hand how little is really known about some of these complex medical conditions. Before Ivan, I had always assumed that even if they were incurable, most diseases were correctly diagnosed, their causes were understood, and medicines could always be prescribed to ease at least some of their symptoms. But in this case of severe epilepsy, the doctors didn't know the cause, and even if the medicines did (briefly) work, they didn't really know why. They were basically changing dosages, hoping to make progress but with little understanding of what might work and what might not.

Wanting to know whether we could have other children, we signed up for 'genetic counselling', which in 2003 was very much in its infancy. This was another field in which we discovered how little is actually known. To start with, no one had any idea whether Ivan's condition was inherited or not. If it was, there might be a one in four chance of it happening again. If it wasn't, it was one in many thousands. So we were offered a sort of 'blended probability' of one in twenty. Remembering how few of my father's 20–1 shots ever came in at the races, we decided to risk it.

It was one of the best decisions we've ever taken.

Nancy arrived in 2004. We were so worried something might be wrong that every movement she made was carefully watched and analysed. We needn't have worried: she was the easiest of babies, and hit every milestone on time.

Above all, we saw the compassion that there is in the NHS. I lost count of the nurses who went above and beyond. Who would stop at nothing to try to make Ivan comfortable. They tried so hard to look after us, as well as him.

A perfect example was when Ivan went for an operation to have a feeding tube – basically a small plastic plug – inserted into his stomach, because his weight loss was getting so severe, and delivering the medicines had become so painful and so difficult. The sight of your little boy about to go under the knife, even for a relatively straightforward operation like this, is hard to bear. I'll never forget the warm-hearted nurse, originally from Zambia, who held my hand as I watched Ivan go under the anaesthetic, tears streaming down my face as I wondered if he would ever wake up again. The tube feeding helped us control his weight and measure the drugs more precisely. Sam and I became expert with the tubes, valves, syringes and measurements.

We were always determined not to hide Ivan away. While he could never tell us his likes and dislikes, we sensed that he liked the stimulation of being out and about in the fresh air. So he would be fed on trains and planes, in pubs and restaurants, usually with a gaggle of other people's children watching. Occasionally one of them would ask if the tube was there because he had been naughty and not eaten his tea.

Just as we experienced a new world of hospitals and tests, so we had to build a new and very different life at home. Looking after someone with Ivan's condition – unable to move or communicate, doubly incontinent and prone to massive and prolonged seizures – meant huge changes. We needed a hospital bed, syringes, tubes, oxygen, suction pumps, sterilisation equipment and a range of controlled drugs, including powerful benzodiazepines and barbiturates. But above all we needed Olympian levels of stamina, patience and love. We did our best, but after a few months we were close to collapse. We tried to cope mostly on our own, but we simply couldn't.

I found the phone number of Kensington and Chelsea council's social workers, and soon, to my great relief, one of them was sitting in our kitchen, notepad in hand, talking about the help that was available.

The list of people who assisted us, in both London and Oxfordshire, is a long one. Children's hospices like Helen House and Shooting Star, and dedicated public servants like the community nursing team, who Samantha would say did more than anyone to save her life and her sanity.

At the moment of greatest crisis, when we were near to breaking point, I found someone who would become very special in the life of our family. Gita Lama, a young Nepalese woman, had worked for a diplomatic family in London and subsequently registered with an organisation that represented domestic workers at risk of abuse and helped them find new work. She became Ivan's night carer, and would later help us to look after him at the weekends at Dean. She loved Ivan as if he were her own, and went on to look after our other children in Downing Street. Now with a son of her own, she remains a good friend of the family.

Kensington and Chelsea were incredibly helpful, and gave us carers who stayed in with Ivan several nights a week. Again, these amazing women – the main two were Shree and Michelle – became devoted to him, and close to us.

Yet for all this help, the emergencies continued. We would often exhaust the range of drugs we were allowed to administer at home, and have to drive at breakneck speed to hospital. Children's A&E at St Mary's became something of a second home: we would arrive and say a familiar 'Hello' to the doctors and nurses. Then the desperate ritual of what became known as 'the protocols' – the administration of a range of ever-stronger drugs to control the seizures – would begin.

The last-but-one stage was a drug called Phenytoin, which was administered rectally. The chemical smelt so strong, you could hardly breathe. A glass test tube had to be used because it could melt plastic. What it did to our little boy I could hardly bear to think of, but it worked. From violent spasms, he would go limp and floppy, and we would hold him in our arms, thankful that the ordeal was over.

The final protocol was for him to be rendered entirely unconscious and put on a ventilator. Once this happened there was no guarantee he would regain consciousness. While we came close at times, we never reached this stage.

We learned a lot about navigating the system to try to get the best for your child. When dealing with epileptic seizures in the A&E department, watch out for the four-hour waiting target: there is a danger of an entirely unnecessary hospital admission. (Once you get close to the deadline the staff, quite understandably, want to shunt you onto a ward, whereas it may be that after just a few more minutes in A&E things will be good enough for you to go home.)

Once your child is in a hospital ward, try to order your next batch of drugs hours before you're due to leave, as they take forever to come. (I used to joke that hospitals were easy to get into, but impossible to get out of.)

When the doctors begin their ward rounds, never leave your child's bedside; it is the only time you have a real chance to find out what on earth is going on.

Nowhere was parental navigation more essential than in the highly charged world of special-needs education. I had already seen as a constituency MP that special schools were struggling, partly because of their high costs, but principally because of the doctrine of inclusion. At its most extreme, this held that all children, whatever their needs, whatever their disability, should be taught in mainstream schools. Of course it is right that children with special needs who *can* be integrated into mainstream schools should be able to be, but some children are undoubtedly better off in a special school. In any event, parents should be able to make informed choices. Far too often they simply weren't being told about what was available. Even though I had seen this happen to others, I rather irrationally didn't see it coming. But of course it did.

We had heard about an amazing special school called the Cheyne Day Centre, attached to the Chelsea and Westminster Hospital. But when the education adviser from the council came around to talk about Ivan's schooling they failed to mention it. We then began a battle to get him in; and once he was, we found ourselves having to fight another battle to keep it open. For a time we were successful, and he received the best possible start. Care, stimulation, therapy and education, all in a place where we knew he was safe and where the staff could cope.

After his fifth birthday Ivan needed to move on. While we had fought valiantly, the cost of Cheyne was too great, and a new special school was being built next to Queen's Park Rangers' Loftus Road ground, which was near where we lived. We accepted the inevitable and agreed to a place at this school, Jack Tizard, which in the end turned out well.

My friends say that the experience of having Ivan and helping to care for him changed me a lot. I am sure they are right. A world in which things had always previously gone right for me suddenly gave me an immense shock and challenge. I tried to rise to it, but am very conscious of the ways in which I failed. I was always there for the emergencies, good at the technical things, never one to hold back when nappies needed changing or drugs delivered. And I loved Ivan with all my heart. I adored bathtime, bedtime, walks, wheeling him everywhere and nowhere. As he got older I would throw him over my shoulder and make sure he was part of everything we did together as a family. But I know that I lacked the real patience and selflessness that are required to be a truly great carer. And that is the truth about accidental carers: we are not perfect, and there is a lot of muddling through. No wonder so many marriages break down when challenges like this come along.

Yet perhaps that was the greatest discovery of all. While I can think of ways in which I failed, I cannot think of a single way Samantha did. I still marvel when I think of how she managed and cared and loved and coped, not just with Ivan but with the rest of our growing family.

The end is almost too painful to relate, even to recall.

We had had some scares and close shaves. Seizures that never seemed to end. Chest infections that he would struggle to shake off. And then one night, 24 February 2009, Shree woke us to say that Ivan's stomach had become badly swollen and he was in terrible pain.

This time Sam said she would take him to hospital, and I should stay with the other children. I will never forget holding Ivan in my arms in the cold night air as Sam threw some clothes and blankets on the back seat and started the car.

As soon as they were gone, I started worrying that this time it was different. So I too dashed to the hospital. When I got there the situation had deteriorated badly. A team was standing over Ivan in the emergency room, working desperately to resuscitate him. But he had gone. Adrenalin injections. Defibrillator pads. Nothing worked. He had suffered a massive organ failure. Sam and I were left holding him as the team, visibly moved, backed away to give us some space. We had always known this might happen, but nothing, absolutely nothing, can prepare you for the reality of losing your darling boy in this way.

It was as if the world stopped turning. Explaining what had happened to the children was so hard, because they were so young. And I had to call Gita, who was visiting her family in Nepal; she was desperate to be there with the child she loved so deeply. I called Ed Llewellyn and told him what had happened and that I would be staying at home. I was leader of the opposition at this point and, as it was a Wednesday morning, I was meant to be at Prime Minister's Questions. What happened later, when Gordon Brown led tributes and the House adjourned for the day, meant a lot to us. It was much more than I had expected, and it showed the real warmth and humanity of Gordon Brown, who had of course suffered in a similar way with his daughter Jennifer Jane, who died shortly after she was born.

The next few days before the funeral were a blur. At least we had to focus on the songs and poems we wanted to remember him by. A friend of Sam's called Damian Katkhuda, who had a band called Obi, sang and played his guitar in St Nicholas church, Chadlington. It was a beautiful service, with our closest friends and family around us. But there was nothing but darkness for us.

You never fully recover from the loss of a child. But you can steadily learn to cope. I threw myself back into my work as a way of trying to manage. When I look back, I realise that I started working again too quickly. For a while I was too fragile and not in the right state of mind to make decisions. Nothing else seemed to matter alongside what we had lost.

But what is often said about grief I found to be true. While at first you think the gloom will never lift, there comes a time – and for me it was many months later – when some of the happy memories start to break through and you remember what you had, not only what you have lost.

And having Ivan taught us so much. About unconditional love. About our total devotion to each other. About the extraordinary compassion in our health service and the lengths that people go to in order to help. We learned about our strengths, but also our limitations.

Ivan lies buried opposite the church in Chadlington. We take the children there, and tell him how things are going and how much we still miss him. Sam found an inscription from Wordsworth for the headstone that sums up so much of what we feel.

I loved the Boy with the utmost love of which my soul is capable, and he is taken from me – yet in the agony of my spirit in surrendering such a treasure I feel a thousand times richer than if I had never possessed it.

8

Men or Mice?

At the time, Michael Howard's 2005 general election campaign was seen as slick and professional. But it was also too right-wing and rather mean-spirited, putting people off rather than turning them towards us. It resulted in another disastrous defeat for the Tories.

I had been responsible for policy coordination, writing the manifesto and acting as one of the party's principal spokesmen around the country. I saw the campaign close-up. Yet just a few weeks after it was over, I was planning an aggressive leadership campaign in favour of a more modern and liberal Conservative message.

How does all that make sense?

The short answer is that in modern politics the tone and content of a manifesto and a campaign are predominantly set by the party leader. Michael Howard was sure that if we were robust and effective, we could make a fairly traditional Conservative message work. He also felt he had to be true to himself. I was already convinced that we had to change, but I understood Michael's position. I owed a lot to him, and wanted to help him make his chosen strategy as successful as possible.

The manifesto itself was short and focused, but it was lacking in policy detail. With Michael's permission I drafted in Michael Gove – who I had helped to persuade out of journalism and into politics, and who was standing in the super-safe Conservative seat of Surrey Heath. We sat in my pokey House of Commons office for several days, dividing the chapters up between us and writing one each before passing what we had done to the other for polishing and improving. We were already friends, and this work brought us closer.

The policies may have been rather workmanlike, but they did actually work. We know this because, while Labour derided our manifesto at the time, they copied and implemented many of its most significant proposals straight after the election. The points system for immigration; the proposals on school discipline. Tony Blair pursued his usual tactic of trashing his opposition, and then coopting any idea that was halfway sensible.

But in modern elections the campaign itself is what matters, and the tone of ours was set not only by Michael Howard, but also by someone I've come to admire as one of the great political campaigners: the Australian Lynton Crosby.

Lynton's hard work is combined with great leadership skills. Twice – in 2005 and 2015 – I've seen him build the happiest, most cohesive, most hard-working teams in Conservative Central Office that I have ever known. His strongest weapon is plain common sense. What's the target? What are your strengths and weaknesses? What are those of your opponents? What, given those things, is the best route to victory? Above all, what's the plan?

In 2005, Lynton came in at a relatively late stage. His view was that the best chance Michael had to win the election, or at least to deprive Tony Blair of another massive victory, was to focus on some straightforward issues that people cared about, while encouraging them to take out their frustrations with the government by voting for the Conservatives.

The famous poster slogan 'Are you thinking what we're thinking?' fitted with this strategy. It was punchy, and it channelled frustration with Labour. It focused minds on down-to-earth-issues: clean hospitals, more police, 'It's time to put a limit on immigration,' and so on. But the tone reinforced the problem with the Conservative image. It was mean-spirited. Too many people answered the question 'Are you thinking what we're thinking?' with 'Well, even if I am, I'm not voting for you lot.'

Added to that, in my view the campaigning on immigration went too far. The message wasn't an unreasonable one. Indeed, I was a strong supporter of immigration control, and had been closely involved in drafting the proposals we put forward. And you could argue that, in the light of what subsequently happened, the decision to make this issue a central one was prescient. But its domination of the early part of our campaign was too much. It felt wrong. It appealed to voters we already had, but made some of those we needed to attract feel uncomfortable – even those who agreed with the policy itself.

The result was the fourth-worst Conservative performance for a hundred years. While we gained thirty-three seats, we only increased our share of the vote by 0.7 per cent, a smaller increase than William Hague had achieved in 2001. Overall, we got fewer votes in 2005 than we did in 1997 – 8.8 million versus 9.6. We won some of our target seats, but even then more than half of those only came to us because of Labour voters switching to the Liberal Democrats, rather than directly from Labour to us.

One other polling figure tells the true story. When people were asked whether a party 'shares your values', the Conservatives came off worst, at around 36 per cent, while Labour and the Lib Dems were at around 50 per cent. Maurice Saatchi put it crisply when he said: 'More anger at the problems of the world we live in is not enough to convince voters that the Conservative Party is fit to solve them.' The problem went much deeper. We needed to change.

Michael announced that he wouldn't stand down until there had been a review of the leadership rules. He favoured a system where if more than half of the parliamentary party settled on one candidate, there would not be a vote of the party membership. In the event this proposal went down badly with both the membership and a significant number of MPs, and wasn't adopted. But the delay in the leadership election that it caused would make all the difference.

If it had taken place sooner after the general election, there can be little doubt that the favourite, David Davis, would have been elected. He had a machine in the parliamentary party, and something of a public profile. There wasn't an obvious challenger. Before one arose, the contest would have been over.

Instead, the party would wait until just before the party conference in the autumn before candidates' declarations were made. A formal campaign would then be held during and after the conference, with the results in December.

But before any of this got under way, Michael needed to appoint a new shadow cabinet. He wanted to give newer MPs a chance, and sounded out both George Osborne and me about what jobs we most wanted to do. I was in no doubt: I wanted to be the shadow secretary of state for education. It might not have been seen as one of the 'big jobs', but for me it stood out above all others. So much depended on it: the life chances of our young people, the future of our country. Our party's prospects too rested on the answers we came up with on such policy challenges, and I wanted to be one of the people driving them.

But choosing the education role wasn't, of course, the most important decision I took after the election.

Slightly to my surprise, and certainly to the surprise of many others, I found myself running for the leadership.

Perhaps for others, deciding to run for such an office comes swiftly, and with few doubts. That is not how it happened for me. Everyone said that I was too young. That I had no ministerial experience. And that I had only been in Parliament for four years. I could be a candidate, maybe a credible candidate, but would I be a credible leader? Would I be part of the party's problems rather than a solution?

During those pizza evenings in Policy Exchange before the election, one of the things our small group of modernisers had discussed was how we might persuade our future leader to act. But nothing we came up with had seemed convincing. We knew, partly from experience with Michael Howard, that it wouldn't be enough to persuade a new leader to mouth words about modernisation. We needed someone who really believed in it, and embodied it in the way they talked and acted and felt.

Gradually some of the group began to feel that maybe the answer was to try to capture the leadership rather than merely influence it. We didn't spend a lot of time on what, at that stage, seemed a little presumptuous and some way off. The moment the election was over, however, it all suddenly seemed more real, and more possible. But was it right?

George's wife Frances was particularly outspoken. The daughter of Margaret Thatcher's cabinet minister David Howell, she knew the brutality of modern politics, but wasn't in any doubt. The four of us were having dinner together at our house in North Kensington shortly after the general election when she looked at her husband and me and asked, 'Well, are you men or are you mice?'

From the moment I really looked at it properly, I thought that I could win. Not because of any special brilliance or powers I possessed; I just saw that all the other potential candidates had flaws that made them eminently beatable.

Ken Clarke had popular appeal, but as the Conservative Party had become a Eurosceptic party, he would find it very hard to win.

Liam Fox, a strong speaker and media performer, was, when you scratched the surface, a pretty unreconstructed Thatcherite. I was fairly sure the party was looking for something else.

That left the favourite, David Davis. He had a great back story, growing up on a council estate, brought up by a single mother and making his own way through business and the Territorial Army to Parliament. He was Conservative aspiration personified. Yet he was another relatively unreconstructed right-winger who would think that the combination of his candidature and another coat of paint would change the party's fortunes.

I knew this wasn't the case. Davis wouldn't be the one to get the Tory car out of the ditch, and I thought the party agreed with me. He had surrounded himself with a rather thuggish crew of former whips from the John Major era, and I rather suspected that a Davis leadership would be like life in a Hobbesian state of nature: 'nasty, brutish and short'. While he was the front-runner he could win adherents who feared being on the wrong side of him, but the moment people began to suspect he might not win, that fear would go, leaving him a much less formidable candidate.

But I still had doubts. Not so much about getting the job, but about myself and the pressures it would bring, not just on me but also on my family. I was still very young, with much to learn. Could I really do all the different elements of the job? Decisions that would put people's lives at risk. Coping with the pressure. Prime Minister's Questions. There were many moments of indecision before the choice was finally made.

My old friend Andrew Feldman played a key role. He told me that I could and should do it. There were also one or two – and it was pretty much one or two – MPs who were similarly convinced. Former SAS officer Andrew Robathan appeared in my office and said that I had to do it. He knew the parliamentary party, and said he was looking at a winning candidate with a winning strategy. Greg Barker, the MP for Bexhill and Battle who had come into the House with me in 2001, was similarly enthusiastic. So was my friend Hugo Swire. Boris was also keen, and generous in coming out for me quite quickly. In a characteristic intervention, he told the newspapers: 'I hope that David Cameron removes his hat from wherever he has got it, and chucks it firmly in the ring. That hat has got to simultaneously decapitate his competitors and land in the ring.'

Yet in the end it was those closest to me who were the most influential in helping me make up my mind. Most friends were enthusiastic. They could all see that the Tory Party needed a new approach, and they thought I should go for it. The only exception was Michael Gove, who called me one weekend at Dean and pleaded with me not to do it. He was worried about the effect on me, on Samantha and the family. For all the subsequent drama in our relationship, I think he had nothing but the best of intentions in making the call.

My mother and father were nervous. I don't remember them ever saying 'Don't,' but my dad in particular was not an enthusiast. He was delighted that I was doing the education job, and thought that I should take one thing at a time. But my brother Alex told me to go for it. This meant a lot.

The most important, of course, was Samantha. Just as she had been worried about the effect on our life of me becoming an MP, she was worried about what being leader would mean. She could see why that side of it worried me, but she was also in many ways the ultimate Tory moderniser. It was a crisp spring day in the garden at Dean when she said words to the effect of, 'What is the point of spending your life in a Tory Party that can't achieve any of the things that you believe this country needs to do?' That was what I really needed, and after her words the decision was made. I was running.

To start with, things came together well. George and I met and talked frankly about the situation. He was being encouraged to consider standing, but he thought he was too young, and hadn't had enough time to develop the sort of story and profile he'd need to succeed as leader if he won. And anyway, his new job as shadow chancellor was a huge challenge. At just thirty-three years old, he was the youngest person in history to hold that role, and he didn't want to be distracted.

But he did offer to run my campaign. There was no pact, no deal, no agreement about anything, including future jobs; but there was something much stronger. A shared view of the challenge, and an understanding that we would stand together and work together come what may.

The rest of the team was small but professional.

Andrew Feldman was the natural treasurer, and he set about raising the necessary funds, starting with the businessman Phil Harris. My old Carlton boss Michael Green chipped in. We wanted a good range of donors, not to rely too much on any one individual.

Ed Llewellyn, who was working in Sarajevo at the time, took unpaid leave to come and lead my team. Kate Fall, who had worked for Michael Howard, came to work as his deputy. They teamed up with my press officer Gabby Bertin and an events team led by Liz Sugg. All would still be with me when I left Downing Street eleven years later.

Steve Hilton, who had been running his own business after leaving Central Office, and had then gone to Saatchi & Saatchi and M&C Saatchi, would play a key role in working with me to put together the case for change.

Meanwhile, in the House of Commons, I started to sound out MPs. The good news was that the early adopters were just the sort of people I wanted: bright, sane, forward-looking, and popular with other colleagues. The less good news was that there weren't very many of them. When we first got together in my office in 343 Portcullis House on 13 June there were just fifteen MPs present: Greg Barker, Richard Benyon, John Butterfill, Michael Gove, Boris Johnson, Oliver Letwin, Peter Luff, George Osborne, Andrew Robathan, Hugh Robertson, Nicholas Soames, Hugo Swire, Ed Vaizey, Peter Viggers, and of course me.

We agreed to spend the summer setting out policies and ideas: we could only beat the Davis bandwagon if there was real substance in what we were saying. When it came to parliamentary colleagues, no jobs would be offered, no future roles dangled in front of them as inducements for support. And we would be unfailingly polite and correct. This was a complete contrast to the Davis operation, which used a combination of brutal arm-twisting ('Support the front-runner or your career is over, matey') and ludicrous promises (by the end, I heard, he had amassed several chancellors and foreign secretaries).

The early campaign was very heavy going. We couldn't get more MPs to declare their support. None of the newspapers were backing us. And I was worried that my freshness, a central part of my pitch, might go stale.

The first parliamentary hustings in July didn't go all that well. The star performer was Liam Fox, who spoke forcefully about the need for change in Europe. And a new issue emerged that was to last throughout the contest – drugs. The MP Mark Pritchard was persuaded (by someone in the Davis camp, we were told) to ask one candidate – Ken Clarke – directly, 'Have you ever taken class-A drugs?' Naturally the spotlight fell on the rest of us to answer. I declined to do so, and while many colleagues groaned when the question was asked, there probably was some damage done.

The *Daily Mail* became quite hysterical about it, publishing a full-page editorial: 'David Cameron, Drugs and the Truth'. I refused to yield, and declined to answer the question about drug use in the past, saying that 'Everyone is entitled to a private past.' My stubbornness won some admirers, and proved that I wouldn't be pushed around.

On *Question Time* I answered a question about whether I had ever taken drugs as an MP by saying, truthfully, that I had not, because 'law-makers shouldn't be law-breakers'. The more difficult question was whether I had ever done so when I was a special adviser, or between being a special adviser and becoming an MP. I simply didn't answer it. Frankly, I didn't want to tell a lie by saying no. Stories began circulating that I had avoided the question because drug use among my friends was commonplace and excessive. This was nonsense. But had I smoked the odd joint with Sam's friends before being elected? Yes. Not at all frequently, but yes.

All in all, it felt as if the campaign was stuck, and outside our small core there were few who thought we could win. But I knew we had one weapon more powerful than those possessed by any other candidate. A clear, powerful and persuasive political message that I was sure the party was ready for: Change to Win.

This oughtn't to have seemed as radical as it sounded. After all, the essence of conservatism, and central to the success of the party, is that it adapts. Far from being the ones trashing the Conservative brand and the Conservative Party, we were absolutely convinced that we were the ones who could save them.

Our goal – which became my mantra – was a modern, compassionate Conservative Party. Modern, because we needed to look more like the country we aspired to govern. Compassionate, because our politics was about extending opportunity to those who had the least. And Conservative, because we believed that timeless Conservative principles – strong families, personal responsibility, free enterprise – were as important as ever.

The speech I made that June, effectively starting my leadership bid, included a strident defence of families and marriage. Some saw this as rather an old-fashioned note in an otherwise modernising score. I saw it as essential to building a stronger and more compassionate society.

In August 2005, I delivered a comprehensive speech on the right approach to tackling the rise of Islamist extremist terror. I made the case for tougher security measures, including action to deport hate preachers and potential terrorists, and arguing that, if necessary, we would have to leave the European Convention on Human Rights. But the real point of the speech was to make clear that I believed we were involved in an ideological struggle that could last for a generation or more. There was no point trying to tiptoe around what we were up against.

And then there was Europe. I thought, naïvely perhaps, that I had the right formula. In line with my own beliefs, we would be genuine Eurosceptics. Not arguing for Britain to leave the EU altogether, but arguing consistently and cogently for reform. Integration had gone too far. Brussels was too bureaucratic. Britain needed greater protections. Far from rejecting referendums on future treaties, the public should have its say.

Crucially, we had to get away from the 'doublespeak' of the past. Margaret Thatcher had railed against Brussels, yet took the country into the Exchange Rate Mechanism. John Major had attacked the single currency, yet said he wanted Britain at 'the heart of Europe'. The Conservative government had opposed a referendum on the Maastricht Treaty, yet most ministers privately prayed for the Dutch and Danish populations to reject it when they were given the chance in referendums of their own. So, above all, we needed to be clear and consistent.

To me, it followed logically that the Conservatives couldn't continue to sit as part of the European People's Party (EPP) group in the European Parliament. The EPP wanted more integration and more political union; the Conservative Party wanted less of both. Yet William Hague, Iain Duncan Smith and Michael Howard had all fudged this issue. In my view we had to act and speak in the same way whether we were in London, Brussels or Strasbourg. Thus my pledge to leave the EPP and establish a new centre-right group.

Some have said that this pledge was made purely for opportunistic reasons – to win the support of Conservative backbenchers. Others that I did not fully believe in what I was doing. I totally reject those criticisms. I was in Conservative Central Office when Margaret Thatcher was persuaded to join the EPP. I thought it was wrong at the time, and I never changed my mind. We agreed with the mainly Christian Democrat parties about many things, but not the future direction of Europe. That should have been a deal-breaker right from the start.

But would leaving the EPP do us harm with our allies? This was a stronger argument, and I suspect it was the one that encouraged my predecessors to pull back from making the full break. But our allies needed to know we were serious about reform. And I was convinced that it would be better for all of us if we were outside the EPP, while cooperating with its members on shared endeavours. This indeed turned out to be the case – we established what rapidly became the third-largest group in the European Parliament after the socialists and the EPP: the European Conservatives and Reformists Group.

Undoubtedly the move helped me win support from Eurosceptic MPs. But many of them had few other options. When it came to Europe, Ken Clarke was already the Antichrist to many. And David Davis was the Maastricht whip who had twisted arms and made MPs vote for a treaty they hated. Meanwhile Liam Fox's bandwagon had limited momentum. So I didn't need to make pledges on the EPP to win the leadership, or even to win the votes of the bulk of the most committed Eurosceptics. What I said to those MPs was what I believed – and I delivered the promise that I made in full.

But still there wasn't enough support. It felt as if there were only two people – David Davis and Ken Clarke – in the race. On one occasion in my office just before summer recess, George said I needed to start thinking about packing it in. He was frank, as always: 'Look, I don't think you're going to win. You've had a good run, made some good points, put down some strong markers – why not leave it at that for now?'

But I still thought the contest was wide open. I was more certain than ever that the party needed to change, and that change wasn't being offered by anyone else. Yet I had a sense that for all my hard work, perhaps I was holding something back. Perhaps I was still trying to temper my radical aims, for fear of scaring too many people off. I knew now that the only way I had a hope of winning was by being true to myself, getting everything out there and going for broke on modernisation.

We had £10,000 left in the kitty, and we would blow it all on the launch, at which we would set out in even clearer terms what was on offer and what was at stake. At least then, even if I lost, I'd have nothing to reproach myself for.

Launch day turned out to be a day that changed my life.

Steve and I spent a lot of time thinking and writing, and then polishing and rewriting, the speech. Steve also spent a lot of time getting the look and feel of the launch right: he wanted it to be as different from the usual Tory leadership launches as it could possibly be. These tended to take place in a House of Commons committee room, or at least in a room that looked like one. They would involve lots of men in suits standing around, sometimes looking faintly deranged and saying 'Hear, hear' too loudly whenever their man (and it usually was a man) said something vaguely right-wing.

We picked a date for our launch, but soon found out that it was the same day the Davis camp had chosen. Instead of changing the date, we hoped that the contrast between the launches would demonstrate new versus old, change versus more of the same. And that's pretty much what happened.

Sure enough, the Davis launch was in an oak-panelled room. Veterans of past Tory leadership elections said that they felt they'd seen and heard it all before.

We rented a bright and open space, with a stage and no lectern. Instead of journalists and MPs we invited friends and supporters. And instead of tea and biscuits it was fruit smoothies and chocolate brownies. Sam asked lots of our friends, some of whom, like her, were pregnant or had recently had babies.

As I stood before the crowd, I felt that this was my chance to say as directly as possible what I wanted to do, and why. I might have been timid at the start of the leadership campaign, but I would be bold when it mattered. Everything had to change. It wasn't enough just to oppose Labour with more vigour. Nor was it enough to produce even more radical policies and push them with even more gusto. 'We can win,' I told the audience. 'We can make this country better, but we can only win if we change. That's the question I'm asking the Conservative Party. Don't put it off for four years. Go for someone who believes it to the core of their being. Change to win – and we will win.'

In one step I had gone from being the outsider to a real contender. And the stage was set for the party conference in Blackpool in just four days' time.

The attention of the press, which had died off over the summer, was suddenly intense, and Gabby Bertin went into overdrive fixing interviews and profiles. She was joined by George Eustice, who came highly recommended having worked for the organisation Business for Sterling, which campaigned against the UK joining the euro. He was a gentle, thoughtful strawberry farmer from Cornwall, as keen as the rest of us to see the Conservative Party change. We quickly became good friends.

The week in Blackpool was undoubtedly one of the most exciting of my life. I could feel the momentum. Every day we were winning more support from MPs and candidates. Every party or event we held or at which I spoke saw more and more people turn up.

Standing in the wings of the Winter Gardens waiting to make your speech is an extraordinary feeling. Even back in 2005 the place was crumbling, but it still had some of its old magic. The acoustics were good, the hall was packed, and the audience was close to the stage. The atmosphere and the potential were tangible.

My speech was not as good as the one at the launch, but many more people saw it, as it was carried live on television and reprised on the evening news. What impressed many people was that I delivered it without notes, having memorised it as we drafted it. Watching it now I find it rather wooden, but it worked.

Within a single day, the polls were transformed: support for me surged from 16 to 39 per cent, while for Davis it collapsed from 30 to 14. Between the conference in October and the ballot in December there appeared to be nothing that might shift the dial back in Davis's favour. And I was going to make sure of it. I resolved to go to as many places as I could, and speak to as many members as possible. For five weeks, life for Liz Sugg and me was spent on the road. Speaking at members' meetings, sometimes with only a dozen people in the room. McDonald's drive-throughs for lunch, a cigarette and a glass of wine for dinner at whichever Travelodge we were staying in.

Politically the only events that came near to attaining significance were television encounters. There were my first two TV debates, one on ITV, the other on the BBC. And I think it is fair to say that I lost both of them.

The face-off with Jeremy Paxman was, by contrast, something of a triumph. I enjoyed his books, his humour, and watching the spectator sport of his political interviews. But as an interviewer I thought that most of the time he was a self-indulgent monster. He wasn't trying to get answers or inform viewers, he was just trying to make his victim look like a crook while he looked like a hero. To reverse Noël Coward's dictum about television, I thought that *Newsnight* was a programme for politicians to watch rather than to appear on. Not least because hardly anyone actually watched it. I could see absolutely no point in doing an interview with Paxman: it would never be an attempt to examine policies or priorities, just an opportunity for him to show off and try to take me down at the same time.

This infuriated my team. There is nothing a press officer hates more than their boss refusing to do an interview. Eventually they wore me down, and I relented. But I was prepared to turn the tables on Paxman.

In spite of endless promises by the BBC about a neutral venue, the interview was staged at some lush wine emporium. And it soon became clear that the whole thing had been set up to try to make me look like a rich, spoilt child of Bacchus. I was a non-executive director of a company, Urbium, that owned and ran bars and nightclubs. And I should have predicted what was coming.

The first question was, 'Who or what is a Pink Pussy?'

I paused and gulped. The only 'Pink Pussy' I had heard of was the notorious nightclub in Ibiza. In a split second I decided – thank God – that no answer was best.

'What about a Slippery Nipple?'

Now I knew where he was going: Pink Pussies and Slippery Nipples were both cocktails. He wanted to get stuck into outside interests and the responsibility of drinks companies. But before he had the chance to get going, I decided to unleash my own Paxman-like rant.

'This is the trouble with these interviews, Jeremy. You come in, sit someone down and treat them like they are some cross between a fake or a hypocrite. You give no time to anyone to answer any of your questions. It does your profession no favours at all, and it's no good for political discourse.'

That, combined with teasing him about interrupting himself, put Paxman off his stride. He got nothing out of me, and I avoided interviews with him for the next five years. I was happy to leave it at played 1, won 1.

And then the campaign was over.

On 6 December 2005 I made my way to the Royal Academy of Arts on Piccadilly for the announcement of the count. The results were read out at 3 p.m. – and the victory was comprehensive. I had won over twice as many votes as my opponent. I had the mandate. I could get down to work.

I went straight from a celebration party for MPs at the ICA on The Mall to the rather grim green office occupied by the leader of the opposition. By the desk at one end of the room is a pair of double doors leading out to a small balcony. I sat down on the ledge and smoked a cigarette as I thought about the day ahead, which would, dauntingly, feature my first Prime Minister's Questions.

The team met to discuss the task. We had talked a lot about supporting the government when it did the right thing, so I was fairly sure that I should make a start on education, promising to support Tony Blair in his desire to give schools more independence, particularly if he faced down the union-inspired opposition on his own benches.

I suggested that if he brought up our approach in the past, I would say, 'Never mind the past, I want to talk about the future. He was the future once.' George said, 'Never mind what he says, just say that line – it's brilliant.' I did. I had only ever spoken from the despatch box three times in my life. The backbenchers cheered behind me.

First hurdle jumped. Many more hurdles to come.

9

Hoodies and Huskies

It was minus 20 degrees. All I could see for miles was snow. Standing on a sled, I clung to the reins of several barking huskies. 'Mush!' I shouted, and we hurtled across the glacier.

It had been four months since I'd taken the reins of a rather different beast. And I had decided to make Svalbard, a Norwegian archipelago in the Arctic Ocean, the destination for one of my first foreign trips as Conservative Party leader. It was dismissed by many as style over substance. But, like all the significant decisions during those early days of my leadership, it was part of a serious, thought-through political strategy.

We wanted to demonstrate in the clearest possible way that this was a new leader, a changed political party, and – above all – that the environment and climate change were issues we were determined to lead on. They were personally important to me, but they also helped to define my sort of conservatism. Concerned about preserving our heritage, aware of the responsibilities (not just the limits) of the state, able to talk confidently about new issues that might not have arisen in earlier general elections, and respectful of scientific evidence.

Yet in opposition it is hard to get across who you are, and to talk about the things you want to talk about. The government can just waltz onto the 10 o'clock news and talk about its latest plan of action, while you have to work relentlessly to try to set the agenda – but with what? Something you *might* do, *if* there is an election, *if* you win it and *if* the issue is still relevant in *n* years' time.

So we were prepared to take risks. And Svalbard really was a risk. For a start, it nearly resulted in images very different from the photos of me gliding along behind the huskies. I was given a whole load of instructions about how to operate the sled. I ignored all of them, and disaster nearly struck. The cameras were set up for a dynamic, fast-moving shot of me steering the sled. I managed to turn the whole thing over at high speed, and collapsed in a ball of snow, ice and, from everyone around me, hysterical laughter.

These weren't quite the pictures we wanted – I kept thinking of another opposition leader, Neil Kinnock, falling over on Brighton beach. Mercifully, these career-maiming shots never made it onto viewers' screens.

Later, as we clambered into a cave, everyone was asked to wear protective helmets. I resisted, remembering William Hague's baseball cap embarrassment as leader of the opposition. As a politician, you're haunted by the ghosts of gaffes past.

It wasn't long before I patented my own.

A leader of the opposition has a car from the Government Car Service to ferry them around, at least partly because they have a number of official responsibilities, and a big case of confidential papers to carry with them. I was allotted Terry Burton, who had driven some of my predecessors.

For years as an MP I had cycled to Parliament, often with George. I didn't want to stop now that I was party leader, and very occasionally Terry would bring this case, and sometimes my work clothes, including my shoes, in to the office for me. Soon the *Daily Mirror* was onto me, exposing the eco-mad Tory leader's 'flunky following behind in a gas-guzzling motor'. The *Guardian* dubbed Terry a 'shoe chauffeur'. I was truly sad that the episode had tarnished our genuine 'Vote Blue, Go Green' message, our slogan for the local elections taking place that very week. And I've never lived it down.

Presentation *is* important, but prioritising the environment through my trip to the Arctic really was as much about substance. We had to take a boat to visit the British Arctic Survey team, and I asked one of its members why they'd put their station somewhere that was surrounded by water. 'Well, the water wasn't there until last year,' he said. It was a profound moment. Global warming was real, and it was happening before our very eyes.

So what *was* the governing philosophy of my leadership of the Conservative Party in opposition?

Two big things had changed.

First, at the time it seemed as if the great ideological battle of the twentieth century – right versus left, capitalism versus communism – was over. We had won. Labour now accepted the need for a market economy to help deliver the good society, and it appeared that full-blooded socialism was dead. The Conservative Party needed to take a new tack. We shouldn't give up on our belief in enterprise and market economics, but it was time to bring Conservative thinking and solutions to new problems.

The second thing that had changed was the electorate. Over the previous twenty years Britain had become more prosperous, somewhat more urban and much more ethnically diverse. Gay people were coming out, more women were going to work and taking senior jobs, social attitudes and customs were changing. And all of this, it seemed to me, had left the Conservative Party, one of the most adaptable parties in the world, behind.

I saw myself, however new and inexperienced, as inheriting the mantle of great leaders like Peel, Disraeli, Salisbury and Baldwin, who had adapted the party. To achieve that, I wanted *Conservative means* to achieve *progressive ends*. Using prices and markets, and encouraging personal and corporate responsibility, could help our environment by cutting pollution and greenhouse gases. Stronger families and more rigorous school standards could help reduce inter-generational poverty. Trusting the professionals in our NHS, rather than smothering them with bureaucracy, could build a stronger health service.

The Conservative Party, in my view, had got into a rut of tired and easy thinking. We had a tendency to trot out the same old answers. Want social mobility? Open more grammar schools. Want lower crime? Put more bobbies on the beat. Want a more competitive economy? Just cut taxes.

We had another, even more profound, problem. People didn't trust our motives. Whenever we suggested something, people seemed almost automatically to add their own mistrustful explanation of our motives. When we said, 'Let's reduce taxes,' they added, 'to help the rich'. When we said, 'Let's start up new schools,' they added, 'for your kids, not ours.'

Part of this was a hangover from the end of the last period of Conservative rule, when Tony Blair and New Labour had caricatured

Conservatives as uncaring. But some of it was our own fault. It was part of what I called – or more accurately what Samantha called – the 'man under the car bonnet' syndrome. We approached every problem or issue with a mechanical, process-driven response rather than a more emotional, values-driven answer about the ends we were aiming to achieve.

At the same time as the new approach and new policies, I was determined that the Conservative Party should make its peace with the modern world. Our opposition to, or sometimes grudging acceptance of, a whole range of social reforms, from lowering the age of consent for gay men to positive action to close the gender pay gap, made us look and sound like a party that was stuck in the past, and didn't like the modern country we aspired to govern.

I wanted the Conservative Party to be more liberal on these social issues. I felt passionately that morally it was the right thing to do, and I thought it would help us to get a hearing from some people who had written us off. It seemed to me an embarrassment, really just awful in every possible way, that someone who shared our values might be put off voting Conservative because they thought we disapproved of their sexuality, or looked down on their ethnicity, or didn't want them to achieve because of their gender.

Part of the problem was our personnel. We were the oldest political party in the world – and we looked it. Just seventeen of the 198 Tory MPs elected in 2005 were women. That was an improvement of four. *Since 1931.*

Totally unacceptable. We were, after all, the party of the first woman MP to take her seat in the Commons. We gave the country its first female prime minister. Up and down Britain, women were among our finest councillors and our fiercest campaigners. But it just didn't show on our green benches, which were, by and large, male, middle-aged, southern, wealthy and white.

By day four of the job I had appointed all my shadow ministers. I thought it was important to bring my leadership rivals into the fold, so David Davis and Liam Fox shadowed the home and defence departments. I thought I'd got a good mix, but I ended up with more people called David in the shadow cabinet (five) than women (four). There simply wasn't the range from which to choose.

Come day seven I was at the Met Hotel in Leeds unveiling a plan to elect more women and ethnic minority MPs (of whom we had, shame-fully, just two). It was imperative that we started to look more like the country we hoped to govern.

The candidates' list was immediately frozen. A new Priority List of 150 candidates, people we thought the cream of the crop, and better reflecting the make-up of modern Britain, was drawn up from the larger main list. All associations in winnable seats would have to choose from this so-called 'A-List'.

It caused uproar. Uproar so furious and so persistent that a year later I ended up agreeing that associations could pick their candidates from the full list, but half of the interviewees had to be women, thereby superseding the A-List.

But the ambition never wavered. We carried on exerting pressure more informally, promoting the candidates we wanted. I knew this required action at every level. More women applying to be candidates. More women getting interviews in safe seats. More procedures during the selection process that emphasised the full set of skills required to be an MP, not just the big speech in front of the full membership. All this was very much driven from the centre.

One of the greatest things about our election victory in 2015 was the seventeen non-white and sixty-eight women MPs elected to our benches, quadrupling the intake of a decade earlier. Indeed, as I write, there are six women MPs in the cabinet, four of whom were on that original A-List.

It was worth the row.

I was learning a great deal on the job. But as I cleared each hurdle – the hiring and firing of shadow ministers, the weekly bout of PMQs, the response to the Queen's Speech – there was one that loomed larger than all the others: party funding.

Long before we inherited a country in debt, we inherited a party in debt by £30 million, largely as a result of the 2005 general election campaign. The funding crisis had a wider significance. Before they let you run the country, people want to see that you are able to run your party.

While donations to political parties had to be declared publicly, loans did not. So wealthy individuals preferred to make loans, and both the Labour and the Conservative parties succumbed to the temptation of this route. This led to the so-called cash-for-honours scandal, and Tony Blair being interviewed by police. Those responsible for Conservative fundraising were called in too. The case for the defence was clear: taking loans was within the rules, and there was a proper vetting process for awarding life peerages. Contributors to party funds shouldn't be excluded, but it should never be the reason for their appointment. The problem was that while the vetting body – the House of Lords Appointments Commission – was told the details of the loans, the public and the media had not known about them.

I resolved that we should stop taking these loans, and should pay off, or convert to genuine declarable donations, those we already had. I also decided that we needed to stop being so reliant on a small number of wealthy individuals. Even if they didn't exercise undue influence over the party – and as far as I was concerned they didn't – it would always look as if they could. For a time I even flirted with the idea of increased state funding for political parties, in some form or other. While I instinctively disliked the idea of taking more taxpayers' money, there seemed to be a recurring problem with our system.

Apart from big individual donors, of course, the whole system of trade union funding of the Labour Party was antiquated and wrong. Whatever people might say about the closeness of business or wealthy individuals to the Conservatives, the unions' funding of Labour gave them votes at the party conference, votes to choose candidates and the leader, and votes to determine policy. They owned Labour lock, stock and block vote.

Throughout the time I was party leader and prime minister there were talks between the parties to try to find a solution. I was prepared to go along with a cap of £50,000, or possibly less, on donations from individuals, as long as it was accompanied by a cap on union donations and the reform of Labour's union links. I supported the idea of tax relief on donations, to ensure that parties had to fundraise properly and listen to their members, not just wait for the next dollop of taxpayer cash to arrive. But the talks always broke down. The caps we were prepared to accept were seen by the other parties as too high, and Labour was never truly prepared to break the union link.

In any event, we were proving, step by step, that party funding through donor clubs, big one-off events and the party conference was possible. We established the 'Leader's Group' of large donors, each committed to giving the party £50,000 a year. While this is a huge amount by any normal measure, it was a great improvement on passing the hat around to a very small number of multi-millionaires for a few massive, often multi-million-pound, donations. At its peak, the Leader's Group grew to over two hundred people, and became the mainstay of our funding.

While the press was determined to paint it as a 'cash for access' organisation, I was very proud of what we had built. We had shown that, even without extra state funding, our party could be properly funded. There were enough members for it to be clear that no individual would have undue influence. The dinners we had were informal and fun. And while there was no improper influence, as the financial and economic crisis hit, we had instant access to some of the best financial brains in the country.

With Andrew Feldman as chief executive and then chairman, we bridged the gap between the person who raised the money and the person who decided how it should be spent, ensuring real commercial control; and from 2006 onwards the party never ran a deficit, and even had a surplus after both the 2010 and 2015 election campaigns, something which is unprecedented in modern party history. We sold our historic headquarters in Smith Square, and even the loss-making annual party conference started to make money: by the time I left office it was making close to £2 million a year. The party was debt-free, and there was around £2 million cash in the bank.

Of course, the most important question in terms of preparing for power was what to do about our policies. A new focus on the environment was one important element. Mending our broken society would be another. On my first full day of leadership I launched one of our new policy review teams alongside Iain Duncan Smith, whose Centre for Social Justice (CSJ) think-tank was pioneering a radical approach towards tackling the cycle of social deprivation. IDS's review, and a speech I delivered on it a few months later, would prove the most controversial of the period. I wanted us to admit that although we had talked about aspiration a good deal, Conservatives had not done enough thinking about those for whom the bottom rungs of the ladder of opportunity just weren't there, or had been smashed before they'd had a chance to climb them.

The speech I made at the CSJ reasserted the Conservative mantra, which I fully subscribed to, that poverty or deprivation were never an *excuse* for crime. But, I added, there was a context, a background, that we needed to understand better. So, as I put it, when people crossed the line and committed a crime, the response needed to be rapid and tough. But to help more of them stay inside that line, we needed more understanding, more help – even more love. I homed in on 'hoodies', the name for both the hooded sweatshirts teens wore and the teens themselves: 'When you see a child walking down the road, hoodie up, head down, moody, swaggering, dominating the pavement – think what has brought that child to that moment,' I said.

We needed to deal with the background issues that led some towards a life of crime, like family breakdown, unemployment, drug addiction, children growing up in care, and educational underachievement. It was a classic compassionate Conservative speech and series of remedies. But the combination of hoodies and love outraged some in the press: 'Hug a Hoodie' was the *News of the World*'s take on the intervention.

I don't regret the speech. It set the context for a new approach: committed to backing the police and supporting tough penalties in our courts, but tackling the failures of the care system, reforming adoption, targeting family breakdown and chaotic families, and beginning the long process of reforming our prisons. These were to be some of our most important achievements in government, and their genesis was in a speech that many at the time said would herald our defeat.

As part of the same train of thought, even before I became party leader I had been developing the idea of a school-age programme that would help our children – all children, not just a privileged few – gain the skills they would need for adulthood, such as resilience, confidence, teamwork, respect and responsibility.

I came up with the idea after talking to those who had taken part in National Service, the period of compulsory post-war service in the forces which ended in the 1960s. The main thing that came across from those conversations was that everyone had been in it together. It didn't matter who you were, rich or poor, white or from an ethnic minority, academic or not – you forged a common identity. That's why I wanted there to be a residential element in this new programme, to take teenagers out of their comfort zones and put them into groups with others of different backgrounds, and also a volunteering element, teaching them the value of putting something back into their community.

National Citizen Service was, I believed, the answer to many questions of our age. The education system was failing to equip children with the skills for adulthood; NCS could help fill in the gaps. Our society was broken; NCS could teach the respect that was so lacking. Integration hadn't worked – we were still too segregated, too suspicious of each other; NCS would bring people together, and prove that ultimately we had so much in common. Although it was never made compulsory, NCS would end up as a rite of passage for every teenager who wanted to take part. Today, more than 500,000 have done so, and it is the largest and fastest-growing youth volunteering project of its kind in Europe.

As we developed individual policies, a theme was emerging. This was helped along by another moment that would have a profound impact on me, and as a result, on the future direction of the party.

Balsall Heath was a neighbourhood in Birmingham that had been blighted by crime, prostitution and antisocial behaviour. House prices fell. The middle classes moved out. But a group of people who remained had got together and taken matters into their own hands. They tore down the escorts' fliers, harassed kerb crawlers and reported the drug dealers to the police. They started taking better care of the parks and public spaces, planting shrubs and trees.

I was so taken by this story that I went to stay with one of the residents, Abdullah Rehman, and his family. I ate with them, slept in their spare room, and walked their children to school with them. Interestingly for a British Muslim family, they had chosen the King David Jewish faith school, on the basis that it had a good ethos and understood the importance of faith. 'We all believe in Abraham,' Abdullah told me as we dropped the children off, before showing me around the community he had helped to transform.

Here, in this Midlands suburb, society was proving more effective than the state. Bit by bit, the idea of government nurturing a stronger, better, *bigger* society was forming in my mind.

So in those first few months there was a lot to sort out: the political strategy, the governing philosophy, the personnel, the purse strings and the policies. But those aren't the only demands on a new opposition leader.

If you have any hope of being an effective prime minister, and of looking like a credible candidate for the job, you need a crash course in diplomacy, and foreign and security policy. My early overseas trips did a lot to shape my world view.

The first was to Paris to see Nicolas Sarkozy, before his run for the French presidency. He was the interior minister at the time, and famous for his fiery personality. My first taste of this was waiting outside his office door with Ed as he shouted at someone. '*Imbécile! Imbécile!*' was all we could hear.

Sarkozy was captivating – small, wiry and full of energy. He was always accompanied by an equally energetic translator, who spoke at a hundred miles an hour. He told me how he admired the British economic reforms, and wanted to be the Thatcher of France. He clearly believed in the 'great man' theory of history – muscular leaders making bold decisions and changing the world – and wanted to be one of them. I later came to feel that Sarko, as he was known, was less radical in reality. But an incredible act of kindness towards me in later years would make me grateful to him for the rest of my life.

I first saw Angela Merkel at an election rally in Stuttgart, when she walked on to the stage to the Rolling Stones song 'Angie'. In her speech she complained about the interference of the European Commission, which had told barmaids in Bavarian beer cellars what they could and couldn't wear. I would use this for years afterwards to persuade her that there was a Eurosceptic lurking inside her too.

My decision to leave the EPP rankled with her, but it didn't affect the close partnership we went on to form. While she profoundly disagreed with the move, she could see that I was a conservative who took a different view to her on the vital issue of European integration.

When we met I could see that she was, as Margaret Thatcher had been, the best-briefed person in the room, able to work out in advance other people's negotiating needs and strategies. I immediately saw that she was someone I could work well with. She has a sense of humour, and is an anglophile. From behind the Berlin Wall she had admired British science and British democracy. She saw us as natural allies when it came to vital issues such as support for NATO, backing fiscal prudence and a belief in free trade. Above all, I liked her down-to-earth, straightforward manner. There was no flummery or flattery – she liked to get on and talk about the things that mattered. And, again like Thatcher, she used her charm to get her own way. But Merkel is not a Thatcher. Her favourite expression is 'step by step'. This was to be disastrous for the Eurozone, which needed bold reform but got incrementalism.

It was in America that I met the forty-third president, George W. Bush. He was charming, intelligent and conviction-driven, quite unlike his caricature, and I admired what he was doing in the fight to combat AIDS and malaria. Yet I had tried to set myself apart from his neo-conservatism in a way that maintained Britain's strong bonds with the United States. On the fifth anniversary of 9/11 I made a speech whose most reported line was that liberty couldn't be dropped from the air by an unmanned drone. This was a criticism of unbridled neo-con interventionism, not a call for the unbridled American isolationism we are seeing a decade on. I didn't believe you could have global US and UK leadership if you point-blank refused to intervene anywhere.

While these were all standard stop-offs, I also strayed dramatically from the path usually trodden by party leaders: India.

As I said in a blog I wrote at the time, we couldn't afford to carry on obsessing about Europe and America while ignoring the fresh new forces that were shaping our world. It was an amazing visit. I travelled around Delhi in a tuk tuk, and walked through the Mumbai slums in the pouring rain to visit a community project, shocked at how starkly poverty and wealth sat side by side. While Tony Blair was fending off an attempted coup at home, I looked as if I was on a prime-ministerial visit. The contrast was helpful.

Sudan was a trickier visit, for here was the humanitarian crisis of our time. In Khartoum we met President Omar al-Bashir, a pariah who was later indicted by the ICC. When I mentioned an attack on a town in Darfur, in western Sudan, he claimed that it had actually taken place in the neighbouring country of Chad. Infuriated, I told him to look at a map. It was my first experience of how some of these leaders brazenly just lie.

The refugee camp itself was unforgettable. The sight of tents and huts stretching for miles, a city in the desert. The families who had lost everything, and had seen loved ones mown down by the Janjaweed militia as Sudanese soldiers looked on. The women, many of whom had been raped, telling me their harrowing stories. The only light relief came when we were sitting around talking through a translator, me bouncing one of the babies on my knee, and the baby decided to wee on me. Everyone laughed. Some things are universal.

In the middle of this hell was a literal oasis – a fifty-foot corrugated-iron tank, providing clean water for thousands of refugees. British aid sustaining and saving people's lives.

Much of my approach towards development in later years could be traced back to that time, and to the pride I felt in the aid workers from the charity Oxfam – based just down the road from my constituency – who we stayed with during that visit.

While some of these visits broke with tradition, my next, the following year, broke with much of the international community.

In August 2008, Georgia, a sovereign country that had every right to regard its borders as inviolable, had been invaded by Russia on behalf of two Russian-backed but unrecognised statelets, South Ossetia and Abkhazia. It was a clear case of illegal aggression and occupation, and I believed the world's oldest democracy had a duty to stand with one of the youngest and say so. I went to see President Mikheil Saakashvili, who I had met before and who I admired for his efforts to eradicate corruption, attract investment and get people to pay their taxes, a problem many leaders fail to crack.

He was under huge pressure, but was just about coping. There was tension in the air. Russian tanks were just twenty-five miles from the capital, Tbilisi. No one was quite certain if the ceasefire would hold, or the Russian tanks would start moving again.

'History has shown that if you leave aggression to go unchecked, greater crises will only emerge in the future,' I wrote in one article. 'Today, Russia says it is defending its citizens in South Ossetia. Where tomorrow? In Ukraine? In central Asia? In Latvia?'

They say you shouldn't make predictions in politics, but sometimes you do without realising it.

While modernisation was still being criticised by some in the press and the party, the public gave its verdict at the ballot box.

In the 2007 local elections we gained nearly a thousand new councillors and thirty-nine new councils. That represented 40 per cent of the vote, with Labour and the Lib Dems on 26 and 24 respectively. We were on track, edging closer to power. But there were rows ahead that threatened to throw us off course.

David Willetts, my shadow education secretary, whose vast intellect led to his nickname 'Two Brains', had given a speech on freeing schools from local authority control. In an aside, he talked about the evidence against grammar schools aiding social mobility, and said that a Conservative government wouldn't open any more of them. Fine – that was our stated policy. I had said from the outset that there would be no going back to the 11-plus on a national basis. I was happy for the 164 existing grammar schools to continue, and to be allowed to expand, as we wanted other good schools to be able to do; but our focus was on improving standards for all 3,000 state schools.

Cue unprecedented uproar when the *Today* programme covered the speech. Shadow Europe minister Graham Brady was enraged. The *Telegraph* was incensed. The 1922 Committee was in revolt. Meanwhile, I was in Hull, spending three days at a school as a teaching assistant, and hearing all this down the phone from Ed.

On the subject of grammar schools, I reached for a new medium to set the record straight. I wasn't just a blogger, I was a vlogger, recording a series of 'WebCameron' videos that were uploaded online.

I felt that the call to 'bring back grammars' was an anti-modernisation proxy, and I wasn't going to stand for it. I looked down the lens and said: 'It is a classic example of fighting a battle of the past rather than meeting the challenges of the future . . . The way to win the fight for aspiration is to put those things that worked in grammars – aggressive setting to stretch bright pupils, whole-class teaching, strong discipline, to name but three – in all schools.' In fact my position was more nuanced than I made it sound. I still believed existing grammars should be able to expand, and in the same vein, that new ones could be built in areas where they were already established and population growth required it. I clarified this, but it looked like a climbdown.

And it came at a bad moment. We were just about to have a change of prime minister. Within a few days of the grammar school row it was Tony Blair's final PMQs.

After he had spoken his final words from the despatch box, the Labour benches stood and applauded. I too stood up, and gestured to my own side to join in. They did.

Cherie Blair came and thanked me afterwards. She is another person who is quite unlike her public caricature. I'll never forget, when I took Ivan to the premiere of the children's film *Ben 10*, Cherie bending down to his wheelchair, looking him in the eye and speaking to him with great kindness and compassion.

I thought it was important to pay tribute to her husband in his last Commons appearance. For good and ill, he had changed British politics forever. And as I applauded, I felt a small inner thrill at the knowledge that a big obstacle on our path to victory had toppled. We were on our way.

But of course, it wasn't to prove that simple.

10

Cliff Edge, Collapse and Scandal

It's June 2007, Gordon Brown is prime minister, and it does not stop raining.

There was something apt about the ex-chancellor's premiership beginning with the wettest weather in decades.

I had – and still have – huge respect for Brown's intellect and his appetite for hard work. And mutual friends have told me how charming and entertaining he can be in private. But in public he seemed to have only one character setting: dour.

And when it came to Parliament, he had only one political setting: everything was about killing the Tories. While other Labour frontbenchers would build relationships with their opposite numbers, Brown would have absolutely nothing to do with his. The one time he did reach out to his shadow George Osborne, George and I were having dinner in Pizza Express in Notting Hill Gate. Brown wanted to 'pair' – i.e. agree that neither of them would vote in an important forthcoming debate. When George very politely explained that he couldn't do this without consulting our chief whip, Brown simply shouted and swore at him, before slamming down the phone.

So when he succeeded Tony Blair, I was rejoicing. We were ahead in the polls. And I was up against someone who hadn't been elected, who had some real flaws – and who I thought it was possible to beat.

But initially things didn't work out that way. As ever, 'events' intervened.

On Brown's second full day in the job, there was an attempted bomb attack in London's Haymarket, and then, the day after, terrorists drove a jeep laden with gas cylinders into Glasgow Airport. Brown reacted swiftly and effectively – and struck exactly the right tone about the threat we faced and how we should meet it.

The non-stop rain led to non-stop floods, affecting first one part of the country and then another. Brown immediately toured the affected areas, pledging money to flooded-out communities and families.

Then, after plagues of fire and rain, came disease. Foot-and-mouth was discovered on several Surrey farms. Having spent little more than a day on holiday, the new prime minister darted back.

And as his side of the political seesaw rose, mine began to sink.

First, Quentin Davies, a pinstripe-suited Tory MP, defected to Labour with a resignation letter of pure vitriol. His criticism of the modernisation project was very personal.

Then came a by-election in Ealing Southall. Our candidate was a successful and engaging British Sikh called Tony Lit, and although we were never going to win in the London borough, we wanted to put up a good fight. But in doing so we ended up setting expectations in the wrong place. I had also agreed to the idea of the candidate running on the ballot paper under the description 'David Cameron's Conservatives'. This looked arrogant and hubristic. I campaigned hard, visiting the seat five times – and we came a dismal third.

I had a chance to seize back the initiative. Social action – our policy of backing volunteering at home and abroad – was a strand of modern, compassionate conservatism we were determined to demonstrate. Project Umubano, led by the MP Andrew Mitchell, was to bring together forty enthusiastic party volunteers in Rwanda that summer, and I was to join them for a night.

The problem was that parts of Witney were still flooded. But I had visited the flood victims, and I was absolutely determined that the Conservative Party would not be a follower on overseas aid, but a leader.

Nevertheless, the visit was dogged by questions about why I was in Africa when my own constituency was under water. That night I looked out from the Christian mission where we were staying, gazing over the lights of Kigali, reflecting on the critical coverage. I knew that it had been a mistake to come. But sometimes there are mistakes in politics you're glad to have made, and this was one of them.

When Brown overtook us in the polls, rumours began swirling around about an impending vote of no confidence in my leadership. It really was personal.

Brown summed up the mood at PMQs with a rare quip (that's how bad things were – Gordon Brown was making effective jokes): 'The wheels are falling off the Tory bicycle, and it is just as well that he has got a car following him when he goes out on his rounds.'

William Hague was emphatic that if Brown was thinking straight, he would call an immediate general election, before the party conference season even began. That way, he would give us no chance to make up the ground we'd lost. I knew that we had just one chance: we had to deliver a Conservative Party conference in October that would meta-phorically blow the doors off.

Though our policy-review teams hadn't even reported back yet, we cobbled together a bumper series of announcements for each day of conference, from cutting stamp duty to introducing new cancer treat-ments. The Friday before conference, the whole lot – every single policy – was emailed to George's chief of staff, Matt Hancock.

But Matt's email address included his middle initial. We had inadvert-ently sent the full Tory plans to eccentric Lib Dem MP *Mike* Hancock.

The sender was mortified. The press officers were up in arms. I, however, was sanguine. 'They're great policies,' I said. 'If they leak, they leak. I'm off home.' So many things in politics are seen as a calamity. Very few actually are.

However, we would spend the whole conference somewhat on tenter-hooks, wondering on which day our precious policies were going to be published before we announced them. To this day I still don't know why they weren't.

Labour had a successful conference in Bournemouth, where Brown's chief bruiser Ed Balls was briefing that there would be an election.

Then came our turn in Blackpool. A cliff-edge moment for our party – and for me.

William opened with a cracker of a speech, chastising Brown for host-ing Margaret Thatcher in Downing Street the previous month (a move he must have hated but which made him look both magnanimous and bold).

George then unveiled what I termed his 'hammock idea', the confer-ence announcement he'd always dream up while reclining somewhere hot over the summer. This year was the biggest yet: raising the inher-itance tax threshold to £1 million. It was deeply Conservative, rewarding people who worked hard, saved and wanted to pass something on.

The finale of the conference, as always, was the leader's speech. It would be back in the Empress Ballroom in Blackpool's Winter Gardens where I'd delivered that leadership-winning, no-notes speech two years earlier. I had been pondering whether I could repeat the feat, not as a stunt, but because I was genuinely frustrated by my seeming inability to get across who I was, what I thought and what I wanted to do for Britain.

The lecterns I spoke behind felt like a barrier between me and the audience, distorting what I was saying and what people were hearing. Steve Hilton agreed. Sam told me to go for it. But last time was just ten minutes, I said. This is an hour. I have to cover everything. And it's my political life on the line now.

But I knew what I wanted to say. It would be me up there, no artifice, no barrier. So in the run-up to the conference I was not just working on my speech with Ameet Gill, but secretly learning its structure, key points and key phrases as we went along.

Come the morning of the speech, I had rehearsed sections but never practised the whole thing in one go. Sam and I snuck out early for a walk on Blackpool beach. I bounded back, full of vim.

As I walked out onto the stage, I knew it was do or die. 'It might be messy, but it will be me,' I told the packed hall. As well as being 'me', it was terrifying, exhilarating – and knackering. After an hour, I reached the peroration. 'So, Mr Brown, what's it going to be? Why don't you go ahead and call that election? . . . Let people decide who can make the changes that we really need in our country. Call that election. We will fight. Britain will win.'

I wish I could say I owed it to Cicero. In fact, it was inspired by the moment that David Niven loses his temper with Gregory Peck at the end of one of my favourite films, *The Guns of Navarone*. All that classical education gone to waste.

Before the conference began we had commissioned a 'tracker' or daily poll to see if anything we were doing was shifting the dial in terms of what the public thought. And we decided to continue the poll as the conference came to an end. It was money well spent. Our poll ratings ticked up daily through the conference – and then shot up at the end. I watched the news that evening and thought that I could see – for once – that I had really made that vital connection: from the hall, through the television, to the viewer at home.

But the country's cameras were now trained once again on Gordon Brown: will he or won't he?

The next day we were straight back into election planning meetings, as the tracker revealed we were neck and neck with Labour.

Then on Friday, as I drove to Dean, Andy phoned to tell me about a significant opinion poll which would be in that Sunday's *News of the World*. It had been carried out only in marginal seats, and it showed, pretty comprehensively, that Labour would not win an election. Far from extending their majority, they would be losing seats to us. It was the final – and in my view, the key – factor that caused Gordon Brown to decide not to hold an election.

Brown argued that his decision had nothing to do with the polls. This enabled us to get the narrative going that as well as being indecisive and temperamental, he was taking people for fools. Andy came up with the refrain 'Brown's bottled it', and we even had bottles of Brown ale made.

A word on being indecisive. The previous year, February 2006 had brought Elwen into our lives. Like Nancy, he was born under C-section at St Mary's, Paddington.

Normally, parents can discuss baby names at their leisure. But we didn't have that luxury. Gabby burst in soon after the birth telling us we had to come up with a name *now*, otherwise I'd look indecisive. I liked Arthur. Boring, said Sam. She sent me out to buy a book of names, and decided on Elwen – not the Welsh Elwyn, but the J.R.R. Tolkien version, meaning 'friend of the elves'. So Elwen he became (but Arthur Elwen on his birth certificate).

Everyone who was there during the summer and autumn of 2007 remarked on how calm I was. Calm on the eve of the make-or-break conference. Calm when I was told about the accidental email leak. Ed found it infuriating that, just as I didn't overreact to bad news, I was often disappointingly unimpressed when he brought me good news – treating triumph and disaster just the same.

People may interpret that as being indifferent, or 'chillaxed'. It's not. It's because I know that bollocking people, blowing your top, throwing tantrums, doesn't get you anywhere. It didn't help Gordon Brown.

But Brown had helped us. By flirting with an election, then pulling out, then denying his reasons for doing so, he exposed his weaknesses. At the same time, he had brought out our strengths – our ability to refuel, to recalibrate, to come together as a team when we were under assault, to stick to the course even when events were trying to divert us. And the fact that our modernisation was working.

Brown continued to demonstrate his tin ear when he stuck with his plans to abolish the 10p rate of income tax for some of the lowest earners in Britain, in order to reduce the overall rate from 22p to 20p. Labour MPs were in full cry on behalf of all those who were going to lose out. There didn't seem to be any way of compensating them without either reversing the policy in its entirety or spending a vast sum of money. It was to have a big impact on the electoral battles ahead.

Conventional wisdom holds – and my experience so far had proved – that there are two days that matter more to an opposition than all the others: local election day and party conference speech day. These are the moments – sometimes the only moments – when the searchlight beam catches you, and people focus briefly on politics and consider whether your party is up or down, and whether you look like a prime minister or a duffer.

I became increasingly obsessed with this theory, and knew that the London mayoral election in May 2008 was another such moment. We would only win in London if we could find a candidate who could reach out beyond our Conservative-voting base.

Boris Johnson likes to say he was my last choice, but it's not true. George and I were keen to persuade him, and we worked hard to do so. One of the promises we made was that we would do everything to help him run the best-financed and organised campaign that money could buy. We made good on this promise by delivering to him the best campaigner on the planet: Lynton Crosby.

On election night, when it became clear the Conservatives were going to have their first London mayor, Boris arrived at the party at Millbank. As we walked in together, we joined hands and raised them in the air. A great pic. But Boris didn't let go. So, rather strangely, we walked into Millbank Tower hand in hand. 'I told you: hold, lift, drop,' Andy chastised us. 'Where was the drop?' Of course, the drop came much later.

A fortnight afterwards we faced a by-election in Crewe and Nantwich, following the death of Labour stalwart Gwyneth Dunwoody. I threw myself into the campaign, visiting the constituency several times. Standing on a bench in the high street, giving an impromptu speech, I looked around at all the support – sometimes you don't need tracker polls, you can just sense it – and I thought: we're going to win this.

We did – the first by-election win from Labour since 1982. Ed

Timpson became the first Tory in Crewe since the 1930s. The tide was turning.

But despite being on a roll, news of another by-election was much less welcome.

The shadow home secretary David Davis had decided, bizarrely, to force a contest in his Yorkshire constituency in protest at Labour's increase of the maximum period of detention without trial from twenty-eight days to forty-two. This was a policy our party was vigorously opposing in Parliament, so the only conclusion I could reach was that the whole thing was a vain – and potentially damaging – ego trip.

William – yet more William wisdom – made me *promise* that I would not guarantee David his job back once the by-election was over. 'It's a team game, and he's decided to leave the team,' was his blunt assessment.

I called David and explained that if he insisted on the by-election, we would support him in the campaign. But I needed a full-time shadow home secretary, and could not guarantee him a return to the role. The truth was that I was delighted to have this unexpected opportunity to dispense with him, without anyone being able to say I was to blame.

I played it safe with his replacement as shadow home secretary, appointing Dominic Grieve, the shadow attorney general and a top Commons performer. While his views on combatting terror were similar to David's, they were a little more nuanced, and I felt he would give more priority to concerns about national security.

In the end, the division between civil liberty Conservatives like him and national security Conservatives like me was to prove fatal, as we were later to clash over the European Court of Human Rights. These tensions paved the way for Theresa May. One of the reasons I thought she'd make a great home secretary was that we agreed on these issues and many more.

All my woes during the beginning of Gordon Brown's premiership were about to pale, however, as two meteorites hit British politics in quick succession: the financial crash and the MPs' expenses scandal. Both would shake people's faith in the establishment, shape politics – and in the case of the crash have a huge impact on people's lives for many years to come.

Brown was right to say that the economic crisis 'started in America', because it was there that subprime mortgage lenders had been providing credit to people who hadn't a hope of paying the money back. Other financial institutions sliced and diced these loans into toxic bonds that were bought worldwide as investors searched for high yields.

And, of course, it was in 2008 that American investment bank Lehman Brothers fell, dragging the world's financial markets down with it. But it was a crisis to which Britain was particularly exposed. Our lending and banking practices had been infected with similar over-exuberance. One of our largest mortgage lenders, Northern Rock, was among the first victims of the credit crunch, and faced collapse in 2007. Our banks were some of the most over-leveraged (indeed, the later bail-out of the Royal Bank of Scotland remains the biggest rescue of a bank ever). And – absolutely crucially – our economy was built on a mountain of debt. Not just private sector debt, but rising government debt from an administration that hadn't adequately used the good years to run surpluses and pay down debt.

So yes, the fire began in America. But Britain had been piling up kindling for many years.

Being the opposition party at this moment left us with a difficult balancing act. Hold the government to account, but don't damage the national interest. Support the government in its necessary action, but make sure you don't become an irrelevant echo. Think through the policies needed for the future in a way that convinces people, while avoiding populist kneejerks. An additional complication was that we were the party that had championed the deregulation that some were arguing had allowed the bad banking practices to take place. We were up against a prime minister who had been chancellor for a decade, and who believed he understood the complexities of the international financial system better than anyone.

And then there was the most difficult thing. We had agreed – and announced back in September 2007 – to match Labour's public spending plans.

Governments determine the base line of arguments about tax and spending. If you depart from it, you end up vulnerable (as we had been in 2001 and 2005) to being described as vicious cutters or, as in Labour's case, big taxers.

Labour had solved that problem in 1997 by offering voters a period of stability in which they would match our plans. After that, all bets were off. We had been critical in the 2005 election of Labour's borrowing and spending, and remained critical, but we had lost the argument.

We had had to make a decision when a possible 2007 election loomed, and had decided to use Labour's 1997 technique. We would match their plans for a couple of years, allowing us the freedom to impose better control after that.

In the light of the 2008 crash, this was clearly a policy mistake, if not a political one, and we needed to change our approach. So we tried to do three things in framing our response.

First, we would be constructive. As Her Majesty's Loyal Opposition, moments of national crisis demand that you put the emphasis on the word 'loyal'. Over in America we were seeing the damage that could be caused by political wrangling, with the rejection by Congress on 29 September 2008 of the Troubled Asset Relief Program. It sent the markets into free fall.

I was in Birmingham at our party conference, and decided to make an emergency statement on the penultimate day. In that, and in my main speech on the final day, I struck a constructive tone. Not only should we be working with the government, but with the financial services industry. I knew instinctively that this was what was needed to meet our short-term priority: preventing a rapid banking collapse and thereby protecting people's jobs, homes and businesses. And I knew it was necessary to meet our long-term aim: fixing the free enterprise system so that never again could it inflict this damage.

That's why we supported Brown's plans for the recapitalisation of the banks, for example when the government bought 58 per cent of RBS in November 2008. There was a strong argument for stripping the most damaged assets out of the banks and creating a 'Bad Bank', as other countries had done in previous crises, but ultimately we backed the injection of public funds to prevent their collapse.

Second, we took our time. We formed a council of advisers, comprising former banking chiefs, top civil servants, Conservative chancellors and others, to guide our approach. Sir Brian Pitman, former head of Lloyds Bank, who I had got to know when I was at Carlton, was a regular visitor. Terry Burns, former Treasury permanent secretary, was key, as was Ken Clarke, who we soon brought back into the fold as shadow business secretary.

They were unanimous that, while it felt as if we were facing a totally new and potentially terrifying set of economic circumstances, there were lessons to learn from history. The Wall Street Crash hadn't caused the Great Depression, it was the banking crisis that came after it, and the policy response to that crisis, which let bank after bank close, taking with them savings, credit and any chance of recovery.

Those who argued that all we needed was tighter financial controls and more government spending were wrong: this was a *monetary* crisis, and the most important part of the solution was *monetary* action: flooding the system with liquidity, preventing the collapse of systemic financial institutions, establishing new sources of finance – government ones, if necessary – to lend money to small businesses now starved of cash.

Confident of this analysis, the third thing we had to do was to be bold. In November 2008, we announced that we would move away from Labour's spending plans. Championing prudence was particularly brave at a time when the whole world was fixated on a Keynesian 'spend, spend, spend' solution to the crash. But we genuinely believed that the government's fiscal position was so precarious that it could not afford to go beyond the 'automatic stabilisers' of higher benefit bills and lower tax receipts that in any event push up the budget deficit when the economy stops growing.

Discretionary increases in government spending and tax cuts were all right for those countries that could afford them; those that couldn't were playing with fire. So, in another bold step – particularly for a party that prided itself on supporting low taxes – when Labour announced a temporary cut in VAT, we voted against it.

The real boldness, however, was in directly advocating a policy of austerity in terms of cutting government spending for the future. After all, what party goes into a general election talking about cuts? And we were using that crucial word: cuts.

This caused more trouble for Gordon Brown, who, after having mistakenly declared himself to have ended the b-words – boom and bust – simply refused for weeks and weeks to use the c-word.

Some critics say that we were as naïve as Brown – and that we never saw the bust coming. But it was *before* the crunch and crash that I'd given a speech at KPMG warning about Labour's unsustainable deficit and debt. We knew their overspending would come to bear on us all. We knew the economy was built on sand. We just didn't know the meteorite would hit when it did.

Other critics say that we were desperate to cut public spending in order to dismantle public services. Well, since we'd promised in 2007 to match Labour's spending plans, clearly that wasn't the case. The reality was, in the phrase George coined and then made famous through endless repetition, they hadn't fixed the roof when the sun was shining.

The reason for cutting was therefore the total opposite. It was to *save* public services. The greater the debt, the more money we would be spending on repayments. The weaker the economy, the less to spend on public services. We saw this clear as day, and I suspected that working people would see it too. They knew the UK hadn't been living within its means, and that that needed to change.

We were making some tangible policies in order to prevent such a situation occurring ever again. That included another bold step, which was to give away a power that chancellors had long held.

I had some experience of the stringency with which annual accounts and results were published when I worked at Carlton. Lawyers and auditors would pore over every word to check for accuracy. But that was business. In politics, it seemed to be totally different. It was up to the Treasury to predict future growth on which its spending plans would be based. That gave it the opportunity to manipulate the growth forecasts and fiddle tax receipts, which Labour took full advantage of.

Gordon Brown and Alistair Darling's forecasts became works of fiction. At every Budget and Pre-Budget Report they would become increasingly optimistic. When better estimates or true figures emerged, there were wild disparities. George and I proposed that an independent Office for Budget Responsibility (OBR) should make those predictions instead. Of course it wasn't easy for anyone to guess how much the economy would grow (or, at this point, contract). But by removing the potential for bias, we could prevent figures being massaged to justify spending increases our country couldn't afford.

That Christmas of 2008, shops like Woolworths and MFI were disappearing from our high streets. The New Year, 2009, brought confirmation of Britain's first recession since 1991. What concerned people wasn't just how the deficit would affect them today. It was how this 'spend today, pay tomorrow' culture was saddling the next generation with debt. How wrong it was – morally – to make our children pay for our excesses. This was perfectly captured in our poster that January – a picture of a baby with the line: 'Dad's nose. Mum's eyes. Gordon Brown's debt.'

Brown's favourite insult to hurl at me was 'This is no time for a novice.'

I worried deeply that, at this time of great financial turbulence, it would become especially potent. After all, George and I had never held ministerial office. And we did make mistakes.

But we had carefully thought out our strategy. We stuck to our instincts – instincts we believed the public would share – that a crisis caused by recklessness and secrecy should be met with prudence and honesty. And we ended the period with more people trusting us than the government on the crucial issue of managing the economy – the first time we had had a consistent lead over Labour since the 1990s.

But the second meteorite was about to hit.

Great political controversies tend to drive us all to think in simple headlines. This one seemed like a straightforward story: brave campaigner sets out on a mission to unearth grave wrongdoing; Parliament resists releasing information it ought to; the information comes out anyway, and it reveals appalling practices, illegality and corruption.

All those things are true of the expenses scandal, but the full truth is more complex. And I didn't just have a front-row seat, I was on the judge's bench – and in the dock myself. And what I saw from that unique position was, yes, wrongdoing that needed unearthing, but also unfairness, heartbreak and lots and lots of grey areas.

The context was this. MPs were entitled to claim an 'Additional Cost Allowance' (ACA) of up to £24,000 per year for running a second home, because they have to be based in Westminster for part of the week and in their constituency for the rest of it. There was also £22,000 of 'Incidental Expenses Provision' for office expenses, over and above the salaries for staff. Rules about employing relatives as members of staff were virtually non-existent.

The ACA in particular was treated by many as if it should be claimed automatically. Over many years the salaries of MPs had been held back – usually for political reasons – and their allowances increased instead. Often, MPs would just send a load of receipts in to the Commons Fees Office and let them decide what household expenditures or repairs should qualify.

Our party had a taste of what was coming when it was revealed that there was little evidence of what MP Derek Conway's researcher, his son, had done to earn the thousands of pounds he'd received at the taxpayer's expense.

Labour experienced its own preview of the scandal when some receipts for which home secretary Jacqui Smith had been reimbursed were published. They included two pay-per-view pornographic films, which her husband soon owned up to.

I knew I had to act fast. I made sure all our MPs filled out a 'Right to Know' form that provided the basic details of their expenses claim and whether they employed any relatives. These would be made public.

Labour's reaction was to continue to attempt to keep the problem under wraps. Leader of the House of Commons Harriet Harman wanted the House to vote to ensure that MPs' expenses were exempt from the Freedom of Information Act. But it was too late. The *Daily Telegraph* bought a stolen disc with every MP's expenses claims set out in full, receipts and all. Day after day it published the details. Determined to remain ahead of the game, I called a press conference at the St Stephen's Club. My response included an apology and a roadmap – and I didn't conceal my anger about what had been going on: 'Politicians have done things that are unethical and wrong. I don't care if they were within the rules – they were wrong.'

I set up an internal scrutiny panel, a so-called Star Chamber, including my aide Oliver Dowden, known as 'Olive', who I also called 'the undertaker', since he so frequently brought me the bad news. The panel, assisted by a team that was combing through all the information, would examine the expenses claims of every Tory MP, and would decide who had to pay money back. Anyone who refused to comply would lose the whip. This was faster and firmer than the Parliament-wide independent inquiry that Brown would set up, and showed that we had understood the severity of the situation and had gripped it early.

The whole thing became a daily ordeal. The *Telegraph* would call up in the morning, give us details about whose expenses they were going to expose the next day, and allow us until 5 p.m. to respond. Ultimately the call – both the judgement call and the phone call to the MP – had to be made by me.

The calls included some of the strangest conversations. 'Your entire family were working for you?' 'Why do you need a ride-on lawnmower?' 'What does your swimming pool have to do with your parliamentary duties?'

As before and since, I always tried to give people a chance to explain their situation, rather than being driven by arbitrary deadlines. But it was hard. Some colleagues didn't help themselves by taking to the airwaves. Anthony Steen became one of the most famous examples. Having claimed over £80,000 for the upkeep of his country house, including rabbit fencing and tree surgery, he insisted to BBC radio that he had behaved impeccably: 'I've done nothing criminal, that's the most awful thing, and do you know what it's about? Jealousy. I've got a very, very large house. Some people say it looks like Balmoral . . .'

He had to go.

I had to deal with Peter Viggers, who had claimed, among £30,000 of gardening expenses, for a £1,600 'floating duck island'. With Michael Ancram, whose swimming-pool boiler was serviced at taxpayers' expense. With Douglas Carswell, who put a 'love seat' on expenses. With John Gummer, whose moles were removed from his lawn using public funds. It just seemed to go on and on, and to get weirder and weirder.

Then there was Douglas Hogg, who was, according to reports, reimbursed by the taxpayer for having his moat – yes, moat – cleaned. To be fair to him, he had never actually claimed for this directly. He had given all his receipts to the Fees Office and, satisfied that they added up to far more than the ACA, they had just given him the full amount. I could see his point. He was doing what he had been told to do. It was all within the rules. But, as Andy put it, the point was that he had a bloody moat.

There were some heart-rending examples at the other end of the spectrum. One MP who was asked to pay back thousands of pounds was desperate not to do so, both because it would look as if he was admitting guilt when there was no impropriety, and because as a young MP with a large family he genuinely couldn't afford to.

The sheer, granular detail being unearthed made the scandal run and run. Receipt by receipt, the *Telegraph* gave an insight into MPs' lives – and revealed a class that seemed completely out of touch with normal people. Never mind that most MPs had claimed only for rent or mortgage interest payments or the odd piece of IKEA furniture, and had been urged by the authorities to claim even more. The colourful examples stood out, showing a world of ride-on mowers, moats and mole-free lawns.

Of course, I had a colourful example of my own – violet-coloured, to be specific.

I had only ever claimed for the mortgage interest on my constituency home. It was a simple approach, specifically permitted by the rules, and seemed to me to be clearly within the spirit of what was intended. But one year I had an extra bill, which I handed to the Fees Office. Ordinarily this would have been logged as 'maintenance', and would have attracted absolutely no attention. Except, like a good West Oxfordshire tradesman, my builder had detailed the work he had done: 'Cleared Vine and Wisteria off of the chimney to free fan.' As with so many other claims, it was the detail that did for me. People still ask me how my wisteria is doing today.

Every party was embroiled. Which meant we could only fix the broken system if we worked together.

In fact, before the scandal broke I went to see Gordon Brown in his Commons office with Nick Clegg to see what the three main party leaders could do. He gave us a take-it-or-leave-it proposal: a per-day allowance for MPs – not all bad, but far from right.

Had we been with Tony Blair, the three of us could have thrashed out something workable. With Brown, it was pointless. He was sullen and stubborn, and couldn't hide his contempt. Clegg and I both concluded that it was impossible – *he* was impossible.

What were the long-term implications of the expenses scandal?

We lost a lot of good MPs, as people who weren't even guilty of any wrongdoing, such as Paul Goodman, left Parliament.

The British Parliament is one of the least corrupt in the world, but it would be forever tainted. I believe deeply that people go into it to make a difference and to serve, not to see what they can get out of it. Yet ever since 2009, my postbag has been full of letters about how venal our MPs are.

It left many Tory MPs feeling aggrieved with me. They felt that the system I had set up to clean house made them look as though they had done something wrong, when they felt they hadn't. At one point Andy walked into my office and said there was a serious chance of rebellion. I said I didn't care. 'This is the right thing to do – if it's going to take me down, then so be it.'

So while it stored up bad feeling between me and some in the party, I don't regret the position I took. Politics ended up with a model that was more transparent and that cost far less, and our party drove that.

In the normal course of things the searchlight might land on politicians once or twice a year, but its harshest glare is saved for momentous events like the financial crash and the expenses scandal. Under that intense scrutiny, I thought we had demonstrated that we were the party with the answers not just to a broken society, but to a broken economy and broken politics too.

But we were also suffering from our own broken – a broken promise.

In 2004, Tony Blair had pledged to hold a referendum on a proposed European Constitution, but it was rejected at plebiscites in France and the Netherlands.

The European powers went back to the drawing board and came back with the Lisbon Treaty, which retained many of the elements of the constitution – creating an EU Council president, foreign minister and diplomatic service, eliminating national vetoes in many areas, and paving the way for more vetoes to be eliminated.

We argued straight away that if the government had said it was going to have a vote on the constitution, it *must* have one on this treaty – especially since it was more significant and far-reaching than its immediate predecessors, Nice and Amsterdam. I was clear about the lessons from Maastricht: it was right to give people their say on such important changes.

I thought – I still think – that Labour's failure to hold the referendum it had promised in 2004 was outrageous. It had managed to avoid all questioning on the European constitution during the 2005 election campaign by saying it would be subject to a separate vote. And then it didn't hold one. So in the *Sun* in 2007, as the treaty was still being nego- tiated, I gave a 'cast-iron guarantee' that a Conservative government would hold a referendum on any EU treaty that emerged from the negotiations.

In 2007 it seemed entirely likely that we would be able to fulfil this if we entered government, since we all thought that the Parliament wouldn't run its full course to 2010. But as Brown delayed, member states had the time to ratify and implement the treaty – including the UK, which did so in July 2008.

By 2009, our last hope was the Czech Republic. I wrote to the presi- dent, Václav Klaus, pleading with him to wait until after a UK general election before he ratified the treaty. But he replied that he could not hold out that long – another eight months – without creating a constitu- tional crisis. The Czech Constitutional Court ruled against the one remaining challenge to the treaty, and Klaus signed it that November. On 1 December the treaty was passed into law. Our promise to hold a referendum on it was redundant.

I gave a speech declaring that a Conservative government would never again transfer power to the EU without the say of the British people, and that any future treaty would be put to a vote. (True to our word, we made this 'referendum lock' law in 2011.) In that speech I talked about 'the steady and unaccountable intrusion of the European Union into almost every aspect of our lives'. I said: 'We would not rule out a referendum on a wider package of guarantees to protect our democratic decision-making, while remaining, of course, a member of the EU.'

I could feel the pressure on Europe quietly building. The anger at the powers ceded at Maastricht and since was reawakened by the denial of a referendum on Lisbon. The anger was not just coming from the usual suspects and the Eurosceptic press, but from constituents and moderate MPs. I felt it too. I was thinking intensively about the issue, and about how to make this organisation work better for us. And I was clearly stating that a referendum of some sort might be on the cards at some point in the future.

11

Going to the Polls

There haven't been many general elections in this country at which voters have shifted en masse from one party to another. The Liberal landslide of 1906, Clement Attlee's triumph in 1945, Margaret Thatcher's victory in 1979, the rush to New Labour in 1997 – these are remembered as some of Britain's great swing elections, whose winning governments went on to change the course of our history.

As 2010 approached, I knew I needed to perform a similar feat. It wouldn't be enough just to take a few extra constituencies. We would have to win 120 more seats, and retain all our existing ones, if we were to have any hope of a majority. No Conservative had achieved anything like it since Churchill in 1950, and even he fell short of an outright majority and needed another election to return to Downing Street. We had a mountain to climb.

I may have had Everest in front of me, but I had the best sherpas by my side.

Thanks to Andrew Feldman we were going into the election in the strongest financial position in our history.

Stephen Gilbert, the thoughtful and reserved Welshman and campaigning powerhouse, already had our target seats identified, our candidates selected, and the volunteers geared up, ready to fight the ground war.

Andy Coulson had transformed my relationship with the media and translated the theory of modern, compassionate conservatism into something tangible and exciting.

Ed and Kate remained by my side. Gabby was joined by Caroline Preston and Alan 'Senders' Sendorek handling the media. Oliver Letwin was still coordinating policy, and Steve still adding his brains and buzz.

Meanwhile, Liz Sugg was poised to turn our plans into practice. At the drop of a hat she could pull together visits, rallies, interviews, drop-ins, walkabouts, anything. Haranguers were kept at bay, staff were all in position, the speech was on the lectern, and there was no danger of me being snapped in front of words like 'exit', 'closing down' or 'country' (there's always the danger of blocking out the 'o' with your head . . .).

But what was the message we should take to the country? We didn't have just one answer – we had several. One focused on fixing our broken economy. Another on mending our broken society. A third was to re-emphasise how much the Conservative Party had changed. And on which of those answers should be given priority, the team was split.

George, de facto campaign chief, and I had been making a series of speeches on the dangers of debt and the need for a new economic policy. The theme was clear: the principal task of our government would be an economic rescue mission.

Combining this with our strong attack on Labour was a single, clear message: only by removing a failing Labour government could we restore Britain's economic fortunes. This was what Andy, running the all-important communications operation, wanted front and centre.

But it was all a bit black-skies. I had been working since 2005 with the famously blue-skies Steve on a sunnier, more optimistic focus – on how we could deliver stronger public services and a fairer, more equal society. We called the organising idea 'the Big Society'.

I loved it. Conservatism for me is as much about delivering a good society as a strong economy. And building that good society is the responsibility of everyone – government, businesses, communities and individuals – rather than the state alone.

This philosophy – Social Responsibility rather than Labour's approach of State Control – was summed up by Samantha when we were mulling over these concepts one evening in the garden in Dean. 'What you're saying is there *is* such a thing as society,' she said, referencing Thatcher's famous (but often misinterpreted) quote. 'It's just not the same thing as the state.' That summed up the theory perfectly. And in practice it fell into three broad categories.

First: reforming public services. We were absolutely not, whatever our critics alleged, going to dismantle taxpayer-funded public services. But to improve those services we wanted to empower the people who *delivered* them – trusting teachers to run schools, doctors to run GP surgeries, new elected police and crime commissioners to run police forces, and so on. Bureaucracy and centralised control would be out, local, professional delivery would be in.

This would only deliver better results if at the same time we empowered the people who *used* those public services, and gave patients, passengers and parents real and meaningful choices, including the ability to take their custom elsewhere. Otherwise we would simply be swapping one monopoly for another.

Just as the Thatcher governments transformed failing state industries into successful private-sector industries, we wanted to bring the same reforming vigour to enable not just the private sector but also charities, social enterprises, individuals, and even cooperatives, or mutuals, to deliver public services. That went right down to local people being able to take over community assets like post offices and pubs.

The second element was about finding new ways to increase opportunity, tackle inequality and reduce poverty.

Since the 1960s, and particularly after 1997, the size of the state had ballooned, spending had surged, more and more power had been centralised, yet the gap between the richest and the poorest had actually increased.

In a number of important ways, the Big State was sapping social responsibility, and as a result exacerbating the very problems it set out to solve. The development of the welfare system was the classic example. Some of the interventions to tackle poverty had had the opposite effect. There were perverse incentives that deterred people from finding work or from bringing children up with two parents.

It took away people's agency. Drug-addiction programmes, for instance, focused on replacing one addictive substance – heroin – with another – methadone – rather than encouraging addicts to go clean.

We wanted to unleash the power of what we called 'social entrepreneurs', usually charities and social enterprises, to tackle some of our deepest problems, from drug addiction to worklessness, from poor housing to run-down communities.

I was inspired by people like Debbie Stedman-Scott. Debbie had come from a tough background, and went to work for the Salvation Army across Britain before setting up an amazing employment charity, Tomorrow's People, which helped people in the most deprived communities to find and keep a job. I visited it several times.

I was also inspired by Nat Wei, who helped create the Future Leaders programme, which sought the best teachers to lead inner-city schools.

And there was also Helen Newlove, who had campaigned tirelessly for the victims of crime since the murder of her husband Garry in 2007. They were amazing people. The Big Society was about empowering them.

The third element was about a step change in voluntary activity and philanthropy.

We proposed that the state should act as a catalyst, boosting philanthropy and volunteering and, for example, encouraging successful social enterprises to replicate their work across the country. That is where programmes like National Citizen Service (NCS) and training a network of community organisers came in.

In the past, we claimed over and over that Labour was the big-state party and we were the free-enterprise party. But we didn't have enough to say about how free enterprise, or indeed any of our other values – responsibility, aspiration and opportunity – could deliver the non-economic things people needed. About how we could provide better schools. Or help people off drugs. Or transform their neighbourhoods. About how Conservative means could achieve progressive ends. The radical reforms that came under this Big Society umbrella had the potential to change all that.

Like all radical proposals, it came in for criticism.

Andy feared that the combined austerity/Big Society message sounded as if we were saying to people both 'Let us cut your public services' *and* 'Get off your arses to deliver those services yourself' – a miser's mixture of 'Ask what you can do for your country' and 'On yer bike.'

Other critics said I was drawing too much on my rural, upper-middle-class upbringing by advocating the Big Society. Well-off people have the time, money and inclination to dedicate themselves to local causes. Those on minimum wage who are juggling two jobs and several children do not. Yet, as I had seen, some of the most deprived neighbourhoods had remarkable social entrepreneurs and community spirit, from volunteers cleaning parks in Balsall Heath in Birmingham to mothers combatting gang culture in Moss Side in Manchester.

I thought the Big Society became more, not less, necessary in a debt-ridden world. Every government in the developed world was having to learn to do more for less – and fast. We could lead the way, and by reducing the long-term cost of social failure, we could drive down the deficit in the process.

Our failure to choose between this theme and the others could be seen in our advertising, specifically our posters. 'We can't go on like this,' the billboards across a thousand different marginal seats said, next to a giant headshot of me, wearing an open collar and a serious expression. 'I'll cut the deficit, not the NHS.' The message didn't land well, because it was a sort of two-in-one. Even worse, my photo had been altered so much that I ended up looking like a waxwork.

It provided an ideal canvas for idle hands. On one Herefordshire hoarding I was spray-painted with an Elvis-style quiff, and 'We can't go on like this' was followed by 'with suspicious minds'. A website was set up for people to produce their own spoof versions. Thank God my children weren't very old at the time. They love teasing me, and they'd have made one for every day of the week.

Yet for all the derision, it was, unlike most election posters, *true*. In government, we did cut the deficit. We didn't cut the NHS.

The disagreements between the team – particularly between Steve and Andy – were never fully solved. By this point the fire-and-ice pair were deliberately assigned a shared office in the middle of the open-plan Conservative HQ, dubbed 'the love pod'.

Sadly, uncertainty and some unforced errors were to continue, and then came a jolt from the polls on 28 February 2010. The *Sunday Times* front page read: 'Brown on Course to Win the Election.'

As an opposition leader, you embark upon the final few weeks before a general election – the so-called short campaign – with exhilaration and dread.

The dread comes from the constant possibility of screwing up. The whole process of a campaign is very presidential, and the result very personal. The exhilaration comes from the fact that you're able to break out of the media cycle and parliamentary timetable and get on an equal footing with the government.

For five years the media cycle had been a source of great frustration. In opposition, we worked hard to research and develop strong policy. If it was about schools, for instance, we would meet and talk with heads, teachers, governors, parents, academics and think-tanks. We would research what had been tried overseas, prepare policy papers, lay out the costs and the sources of funding – and then perhaps arrange a visit to something equivalent that already existed in order to accompany the announcement. It's really strong, exciting stuff – and you set out the steps for how it's going to change the country.

Then you switch on the news that evening and find you've been given twenty seconds to explain it. What follows is analysis – often reporters interviewing other reporters – not about the policy itself, but about what political advantage you are seeking by coming up with this new idea and whether or not it's an election winner. And, of course, all this is combined with the latest plot twist of who is up or down in the great parliamentary soap opera.

On 6 April, Gordon Brown fired the starting gun for a 6 May election. I was sitting in my glass-walled office in CCHQ after our first 7 a.m. daily election meeting. Officers from the Met were on their way to give me police protection throughout the campaign. If we won, they'd probably be with me for the rest of my life. (They are the most wonderfully kind and dedicated people. And they do try to give you as much personal space as possible. A week after the election, Sam and I were out for dinner and she leaned over to me and whispered, 'Those people on that table there – I've seen them before.' 'Yes, darling,' I said. 'They're from the protection team.')

Off we went to Birmingham, then Leeds, then seventy-three other constituencies in just four weeks. For years I'd done my Cameron Direct events, letting the public fire questions at me on any subject. It was exactly where I wanted to be, on a little stage we were carting around the country, not much more sophisticated than the soapbox John Major had taken around eighteen years earlier.

What Major didn't have, though, was a man dressed as a chicken following him everywhere. Tony Blair did – one of CCHQ's apparently. But what goes around comes around, and now I had a chicken of my own, this time from the *Daily Mirror*.

To begin with it was funny having this birdman on my tail. But the novelty wore off, and I finally decided to confront the stooge, unmasking him by lifting the head off his costume. It turned out that he was called Tristan, and he was left completely speechless when I asked what it was he was so keen to ask me. The next day, in Saltash in Cornwall, I was hit by an egg, enabling me to finally answer the question of which came first.

By this point I felt we were really getting somewhere with our economic message. Leaders from great British brands like Marks & Spencer, Sainsbury's and Corus steel were coming out to condemn Labour's proposed National Insurance rise, branded a 'jobs tax'.

It's important to emphasise what a shift this was. Since Black Wednesday, New Labour had courted Britain's businesses effectively. Now we could claim to be the party of business once again.

We were making progress on our society messaging too. With the help of Michael Caine we launched the centrepiece of the Big Society, the National Citizen Service, or NCS. Expressions of interest in setting up Free Schools were coming in from around the country. And our pitch to public-sector workers about cutting bureaucracy and enhancing local control was a vast improvement on our efforts in 2001 and 2005.

A few days later, I launched our manifesto at Battersea Power Station. The manifesto was a blue hardback book titled 'An Invitation to Join the Government of Great Britain', which emphasised the Big Society theme.

But all the usual election paraphernalia – posters, chickens, eggs and manifestos – was about to be eclipsed by something completely new.

Ever since 1964, when Harold Wilson challenged prime minister Alec Douglas-Home to appear in a TV debate, there had been a similar call from someone during general election campaigns. In the past, it had always been the underdog doing the calling, and the favourite refusing (and in recent years that favourite had been followed around by a chicken – because they'd 'chickened out').

Until now.

I had decided back in 2005 that I wasn't going to fit into the normal pattern of resisting debates if I was in the lead, or of calling for them if I was falling behind. I was going to go for it. I liked TV. I liked debating, although perhaps I hadn't paid enough attention to the fact that when I'd debated on TV during the party leadership campaign it hadn't gone well. Anyway, I always felt that TV debates were coming. The UK's first general election leaders' debates would take place in 2010 because, for the first time, the front-runner was calling for them.

Bill Knapp and Anita Dunn, the US experts I had hired to help me prepare for the coming ordeal, were brutally frank about the reality I was about to encounter: to my disappointment they told me that these wouldn't really be debates at all. You don't want to engage with your opponents' argument, you just want to put your own point across. You should focus your efforts on delivering your pre-prepared soundbite down the camera lens. Avoid too much spontaneity in taking apart opponents' arguments; it's far too risky. Just get your 'zinger' – a one-liner destined for the headlines on the news programmes after the show – ready beforehand, and deploy it as soon as you can.

My disappointment quickly turned to worry. We did some practices in Millbank, with Damian Green (and sometimes Olive Dowden) playing Gordon Brown, and Jeremy Hunt as Nick Clegg. Halfway through, I threw down my notes. 'It's hopeless. Clegg will win hands-down. It's easy. He can just say "A plague on both your houses."' Even if I'd been Demosthenes or Cicero, he was going to win.

Before that first debate, history in the making, I'd never been so nervous in my life. The news channels covered the build-up as if it was England in the World Cup final. Brown, Clegg and I stood on a primary-coloured set like gameshow contestants. As predicted, Clegg was painting the blue and red parties as the old guard, and himself as the new kid on the block. It seemed a breeze for him. He was even using the same phrases that Jeremy Hunt had as his stand-in during our mock debates – 'two old parties', 'more of the same', 'there is an alternative'. Nick had prepared well – and he was good.

I was bad. Not switch-off-the-telly, hide-behind-the-sofa bad. But aloof and stiff. Lacking passion. Anecdotes that were too contrived. And one bit of absolutely essential preparation that I failed to put into practice was properly looking down the camera when I spoke. Colleagues and friends were polite afterwards, because while Clegg had undoubtedly won, at least I hadn't lost (that honour went to Brown).

But Samantha was brutal. 'You were hopeless – and you've got to watch the whole thing through all over again to see just how bad it was.' She was right – and I did.

The Lib Dems surged ahead in the polls – into the lead in some. There was even a poll that said their leader was nearly as popular as Winston Churchill. Britain was in the grip of a new phenomenon: Cleggmania.

And I took it hard. I was the one who had wanted to do these debates. I hadn't prepared properly for them. I'd let everyone down. I lay in bed, running through a list of people in my head, friends who I thought were going to lose their seats because I'd screwed up. The feeling was worse than fear or disappointment. It was guilt.

The second debate, in Bristol, went much better. It was on foreign affairs, and Clegg was vulnerable here. His party manifesto rejected 'like-for-like replacement of the Trident nuclear weapons system', thus putting our deterrent at risk. It didn't do him too much damage though – the two of us drew in the opinion poll afterwards.

Arriving in Birmingham for the third and final debate, my anxieties reached a new high. So much was riding on my performance. The future of the country. The future of my party, my team, my friends and family. My future.

What happened next wasn't planned or predicted, but I suppose it was inevitable. Brown and I had clearly both gone away and done the same thinking about Clegg. This guy was – according to the polls – running away with the election. Yet he was inconsistent. His policies had never really been subjected to proper scrutiny. The numbers didn't add up. His manifesto included some seriously odd ideas.

We took him to bits, starting with his pledge of an amnesty for illegal immigrants. Months afterwards Nick told me that if he'd known this policy would be so contentious he never would have let it into the manifesto.

Afterwards, I bounded back into the hotel room. A poll there and then showed that I had won the debate. The relief was enormous. Finally I could get back to the real campaign, which culminated a few days later in a twenty-four-hour sprint to the finish.

My top-to-toe tour of Britain began in a hi-fi factory in East Renfrewshire in the evening. By 10 p.m. I was in a Carlisle fire station, with officers clocking on for the night shift. At 1 a.m. I was wandering around a factory in Darwen, Lancashire, before crossing the Pennines for a 3 a.m. tour of a Morrison's depot in Wakefield. Next was the Grimsby fish market at 5 a.m. as trawlermen delivered their morning catch, followed by the first lesson of the day at a school in Nottingham, and then an ambulance station in Dudley.

Life on the campaign bus (which is rather like a band's tour bus, but with less booze and more journalists) tests your senses as well as your stamina. After we had boarded with wet shoes from the Grimsby fish market, Sky's Joey Jones decided to put some roast beef in the oven that was on the bus. As we wended our way round the windy Welsh roads towards a school in Powys, surrounded by the inescapable smell of fish and beef, everyone began feeling sicker and sicker.

At long last we reached our final stop, Bristol, where supporters including Sam had gathered. I was wrung out, but I had to give a rallying cry with what felt like the last breath I had in me: 'I want a government that makes us feel good about Britain again – all that we are, all that we've done, all that we can do in the future . . . a government that's about hope, and optimism and change in our country, not the doom and the gloom and the depression of the Brown years, which we can, tomorrow, put behind us – forever.'

But the news bulletins were focused on something else. Athens was ablaze, and several people had been killed after protesters reacted to planned austerity measures. Economic volatility and the vulnerability of countries like Britain returned to the foreground. The world was in turmoil. Who were the public going to ask to run Britain in these uncertain times? A Conservative government? A Labour government? Or something else?

As on every previous polling day in my adult life, I got up early on 6 May to go to cast my ballot. Sadly, my early voting was delayed by four whole hours, as two jokers had scaled the roof of my local polling station, Spelsbury Memorial Hall, to erect 'Vote for Eton' signs and swig champagne.

It is strange voting for yourself to become an MP. It is even stranger voting for yourself to become prime minister. And whereas at previous elections I would have visited polling stations and committee rooms in my own constituency, this time I had sent all my party workers to neighbouring marginal seats. So I spent the hours that followed fiddling around in my vegetable garden and chopping logs for our fire – two of my favourite ways of dealing with stress.

Later that night, Sam, my close team and I gathered in the sitting room at Dean to watch the exit poll. I felt a mixture of fear and hope – fear that we'd fail totally, and hope that we might be about to defy expectations. But I was left with a strange, in-between feeling when at the stroke of 10 p.m. David Dimbleby announced the result of the exit poll: 'It's going to be a hung Parliament with the Conservatives as the largest party.'

Some of the results that followed started to point towards a majority – the swings in Sunderland, the seats we picked up in the south, our better-than-expected performance in the south-east. At 1 a.m. we won Kingswood near Bristol – and that was 135 on our target list. But the north of England and London weren't going as we needed them to.

As the voters' verdicts unfolded across the country, I went to Windrush Leisure Centre in Witney for my own count. Though my eyes were also on the 649 other contests taking place, I was still eager to succeed in the patch I loved so much.

I was also reminded why it's so important that our prime ministers are also MPs. Only in Britain would the person bidding for the highest office be sitting on a plastic chair watching his party's fate unfold on a crackly TV. It is humbling and grounding to be accountable to your own constituency. And it was with genuine pride that I increased my majority.

By 3 a.m. it was confirmed there *would* be a hung Parliament. I took a call from Arnold Schwarzenegger, the governor of California, congratulating me on my win. But Arnie, I said, it wasn't a win. In what was the most surreal moment of the whole election, there I was, in a leisure centre, in the middle of the night, explaining the first-past-the-post electoral system to the Terminator.

When I got back to London I had to go to the CCHQ party in Millbank. The atmosphere was jubilant, but I was cautious – I could see from what was on the boards of 'results in' and 'results to come' that we were unlikely to make it.

Over at the Park Plaza Hotel I attempted to grab an hour's sleep. As I closed my eyes, I lay there pondering it all. It was looking like being the most successful Tory result in eighteen years. But I was surprised and confused about the Lib Dems. Cleggmania had well and truly faded. They had lost seats. Yet – another odd feature of our politics – he might now be kingmaker.

I steeled myself. This had been the hardest slog of my life. But what was to come might be even harder.

I am clear what it was that produced the great swing in our favour. We had changed the Conservative Party, making it appeal once more to Middle England and making people in urban, liberal Britain feel that they could vote for us.

I am also fairly sure why we didn't win outright. There was too much 'and' in our campaign – the Big Society *and* austerity; cutting some public services *and* increasing others; continuing to modernise *and* hammering Brown and Labour.

As for the debates, they didn't have a dramatic impact on the outcome. The Conservative share of the vote was close to where the polls were at the start of the formal campaign. The Lib Dems also ended the campaign close to where they began, though Nick Clegg gave his party a tremendous boost where otherwise it may well have been squeezed by the two bigger parties and lost even more seats.

The truth is the real benefit of the debates to the Conservatives was elsewhere, and is often overlooked. By sucking the life out of the campaign, the debates meant Labour was never able to get its powerful anti-Tory cuts campaign off the ground.

Looking back, my view is simple: in those desperate economic times, even after the changes we'd made, the Conservative Party hadn't quite sealed the deal with the electorate. People were still uncertain about the Conservatives.

This was even more true at a moment when budget cuts were needed. As I've said, we'd intended to respond to voters' concerns by matching Labour's spending in the first two years, and by promising to share the proceeds of growth between more spending and tax cuts. This formula was easy to understand, and allowed us to reduce the relative size of the state while still increasing the amount spent on essential services.

Then came the financial crash, and these reassurances weren't possible any more. So people were uncertain. And we had been reflecting their uncertainty rather than allaying it.

Remember also that it was Everest we were trying to climb. We were trying to win a historic number of seats, while the electoral geography massively favoured Labour.

The data shows why. In 2005, a 35 per cent share of the vote had given Labour a majority. With our 36 per cent share in 2010 – and two million more votes than Labour – we didn't clinch it.

So how did we measure against those great landslides of political history in the end?

The swing from government to opposition was less than Attlee had managed in 1945 or Blair in 1997, but it was on a par with Margaret

Thatcher in 1979. And we had gained more seats – ninety-seven – for the Conservatives than at any election since 1931.

Yet it was what would happen next, the relationships I would forge and the decisions I would take, that was to make more significant political history.

12

Cabinet Making

'David, congratulations!' came the voice down the phone. It was President Barack Obama, and this was my first evening as prime minister after five tumultuous days of negotiations in May 2010. 'Enjoy every moment,' he said, 'because it's all downhill from here.'

I would often tell the story – and would use the same line when ringing other presidents and prime ministers after their election victories. But it wasn't entirely true. The early days and weeks in government went extremely well, in a way that confounded many people's expectations. Some thought a coalition would be unstable and prone to early collapse. In fact, at a time of great difficulty, when markets were fragile and protests were breaking out across Western capitals, the administration that I put together with Nick Clegg was to prove one of the most stable governments in Europe.

I believe the coalition succeeded in those first few weeks and months in part because our party had spent five years in opposition preparing for power. I had thought a lot about how to do the job of prime minister. I knew that the 'who', 'what', 'where' and 'why' questions would hit me the moment I walked through that big black door. And I understood that the mechanics mattered. Because of all that preparation, as well as feeling daunted, honoured and excited by the prospect of being PM, I felt ready.

In opposition we had developed a good system of short and focused daily meetings to bring the top team together and despatch the business of the day, chaired by George in my absence. The civil servants were doubtful that the routine would last, but six years later we were still assembling – PM and chancellor – for the daily '8.30' and '4 o'clock', as we called them.

I've always believed that ministers who automatically purge their predecessors' staff are cutting off their noses to spite their faces. So

I decided to keep the private secretaries I had inherited from Brown, including the smart and sardonic James Bowler as my principal private secretary, as well as cabinet secretary Gus O'Donnell and No. 10 permanent secretary Jeremy Heywood – both former colleagues from my Lamont days. Their expertise would be invaluable. 'Your job is not to tell me when I get it right, but to tell me when I get it wrong,' I told them.

The integration between my staff and George's was to continue in government. We were one team, and I believe that became one of the secrets to our success, particularly as our driving mission was economic rescue.

The 'one team' spirit also applied to the No. 10 operation, where I wanted the political appointees and the civil servants to work together. And I wanted that open, trusting, collegiate atmosphere to flow through the coalition too. That meant, rather controversially, that our spads would work side by side, sharing offices. Sometimes people would walk into a room and find it difficult to tell who was the Tory, who was the Lib Dem and who was the civil servant.

We didn't get it all right. The Conservative Party in opposition had tended to criticise the growth of No. 10 as making the PM's office too 'presidential', and, in line with that thinking, I scrapped the PM's 'delivery unit'. This was a mistake – and we reversed it over time, building a similar team focused on the implementation of government policy.

Another early error was running a joint Conservative–Lib Dem Policy Unit. It soon became clear that this would be very difficult when everyone involved had loyalty to different leaders and their eye on the next party conference or general election, at which point we would be competing, not collaborating. The Policy Unit was split in April 2013.

The next question was where would I base myself in Downing Street. Margaret Thatcher had what was called her study on the first floor, by the stairs that led to her No. 10 flat. Tony Blair had his 'den' at the bottom of the main staircase, whose yellow walls are tiled with pictures of his – our – predecessors. Gordon Brown opted for something different – something that resembled a trading floor or newsroom.

In the end I chose the room that had been Blair's. It was close to everyone else on the ground floor – the private secretaries, the duty clerks who staff the place night and day, the 'garden rooms' teams who support the PM wherever he or she is in the world, the press office, speechwriters, and events and visits team. But you could also shut the door, hold very private meetings and work, write or make telephone calls without being disturbed.

It was in that office, on my first evening, that I sat and read the letter Gordon Brown had left for me. Tony Blair's letter of congratulations came soon afterwards. One of his pieces of advice stuck with me: however tough the job gets, remember that the British people have a grudging respect for whoever is trying their best to do it.

The one issue Sam and I had not settled was where we would live. Sam wasn't sold on the idea of moving into either the flat above No. 10 or that in No. 11 – it was something we would have to work out. So that night, my first as PM, I went home to North Kensington.

Leaving the next morning for my first full day in my new job, I stepped out of our terraced home and into an armoured Jag. We sped off towards central London, with eight motorcycle outriders around us. Some would split off to block the junctions that fed into the main road, clearing our way. Which was incredible – until I saw the tailbacks I had caused. I felt like President Mugabe. So that was it – day one, executive decision: no outriders, except for emergencies. (I did use the Special Escort Group more and more as time went on. And they really are the most professional elite officers, protecting everyone from the royals to visiting PMs and training other forces from around the world.) The experience also made me wonder how practical our west London home might prove to be.

Then, of course, came the crucial question of *who*: who would I appoint to each department of government?

The art of the shuffle – and the reshuffle – was something else I had learned a lot about in opposition. But the pressure and media attention in government is many times greater. Basically, reshuffles are a nightmare. You are dealing with egos, big ones. Every move is scrutinised. Any delay is dithering. Any job rejected is a snub. End up with the wrong balance of left and right, male and female, and you are either hopelessly politically naïve or absurdly politically correct – or if you are unlucky, both at the same time.

Friends in business used to say, 'We all have to take tough decisions to get the right top team – why all the fuss about political reshuffles?' To which I would reply, 'Yes, but you don't have to appoint your entire team all on the same day, in full view of the world's television cameras. And the ones *you* sack go away. The ones *I* sack sit behind me and plot my downfall.'

The best way is to plan reshuffles like a military operation. You need to have a strategy and stick to it. Every move is timed and scheduled. And whatever the resistance, even if you are forced to make small tactical retreats, you need to keep advancing.

Appointing your first government, though, is, relatively speaking, a pleasure. After all, you are helping colleague after colleague achieve one of their political dreams: taking office.

The first moves were obvious: making George chancellor of the exchequer and William Hague my foreign secretary. We had worked closely together over the previous five years, and I had assured them that the jobs they had in opposition would be the jobs they would take in government. I made William first secretary of state, essentially my deputy, so he could chair cabinet and take PMQs in my absence.

On that first night in Downing Street I also confirmed to Patrick McLoughlin that he would be the chief whip, and could work with the team at No. 10 to plan the cabinet and government formation that would happen the next day.

I wanted it to be straightforward. I had long believed that secretaries of state were shunted from job to job too often. Labour's ministerial musical chairs had become increasingly absurd – they had a new home secretary every two or three years.

Most members of my shadow cabinet knew what jobs they would be doing in government because they had been shadowing them for much of the past five years. Andrew Lansley went to Health, Liam Fox to Defence, Eric Pickles to Local Government, Owen Paterson to Northern Ireland, and Andrew Mitchell to International Development.

But after the shoo-ins there were some surprises.

Theresa May stepped into the Cabinet Room and sat opposite me across the green table. She had huge experience in shadowing government departments, having covered the Education, Transport, Culture and Families portfolios over the years, as well as being the Conservatives' first female chairman. Most recently I had appointed her to shadow Work and Pensions – and I am sure that's what she expected to be offered.

'I want you to be home secretary,' I said to her. Theresa is not always the most expressive person, but she looked genuinely surprised and delighted. This may have been one of the least expected appointments – but it would turn out to be one of the best.

Another surprise was my choice for Work and Pensions. George – ironically, given their later battles – persuaded me that right-winger Iain Duncan Smith would provide balance in our cabinet.

Then there was the choice of party chairman. I wanted someone who was a loyalist and who would continue the work of delivering a modern, compassionate Conservative Party that reflected and represented all of Britain.

Sayeeda Warsi had joined the party as an adviser to Michael Howard after setting up a legal practice in her native West Yorkshire. She had been by my side through my leadership (halfway through I'd appointed her to the Lords), speaking with a no-nonsense conviction on issues domestic and foreign that I found refreshing and impressive. I told her I wanted her to be joint chairman with Andrew Feldman, who I hoped would continue his great work. She couldn't believe it. The daughter of a textile worker from Pakistan had taken her place in a Conservative British cabinet, the first Muslim woman to do so. The partnership was a statement about how far we had come, in that we had a Muslim and a Jew jointly chairing our party.

The cabinet jobs lost to Lib Dems would upset some, but there were fortunate solutions for most of them. Ken Clarke, for example, had been shadowing Business, but Justice was a good alternative, given his legal background. Francis Maude and Oliver Letwin were happy – and extremely effective – in the Cabinet Office, which would become an engine room for both delivering savings in government spending and making the coalition work.

Of course there were some people who were disappointed. Chris Grayling and Theresa Villiers had been in the shadow cabinet. Neither made it into the cabinet at the start of the government, but both took their lesser appointments with good grace and worked to prove themselves in their new roles. Theresa later made an excellent Northern Ireland secretary, and Chris served as lord chancellor.

But there were those who were more openly upset that their shadow portfolios were taken by Liberal Democrats. There were some awkward conversations, and in one or two cases it created simmering resentments. I understood how frustrated some colleagues were. Some had shadowed a government department for years on end, asking questions in Parliament and building up their expertise, only to have their dreams of office snatched away by the arrival of an unexpected coalition.

Appointing 118 ministers across just two days, there are bound to be some slip-ups – and I made a couple of errors.

The number of full cabinet members is strictly limited – to twenty-three – and, as other PMs have done before me, I tried to make the numbers add up by having some ministers 'attending cabinet' but not paid the full salary. Foolishly, I included Tom Strathclyde, leader of the House of Lords, in this category.

Tom explained – rather forcefully – that it wasn't about the money, it was that their lordships simply wouldn't stand for their champion in cabinet having 'lesser status'. He was right. The problem was that when he told me this I had already filled all the other posts.

Most secretaries of state have to be in the cabinet, so the only solution was to pick on the most agreeable Member of Parliament that I had already appointed and ask them to accept a downgrade. There was one outstanding candidate for this honour: my leader of the House of Commons, Sir George Young, a veteran of the Thatcher and Major governments and known as 'Gentleman George'. He lived up to his name, and couldn't have been nicer about the whole thing, pay cut included. All too often in politics the stubborn get rewarded and the good guys get shafted. On the other hand, George's grace is among the reasons he has had one of the longest, most extraordinary ministerial careers of modern times.

We agreed that the Lib Dems would have five cabinet positions and a number of ministers that corresponded to their proportion of MPs. And that led to the question of what job Nick Clegg should do. I was keen that he consider holding a major office of state, such as the Home Office or Environment. But he wanted instead to be deputy prime minister.

I didn't feel strongly about it. My priority was making the government stable at a time of such uncertainty, and I was perfectly happy with any solution that did that.

As for his choice of ministers, I insisted on formally appointing each of them as I did those from my own party. Again, I wanted to emphasise that this was one team under one prime minister.

Some of the Lib Dems on the left were sceptical but willing to make it work, like Vince Cable. Others towards the right of the party were enthusiastic, like David Laws. Another even said to me: 'You don't have to worry about me, I'm basically a Tory anyway.'

A few were quite emotional, even tearful, like Tom McNally. He had been elected a Labour MP in 1979, having previously worked in Jim Callaghan's No. 10. Now a Liberal Democrat, he was the party's leader in the Lords in a Conservative-led government. Although a left-leaning Liberal, he was determined to make it work, and did a great deal to ensure it did.

To begin with George and I were highly sceptical about whether the Lib Dems should get two big economic portfolios: chief secretary to the Treasury and business secretary. But in the end we agreed that it would be helpful to make sure there was a Lib Dem in the Treasury, and seen to be as responsible for the cuts programme as we were.

With a ministerial team in place and the mechanics whirring away, it was time to answer the most important question: *why*. Why was I in government, taking this approach, joining forces with a rival party? This was the most straightforward question of all to answer: we were engaged in a full-scale economic recovery job, to turn around the British economy and deal with what had become one of the largest budget deficits in the world, forecast to be over 11 per cent of our GDP.

And we were going to do it in a way that was true to the modern, compassionate Conservative message we had developed.

On our first full day, the stage was set to deliver the message. The chairs were laid out in rows on the Downing Street lawn, with an aisle down the middle. Cameras clicked, there was an expectant air, as Nick and I emerged from the Cabinet Room to stand before the gathered crowd.

The marriage metaphors used to describe the 'Rose Garden' press conference weren't that far-fetched. It did look and feel a bit like a wedding. And, like a groom, I was nervous, aware of the immensity of the occasion and the need to rise to it. 'We mustn't come up short here,' I said to Clegg in the Cabinet Room before we stepped outside. 'It is one of those times when we need to give it 20 per cent more than feels appropriate.' And that's what we both did.

'Today we are not just announcing a new government and new ministers,' I said. 'We are announcing a new politics. A new politics where the national interest is more important than party interest. Where cooperation wins out over confrontation. Where compromise, give and take, reasonable, civilised, grown-up behaviour is not a sign of weakness but of strength.'

The tone was meant to be historic, but on rereading it feels a little histrionic. These were unprecedented times, however, and I was trying to rise to them.

One consequence was that those who were disappointed that we hadn't achieved a majority felt we were rubbing their noses in it. And that I was enjoying myself with the Lib Dems a bit too much.

But while I understand this criticism, overall I don't regret the Rose Garden performance. The banter and bonhomie *did* help to set the tone for what we were about to embark on. They showed that Nick and I were confident we could work together and were clear about our task: to confront the economic challenge ahead of us.

Britain – this proud, prudent, economic powerhouse – was in deep economic trouble, and it needed a government that people had confidence in. One that would persist and could weather political storms, one that would give creditors confidence in the country they were being asked to lend to. Forming such a government was part of our solution to the problem.

The urgency of our task had been underlined on our first full day in office, when news emerged that unemployment had hit its highest level since 1994. We started work straight away on the £6 billion of in-year cuts that we had promised and that our coalition partners had now agreed to.

As politicians, we took the lead ourselves, cutting ministers' pay by 5 per cent and freezing it for five years. We also scrapped ministers' personal drivers.

We agreed on the importance of cutting taxes for the lowest-paid. So another early action was raising the threshold at which people started paying tax to nearly £7,500 – meaning that nearly a million more would no longer pay any income tax at all.

All this was underpinned by the 'Coalition Agreement' and the 'Coalition Programme for Government', both hastily compiled but surprisingly comprehensive combinations of our two manifestos, which became our blueprint for government. They stood the test of time.

I didn't feel we were being constrained by the burdens of office or the addition of the Lib Dems. On the contrary. I felt liberated by office. We were finally doing the things we wanted to do, not just talking about them.

While economic policy would prove the central challenge, it naturally wasn't the only one. Although I didn't appreciate at this point quite how much of my time as prime minister would be consumed by foreign and security issues, I was persuaded that we should prepare properly for the work that would need to be done.

One of the big changes to the machinery of government was the inauguration of a National Security Council, which came out of our policy review, chaired by Pauline Neville-Jones, in opposition. The rationale was simple. It no longer made sense to consider foreign policy on its own. The challenges we faced required a response from across government, not just the Foreign Office. Particularly with the rise of threats from what the experts like to call 'non-state actors' – basically terrorists – we needed to combine diplomatic, military and counter-extremist thinking.

Afghanistan was the classic example. We could only make progress if we could deal with the poor relations between Afghanistan and Pakistan (Foreign Office), assist in the fight against the Taliban (Ministry of Defence), deal with the flow of drugs from the region (Home Office), improve the country's potential for economic development (Department for International Development), while all the time working on the vital issue of countering Islamist extremism (Home Office again).

The National Security Council brought all of these departments together, combined with our intelligence services, MI5, SIS and GCHQ, and the armed forces, as represented by the chief of the defence staff. I appointed Peter Ricketts, who was the permanent secretary at the Foreign Office, as the first national security adviser. Coming straight from the FCO, he secured that department's cooperation with the new arrangements and carried out his new role with huge ability. The NSC is now a vital part of the UK government – and I believe will remain so.

Of course, there was another early national security question to be settled – this one by me alone.

Every PM must decide what set of instructions to send to the commanders of the Trident submarines for use in the event of a nuclear attack on the UK that has rendered other means of communication redundant. These are the so-called 'letters of last resort'. A senior naval commander comes to your office to brief you on the options and the process. You're then left alone with a series of alternative letters, and you decide which instructions to give. The others you shred in a giant, industrial-sized shredder which seems to appear in your office that morning.

It is the moment when the full seriousness of your responsibilities as prime minister come home to you. I had spoken with John Major about the approach he had taken, and I had decided on what I believed to be the right course of action. But even so, I sat and stared at the cold words on the page, trying to imagine the scenario in which one of our submarine commanders had to open one of my letters.

As I handed over the chosen letters to the officer – letters I prayed would never, ever, have to be opened – one of the envelope's seals popped open. A call for Pritt Stick and Sellotape was rapidly answered. An absurd moment in such a solemn process.

So what were my perceptions of office as the sun shone on us in those early days?

All in all, I felt we were successfully setting the scene for a long-lasting coalition, and for turning around our economic fortunes.

By the end of the first month Sam and I decided we and the children would move into Downing Street rather than staying in west London, and we brought across the entire contents of our home – bikes, beds, beanbags and, after a few months, our new baby daughter.

When we departed six years later, we left some of the furniture behind. This included an IKEA kitchen cabinet which I had assembled in the days just after Florence had been born. Nick Clegg had needed to see me, and found me in the kitchen surrounded by pages of instructions, wooden panels, nuts, bolts and screws. He immediately helped out, and we joked as we assembled the 'coalition cabinet'. Samantha commended us on our work, but pointed out that the two doors did not quite align with each other.

For me, living in Downing Street was perfect. Whether I was upstairs in the flat or downstairs in the office, I was never far from my two enormous responsibilities, to my family and to the country.

13

Special Relationships

Within minutes of arriving at the Renaissance palace in Rome that was Silvio Berlusconi's official residence, I was in his bedroom. The Italian prime minister was showing me an ancient two-way mirror. 'They didn't have porn channels in the fifteenth century,' he explained.

During your first few weeks and months as prime minister, you must begin forging the relationships that will help advance Britain's interests around the world. Personal bonds are vital; relations between countries really can be enhanced by the rapport between their leaders or jeopardised by the lack of it.

In the digital age, the old ways of doing things – messages passed through ambassadors or fixed times for formal telephone calls – are being augmented with new methods. I had a pretty regular text relationship with the Dutch and Swedish prime ministers and the crown prince of UAE, for instance, and I also exchanged communications with the Australian and New Zealand prime ministers.

But the traditional methods, including phone calls, formal diplomatic visits and international summits, do still matter. I had spent years laying the groundwork for these relationships. However, the amount of time I had to devote to foreign affairs as PM still surprised me. We once did a calculation which showed that a third of my time was spent on trips overseas, foreign policy meetings, hosting foreign leaders and National Security Council meetings.

My approach to foreign policy began with what I suppose might be called a patriot's view of British history: not one that ignored its flaws, but nevertheless one that felt great pride in our role and contribution. I didn't accept the idea that Britain was facing inevitable relative decline. Previous predictions of our demise in the 1960s and 70s had been defied by the economic success of the Thatcher years, with their global exports of privatisation, shareholder capitalism and the rule of law.

It saddened me to see some commentators talk about an inexorable waning of our influence. I understood that with the rise of India and China power was moving south and east, but I didn't accept that Britain couldn't forge its own important role in the world.

We still had some great advantages: our time zone placing us between Asia and the Americas, English as the global language, our universities and science base, expertise in aid and diplomacy, widely respected armed forces and an unequalled network of global alliances, including NATO, the EU, the Commonwealth, the G8, G20, IMF, and permanent membership of the UN Security Council.

The Labour governments had had some foreign policy successes. There had been the actions to save Sierra Leone and Kosovo, and Gordon Brown had used the G20 effectively to help coordinate action after the global economic crisis. But at the same time their foreign policy had been disproportionately defined by two relationships: with the US and the EU. Elsewhere, they had closed embassies and downgraded the importance of the Foreign and Commonwealth Office. Far too many old alliances had been allowed to slide.

It was clear to me that, alongside our economic rescue, reasserting Britain's global status would be one of our biggest missions in government. In fact, the two were intertwined. Building a stronger economy relied upon the goods and services we sold abroad and the investment we attracted at home. And our global reputation rested upon our ability to fix our economy.

I approached our foreign policy challenges as a 'liberal conservative', not a 'neo-conservative'. While I understood and sympathised with the doctrine of Bush and Blair – that spreading democracy around the world helped peace and prosperity – I felt that their rhetoric and actions didn't reflect the difficulties of achieving such change. I wanted a foreign policy that was practical, hard-headed and realistic.

My practical approach made me sceptical of the view of some in Whitehall that the job of the politicians was simply to set the overall strategy and leave its implementation to officials. Some of the military top brass were particularly keen on this, and thoroughly disliked it when

I interfered in deployments of ships, troops or submarines. But you cannot always separate tactics and strategy: the politicians need to be intricately involved in both.

The hard-headedness came from my belief that we should never be ashamed of using foreign policy to help generate prosperity at home. For too long, the FCO had neglected its economic responsibilities. India was one of the world's fastest-growing economies, yet we had fallen from their fourth biggest source of imports to their eighteenth in just one decade. We still exported more to Ireland than to the BRICS economies of Brazil, Russia, India, China and South Africa.

I saw an opportunity to reposition our diplomatic network as a generator of trade and inward investment. Some sneered at 'commercial diplomacy'; I made it a key plank of my foreign policy. We had a presence in 196 countries across the world. That was 196 potential shop windows in which to advertise Britain.

The realism was evident in our recognition that democracy wouldn't take root at the drop of a hat – or bomb. Its components – the rule of law, a fair judiciary, property rights, a free press – take time to embed. This was not a disavowal of military action. Instead, I saw it as a course correction from the excessive interventionism of the New Labour years.

I admired Tony Blair's passion for trying to solve problems and to intervene in difficult situations. But his foreign-policy actions lacked the consideration of a good driver – too keen to keep his foot on the accelerator and insufficient in his use of either brakes or rear-view mirror. At the same time, what I didn't want was for the pendulum to swing too far in the other direction and rule out intervention altogether. There are occasions when you have to intervene. The failure to intervene in Bosnia and Rwanda in the 1990s influenced the rush to intervene in Iraq. I didn't want Iraq to stop us intervening somewhere that it was desperately needed in the future. Every case and conflict is different.

When it came to the 'special relationship' with the USA, I had always felt it was special to me personally: I had huge admiration for the United States, loved spending time there, and as a child of the Cold War I was clear that it was our best friend and ally. It seemed obvious that we should want to maintain, and where possible enhance, the special relationship. And I believe that word – 'special' – is merited. In my view, the intensity of the partnership between our intelligence services justifies the title on its own, quite aside from the powerful ties of history, language, culture, trade and values.

But it shouldn't be our *only* special relationship. I wanted to carve out privileged partnerships with global giants like China and India, and to restore bonds which I felt had been weakened with Australia, Canada, New Zealand (no recent Labour foreign secretary had been to Australia, for example) and our other Commonwealth allies, like Malaysia and Singapore.

In the Gulf there was a whole series of countries that saw Britain as their oldest ally and friend – we needed to show that meant something to us too. Then there were the South American countries, many of which had long-standing historical relationships with Britain, such as Chile and Colombia, which were on the rise again, but with which Britain was not engaging properly.

The relationship-building with foreign leaders starts with the phone calls. The call with President Obama was followed by calls from all the foreign leaders over the next few days. Such conversations are useful, but it is the one-on-ones you have in which relationships are really cemented. I look back on those meetings now, and on what I was trying to achieve, and I can see them falling into four categories.

There are the agendas you began and which yielded real and lasting results. There are those you started but which got nowhere. There are those you wish you had pushed harder. And there are some you wish you had never started at all.

My first visit – and it is important which country you pick first – would be to France. With the Entente Cordiale and the fact that I knew Nicolas Sarkozy, it made sense.

There was one man who would prove essential to all this: my foreign affairs private secretary, Tom Fletcher. Tom had worked for Gordon Brown, and became my support, sounding board and source of information about virtually every country on Earth. He correctly warned me before this first trip that Sarko would be my 'best friend and biggest rival'.

Sarkozy made great statements publicly and privately about our friendship and how well we would work together. Knowing my love of tennis, he presented me with two state-of-the-art Babolat racquets, one yellow, one blue, reflecting the coalition colours.

The truth of our relationship was slightly different. Although we were both conservative politicians, and similar in many ways, even at this early meeting I could see how divergent some of our ideas were. His criticisms of the European Union tended to be that it wasn't doing enough, whereas in most areas I thought it was doing too much. He had the traditional Gallic and Gaullist distrust of finance and a suspicion of the City of London. I wanted a more balanced economy, for sure, but was totally opposed to harming what was a vital British asset, or sanctioning even more European action in this area.

With the warm atmospherics but the more difficult policy obstacles, I saw that our relationship was going to be strongest in terms of defence, security and countering terrorism – and therefore that we should focus as much as possible on those things. The 'Lancaster House Agreement' we would sign within the year was a step change in defence and security cooperation between the UK and France, including collaboration over the most sensitive of all issues, our nuclear deterrents. But this is an agenda where I wish we had gone even further.

France and Britain both share a global reach and a global ambition. We are Europe's only nuclear powers, and have Europe's only two properly capable armed forces. We're both permanent members of the UN Security Council. Our relationship within NATO became stronger with Sarkozy's brave decision to take France back into the alliance as a full political and military member. If we could bury some of the mutual suspicions about each other, the military and security cooperation could be far deeper, and could lead to great economies of scale.

Relations between us, having started so strongly, would dip during my first European Council meeting the following month. Sarko was complaining about the lack of effective action in the EU, whereas when it came to my turn to speak, I bemoaned the ongoing transfer of powers to Brussels. I could feel the room collectively sighing – particularly Sarko. It was a taste of things to come.

Contrast this with the bonhomie on display when he came to visit us later at No. 10 with his wife, the model and singer Carla Bruni. Waiting behind the famous black door before their arrival, I asked Tom, protocolwise, whether I should kiss Carla. 'Definitely,' he replied. 'How many times?' I asked. He left his answer until the moment the door opened: 'As many times as you can get away with, Prime Minister.' I met them both with a huge smile on my face.

With Angela Merkel, the tone was less emotionally charged, though she put on a full military welcome for my first visit to Berlin as PM, straight after I'd been to Paris. I found her logical, sensible and focused – clear about what you wanted and what she wanted. But at the same time she was fun, and often had a mischievous twinkle in her eye.

Although we had made up over the EPP, it was still an issue of concern for her. She wanted the strongest possible conservative bloc in the European Parliament. I responded (again) that my new group would be allies rather than enemies of the EPP on most issues, while taking a different view on EU integration.

In the end, I would have a much better relationship with Merkel than with Sarkozy. I was fascinated by the dynamic between them. Despite their nickname, 'Merkozy' – a press confection expressing the idea that theirs was a close alliance – her real attitude to him seemed to me a mild, eye-rolling disdain. I remember suggesting to her that the three of us should do something on promoting more free-trade deals between the EU and other countries and regions. She said, 'Let's do that together and let's get Nicolas along later, because he can be quite histrionic.'

A few months later at the G8 and G20 Merkel and I spent most of our time sitting next to each other. That's where the relationship grew stronger. During one of the economic sessions at the G20 in Toronto, England and Germany were playing in the World Cup. She was getting regular text updates about the game from her team, while I was pushing the 'refresh' button on the BBC Sport website and getting frustrated at the slow wi-fi. At one point she leaned over and whispered, 'I'm very sorry, David, but you're 2–0 down.'

After the meeting broke up, we agreed to go and watch the second half together, along with our teams of advisers. We drank beer and chatted, with one or two moments relieving the gloom of the result (Germany won 4–1). At one stage the commentator said, 'And now England makes an aggressive assault on the German defences,' and Angela turned to me, smiling, and said, 'We're never allowed to say things like that at home.'

While I was to build a strong and trusting relationship with Merkel, I would also put this down as an agenda I didn't carry far enough. Together we would chalk up some important collaborations, opposing attempts to allow protectionism at the G20, fighting off attacks on the need to cut budget deficits and achieving what many saw as impossible:

cutting the EU budget. But in the end the relationship didn't deliver everything I needed.

One diplomatic relationship that never even began was my attempt to appoint a new ambassador to the Vatican, as the current one was retiring. Who better, I thought, than Ann Widdecombe, a former Tory MP and one of Britain's most prominent Catholics?

But when I called her from my office in No. 10, she didn't believe it was me, and said, 'I think this is a hoax call.' On and on I went, trying to convince her. When she finally conceded that it was *probably* the PM she was speaking to, I began to tell her about the appointment I had in mind. But just as I thought I was winning her round, she apologised and said she couldn't – she had committed to take part in *Strictly Come Dancing*.

Diplomatic relationships are also made and nurtured at multilateral gatherings, and there were two major ones during my earliest weeks in office. The annual summit of Canada, France, Germany, Italy, Japan, Russia, the UK, the US and the European Union – the Group of Eight, or G8 – was held in Muskoka, Ontario, in June 2010, followed by the G20 in Toronto.

My overriding aims for my debut on the world stage were simple. I wanted backing for our economic strategy at home, and to defuse the 'fiscal stimulus versus deficit reduction' row that was brewing between the big nations. My view was straightforward: fiscal stimulus – i.e. unfunded tax cuts or additional spending paid for by more borrowing – was fine for those countries that could afford it. Britain couldn't. The stimulus we needed was the monetary boost being provided by the Bank of England and pro-enterprise policies, as well as support for free trade and opposition to protectionist moves internationally.

The discussion I remember best from that G8 is actually a minor bust-up I caused, which had the side effect of reminding the Americans that we were their closest allies. I mentioned that part of the answer to all the things we had been talking about – the Middle East peace process, Iran, Iraq, Afghanistan – was a stronger relationship with Turkey, and harnessing its support as an EU ally as well as a NATO ally.

Turkey. It sent Sarkozy off. Merkel backed Sarkozy, I think because she felt she had to. Obama backed me.

The scenery in Muskoka was incredibly beautiful – pine forests dotted with deep lakes. I have always been an enthusiastic open-water swimmer, and my prime ministerial wild-swimming career would take me

– and Liz, who always joined me – from the Shetland Islands to Australia. On day one of my first G8 I started with a swim before breakfast, with two speedboats full of police bobbing around at either end of the lake.

Berlusconi got to hear of my swim and, determined not to be outdone, turned up at one of our first group meetings showing off pictures of him in his twenties standing on the shore of the Mediterranean in swimming trunks. Obama and Merkel looked perplexed, but were polite. 'Very nice, Silvio,' everyone was saying.

As we all prepared to leave the G8 for the far more urban setting of the G20, I was told that bad weather had grounded our helicopter, which would mean a three-hour car journey instead. My team immediately pulled some strings to get me a lift with Obama in his more robust helicopter, Marine One.

I was driven to the helipad with his team, and I used the time to explain my reasons for setting a deadline for bringing combat troops out of Afghanistan (which I'll discuss in the next chapter). The reaction from one of Obama's advisers was, 'Oh, Afghanistan – we're done with that.'

Once we were on board, Obama and I swapped stories about family and work life. He was fascinated by the fact that we were in coalition. At one stage he said, 'David, if you were an American politician I think you would be on the soft right of the Democrat Party.' I had been a strong supporter of George Bush Senior and a fan of Ronald Reagan. But with the changes in US politics – and the emphasis on 'guns, gays and God' in so much of the Republican Party – I wasn't sure I could disagree.

In many ways the G20 is the more important of the two gatherings, bringing together the G8 nations with the new and rising giants, China and India. But with the extra size comes more formality and the tiresome reading of prepared speeches. Genuine dialogue, interaction and argument have to find their place on the 'margins' of the meeting, where much of the real business is done in bilateral or informal gatherings.

There was one such moment in those margins when Obama, Merkel and I were discussing the handing over of European travel data to the Americans to help in the fight against terrorism, which was being delayed by the fact that a required EU directive was being blocked by the Parliament. Merkel uttered the immortal words to Obama: 'Well, of course we've given the European Parliament far too much power.'

The strength of the G20 has been (and to some extent remains) as a mechanism to agree and process better financial regulation. As with the

G8, much of the discussion concerned how we should deal with the financial crisis.

I talked about the importance in our case of getting the budget deficit down, and I found lots of people willing to listen. Britain was blazing something of a trail – and this formed part of the wider debate at the G20, indeed the wider debate across the world, about deficit reduction versus stimulus.

However, the idea that there was some great division about this at the G20 was overblown. The IMF approach was clear: those countries with high budget deficits whose stability looked under threat needed to deal with their deficits; those that could afford stimulus should consider it. The IMF was more focused on trading imbalances between countries, and indeed between regions of the world.

In a meeting that brought together China, Saudi Arabia and Germany on the one hand with their enormous trade and financial surpluses, and the United States, Britain and much of Europe on the other with their significant trade deficits, it was possible to have a serious conversation about the scale of the financial imbalances in the world. To some extent these imbalances lay behind the financial crisis. The huge build-up of financial surpluses in the creditor nations, combined with the long period of low interest rates globally, had led to the 'search for yield' in which huge risks were taken.

Regulation of banks had become too lax, and leverage ratios too high (i.e. too much lending allowed against too little capital). This had led to the creation of too many inappropriate financial instruments, like the subprime mortgage bonds. When they turned out to be worth so much less than advertised, the world teetered on the edge of financial collapse. From this it followed that we needed, yes, to improve regulation and to ensure that banks had adequate levels of capital; but we should also deal with the underlying problem: the imbalances.

The regulatory changes that followed were sensible, but what I felt was still missing was the need for the IMF when it came to identifying financial imbalances, and for the WTO, when it came to calling out protectionist moves, to act with more vigour. They should lead the debate, and not be frightened to judge countries that were dragging their feet or heading in the wrong direction.

There was also an opportunity in this forum to build a reputation for Britain as the country that is always championing and facilitating free trade. The French were pushing back. The Indians were defensive. The Americans were saying that the Doha Round, the ongoing attempt to reduce global barriers, wouldn't benefit them by any significant margin.

But even if we weren't making the progress I wanted, my idea was to mark us out as the most open, globalised, free-trading, anti-protectionist nation on Earth. I was forming alliances with the Germans, Turks and Mexicans. (The full Doha package never made it, but parts of it were carved out into a Trade Facilitation Agreement that Britain championed and helped to get across the line years later, in 2014.)

Generally the summits had fallen into the 'yielding results' category, though there was one moment of farce, involving the colourful and volatile president of Argentina, Cristina Kirchner. Officials were worried that she was going to corner me at a drinks party and harangue me about the Falklands, and warned me that I should be robust about my deep belief in the islanders' right to self-determination.

But when we came to meet, she didn't even mention the British territory. 'How can you be tough on Latin American countries' budgets when your deficit is so high?' was her line of attack. The only thing she seemed intent on invading was my personal space. We were locked in this Argentine tango where I kept moving backwards as she got closer and closer.

Gabby had already briefed our lobby pack that I would kick back very hard against Argentina if it raised the future of the Falkland Islands. '*Would*' became '*did*' as the press hurried to hit their deadlines, and the next day the *Sunday Express* splashed with 'Cameron . . . Tells Argies: Hands Off Falklands'.

It was one occasion where, following a complete falsehood, neither I nor the press wanted to correct the record. We were genuinely all in it together.

Then, in July 2010, came my first trip to America as prime minister.

I knew that getting my relationship right with Barack Obama was essential. So many of the things I wanted to achieve in foreign policy – from bringing an end to the conflict in Afghanistan to pushing for progress on climate change – would depend on the approach taken by the US.

I had met Obama when he was a senator, and we had got on well. I remember asking him if he thought he was going to win the 2008 presidential election. He said he hoped so, but he thought he had a lot of things to get over – the 'inexperience thing', the 'black thing', the 'Muslim thing'. I admired his frankness and candour, and we had built on our bond at the recent G8 and G20.

The protocol – bilaterals, meetings of our teams, press conferences – of that visit all went to plan, and Obama insisted on giving me a personal tour of his garden and the private apartments at the White House. He described the place as a 'beautiful cage – but still a cage' and showed me a bust of Churchill he had put in pride of place outside his bedroom.

The relationship was plain for all to see in the easy and instant connection we forged – one that was rooted in the values we shared.

Yes, he was a Democrat and I was a Conservative. But when it came to the issues in front of us, we didn't need time to reach agreement. We were both committed to the defeat of the Taliban, but believed the war in Afghanistan shouldn't go on forever. We agreed that lethal force should be used to defeat terrorism, and were passionate about progress on climate change and development. We were also pro-free trade, and wanted to counter the forces of protectionism that risked dangerous trade wars. Indeed, I always sensed that Obama was envious of our political system. I was in coalition, but still had the freedom and power to go further on these issues than he could.

My impressions on that visit were that he was a likeable, decent man with a great sense of humour and, as a former law professor, the ability to provide a brilliant analysis of the most complex situations. What a relief it was, I thought, that someone with good values and buckets of common sense was in the White House, in the most powerful job in the world.

The strength of the relationship also mattered at that first meeting, as there were some particularly tricky bilateral issues to deal with, including the horrific oil spill from BP's Deepwater Horizon rig in the Gulf of Mexico in April 2010. This was being billed as a showdown between the US and the UK, between Obama and me. US officials continued to refer to the company as 'British Petroleum', though it had dropped the name in 2000 after a series of mergers and had, as I pointed out to Barack, the same percentage of American shareholders as British ones.

I accepted – as did the company – that it would pay a hefty fine in compensation, but the US system seemed to be hell-bent on breaking the business altogether. Not only would that threaten the livelihoods and pensions of many people – British and American – it would also leave this important company weak and susceptible to takeover, perhaps by a state-owned or state-controlled business from Russia or China. How could that be in our interests?

There was also an issue of basic fairness. When US-based companies – like Halliburton, which was one of the three companies involved in Deepwater Horizon's operation – were not paying unlimited damages, why should a British-based one be singled out for such treatment?

I made all these points, but it was initially hard to tell how much impact they had. At least Obama could see how strongly we felt about the issue.

The most significant question we discussed in Washington was whether and how to talk to the Taliban. Though I was to make progress on this in the following years, it was another of the agendas I wish I had pushed harder.

After the summits and America came Turkey – not a country that is usually found on the itinerary of new prime ministers' Grand Tours. But I was eager to change that. In a bid to reach out to the powers of the future, I wanted to go there as early as possible.

I had always believed Turkey's success was important. It was a potentially powerful ally to Britain – vital to the Middle East peace process and in halting a nuclear Iran. It was a fast-growing economy of considerable size – 'our Brazil', I called it. But above all, it is one of the world's most prosperous Muslim countries, proof that Muslim democracies and Muslim market economies can work.

For these reasons I believed that, in the long term, Turkey ought to be in the European Union. As I had said in my G8 speech – which apparently drove Sarkozy almost mad – I wanted to 'pave the road from Ankara to Brussels'. I wasn't saying that millions of Turks should automatically be allowed free movement across Europe (a stick I was to be thoroughly beaten with later). I was saying that in a flexible, multi-speed Europe, where membership meant different things for different countries – the Europe we wanted and would argue for – Turkey should be inside the wider tent. The EU should not be reserved for Western Christendom alone. (Incidentally, Boris Johnson thought this too.)

I spent a lot of time with prime minister Recep Tayyip Erdoğan. He was canny and easy to get along with, and I was as frank as I could be. I told him I thought the Turks were in danger of going in the wrong direction. That there was a growth of Islamist sentiment which his party seemed to promote and appease, rather than challenge. If we genuinely wanted to move Turkey closer to Europe, he needed to help his friends by demonstrating the extent of reform and concern for democratic and human rights. And while Turkey had been one of the few Muslim countries to have a good dialogue with Israel, it was now leading the verbal onslaught against it.

Enhancing the British–Turkish relationship was an example of an agenda that ran into the ground. The trade issues progressed well, but the fact was that the mood in Europe was turning against any form of Turkish membership – and Erdoğan's real ambitions were heading in a different direction.

From Ankara I flew to India to lead my first-ever trade delegation. I loaded up a plane with CEOs, cabinet ministers, top officials, heads of museums and galleries, even Olympians Kelly Holmes and Steve Redgrave.

This was commercial diplomacy in practice – and it was an agenda that yielded enormous results. We more than doubled exports to China and South Korea, and became the foreign investment capital of Europe, something that would be a key driver of the explosion in private sector jobs in the years that followed.

When it came to India, I argued that we needed a modern partnership – not one tinged with colonial guilt, but alive to the possibilities of the world's oldest democracy and the world's largest democracy. Many of Britain's most successful business leaders and cultural figures are from the Indian diaspora community and would be our greatest weapons in that endeavour. I was proud to have many of them, like Priti Patel, Shailesh Vara, Alok Shama and Paul Uppal, on the Conservative benches in the House of Commons.

I got on well with prime minister Manmohan Singh. He was a saintly man, but he was robust on the threats India faced. On a later visit he told me that another terrorist attack like that in Mumbai in July 2011, and India would have to take military action against Pakistan.

Then came my curveball: Italy.

'What's the dress code for dinner?' I asked one of my team before we set off for Rome. 'A thong,' he replied.

What a dinner it was. Everything was tricolour – mozzarella, avocado and tomato for the first course, then white, green and red pasta to follow. And so it went on. As did Berlusconi. Outrageous joke after outrageous joke – none of them funny.

Italian politics was a riddle to me. What did they see in a billionaire playboy with apparently no filter on what he said – or did? He once interrupted a long night at an EU Council by pushing the button on his microphone and announcing to the other twenty-six leaders that if the meetings were going to run on like this, 'you should all do the same as me – and take a mistress in Brussels'.

On another occasion he walked into the room where we held the interminable meetings to find me chatting with the prime minister of one of the smaller countries whom he clearly didn't know. When I introduced them he said, 'You must come to Rome and meet my wife – you look just like her lover.' That one, I confess, did make me laugh.

Over time I came to realise that, far from putting people off, Berlusconi's unscripted brashness was part of the attraction. Italy had suffered so much from corruption in its political system that his eccentricities were a permanent reminder that he was different.

For all his loucheness, Berlusconi shared some values with me: he was pro-free market, pro-enterprise and anti-regulation. We tried to unblock a lot of foreign investments, including a BP gas terminal in Italy that had been in limbo for twenty years.

Italy was the fourth-largest European economy, a net contributor to the EU budget and a major player when it came to security. Yet it was often left out of discussions between the 'big three' – Germany, France and the UK. I thought we could rectify that, making Italy a big ally on NATO and on building a more flexible Europe.

There were four Italian PMs while I was in office, and I was much closer to Enrico Letta and Matteo Renzi, who followed Berlusconi and Mario Monti. But I have to accept that, while Italy was a strong NATO ally and helpful during the Libya intervention, my attempt to put rocket boosters under the Anglo–Italian relationship was an example of an agenda that didn't get very far.

Another agenda I hoped would produce real and lasting results was China.

The country had gone from the world's eleventh-biggest economy to the second-biggest in just twenty-one years. Eight hundred million people had been lifted out of poverty – all because this communist country had embraced the capitalist principle of liberal economics.

Not that ours was a likely alliance. For them, we were still their oppressors from the nineteenth-century Opium Wars, and awkward neighbours during the years when Hong Kong was a British territory. For us, no matter how liberal China's economics, it was still a one-party, authoritarian, communist state, with a woeful record on human rights and a tendency to rip off intellectual property, keep its markets closed where we opened ours, censor the internet and spy on just about everyone.

Yet the advantages of a deeper connection were undeniable. Thousands of Chinese students and tourists flocked to Britain each year. Chinese people loved our history and culture. With a burgeoning Chinese middle class, there was a massive potential market there, and although Labour had made some inroads, it remained largely untapped.

So, China: opportunity or risk? The coalition was split. Nick had been campaigning on the human-rights issues for years. And William was sceptical about the Communist Party's intentions. He wrote me a letter during our first month in office with a stark warning about China 'free-riding on wider global public goods' – obstructing international action if it conflicted with its own domestic imperatives.

My stance – and George's – was more pragmatic. It wasn't, of course, that I didn't care about human rights, or that I trusted a party which appointed one 'paramount leader' every five years. It was just that I thought there was a longer game to play, and a better way of getting what we wanted.

Yes, our ambition was largely economic, but an essential outcome of that economic partnership would be the greater political leverage it would give us. In other words, the trust that we could build from doing business together could lead to trust across a whole range of areas.

Naturally, we would always be vigilant when it came to China. We were hard-headed about the threat it posed, instructing our security services to map and counter the Chinese threat. Only by understanding and guarding against that could we trade with the country.

This would be the best, the safest, way to bring China into the rules-based international system – through rules on trade, but also rules on climate change, terrorism and human rights. The more it was part of the UN, the WTO, the G20, the more cooperative our relationship would be. We could influence its views on everything from climate change to Burma to North Korea. Which was vital – China would be a linchpin in all these matters.

On my first visit to China as prime minister in November 2010, with four cabinet ministers and forty-three business leaders, I knew I'd have to walk a tightrope, on the one hand trying to strike trade deals and showcase what we had to offer investors, and on the other raising issues around human rights and democracy.

Nothing was going to stop the Chinese in their pursuit of growth – getting people out of the countryside into the cities, into employment and out of poverty. If you were useful – if you could supply inward investment, exports and scientific knowledge, as we could – then you were considered a partner. 'Transactional' is the FCO jargon for that sort of relationship.

But there was still a tightrope to walk. So, when I went to Beijing University, I decided to make a more wide-ranging speech, focusing on the importance of democracy and the rule of law. Given that you can be locked up in China for so much as criticising the socialist state, going further than most Western leaders in promoting democracy would, I reasoned, more than offset any perceived obsequiousness in our economic dealings.

I didn't quite stir a revolution. But I did provoke something when I said: 'The rise in economic freedom in China in recent years has been hugely beneficial to China and to the world. I hope that in time this will lead to a greater political opening.' The first question from a student after my speech was about what advice I'd give to the Communist Party in China in an age when more countries were having plural politics. An amazing noise went around the room, half admiration and half shock. I gave a measured answer about our countries having two very different systems. But as I looked around at the sea of faces I thought: is this system really going to last? My conclusion was that, in its current form, it couldn't. After all, surely this was now a consumer society in which people had increasing amounts of choice over their lives. How could the ruling party frustrate that when it came to politics?

While my fundamental view hasn't altered – change, in some form, will come – multiple visits to China have led me to a more nuanced view. The primacy given by both government and people to economic growth. The fierce sense of pride and exceptionalism. The attention given by the nation's rulers to emerging trends and problems across the country. All these things mean that China's path to greater pluralism may be a very long one, with a different destination to our own.

Given that so many other countries were trying to align themselves economically with China, this incident, and the unfortunate poppy set-to, which happened on the same visit, weren't exactly welcome. A three-day visit to the UK by Li Keqiang, who was heavily tipped to replace Wen Jiabao as premier, gave us a chance to up our game. Li, who received red-carpet treatment, made it clear that some of my comments had not been welcome. But despite that we would make real inroads.

Finally, there is an agenda I really wish I had never started. That was the British bid to host the 2018 World Cup.

Britain had a strong case: the best stadia, the most enthusiastic supporters, club teams that are followed across the world, and a football-mad culture. We had also learned a lot from our successful bid for the 2012 Olympics.

The biggest barrier to bringing the tournament home for the first time since 1966 was the notorious world football governing body, FIFA, and its susceptibility to corruption. The bid also pitched us against Russia, a country with a government quite prepared to do whatever it took to win.

We threw everything into beating Russia – and Belgium and the Netherlands, Portugal and Spain – to host the 2018 Cup. I had the FIFA president, Sepp Blatter, to No. 10. He even got to hold Florence – a privilege reserved for presidents and monarchs, I joked at the time. Looking back, knowing what I know now, it makes me wince.

Then, in December 2010, I spent three days in Zürich, where the bidding process was taking place for both the 2018 and 2022 World Cups. With Prince William and David Beckham by my side, my confidence grew. We had the best bid – and a dream team to bring it home.

The process was rather like speed dating, with an allotted amount of time with each voting nation's representatives to try to persuade them to back our bid. The three of us pleaded with people from Cyprus to Paraguay. America's representative, a man called Chuck Blazer, was so enormous that as he got up to leave, his chair went with him.

The corrupt undertones were all there, but, typically British, we gave it our best and got through it with jokes. Vladimir Putin's approach was classic. He suddenly cancelled his appearance, claiming that the whole competition was riddled with corruption.

The moment of our presentation approached, and my role was to kick it off with a short, off-the-cuff speech. I confessed to Prince William that

I was nervous. He told me not to be – and just to imagine Chuck Blazer naked.

The three of us stood up and gave our pitch. We were followed by a video accompanied by Elbow's 'One Day Like This'. It was stirring stuff, and we got a strong ovation.

Our confidence grew. Six nations promised that they would vote for us in the first round: South Korea, Trinidad and Tobago, Côte d'Ivoire, Cameroon, USA and England.

How many votes did we end up with? Two.

Russia, whose bid was fraught with problems including racism, won the chance to host the 2018 World Cup. Forty-degree Qatar, hardly a footballing hub, got 2022. Putin didn't need to come. The fish had been bought and sold before we'd even got to the marketplace.

David Beckham was upset and angry. 'I don't mind people lying to me, but not to my prime minister and future king,' he said. Blatter said we were just 'bad losers'.

In the years that followed, a criminal investigation into the way the hosts of the 2018 and 2022 World Cups were chosen took place. Nine of the twenty-two members of the FIFA Executive Committee who awarded them have been punished, indicted or died before facing charges, including Chuck Blazer, who admitted fraud, money laundering and taking bribes on the 1998 and 2010 World Cups. For seventeen years Sepp Blatter presided over an organisation riddled with corruption. He has been banned from football for six years, and his plaque removed from FIFA headquarters.

One issue that proved to be more prevalent than I had expected before I became prime minister was corruption. I kept on seeing it for myself: from Omar al-Bashir's refugee camps in Sudan to Blatter's boardroom in Zürich. Those same forces that had denied Britain the World Cup – bribery, lack of transparency, collusion, fraud – were depriving people around the world of safer, healthier, wealthier lives.

At international summits we focused on everything – security, poverty, growth, aid, the environment. But we seldom said a word about one of the biggest drivers of these things: corruption. I resolved to spend my time in government – and after it – trying to change that.

14

Afghanistan and the Armed Forces

When I took office there were more than 10,000 British troops in Afghanistan, engaged in a conflict that had lasted nearly a decade. That made me the first prime minister to come to power from a different party when the country was at war since 1951 – the year Churchill replaced Attlee while British troops were fighting in Korea.

I spent more time on Afghanistan – visiting, reading, discussing, deliberating, and yes, worrying – than on any other issue. The burden weighed heavily upon me every single day until the final British combat soldier left Camp Bastion in 2014. I still care deeply about Afghanistan's future today. And I will always remember the families of the fallen, and those living with life-changing injuries because of their service.

Many leaders have written about what it's like to send brave men and women into battle. My reflections are about inheriting that responsibility and handling a conflict whose aim had become ambiguous and whose unpopularity was growing.

I supported the decision to send troops to help rebels overturn the Taliban government in 2001. The 'invasion' of Afghanistan was justified. The brutal Taliban regime, which controlled 90 per cent of the country, was harbouring al-Qaeda, the perpetrators of 9/11. It was continuing to train jihadists and plot attacks against the West. When asked to expel Osama bin Laden and al-Qaeda, it had refused. The US had no sensible choice but to act. And we were right to support it. As a young back-bencher, on the day after the invasion I said in the House of Commons that our actions were 'every bit as justified as the fight against Nazism in the 1940s'.

By 2005, when I became Conservative leader, there was a growing sense that while Iraq might be the 'wrong war', as it had little to do with tackling Islamist extremism and might in fact be encouraging it,

Afghanistan was the 'right war'. We were at least trying to deal with one of the principal sources of the problem, and responding to a direct attack.

Come 2010, my message matched what I had said in the House of Commons almost a decade before. Our troops *were* combatting terrorism, with the Taliban beaten back. They *were* defending our security, with plots no longer coming directly out of Afghanistan (although there was still more to be done to address the threat from ungoverned parts of neighbouring Pakistan).

Sadly, like whack-a-mole, the scourge of Islamist extremism and its promotion of terror would rear its head elsewhere. Every broken or fragile state was a potential incubator, and the Afghanistan–Pakistan border was the most virulent region of all. But in tackling it there, we were tackling the motherlode.

In any case, it is true to say that Britain is safer as a result of the hard work and bloody sacrifice of our troops in Afghanistan.

That's why the view of our intelligence experts in 2010 was clear: what we were doing remained justified. The most significant security threat to the United Kingdom remained al-Qaeda. If we left Afghanistan precipitately, it – and its training camps – could return.

But our action came at a grave cost. We were taking casualties almost every day. And, like the vast, sandy plains we were fighting over, the war seemed to have no end in sight.

What helped me now I was PM was that I'd visited Afghanistan more than any other country. I knew the drill: picking up a Hercules turboprop plane or a C17 transport jet in the Gulf to take us to Camp Bastion or the capital, Kabul. Then travelling onwards in the Americans' Black Hawk helicopters or our own Chinooks, where I would sit upfront in the 'jump seat' just behind the pilots; landing to see the small fortified Forward Operating Bases (FOBs) our troops were defending, often against ferocious attack.

I knew the landscape: the way the totally dry desert merged into the lush green Helmand River valley, the beautiful blue-green water fertilising the land; the mud-brick houses that told of a deeply conservative society that hadn't really changed for decades, even centuries.

I had seen some of the worst trouble spots, visiting the Helmandi town of Sangin in 2008 with William Hague. As we sat being briefed on the roof of the district centre under the fierce sun we could hear the crackle of small-arms fire. A local checkpoint was under attack. There was the quietest rush of air as a stray bullet passed overhead. The commanding officer gently ushered us under cover, muttering about the dangers of sunburn. Wonderfully British.

I had got to know a lot of the personnel, too. Hugh Powell, a top-rate civil servant, had served as the head of the Provincial Reconstruction Team in the capital of Helmand, Lashkar Gah, as well as being the senior Brit coordinating the UK counter-insurgency in the south of the country. One time we had breakfast together wearing our helmets because of an imminent security threat. Another, we tried on the full kit, body armour and radio of a lance corporal, jogging with a rifle around the yard of an FOB in the forty-degree heat.

When I first visited Bastion I climbed the air-traffic control tower and looked over a small camp of tents. Since then it had grown to an area the size of Reading, complete with a runway, a water-bottling plant, a medical facility as sophisticated as any district general hospital – and a KFC. Many of the pilots were from RAF Brize Norton in my constituency, and would talk candidly to me about life at the base. I heard about the problems first hand, from frustration at the lack of contact time with home to the potentially fatal shortage of helicopters and delays in getting new body armour.

Sometimes we would stay in Kandahar, the massive, mostly US, airbase covering the whole of southern Afghanistan. It was home to many of our fighter and helicopter pilots and our remarkable troops. Two miles long and right next to the city that is in some ways the spiritual home of the Taliban, it was subject to the occasional rocket attack and repeated ambushes.

I had also seen what ordinary life was like for Afghan people. I'll never forget watching children flying kites across the river from where we were staying, or visiting a school funded by the UK's Department for International Development.

I knew the lingo: a mixture of military-speak and banter. The heat: like a blazing fire that hits you the minute you step off the aircraft, made worse by the altitude that can leave you exhausted. The sand: it got everywhere; you'd come home with it still up your nose. The sounds at night: the constant coming and going of helicopters, the alarms when there was a threat of attack, and the endless sound of the diesel generators.

I even grew used to the tight, corkscrew take-offs and landings designed to evade rocket-propelled grenades and small-arms fire from the Taliban on the ground. Fortunately, I rarely had to contemplate the threat of a direct attack. My only brief taste of one was when our two helicopters were leaving the relative safety of Camp Bastion for a military outpost in the middle of Helmand. I was seated in the body of the second Chinook, looking out of the back, where the ramp is always slightly ajar, at the mountainous expanse below. Suddenly the helicopter turned back. There was good intelligence that the Taliban knew that someone senior would be coming in by helicopter.

That feeling – of being in the crosshairs of the enemy – is what our troops lived with every day. I don't know how they did it.

I had a constant reminder of the tragic reality behind the growing death toll when I wrote by hand to bereaved families. So often were we losing men and women at the start of my premiership that I'd have one to write every few days. It was a task I would carry out at the kitchen table in the flat or in my study at Chequers. I would commend the soldier for their bravery and, carefully reading the citations made by their comrades about their service, I would take points from them. I would also use some of my own experience – particularly to parents – about losing a child. I tended to say that while there was nothing anyone could say to lessen the pain and grief, I knew that over time at least some of the clouds would part, and some of the happier memories from the past would come through. I would try to explain what we were doing in Afghanistan. It was a country far away, but the struggle against Islamist extremism and terror was something that affected us back at home.

It was this, instead of any overarching ideology, which would inform the decisions I would take. My approach was hard-headed and pragmatic. I tried to make the choices that would best guarantee the stability of Afghanistan, the security of our country and the safety of our troops.

As prime minister for less than a third of that thirteen-year war, I took some of its defining steps. There were three key things that I thought necessitated them.

The first was that the overall goal of our intervention had veered too much towards nation-building – 'creating a Denmark in the desert', as some put it – whereas we should be aiming for Pakistan-lite. We needed realism. A failing state would be better than a failed state. As I put it during that first visit to Afghanistan as prime minister: 'We are not here to build the perfect society, we are here to stop the re-establishment of training camps.'

The second thing we needed was time to build up a sufficient Afghan National Security Force and government so they could handle the civil war without our help. (In fact, the size of the ANSF – army plus police – doubled between 2009 and 2012, making it strong enough to manage the fighting largely without NATO from 2015.)

The third thing was the need for a deadline. I could see the case against this. The Taliban could just wait for us to leave. But the counter-arguments appeared more compelling. Our military high command seemed to have settled on the idea of being in Afghanistan almost indefinitely. The Afghans had become far too reliant on our presence. And as we lost troops, public consent was dwindling. A date would force everyone's hand to reach a satisfactory and stable position before support at home disappeared altogether.

One of the early defining steps I took was recommitting Britain to the war, by making sure Afghanistan was our number-one security priority. On my first full day in Downing Street, I convened the new NSC. We were a country at war, I told the assembled ministers, and this would be our war cabinet. We wouldn't just be setting the strategy and leaving the heads of the army, navy and air force to fill in the gaps. We would seek to shape events more directly and take urgent action. We would have monthly published reports on progress and quarterly statements – by a cabinet minister – to Parliament. This would not become a forgotten war. We started with a boost for the troops, doubling the operational allowance they were paid while on tour.

I was clear where action was needed most urgently: Sangin. President Obama had rightly ordered a 'surge' in the number of US forces in 2009, increasing them from 30,000 to 90,000. Britain meanwhile had committed to an increase from 9,000 to 9,500 troops in Helmand, where we had taken over security in 2006.

I didn't object to the increase; I objected to how thinly spread we would end up being in comparison to the Americans. The advice was that our numbers would be sufficient. But I commissioned some figures that revealed that while the US would now have up to twenty-five soldiers per thousand members of the Afghan population, we would have just sixteen.

Sangin demonstrated this lopsided deployment. This small town is where we sustained nearly a *quarter* of all our losses during the conflict. I knew that we had to match America's densities of troops. But we had carried more than our share of the burden – we made up only a tenth of the fighting force in Afghanistan, but suffered a third of the casualties – so the option of increasing UK troop numbers seemed wrong. Taken together, this meant getting out of Sangin. I overruled the advice and secured agreement for the withdrawal.

The US Marine commander who took overall responsibility for Helmand was a splendid man called Brigadier General Larry Nicholson. As if marching straight out of Central Casting, he crushed the bones of my right hand with his handshake and declared that he had come straight from Fallujah – one of the toughest battles in Iraq – and was ready to take the fight to the Taliban. He proceeded to use a dried opium-poppy stick to point at the map on the wall and run through his plans. He recognised that our decision was a reasonable one, and US forces took over what remained one of the toughest jobs in the country.

Partly because of this redeployment, our casualty numbers fell dramatically. In 2010, 103 British troops were killed. In 2011 and 2012, that fell to forty-six and forty-four respectively. For the final three years we were in Afghanistan, the British death toll dropped to single figures each year.

The move also improved our performance in the rest of Helmand. But action wasn't just needed in adjusting the force; it was needed on the state of our equipment, which had become something approaching a national scandal. I knew from my previous visits the key improvements needed: more helicopters, faster casualty evacuation, more rapid improvements in body armour. And, above all, better-protected vehicles.

The Taliban's weapon of choice was the improvised explosive device (IED), which was becoming ever more sophisticated. Every time we increased the armour on our vehicles, they would find a new way of burying more explosives. Every time we developed a metal detector with more sensitive scanners, they would find a way of using fewer metallic components. IEDs became the primary instrument for killing and maiming not just our troops, but local people, including children. Our forces' ageing Snatch Land Rovers were no match for these roadside bombs. To be fair to my predecessor, plans were in place for improvements, but I did everything I could to add to them and speed up their delivery.

The action we took in the NSC – including expanding a new system that bypassed bureaucracy and delivered equipment more quickly – aimed to give our forces what they needed. After a couple of years, no one raised concerns with me about poor vehicles or a shortage of helicopters. Some even said to me that in some regards our equipment was superior to that used by US troops.

The biggest decision, though, would be about our long-term involvement in Afghanistan. After nine years of the conflict – and four bloody years in Helmand – people were rightly asking: when will our work be completed? And when will our troops come home?

To answer those questions, we needed to be far more precise about exactly what it was we were trying to achieve. And while some elements of that nation-building would be important – getting children into school, improving healthcare, constructing infrastructure – we had to show common sense and keep a grip on what was possible.

Early on, I called a seminar in Chequers, packed with experts from the worlds of military, academia, aid and policy. There was a debate between 'The war can't be won' on one side and 'Stick with the mission' on the other. But the seminar did point me towards a third option: concentrate more on training the Afghan army and police as the route for our exit. This was a middle way. Don't leave immediately, job undone. Don't stay indefinitely, a war without end. Set some sort of deadline.

So that was my decision: to withdraw from Afghanistan, but only after giving the military enough time to prepare and to hand over to the Afghan forces. I chose the end of 2014, because I thought that left us enough time to do the work we needed in order to prepare.

I set out the plan at the G8 in Muskoka. No cabinet meetings, no pre-briefings, no public rows between the military and civilians – just private agreement with George, William and Nick, and a statement to the press. To give a strong lead to our troops and change the tone, I knew it was one of the things that needed to be done quickly.

But I didn't want us to leave and that be that. We were committed to financial contributions, both in terms of aid for economic development and funding for the Afghan security forces. Together with Obama, I would lead the charge at subsequent NATO summits – in Lisbon, Chicago, Cardiff and then finally Warsaw in July 2016 – to help secure similar commitments from other countries.

I knew my history. The Afghan government didn't collapse in the

1990s when the Russians left; it collapsed when they withdrew funding for the military. The other way we could help make Afghanistan more stable beyond our military input would be by helping to foster a political agreement that brought at least some of the Taliban into the political settlement.

After all, the big mistake had happened at the beginning, back in 2001, when the West insisted on a settlement that was free of Taliban involvement – whereas my view is that true peace is only possible if it includes at least elements of the Taliban.

In many ways, of course, war suited both sides. The Afghan president Hamid Karzai and his allies could use war to favour their own tribes. The Taliban could use that fact to portray the government as tribalist and anti-Islamic. Plus, talking to the Taliban wasn't exactly easy. (One fact illustrates this perfectly. We were being told repeatedly – right up to 2015 – that their willingness to cooperate depended on the permission of Mullah Omar, whereas we now know that Omar died in 2013.)

In the end it would have to be Afghan talking to Afghan, but we hoped that – with our strong relationship with Pakistan – we might be able to help get the ball rolling.

There was, however, a chronic lack of mutual trust between the two governments. The Pakistanis thought that the Afghans were too close to their rival, India. The Afghans thought that Pakistan was harbouring the Taliban, hedging its bets by allowing terrorists to weaken its neighbour, and still had designs on redrawing the border to unite all Pashtuns inside Pakistan.

I reasoned that one of the best ways to make progress would be to try to strengthen the personal relationships between the leaders and – where possible – include their military and intelligence chiefs in their discussions.

In opposition I had got to know Karzai quite well, and he was the first foreign leader I received at Chequers, shortly after taking office. He was charming and wily, and claimed to have a real affection for Britain. He believed that, because of our history, we had a greater understanding of the situation in Afghanistan than did the United States.

Part of my job, as I saw it, was to help focus him on making his government work and ensuring that the country was run properly, for example appointing governors, passing basic laws and dealing with rampant corruption. Karzai himself had been dogged by corruption claims in both the 2004 and 2009 elections: Hugh Powell had visited a polling station in Helmand with a turnout of three hundred which had reported 15,000 votes for Karzai.

One of the reasons some Afghans turned to the Taliban was because they seemed better at dispensing justice and ensuring order than the legitimate authorities. And corruption was so ingrained in the country's culture that Karzai could never quite accept that we were there because we genuinely believed in the mission as part of an international coalition. In January 2012 I remember him asking me what was it – minerals? mining rights? – that we really wanted from Afghanistan.

He could be hard work. He would criticise the activities of British and American troops, even though they were making extraordinary sacrifices and were essential for his regime's survival. And he found it too easy to play the nationalist card and blame all his problems on Pakistan.

But there was enough there for me to be able to use the relationship I had built up, and the fact that the Pakistanis trusted us more than the Americans, to help build the trust between Afghans and Pakistanis.

The high-water mark of my efforts came at the Chequers summit in 2013 when Karzai and President Asif Ali Zardari of Pakistan spent two days in talks. They slept in adjoining bedrooms, each with a guard outside sitting ramrod-straight and wide awake in his chair. I came downstairs the next morning to find the two presidents joking about which of them had been snoring the loudest.

We agreed a series of small steps to build confidence: publicly praising each other's leadership; agreeing visits to each other's countries; and – crucially – recognising that the security of one was the security of the other.

We aimed to move on to talk about border security and the importance of dealing with terrorist safe havens on both sides of the Durand Line (the controversial border between the two countries). We wanted to get some of their military and intelligence personnel to sit together and work together – even suggesting joint patrols. A further series of coordinated steps would then follow to help deliver a peace process: the release from Pakistani custody of potential Taliban peacemakers so they could carry out talks, hints that Afghanistan would consider constitutional reforms, and so on.

But Karzai couldn't bring himself to trust Pakistan. Zardari was often willing, but as we know, it is really the military that makes the key decisions.

The Americans were supportive and appreciative of our efforts. They took up the cudgels for contacts with the Taliban, but ultimately the distance between the two sides, and their half-heartedness about a compromise, was too great to make meaningful progress.

This is the agenda I most wish had come off. But I am convinced that it remains crucial today, and that it can be done.

So, for all the blood spilt and treasure spent, was Britain's involvement in the Afghan war worth it? Historians say it's too early to say. It is incredibly depressing whenever the country slips back. Sangin is back in Taliban hands. Their flag flies over Musa Qala. Opium fields still stretch across Helmand. These are painful things to write.

But at the same time, in 2014 Afghanistan saw its first peaceful, democratic transfer of power, to the anti-corruption academic Ashraf Ghani. It now has its own police force and national army. And more than that. By 2011, 85 per cent of the country had access to basic medical care, compared to 9 per cent under the Taliban. Seven million more children were in school compared to one million in 2001. A third of them were girls. Not a single girl went to school in the Taliban years. As long as we go on funding the Afghan army and police (and the international community remains committed to this), the Taliban is unlikely to win the whole country, and terrorists cannot get the same foothold they had before.

The agenda is still the same. The Afghan government needs to deliver for all its people. It needs to find a way of bringing at least some elements of the Taliban into the legitimate political sphere. It will only do this if it forges a trusting partnership with Pakistan, where both accept that allowing safe havens on either side of the border for terrorists will end up destroying them both.

The difference now is that our troops are not exposed to the daily risk they once were. Arguably, it will be easier now for some sort of deal to be done because the provocative presence of foreign troops on Afghan soil is so much less.

The prevailing views are that this war was either doomed to fail or that it should have been pushed harder. I believe there is a third category, where you do the right thing and keep doing it, but it takes a very, very long time before you achieve stable and lasting success.

Samuel Beckett said, 'Fail again. Fail better.' Foreign troops could only ever provide a breathing space for a legitimate Afghan government to get its act together and deal with the fundamental issues.

Delivering security and some semblance of uncorrupt administration; getting the relationship between Afghanistan and Pakistan right; achieving a political settlement which demonstrates that all Afghans are welcome. I believe all these things are achievable. The story isn't over.

Afghanistan brought me into close contact with the UK's military chiefs. While there were robust arguments and discussions, they were generally good-natured. However, my relationship with them would come under greater strain when we had to discuss another intractable challenge: how to make sense of the UK's defence budget in an age of austerity.

I had read widely about the history of military top brass interacting with those at the top table of government – particularly the blazing rows between Churchill and General Alan Brooke in the rooms I was now so familiar with in Downing Street. I have huge respect for the chiefs of staff who head up the army, navy, air force and the armed forces as a whole. But PMs need to build up their own expertise. Like all my predecessors since James Callaghan, I didn't have a military background, so I decided to hire a senior military adviser to be in my private office.

Colonel Jim Morris, a Royal Marine, followed in later years by Colonel Gwyn Jenkins and Lieutenant Colonel Nick Perry, advised me in Downing Street, and did so brilliantly. They had all done multiple tours of Afghanistan and Iraq, and were at my side through everything: from hostage situations to those trips to Afghanistan. It was a small change to government that paid dividends – and I believe it will endure.

Meanwhile, the tenure of the chief of the defence staff, Sir Jock Stirrup, was coming to an end, and it was time for me to choose a successor.

General Sir David Richards, head of the army at the time, was a swashbuckling, charismatic, can-do character, whose initiative and daring had helped end the bloody civil war in Sierra Leone. I had met David when he commanded the international forces in Afghanistan between 2006 and 2007. He had done an excellent job, and had the respect of the US. He was famed in Whitehall for coming out with clever but infuriating catchphrases to put the politicians in their place. One of his favourites was 'Amateurs talk tactics, professionals talk logistics.' As I've said, I don't believe you can separate tactics, logistics and timing from strategy, so my reply was, 'Well, I'd better get involved in those as well.'

His promotion wasn't without risk. I was warned by more than one Whitehall mandarin that David would spend more time briefing the press than briefing me, and relations would become strained.

They were right. He was an innovative thinker, yet there was no doubt he was a leaker, and that was damaging to the government. And he'd say one thing in private – we should cut the army to 82,000 – and a different thing in public – nothing less than 102,000 was acceptable. I don't regret his appointment, but if you deal with a swashbuckler, you've got to be prepared to get swashed from time to time.

Another key figure, Liam Fox, had been one of my rivals for the party leadership, and was delighted when I made him shadow defence secretary back in late 2005. Despite being eager to continue the role in government, he occasionally seemed overwhelmed by the decision-making. It left me having to step in a lot myself, especially because the defence budget was a total mess.

An eye-watering £38 billion had been pledged to future programmes, which was actually *more* than the annual defence budget of £33 billion. The previous government had made attractive-sounding pledges for *future* equipment without a realistic prospect of being able to pay for it. And many of the *current* projects were wildly over budget. The only way forward was a complete overhaul. Labour hadn't had a defence review for years, so the changes would be drastic – projects cancelled, spending cut and even jobs lost.

The Treasury had long wanted to take an axe to defence spending, and was pencilling in at least a 15 per cent cut. The chiefs, the MoD and Liam Fox were reluctant. A letter from Liam to me, leaked on the eve of the 2010 party conference, warned of 'grave political consequences for us, destroying much of the reputation and capital you, and we, have built up in recent years' if we made cuts. This would be a very tough fight.

I was ready to be ruthless, and to go into battle on multiple fronts – against the right-wing press, the retired generals brigade and the rather gung-ho usual suspects in my own party. Better to do something drastic but necessary, I thought, than fudge, fiddle and muddle through as governments had done for years. We were Conservatives: we could take tough decisions about defence and finally sort this mess out. I insisted on a re-examination of the government's approach to the whole security agenda. So the Strategic Defence and Security Review (SDSR), as it was known, would be coordinated by the new NSC and the national security adviser.

The review would be characterised in the immediate aftermath as bungled, rushed and short-sighted. But judging it straight away and only in terms of immediate impact – surely *that* is short-sighted. The real test of any defence review is to take a longer view, and ask: over the next five years, did the decisions you made jeopardise current needs or future capabilities? Was anything lacking on your watch or since then?

My answer is 'no'. We brought the defence budget back into the black, while modernising the services. We reduced the size of the army, navy and air force, while dealing with new and ever more complicated threats. We cut spending by 8 per cent – far less than other departments and far less than the original Treasury ask – while engineering it so that the budget would rise in the future. And, in the end, we retained our NATO commitment to allocate 2 per cent of our national income to defence. This was a benchmark few countries were willing to meet, and although difficult, it was important to our standing in the world that we were able to do so. It is one of the achievements of which I am most proud.

Yes, there was a period of retrenchment and redundancies, but what would emerge – the so-called 'Future Force 2020' concept – was forces fit for the future. This was the beginning of the end of a political orthodoxy that correlated spending and success. Over the coalition years we showed that you could save money and improve services – not just in defence, but in local government, policing, environment – if you spent smart.

The bottom line of the review was that we would have to make savings. And in doing so, there was a big choice. Was it better to maintain a balanced, albeit smaller, force, with a broad set of capabilities? Or was it better to focus the valuable resources on one type of warfare?

David Richards championed the latter option. He wanted a smaller navy and a bigger army – for us to be really good at one thing. It was tempting but limiting, and didn't add up strategically or financially. Nor was it any use to prepare for the previous conflict when the next one will be different. And choosing the former type of armed forces – smaller but balanced and flexible – has proven the right choice given the conflicts we have since engaged in.

That's not to say that, along the way, I didn't feel any regrets. I felt two – one technical, one emotional.

The technical regret was that we didn't move fast enough from Cold War capabilities towards modern-day capabilities. The things I wanted more of were the things we were moving towards. More drones. Chinooks that could be instantly packed into transport planes, despatched anywhere in the world and ready to fly on arrival. More measures to thwart cyber-attack. But moving towards isn't the same as arriving.

The emotional regret concerned the saga of the aircraft carriers and their jets – a tale of bureaucracy, bungling, of rocks and hard places.

When I took office, two new aircraft carriers had already been commissioned. The largest ships in Royal Naval history and the size of three football pitches, they would be a proud national asset. They were also too big. Yet the cancellation penalties were so high it would be cheaper to just carry on building them. Their immediate predecessor, HMS *Ark Royal*, was due to be decommissioned anyway. So the decision was taken to continue with both the carriers, since one could be under-going repairs while the other was in use.

That led to a decision over our fleet of fighter jets, some of which used the carriers. We now had four types in service, which was too many. Our aircraft carriers could accommodate the new F35 Joint Strike Fighters, so the prime candidate to be scrapped was the much-loved Harrier, famous for its vertical take-offs. But that then meant there would be a gap in Britain's aircraft-carrier capabilities while the F35s were being built.

Having grown up in the era of HMS *Invincible* and HMS *Hermes*, I had a great romantic attachment to the idea that we could fight anywhere in the world to defend our interests thanks to these vast mobile platforms. It pained me to see the Harrier 'Jump Jets' retired. I remember, aged fifteen, listening on the radio to stories of their crews' great daring in the South Atlantic as we retook the Falkland Islands from the Argentinian Junta.

In practice, there aren't that many occasions on which we need to project power somewhere that we don't have a friendly country nearby to base our aircraft in. The only time the absence of an aircraft carrier affected us while I was prime minister was during the mission in Libya, when we had to use a base in southern Italy. But I remember meeting pilots there at Gioia del Colle on the eve of their mission, and they assured me that the new Typhoons flown from an airbase were the best option anyway. 'Yes, you may have to spend extra hours in the Typhoon. But what an amazing machine to spend them in,' they said.

My defence dramas wouldn't end there. In October 2011 it was revealed that the defence secretary had been taking his friend Adam Werritty into official meetings – often overseas, without security clearance, without any apparent reason. I commissioned a full investigation from the cabinet secretary Sir Gus O'Donnell – and details began to emerge about Tory donors funding Werritty's company. Before the report was even finished, however, Liam Fox called me and said he was resigning.

His resignation led to a mini-reshuffle, and I was convinced that the fair and focused transport secretary, Philip Hammond, could get a good grip on the department and deliver the efficiencies we needed.

One other thing that was stymying us – in fact shaming us as a nation – was the less than adequate treatment of our armed forces. Those people who put on uniform and risk their lives are doing the most any human can do for their country. It is a tacit understanding that we, as a country, do as much as we can for them and their families. For centuries this had been referred to as a Military Covenant, and it came to represent everything we ought to provide – pay, healthcare, housing, schooling, support – for service personnel, veterans and their families.

In opposition we made a small but important gesture towards our service personnel by launching the 'Tickets for Troops' charity which provided them and their families free entry to sporting, musical and cultural events. Since then, it has distributed over a million tickets.

Oliver Letwin and the Covenant Reference Group brought the covenant back to life, making sure we delivered for veterans in the way other countries had done but Britain previously hadn't.

My visits to the state-of-the-art facilities in Headley Court, Surrey, which was home to wounded and recovering soldiers, brought home to me how important this was. Here were young men in the prime of their lives who had just lost one or more limbs. What they wanted was not merely prosthetic limbs to get around on, but the most modern and capable prosthetics there were. They didn't want a quiet life, but a full and active one.

The fact that my dad had lost both of his legs – albeit in very different circumstances – meant that I knew what a mental as well as physical mountain they were having to climb. Talking to former troops made me understand what we needed to deliver for them. They'd fought the Taliban, and we had been by their side during that battle. Now they were engaged in dealing with physical and mental injuries, finding training and employment, and other challenges of civilian life. It was a different sort of battle, but one during which we should also be by their side. The Military Covenant was enshrined in law in 2011.

The following year we were able to go further. When the news broke about the widespread manipulation of the London Interbank Offered Rate – or LIBOR – over several years by multiple banks we launched a parliamentary inquiry. George announced that the fines collected by financial regulators would be given to military charities. This was the brainchild of his spad, Ramesh Chhabra, and it resulted in tens of millions of pounds being given to the most deserving of causes.

The final thing I inherited involved our continuous at-sea nuclear deterrent, Trident – four submarines carrying ballistic missiles that are tipped with nuclear warheads. Tony Blair only got the decision to replace the subs through Parliament because our party had supported it in 2007 (eighty-eight Labour MPs voted against). Once again, its continuation needed to be approved.

Before I did so, I wanted to prove to myself that Trident was the right option for us. Very secretly, and with only the chancellor, foreign secretary and defence secretary knowing, I commissioned a report on the alternatives.

And there *was* an alternative: Cruise missiles tipped with nuclear warheads. The problem was that over their forty-year lifespan there was a risk that they would become comparatively slow as new defence systems were developed. They might have been less expensive than Trident, but by the time the tests had been done to see if they worked, they could well have cost the same or even more. Even though this proved a dead end, it was one that I was glad I explored.

Looking back over all this, I know that we left the nation's defences in a better state than when we found them. Even though there were fewer personnel, there was better kit, better body armour, higher pay for those on tour, and military hardware fit for this century.

A navy built around two brand-new aircraft carriers, with replacement submarines for our nuclear deterrent, a fleet of hunter-killer subs, nineteen frigates and destroyers, and a modern amphibious task force manned by Royal Marine commandos.

An army capable of fighting three conflicts simultaneously – including complex stabilisation and peacekeeping missions – and of maintaining a force at high readiness for rapid deployment and, with sufficient warning, able to deploy 30,000 troops into action with maritime and air support.

And an air force with new fleets of transport planes – the C17s and A400Ms – and a large fleet of fast jets, principally Typhoons, soon to be augmented by the new F35.

The budget fell initially, but when I left office it was rising again. In the military expression, there was now 'more in the tooth and less in the tail'. Or, for civilians, more fighting force and less backroom bureaucracy.

Far from having a smaller and less capable army, we could still field a force similar in size and capability to that deployed in the Falklands War in 1982 or Iraq in 2003. We could still sustain over several years a force similar in size to the Afghanistan deployment while taking on an additional small conflict (like Sierra Leone – or, as it turned out, Libya).

Among those who care about Britain's standing in the world, there are many who think defence spending and army size are sacrosanct. For them, defence would have been completely exempt from the economic rescue we were undertaking – protected from cuts, like the health service or international aid.

But the reality is that a nation's vulnerability is far greater if you have a tanking economy than if you have a temporary gap in your aircraft carrier capability. If you care about Britain's standing in the world, you have to worry more than anything else about the economy and economic success. That is why it was right to ask the MoD to be part of the financial retrenchment – and why there would have to be many other cuts to come.

15

Budgets and Banks

I heard the chants from my office in No. 10 and saw the placards on TV. 'Fight the ConDem Axe Men'. 'Tory Scum, Here We Come'. 'Stop the Cu*ts!'

Before the general election, the governor of the Bank of England, Mervyn King, had warned that the measures needed to fix Britain's economy were so drastic that the winning party would be out of office for a generation.

And now, here I was, stuck *in* my office, as people protesting against those very measures in our October 2010 Spending Review encircled Downing Street, putting us in a 'lockdown', with no one allowed in, and no one allowed to leave.

As a pragmatist rather than an ideologue, I have always thought that the most important job for a prime minister is to do their duty – to tackle the most urgent task in front of them, whatever that might be. In 2010 there was absolutely no doubt what that was: to rescue our economy.

For me there wasn't any doubt about how that could be achieved. It doesn't require a degree in economics to appreciate that if you keep spending faster than the economy grows, and faster than tax revenue grows, eventually you will be in trouble. Which Britain had been doing and now was.

It wasn't just the right economic moment but the right political moment. Cuts are always controversial. You can't get them through if there isn't political assent, and that assent is only forthcoming when people see how serious things are. Which, in 2010, they did.

I believed profoundly that, if done properly, it would work. And I was absolutely certain that far worse than administering the medicine would be failing to take action.

George and I would to and fro endlessly about the measures, but we never argued about the necessity of it all. Not once. After all, the budget deficit (the annual difference between what was coming into the country's coffers and what was going out) was on course to be the highest in our peacetime history. That was adding an unprecedented amount to our national debt (the total outstanding stock of money owed by the government), which was already, again, the biggest in our peacetime history.

The more intelligent opponents of austerity would argue – rightly – that what really matters is the ratio of your outstanding debt compared to the overall size of your economy, the debt-to-GDP ratio. But even on this measure, real trouble was brewing. The debt as a share of our GDP was already forecast to rise from 34 per cent in 2003 to 75 per cent in 2015.

This matters for three reasons.

First, as the debt grows, so does the interest bill you have to pay – and that can squeeze out other vital government spending programmes. Indeed, when we took over, the government was spending more on debt interest every year than on defence. It was by far the fastest-growing government spending programme: £44 billion on simply servicing our debt.

Second, once your debt heads towards 100 per cent of your national income, there are dangers. Investors won't invest in your country, and lenders won't lend, or will ask for much higher interest rates in return for the risk they are taking. The higher rates choke off growth, tax revenues fall and the deficit rises, putting further pressure on interest rates.

In this situation, an economy can enter a form of 'death spiral': higher debt means higher rates means less growth means higher debt.

And third – and in my view the most vital reason – if your debt-to-GDP ratio gets too high, you may not be able to borrow through a crisis the next time trouble comes.

No one can abolish 'boom and bust'. Or recessions. Or financial crises. Of course you try to prevent them, but nothing matters more than your country having finances strong enough to be able to cope – because you don't know whether the next crisis is twenty or five years away. The alternative is collapse and state failure.

So during those early days in office I looked at the projected budget deficit of 11 per cent (it ended up hitting 10 per cent) and compared it with one of 6 per cent when we had gone cap in hand to the IMF in the 1970s – a humiliation seared into the memories of my generation.

I looked at our debt-to-GDP ratio, projected to be nearly 75 per cent, and compared it with nearly 50 per cent in 1976.

We were at greater risk than I had ever experienced or envisaged – and it would be a complete dereliction of duty to fail to act and to leave the country unable to cope with future problems. Something Margaret Thatcher had said kept coming back to me: 'We are in the business of planting trees for our children and grandchildren, or we have no business to be in politics at all.' This wasn't just today's fight; it was a fight for all those future generations.

However, we couldn't just address the debt and the deficit; we'd have to fix the broken economic system that had allowed them to get this bad. That meant picking apart everything that had gone wrong over the past thirteen years – and more.

People might look back at those years of economic growth under Labour and think that surely something must have gone right. On the face of it they would be correct. The number of jobs *did* increase. But over 70 per cent of them went to workers from overseas. If you took away immigration and the public sector, the number of jobs had actually *fallen* since 1997.

The unemployment rate *did* come down. But then it went back up again. And anyway, more revealing was the overall rate of inactivity (i.e. the percentage of people not actually *in* work), which was 21.5 per cent – over a fifth of the working-age population. This was masked by a ballooning welfare system. One in three households – those out of work but also, unsustainably, those *in* work – got at least half of their income from benefits.

Tax revenues *were* flooding in. But largely these were from the City of London and financial services, which, along with corporation tax, stamp duty and bonuses, made up more than 11 per cent of UK tax revenue. They were also coming from banks, companies and households whose spending ability increasingly came from debt.

And there *was* growth. But this was primarily in the size of the state, which increased in a decade from 34 per cent of GDP to 45 per cent, faster than almost any other economy over the same period. And real growth – private-sector job growth – was focused on only one part of the country. A statistic George and I often referred to was that for every ten private-sector jobs created in the south, just one was created in the north and the Midlands.

So it turned out that Labour's economic sunshine was just clever lighting. They had disguised falling employment, a shrinking private sector
and a growing deficit with a ballooning state and welfare system, surging
immigration and the tax revenue of an out-of-control financial services
industry.

It wasn't a placard that encapsulated this, but a note. When the new
chief secretary to the Treasury David Laws looked in his office drawer,
he found an envelope from his Labour predecessor. The message inside
it read:

> *Dear Chief Secretary,*
> *I'm afraid there is no money,*
> *Kind regards – and good luck!*
> *Liam Byrne.*

The note struck a chord with the public. It symbolised Labour's profligacy so perfectly that, five years later, I would carry around copies of it
with me on the campaign trail.

By immediately establishing the Office for Budget Responsibility
under a founding member of the Bank of England's Monetary
Policy Committee, Alan Budd, we were able to make sure all fiscal
policy would be based on independent, public forecasts.

The step was part of a three-tiered response we had been working on
since the crash first hit in 2008:

Monetary activism. Backing the Bank of England's strategy of low
interest rates – and in 2011–12 encouraging it to go further by directly
extending liquidity to banks and building societies, so that borrowing
and lending, including to small businesses and homebuyers, would
actually take place.

If you like, low interest rates *in principle* needed to be backed by low
interest rates and the availability of credit *in practice.* In opposition we
spoke about a massive loan guarantee scheme. We took important steps
in this direction, which I will briefly describe below. But with hindsight
we should have been far bolder, far sooner.

Supply-side reform. Cutting regulation. Encouraging start-ups and
entrepreneurs. Making it easier to hire staff and grow a business. Backing
inward investment. Reforming education. Building infrastructure.

These changes to the so-called 'supply side' of the economy can raise your country's underlying growth rate, particularly when other forms of stimulus are unavailable, because interest rates are already at rock bottom and the budget deficit is so big that tax cuts or spending increases are unavailable.

Our biggest supply-side move was on corporation tax, which would be slashed from 28 to 24 per cent, and eventually 17 per cent – the lowest rate of any major Western economy. By the time we left office there were a million more businesses up and running – and the UK had cemented its reputation as the start-up capital of Europe, the home of international commerce and one of the most competitive places in the world to do business.

Fiscal responsibility. In other words – and we were never afraid of using the word – austerity. Ensuring that Britain could live within its means was vital to giving businesses the confidence to invest and consumers the confidence to spend. And, of course, you could only rely on things like monetary activism if fiscal policy was credible.

The situation in Europe showed how urgent this was. Some countries – especially Greece, but others too – looked in danger of entering a death spiral of low or no growth, ballooning deficits and high interest rates. In fact, one of the first briefings I received in government said just that: there was a risk that the markets would shift their focus from the Eurozone to other highly indebted countries, like us. That could mean much higher interest rates – and real trouble ahead.

As I've said, one of the first decisions we took was to make immediate in-year savings. Consultancy fees, IT costs, first-class travel, unnecessary quangos; they were obvious candidates for the axe. We managed this while paying to scrap Labour's 'jobs tax' – their proposed increase in National Insurance contributions, which we believed would make the situation worse, and had played an important role in our election campaign. Interest rates came down in response to these savings, which were, bit by bit, reducing the amount the government was spending on its debt – some proof that we were on the right path.

But there was criticism too. The one that hit me most was that our action was akin to Margaret Thatcher's in the 1980s – illustrated by one of those placards, which had my face on it and the words 'I can't believe it's not Thatcher'.

I would argue that this was not like the 1980s. Then, the rhetoric was very tough, but the overall cuts were really quite mild. Conversely, what we were attempting in terms of cutting government spending was far tougher in reality, but I wanted to take more of the country with me.

Of course, economic responsibility was our goal. But another value – fairness – would be our guide. I wanted us to be the ones to prove that you could do more for less. To demonstrate that those with the broadest shoulders could bear the greatest burden. To protect the weakest and most vulnerable. To come out at the end of it with not just the books balanced but with a fairer balance between those who were fortunate and those who were not.

'We are all in this together' was the slogan I used in my 2005 victory speech, and which was adopted by George in the years to come. Winston Churchill's grandson Nicholas Soames, the MP for Mid Sussex, had given us the idea: 'It's as if you are leading the country through a time of war,' he boomed down the phone. 'You need people to see that this is a great national endeavour – that we are all in this together.'

There were many who didn't feel part of the endeavour. They thought cuts weren't just unnecessary but cruel, particularly to those receiving reduced public services, who were often the most vulnerable. Did those critics not think, though, about what would happen to those public services if we let Britain's economy tank? Take the moment when Britain teetered on the brink of bankruptcy in the 1970s. A condition of the IMF loan was £2.5 billion of immediate spending cuts – 4 per cent of all government spending. That meant an inevitable fall in funding for areas like unemployment benefits and the NHS – in fact, the only ever fall in the NHS budget since the first few years after it was established.

We knew we could also mitigate the impact on the most vulnerable with our approach towards the regions. This was another way in which we differed from Thatcher. A hit to the regions had been inevitable in the 1980s because of the closure of old and inefficient industries. This time revitalising the regions was a priority, since half of all the country's growth was concentrated in London and the south-east. The Regional Growth Fund started off as a pot of over £1 billion to help lever private investment into parts of the country that relied more heavily on state spending. It would be down to those areas, via local businesses, councils and new metro mayors, to dictate where that money was spent.

However, we couldn't rely on cuts alone to balance the books. There would also have to be tax increases, and a decision about how much to impose of each – cuts and taxes – would have to be made. The evidence internationally was that the best economic outcome would be achieved if 80 per cent of deficit reduction was made through cuts to spending and 20 per cent through taxation. It was the ratio we had promised at the election.

The question was which taxes we could increase. Increasing income tax blunted incentives and put off inward investment. Raising National Insurance was a jobs tax. There were changes we could make to the taxes paid on buying property and capital gains that would make the system fairer and ensure that the rich contributed the most – and we made them – but they couldn't raise the sort of revenue we needed.

Raising VAT, on the other hand, would take a little bit of money off a lot of people, and raise a lot of revenue in the process. The distributional effects weren't perfect, but the rich paid the most because they spent the most, and because some basic goods were VAT-exempt. So after putting the Treasury through its paces, I assented to something I'd made sure we'd never ruled out. Increasing VAT from 17.5 to 20 per cent was the biggest act of our Emergency Budget on 22 June 2010, and it would take effect from January 2011.

As well as raising VAT, we took action on Capital Gains Tax. I have always been in favour of a low rate for this tax. You don't want people holding onto assets simply to avoid paying tax (not least because it holds back the revenue you raise). And setting a low rate is one of the ways a country can build a more entrepreneurial climate. But Labour had left an unsustainable situation. Their main rate, at just 18 per cent, was so much lower than the top rate of income tax (too high, at 50 per cent), that people were finding ever more clever ways of turning income into capital growth and therefore circumventing income tax. Proposed increases included 40 per cent (the rate it had been throughout most of the Thatcher years) and 30 per cent, but I agreed to what the Treasury said was the revenue-maximising rate: 28 per cent was fair.

Our coalition partners were consulted on every cut, at full cabinet meetings and during regular 'Quad' meetings comprising me, George, Nick Clegg and the chief secretary, David Laws. Later, David was replaced by Danny Alexander. Standing over six feet, with ginger hair, Danny was something of a gentle giant. He really wanted to make the coalition work, and was often the voice of reason and conciliation when the arguments became heated. This was a gathering convened for our first Budget, but it lasted the entirety of the Parliament. And there were endless meetings and texts between me and Nick, usually about deals we were having to do.

It wasn't easy getting the Lib Dems to the stage where they saw the need for fiscal consolidation along the lines we wanted. George and I had to give way on a lot of their asks. I was relaxed about this. I knew that overall the government and I would be judged on whether we fixed the problem, not on who got what along the way. A lot of the Lib Dem requests we were more than happy to accept, for example the initial increase to the personal tax-free allowance, which the Conservatives gladly adopted, and which eventually took four million people out of paying any income tax during my six years in office.

The Quad got along well. Nick used to say to me, 'You're so lucky. You've got George as friend, confidant and partner. I don't really have anyone like that.' In time his partnership with Danny Alexander would grow stronger as Danny became more senior, but I could see what he meant. His senior Lib Dem colleagues only seemed his allies up to a point.

Many of the discussions around the economic rescue took place not as a Quad in the Cabinet Room but in the dining room of the No. 10 flat with me, George, Jeremy Heywood and Treasury spad Rupert Harrison, one of the most talented economists of his generation, passing around endless pages of A3 modelling different proposals, while I served up the pasta. The decisions about which departments' budgets to protect, though, had evolved between George and me over several years. In opposition we had decided to protect the NHS, our most precious national asset. I was clear: the NHS was sacrosanct. Reform, yes – but cuts, no.

Protecting the overseas aid budget was much more difficult to get agreement on. In these straitened times there was, unsurprisingly, a great deal of opposition to our commitment to spend 0.7 per cent of national income on development. Even Nick Clegg was prepared to delay it at one stage. I'll discuss later why we did it and why it was so important.

There was another area I had said before the election we would protect: pensioners' benefits, such as free bus passes, TV licences and winter fuel payments. In government I was going to go further, and make sure their pensions saw proper increases. My reason was clear. These people have worked all their lives, and they deserve security in their retirement. They are no longer able, like younger people, to change their circumstances. They can't get a better job or work more hours. So it is right to make sure that they can rely on the state pension they've got coming in each month.

Of course, there are all sorts of arguments about means-testing these benefits, but that process would cost more than just keeping them universal. Over the years the DWP and Nick Clegg kept trying to cut them, and I had to keep stopping them.

So I protected pensioners' benefits and, crucially, brought in what we called the triple lock, which made sure their pension would rise each year by the same amount as the Retail Price Index, average earnings or 2.5 per cent, whichever was highest. Pensioner poverty continued to fall. Means testing was reduced. And then, in a signature reform, we introduced a higher 'single-tier' pension.

Much of our strategy for making savings had been formulated in opposition too, including the controversial step of freezing public-sector pay for two years. George had announced this in his pre-election party conference speech. It was remarkable that we went into an election saying to nurses, police officers, teachers and social workers that they wouldn't get a pay rise – and *still* we won more seats than anyone else. Indeed, a majority of doctors voted for us.

I believe that was because people agreed this was fair. We were only replicating in the public sector what had already happened in the private sector, where many people had seen pay freezes, pay cuts and job losses. We were doing it in a way that protected the lowest-paid, with anyone earning under £21,000 a year exempt. And we were doing it so we could afford to fund those public services in the long term.

When weighing up where to make cuts, we had to look to where the most money was being spent. Without the DWP – which was responsible for a quarter of all government spending – making its fair share of cuts, other departments would face 33 per cent reductions in their budgets, compared to the eventual 19 per cent.

The welfare budget is divided into three parts: pensions and pensioners' benefits; help for the poorest and disabled; and working-age welfare. It was this third element that had seen rapid increases.

Take housing benefit. That was started by Thatcher in the 1980s and cost £2 billion. By 2010 it cost £20 billion – half as much as the entire schools budget. Some families received £60,000, £70,000, even £100,000 to pay for their housing costs each year.

Then there was the 'tax credits' system that was used to top up people's wages, and that had been radically expanded by Gordon Brown. One problem was it was being given to well-paid people, including some earning £60,000 a year or more. Another was that, with its bizarre incentives, many low-paid people who received it would be financially better off working less or not at all.

With such carefree spending, it's easy to see how we had reached a point where the benefits bill actually came to exceed the receipts from income tax. Every penny people paid into the pot from their pay cheque was being spent not on hospitals or schools or roads or railways, but benefits.

I know Iain Duncan Smith shared my outrage at this – not only at the money wasted but at the lives wasted in a system that encouraged people not to make the most of their potential. Iain had immersed himself in the pursuit of social justice, passionately seeking ways to cut poverty and transform people's lives. Now work and pensions secretary, he was able to right these wrongs. He had developed a solution – Universal Credit – with the Centre for Social Justice. It was a simple but profound idea. In the past, if you got a job you tended to lose many of your benefits – not just unemployment benefit, but housing benefit and council tax benefit too – meaning that you could be left worse off. The fear of losing benefits stopped many people from looking for a job in the first place.

Universal Credit was about combining these benefits together (thus the tag 'universal') and paying them to jobseekers and workers on a monthly basis, tapering the payment to make sure work always paid (thus the idea of a 'credit'). But it was, inevitably, very complicated to implement. It was also hard to get Iain to consent to the idea of saving money as well as implementing his reforms.

In the end, we won him round. The Emergency Budget cut £11 billion from the welfare budget through a combination of measures: housing benefit capped at £400 a week; child benefit frozen for three years; tax credits decreased for those earning over £40,000, but increased for the lowest-paid. And benefits would now rise in line with the less generous Consumer Price Index rate of inflation, rather than the Retail Price Index.

Today, Universal Credit is rolling out, slowly. Such a step change is never going to be straightforward, and the issues that have arisen have been with delayed payments. It is vital that process problems do not derail the concept at the heart of UC: that work must always pay.

If we stuck to this course, the OBR forecast that we would eliminate the 'structural deficit' in five years. That was our plan: by 2015 – the next election – Britain would be back in surplus for the first time since the Tories last left government. Even Vince Cable deemed this a Budget we could be proud of.

By October, the Spending Review came to its conclusion. This was what inspired that big march on Whitehall with the placards and chants.

George stood up in the Commons Chamber and went through the details. The Treasury would be asked to make the biggest reductions (33 per cent), followed by Environment (29 per cent) and Communities and Local Government (27 per cent). Half a million public-sector jobs would be lost, which we believed would be made up for by an increase in private-sector jobs. Welfare cuts would be increased by a further £7 billion – something that was much harder to get out of Iain than the first £11 billion.

Then there was the decision to remove child benefit from high-level taxpayers. Why should people on lower wages fund benefits for those on high wages?

Pensioners would keep their benefits and pensions would increase, but in the future would have to meet us halfway and accept an increased retirement age of sixty-six by 2020. We saved £34 billion and demonstrated serious government, contrasted with others, such as the French, who at a time when populations were ageing and money was in shorter supply were proposing to *lower* the pension age.

Social care would receive a £2 billion boost. Free museum entry would be preserved. We would go ahead with the new train line in London, Crossrail, the biggest engineering programme in Europe. We would continue funding the country's largest biomedical research facility, the Francis Crick Institute, and protect the science budget. As we were tough on pay and welfare we were able to be (relatively) more generous on education. We saw this as the next ring-fence to build – if we could possibly afford it.

I've said we needed to pick apart everything that had gone wrong over the last decade and more. One of those things was the banking malpractice that had helped to topple our economy. The system of regulation that the Labour government had left us was muddled and incomplete, and had proved a failure. We needed wholesale reform, of a type that addressed the justifiable anger the country felt towards such behaviour.

First, while there was plenty of financial services regulation, there was no single organisation with the task of assessing the overall risk to the economy of the level of indebtedness. As we had said in opposition, this should be vested in the Bank of England, which would need to be strengthened. This was one of the most enduring reforms we made.

Second, if a bank failed, or looked likely to fail, the system gave the government no realistic choice but to bail it out with taxpayers' money. This probably caused more resentment from the public than anything else. Even though politicians knew that banks couldn't be allowed to fail, because they would drag so much of the 'real' economy down with them, there was no way of enabling them to fail safely. So we legislated to ensure that, in future, banks facing failure would be 'bailed in' by their bondholders, not 'bailed out' by taxpayers.

Third, there was the longer-term, structural question. Given that the collapse of banks had done so much damage to the financial system and the real economy, wasn't it time to recognise that there were different types of banking, with totally different risk profiles, and they needed to be handled and regulated in a different way?

The traditional retail bank took money from depositors and lent it to businesses: a straightforward transaction that was not without risk, but was relatively easy to regulate. 'Global' banks, on the other hand, combined this retail model with far riskier activities such as high-frequency trading, complex derivatives trading, taking big bets on market movements and investment banking operations. We set up the Vickers Review which came up with a ring-fencing proposal, separating the two, which we put in place.

And we needed to address the behaviour of the bankers. They had done some truly dreadful things: taking unsustainable bets, lending to projects they hadn't properly evaluated, selling people products that were completely unsuitable, and arranging their own remuneration in a way that encouraged dangerous risk-taking. There was a complete lack of proper rules in place, and ineffective laws to prosecute offenders. So we fixed that too: we introduced a criminal sanction for the sort of reckless behaviour that had led to the financial crisis.

We also brought in far tougher rules over bonuses, pay and transparency.

I knew my response had to be nuanced: condemning the wrongdoing while continuing to support business, enterprise and go-getting capitalism. In the end, London had some of the toughest rules in the Western world, and still maintained its place as the most successful financial capital in Europe.

I believed there had to be a limit to the 'banker bashing' that Labour – as well as the Liberal Democrats and some in my own party – would have liked to unleash. We needed those banks to support the real economy. For 'monetary activism' to work, the banks would have to lend to small businesses, and get the mortgage and housing markets working again.

There was also the fact that those working in financial services weren't all champagne-swilling, City-working caricatures. The industry employed one in thirty of the people in the country. Many, contrary to popular belief, were outside London, in places we were trying to boost like Birmingham, Manchester, Cardiff, Edinburgh and Glasgow.

My solution was to try to reach something of a 'grand bargain' with the banks over what needed to be done. It would work like this. We would agree that there would be no further special bank taxes, over and above the levy we had established, and they would agree to increase lending to small and medium-sized businesses, while also boosting some other government initiatives, like the world's first Social Investment Bank – Big Society Capital – that we wanted to set up.

There was also agreement about the new rules governing pay, bonuses and the outline of regulatory reform. It became known as 'Project Merlin'.

A by-product of this was the Business Growth Fund, an investment scheme we encouraged the banks to set up in order to put equity into small and growing businesses. As I write it has invested £1.9 billion in 275 businesses, and is a model being followed elsewhere in the world.

One key figure in helping to strike this overall balance between bank reform and bank lending was Mervyn King himself. Quiet and considered, with goggle-thick spectacles, he spoke in measured tones and was very supportive of both the fiscal responsibility part of our agenda and vigorous supply-side reform. But beyond simply keeping interest rates low, he was less helpful when it came to the 'monetary activism' that we believed was so necessary.

In meeting after meeting, George and I would politely point out that we were doing our bit on fiscal policy, getting the deficit down. It was the Bank of England's turn, we thought, to do its bit on monetary policy to get credit flowing.

'It's great that you're keeping interest rates on the floor,' I'd say to him, 'but that's no good to a small businessman or woman who can't take advantage of them because the banks won't lend.'

On this, Mervyn was something of a roadblock. Eventually he produced some good initiatives, like 'Funding for Lending' in July 2012, which linked the availability of Bank of England finance to the ambitions of clearing banks to lend to small and medium-sized enterprises.

But we had to wait too long for too little. Project Merlin had some success, but really it was nine months too late, after the Eurozone crisis had flared up in August 2010.

Today the government has sold all its Lloyds shares, and our financial services industry is healthier, safer and better-regulated. We started to resell RBS in 2013, but it still remains mostly government-owned. As I argued, maybe the approach the Brown government chose, keeping it as one bank and pushing in taxpayers' money to keep it afloat, was the wrong model. Perhaps we should have gone for separation, and put the poorly performing assets in a 'Bad Bank', fully nationalised, while rapidly floating off healthy banks to help get lending going. We didn't do that because I judged it was too late to change course. And at least then we showed strategic patience, giving the management of RBS time to nurse this mega bank back to health. As I write it is still a work in progress.

My role in this was to bring the right parties together. Sometimes it was tricky and intricate. And of course an alternative, bolder approach was possible. But real political life is complicated, and full of compromise and grey areas.

What we got wrong on the monetary-activism part of our strategy is that we didn't go far enough. What we got wrong on the fiscal responsibility part was that it took us a while to work out that the best reforms were those that made a little money from a lot of people, for example freezing pay or aligning welfare upgrades with CPI.

What didn't work were reforms that people either didn't back or couldn't see the logic of. These were the things that Oliver Letwin would caution against as 'tiny atomic explosions' – where the outrage was not worth the saving.

The proposal to sell the Public Forest Estate, which made up a fifth of Britain's woodland, was one such example. Caroline Spelman, the secretary of state at DEFRA, put the plan out to consultation after getting the go-ahead from No. 10. Then came the blowback. The fire was stoked by

38 Degrees, the online lobbying organisation that would jump onto any bandwagon and fill MPs' postbags with endless cut-and-paste objections. It wasn't a bad policy – but it was politically ridiculous. We were meant to be in the business of planting trees for the future, and here we were selling them off. Once again, I was the man under the car bonnet, falling for a rational and clever plan rather than standing back and looking at the big picture.

I was learning my lesson. Years after, when a proposal on diverting Britain's much-loved public footpaths away from gardens, farmyards and businesses appeared in my box, I was very clear in my reply: 'Avoid. No. No. No. No. No. No. No. No. Which genius thought of this right now? NO. Just leave this as it is. DC.'

Looking back over those first few months at the helm of Britain's economy, what do I think? Did we cut too much?

If you'd asked the BBC at the time, they'd have said yes. I am a fan of the BBC and of its news reporters. But sometimes they seemed like the 38 Degrees of the broadcast world, constantly fighting a battle against austerity. I think they were wrong. Because the cuts – £1 in every £100 spent by government – weren't that bad. They weren't as bad as I thought they'd have to be before we took office. They weren't as bad as the protesters' placards said or Mervyn King implied they'd have to be. They certainly weren't as bad as the impression gathered from listening to the BBC.

My assessment now is that we probably didn't cut enough. We could have done more, even more quickly, as smaller countries like Ireland had done successfully, to get Britain back in the black and then get the economy moving.

Those who were opposed to austerity were going to be opposed – and pretty hysterically – to whatever we did. Given all the hype and hostility, and yes, sometimes hatred, we might as well have ripped the plaster off with more cuts early on. We were taking the flak for them anyway. Today, with austerity fatigue setting in, and the effects of that shown in the polls and election results, it's even clearer that we should have taken advantage of the window of public support for cuts when it was open. The job we started still needs to be finished.

16

Nos 10 and 11 – Neighbours, Friends and Families

Downing Street. A strange row of a few tall, mismatched townhouses, cowering in the shadow of the mighty Foreign Office.

A little figure pops out of the shiny black door and sails down the road on a lilac-coloured scooter.

'Hello, Flo,' beams the first police officer she passes, his finger on the trigger of a machine gun. His colleague opens the iron pedestrian gate and the pink figure glides through, passing a photographer who is snapping away.

As she weaves through the tourists on Whitehall, her knackered dad, clutching a baby doll and a little glittery bag, accompanied by a plain-clothes protection officer, tries to keep up with her, before they cross the road and disappear into a side entrance of the House of Commons.

This is Florence Cameron on her daily journey to nursery, and this is the only world she has ever known.

Florence first entered Downing Street as a bump on 11 May 2010. The pregnancy was complicated, because Samantha had a low-lying placenta and a real risk of bleeding and a very premature birth. We wondered about the wisdom of going to Cornwall that August for our annual holiday. But the baby wasn't due until September, and neither of us fancied being cooped up in London.

When Sam began suffering birth tremors we went to the Treliske Hospital in Truro. The doctor was happy to go ahead there and then with the caesarean that Samantha was in any event planning to have. (After Ivan's emergency caesarean, all our children were born this way.)

Our daughter arrived an hour later, weighing six pounds one ounce – and for the fourth and final time I had that amazing feeling of being the first person to hold the new baby, looking back at the team of midwives, doctors, nurses and anaesthetists who'd helped to bring this tiny, precious girl into our lives.

There followed some of the happiest days of my life. I went back to the holiday home to tell the other children the news, and for the next couple of days drove between Trebetherick, where we were staying, and the hospital in Truro. When Samantha and the baby were ready to leave, we were completely unprepared for the arrival of a baby. No cot. No pram. No bottles. No baby clothes.

Nancy, who had inherited her mother's genius for design, had found a cardboard box, decorated it in tinsel and crêpe paper and put a pillow inside as a mattress. This was where our new daughter spent her first days – indeed weeks, as it was so successful that we continued to use it when we returned home to Downing Street.

After our initial scepticism, we had created a great home in No. 10. The biggest factor in this was Samantha. She was the one who did all the work to make our home life there so successful. She was the one who thought so carefully about how to protect our children and keep them grounded with their old friends and in their existing schools. She was the one who looked after me and kept me vaguely sane.

At the same time she fulfilled the role of prime minister's wife in the most brilliant way. It's not easy to do. Simply carry on with your own career, and you are criticised for not helping. Stay totally in the background, protecting your family, and you are accused of shrinking away. Back your husband or wife too vigorously, and you're accused of meddling in politics. And heaven forbid that you speak out and offer a political opinion of your own.

Samantha continued to work for the business, Smythson, where she had been for fourteen years, starting as a window dresser and rising to design director. She held charity receptions in Downing Street every week. Quietly, and without publicity, she volunteered at a homeless shelter, the Passage, and for an organisation that helps to promote women, Dress for Success. She was assisted magnificently over the years by Isabel Spearman, then Kate Shouesmith and then Rosie Lyburn.

I can't recall anyone criticising her for any element of what she did – and rightly so.

What was her secret? As well as having great judgement and being extraordinarily efficient, I think one of the keys to her success was that she didn't change. At all. She was determined that her friends would stay the same. As much of her life as possible would stay the same. She would still go on holiday in the same places and spend time doing the same things with the same people.

And one other thing that never changed was that while she supported what I did, and cared passionately that I got things right, she wasn't at all impressed with any of the trappings of office. She was thoroughly grounded – and hopefully she kept me slightly grounded as well.

Though, of course, I did relish living in that great building. Showing friends or visitors the State Rooms or the Cabinet Room after dinner, padding through the dark corridors trying to find light switches and door keys and to remember the important points about pictures, statues and pieces of furniture – the thrill never wore off.

Because while everyone knows the iconic front door, few people actually know what it's like behind it. No. 10 is not open to the public, and the road is gated at either end. There are only a couple of hundred people working there, and just a few dozen who have ever actually lived there. The house was gifted to the first prime minister, Robert Walpole, by King George II, but Downing Street was thought a rather grotty part of town to live in, and it was only really in the twentieth century that successive PMs would actually make it their home.

What surprises many people is that the famous door is not just reserved for world leaders and photocalls; it's what staff and visitors come in and go out of every day. There is no key; it's manned by the team of 'custodians', 24/7.

The former drawing rooms and studies of the original seventeenth-century house now form a mishmash of offices large and small, and the many extensions and renovations over the years add to that higgledy-piggledy feel. The fact that our country's seat of power has this slightly improvised, rickety sense – that it is both grand and homely – seems brilliantly British.

The big black door isn't wooden, it's metal (and the zero on it is a bit wonky because the original apparently was). Up the famous staircase are the beautiful state rooms, with their extraordinary views of Horse Guards and St James's Park. The State Dining Room doubles up as a dance floor for staff summer and Christmas parties. Thatcher's Study, now home to the vast oak table from the G8 summit, is used for everything from bilateral meetings to small drinks and dinner parties. It would also be the place I'd watch the EU referendum results come in, a whisky in hand, with my small team and Nancy by my side.

Everywhere you walk there are curiosities – fake doors, quirky objects – layered with historical facts and more recent gossip. Churchill's chair (surprisingly small). Pitt the Younger's desk (miniature for someone who was six feet tall). Thatcher's gold-leaf ceiling. Macmillan's coffin-shaped cabinet table. Countryside scenes by Turner not far from Tracey Emin's neon art (an unexpected Tory supporter, she gave me a piece of art for the house – a pink lit-up sign saying 'More Passion' that I chose to go above the Terracotta Room entrance).

Through the arch in the street, opened only on rare occasions, you can walk into the Foreign Office courtyard, and beyond that into the Treasury.

Somewhere in all this there are two flats, behind alarmed doors. Because at the same time as being the heart of government, Downing Street provides a home for not one but two families (who tend to use the back entrance facing St James's Park, avoiding the cameras and crowds).

From the flat above No. 11, a spiral staircase leads down to the garden. Of course the garden has its official purposes. It was the setting for political gatherings, charity receptions and even Shakespeare plays. We used it to promote the Olympics and school sports, and to celebrate the success of national football, rugby, cricket and other teams. From it you can see a reminder of the 1991 IRA mortar attack, whose damage to the brickwork has been left as a reminder.

But to Nancy, Elwen and Florence the garden was a football pitch, a cricket pitch, a playground (we had climbing frames installed) and a place to explore and have fun. I was often rescuing them from trees they had climbed up but couldn't climb down.

Although Downing Street never sleeps – the duty clerks, press officers, custodians, police and switchboard operators work around the clock – it can be surprisingly quiet at night. Except, that is, in the No. 10 flat, which we moved to at the end of May 2010, and where Margaret Thatcher and John Major both lived. The noise of the bell in Horse Guards Parade would wake me at regular intervals.

Of course, for the children there was no distinction between what were home areas and what were work areas. It was all theirs. It was one giant labyrinth to explore, and they loved it. They'd climb across the green baize of the cabinet table and jump onto the chair used by Churchill, only half aware that they were here because Daddy was doing the same job as him. I say half aware, because they were more interested in the fact that there were Fox's Glacier Mints in little bowls on the table. 'Daddy, your office has sweets!' I recall Elwen declaring.

I remember seeing the chief of the defence staff coming through the front door in full-dress uniform, decked out in medals, to be confronted by Florence, sitting on the black-and-white chequered floor of the hall-way, asking him, 'What are *you* doing in *my* house?'

She came to know people in the building well. She knew which desk she could expect a Polo mint, an apple, or part of a bar of chocolate from. She truly was a daughter of Downing Street.

Sam was growing to love it there too. While we were in the No. 10 flat she was preparing a refurbishment of the much bigger one in No. 11, where the Blairs and Browns had lived with their families. When we arrived it felt a bit like a tired London hotel – lots of brown furniture, pink carpets and damask. But beyond that there were beautifully proportioned rooms, high ceilings, tall windows and a sweeping staircase with an Adam-esque dome above.

Samantha was determined to bring it up to date and get it finished over the summer. Heavily pregnant with Florence, she would be marshalling builders, plasterers, decorators and carpenters while at the same time dealing with the No. 10 works department and clearing all the hurdles involved in making changes to a listed and complicated building.

We decided that we would add our own money to the allowance set aside to ensure that the occupants kept the place up to scratch and didn't allow it to become dilapidated. It was well spent, and Sam did an amazing job. It was fresh and cool, but cosy too. The kitchen became the heart of the flat – and it was there that the dramas of family life, with all its tears, tantrums, homework crises, nit combs, art projects, cooking experiments and family arguments were played out over the next six years.

It saw its fair share of political dramas too. I found it very helpful to bring MPs, advisers, my coalition partners and world leaders – including the king of Jordan, the crown prince of UAE and the Chinese premier – up to the kitchen or living room to talk in a more informal setting. It was in our kitchen that Angela Merkel and I sat and agreed that we would do everything we could to stop Jean-Claude Juncker becoming president of the European Commission. Sadly, the informal setting didn't always lead to the formal result that was desired . . .

In spite of the fact that we refurbished Downing Street, it was still infested with mice. There was one occasion when I invited Iain Duncan Smith and Northern Ireland secretary Owen Paterson for dinner in the kitchen to take their views about how to get the party together over Europe. As the two of them were urging each other on to ever greater bouts of Euroscepticism, their flow was interrupted by a succession of mice running across the floor – and by me eventually hurling an empty wine bottle at the creatures. It was a useless attempt at pest control, but it did briefly stop them 'banging on' about Europe.

The rodent problem was so bad that Liz Sugg was charged with a visit to Battersea Dogs and Cats Home. She came back with a chunky tabby who had been found as a stray and named Larry. He was not 'the Camerons' cat', as the press liked to claim, but 'Chief Mouser to the Cabinet Office'. He didn't catch many mice, but he was a fine-looking cat, although not the most friendly. I liked having him curled up in my office – even if he did cover every chair in white fur.

People assume that there are chefs in Downing Street, but there's nothing of the sort.

Since there are no supermarkets around Whitehall, that meant weekly Ocado deliveries and the occasional dash to the Tesco Express at Trafalgar Square for forgotten items or requests. We did, however, have a wonderful cleaner called Connie, who took almost all her holidays in Las Vegas, then Debbie, who cried when we left.

But when I think back on Downing Street, while I think of a family home, of course most of all I think of work.

Every day began at 5.30 a.m., or shortly afterwards, when the alarm would go off and I would try to get out of bed to be at the kitchen table going through my paperwork by 5.45. I have always been a morning person, but I would still need to kick-start my system with a strong cup of instant black coffee. This was the time of day I could focus and get things done.

It all centred around the big red box – often more than one – which would be stuffed with papers submitted by my staff overnight. I would get through reams of them, and shocked them with my continued enthusiasm for this task, and also my eagle eye for detail.

An average box would contain thirty-two notes or briefings – each one of which might be dozens of pages long – as well as four intelligence reports and nine letters to read and sign. I'd also read some of the letters I'd been sent by members of the public, of which there were about 300,000 in my first six months as PM. That was before all the papers in my 'day file', which would contain the relevant information for each meeting, debate or event. A typical 'PMQs pack' alone would be two ring binders.

After that work and some breakfast with the family, I would head down to my office and plonk the heavy box on the desk of Ed Llewellyn or Chris Martin, who replaced James Bowler as my principal private secretary.

When people asked if I enjoyed the job, I was never quite sure how to answer. It was hard, and I had to take decisions that I knew would have a huge impact on people's lives. But I never got over the feeling of it being an immense privilege to have such an opportunity. And I liked it – I liked the intellectual and physical challenge of being a political decathlete, switching from one discipline to the next and trying to give every single one of them your best.

Often Sam would ask me in the evening what I'd done that day, and I'd have blanks about the morning because so much had happened since. A day is a long time in politics for the exhausted incumbent of No. 10.

I was experiencing what I called 'a new form of tiredness', feeling not just physically run down, but mentally as though my brain was close to shutting down. For that reason, I rarely struggled to sleep. PMs' sleeping patterns fascinate people. They think of Mrs Thatcher creeping down to her study at dawn after just four hours' slumber, or of Churchill in his siren suit, lying in bed with a whisky and a cigar in midmorning after a night working on his papers.

Although I had so much on my mind when my head hit the pillow, chronic tiredness would drag me down into a reasonable sleep of seven or so hours most nights. Of course there were times when I knew a military operation was under way, or a hostage rescue was being attempted, when I would sleep with the telephone almost in my hand.

One of the things that helped get me through was the fact that I have always been able to catnap, to shut my eyes and recharge on the road, in the air and sometimes simply on the floor of my office with a book or a cushion under my head.

And one of the greatest advantages of the set-up was having my closest colleague living next door. The Osbornes started off staying at their home in Notting Hill, but in August 2011 they decided to move into the No. 10 flat. Not only were George and I good friends, but Samantha and Frances were close, and our children became close too. Nancy and Liberty Osborne (my goddaughter) would take it in turns to make unbelievable messes in either of our kitchens through their cooking experiments. And Elwen (George's godson) and Luke Osborne would play various sports in the garden. On Monday nights they would have art classes together, something we have continued with since we all left Downing Street.

Did the dads ever argue? Often, but never with anger. Together, we found Downing Street a happy place to live and work. But for me those early days were disrupted by great sadness.

On 8 September 2010, my father died.

He had been an enormous influence on my life, and was incredibly proud that his son had become prime minister, but I think he always worried about the effect it might have on Samantha and me and the family. I remember being so happy that I had got Mum and him to come to Downing Street, and also to Chequers, in August 2010 before we went on holiday, and before they made what had become an almost annual pilgrimage to a very pretty hotel in the south of France each September.

At Chequers we made up a bedroom downstairs, because Dad was finding it increasingly difficult to move, but he was determined to see all of the house. So three of us helped to get him up the stairs and into the Long Gallery, where I could show him Oliver Cromwell's sword and Nelson's pocket watch. I know it meant a lot to him to have time alone with me then, and to talk about everything I was doing.

By then Dad had had both his legs amputated, and had also lost his sight in one eye. He was becoming increasingly infirm – and it troubled him. He knew he was already a difficult person to care for, and that he would be an even trickier invalid. He said to me at the time, 'I'm feeling old and tired, nothing really works any more and I feel I'm going a bit dotty. The list of things I can do is getting shorter.'

It wasn't a 'goodbye' moment, because I didn't know he was going to die, but I did feel that he was on quite a steep downhill track. Anyway, off we went to Cornwall, and before we got back with the newly arrived Florence, off they went to France.

Early on the morning of 8 September I was getting ready to prepare for Prime Minister's Questions when the phone by my bedside rang. It was my mum telling me that Dad had been taken ill – possibly some sort of stroke – and she was very worried. I immediately feared the worst, and thought I must get over there with the rest of the family. Phones buzzed between all the siblings. Alex, Clare and I agreed to go. I called Ed Llewellyn to ask Nick Clegg to stand in for me at PMQs and rang Liz Sugg to see what flights were available to the south of France.

The family converged on City Airport, and we were all about to board the BA plane when we rang Mum one more time to find out the latest news. She thought things had stabilised; he had had a stroke, but he wasn't getting any worse.

At that moment I thought it best if I stayed behind, and I said goodbye to the others. But as I headed back towards Downing Street and PMQs there was an amazing intervention. Nicolas Sarkozy came through on the car phone to tell me that he had heard my father was unwell, and his office had spoken to the doctors concerned. They were worried that the stroke was potentially fatal. He said, 'Whatever you do, David, turn around and get back on the plane, and I will get you to your father.'

And that's what happened. As we stepped off the plane at Nice Airport a helicopter was waiting to take us to the roof of the hospital where Dad had been taken. We all got to see him, but he was in a pretty desperate condition, having had what turned out to be both a bad stroke and a substantial heart attack.

We were able to briefly grip his hand before he was wheeled into one of those large lifts to be taken to an emergency room. At the very moment the lift door closed it seemed as if he suffered a further seizure. I remember Mum saying, 'Poor Ian. I hope they don't try and save him. I know that he's going.' When you have been married to someone for forty-five years, I suppose you're so close you even almost know when they want to die. It was extraordinary.

Sarkozy then called to insist that we all stay at his official house, Fort de Brégançon, which is on a small and beautiful isthmus of its own fifty miles from Marseille.

We were cooked a delicious dinner, and sat reminiscing about Dad, drinking some of the president's best wine and brandy. We drank long into the night, telling more stories about our father, and telling each other that such a quick departure when he was happy and in a place he loved would have been what he wanted. Mum pointed out that while us remembering him over drink and food was 'what he would have wanted, he would have loved to have been there too'.

In the bay outside, two French navy patrol vessels stood guard, as did French soldiers over my father's body in the mortuary.

These were extraordinary gestures, which I will never forget. Sarkozy and I were to have some great collaborations, particularly over Libya, and some ferocious rows, almost always over the EU. But without his intervention I wouldn't have seen my dad before his funeral. I will never forget that act of compassion.

17

Progressive Conservatism in Practice

From the moment I entered No. 10, my life was dominated by Britain's two most urgent challenges. The economic crisis occupied my days. The war in Afghanistan filled my head at night. But it was the prospect of radical reform that motivated me.

I know that some thought my modernising mission was just a paint job – a way of making our party more attractive to more voters by adopting a wider, more progressive range of concerns. Many in the press and in my own party thought that 'Vote blue, go green', 'modern, compassionate conservatism' and 'the Big Society' had just been slogans, not serious intentions. That I had ridden to the centre ground on a husky-powered sled, but now, in office, I would tack back to the Conservative comfort zones of the economy, defence, crime and Europe.

I scorned that assumption. Granted, not having the money to spend would make every part of our project more difficult. But the challenge of doing more with less, not least by reforming public services, was one that every Western government needed to meet. It wasn't the only challenge we shared. How to make our environment greener. How to make the most of the opportunities offered by technology. How to put more power in the hands of a public that had been empowered by this technology and was increasingly disillusioned by centralised politics and the establishment. I was determined that, in all these areas, Britain would take the lead.

To achieve this, we alighted on 'progressive conservatism' – the idea that progressive ends are best achieved through Conservative means. It might have been the least well known of our phrases, but it was one of the most important.

It frustrated me greatly that so many of those progressive ends – a fairer, more equal society; a greener environment; more responsive public services; and more power for people – were most commonly associated with the left. How could Labour claim to own social progress, I asked, when it was Conservatives who had made such strides in cleaning up the factories and fully legalising trade unions (Disraeli), brought in free universal secondary education (Balfour and Rab Butler) and built houses for the many (Macmillan)? Few acknowledged that it was Churchill who was one of the first to endorse the idea of a National Health Service, or Thatcher who championed action on climate change. John Major should be cited as the leader who outlawed disability discrimination, but rarely is.

Many failed to understand that competent economic management – for which the centre-right *was* given due credit – underpinned all this. With forefathers like Burke, Adam Smith, Shaftesbury and Wilberforce, the centre-right did not deserve its reputation for selfish individualism. I was determined to resurrect the spirit of these great Tory thinkers by leading a generation of great Tory doers.

I wanted to show that the Conservative means that had worked so well in delivering prosperity in our economy – decentralisation, choice, competition, transparency, accountability – could deliver better public services and a stronger society too.

I firmly believed that we could smash through the old divides and become the party of economic competence *and* a first-class NHS; traditional values *and* equal rights; strong defences *and* a greener environment. Because these things weren't the either/or binaries they were assumed to be.

Someone who was vital to all this thinking was Oliver Letwin. Oliver combines intellect with a massive capacity for hard work. His influence over our policy programme in opposition and in forming the coalition was huge. Now we were in government he remained instrumental. In many ways, he was another deputy prime minister.

And he was the person who, during our first days in government, set out to wary civil servants what 'progressive conservatism' would mean in practice. Businesses and charities having the chance to deliver public services. Remuneration linked to results. Devolving power from Whitehall to cities, communities, families and individuals – a great power giveaway.

The approaches of Blair, Brown and Thatcher proved my theory that the right had focused too much on markets; the left had focused too much on the state. Both forgot the space in between: society.

Far from feeling outmoded by 2010 – a sunny idea in a dark time – the Big Society felt to me like an idea whose time had come. Not only were we having to do more for less. Disillusion with top-down, big-state solutions was setting in. People talk about populism as if it was invented the day Britain voted to leave the European Union and patented when Donald Trump was elected. But it was already a growing force by the time we took office.

One example of the philosophy in practice was an outrage I felt viscerally: the fight against poverty.

I don't doubt that Labour's changes in this area were genuinely intended to help the poorest. They topped up older people's state pensions with pension credit. They supported families with the mass expansion of working tax credits. The problem was that the benefits bill and the number of people reliant on welfare ballooned. It became disadvantageous for older people to save money if, as a result, they would lose their pension credit. It became self-defeating for parents receiving working tax credits to take on more hours or get better-paid jobs. The safety net became more of a spider's web, trapping people, sometimes for life.

What made it worse was that the political argument about poverty had settled into a sterile debate between those who saw it in absolute terms and those who saw it in relative terms.

The left favoured a relative measure: you were poor in relation to the median national income. This produced the odd effect that as the economy grew, poverty often appeared to get worse even though poor people were actually better off. And Labour locked in this interpretation with the Child Poverty Act.

On the right, the flaws in this seemed obvious. Relative measures might deem someone without, say, an iPad and an annual foreign holiday 'impoverished' because most middle earners have these things. Focusing on achieving some wider parity of lifestyle could sideline the real, urgent battle with grinding destitution.

I didn't think either of these approaches was correct. I shared the view on the right that what mattered most of all was people who had been stuck in poverty. But I also felt that relative measures did have some role. The income gap does matter. 'Even if material want did disappear, that would not be the end of the line for poverty,' is how I put it during a speech in 2006.

What we needed to focus on was how an array of life problems could cause persistent poverty – the kind of disadvantage that imprisoned whole families in both absolute *and* relative deprivation. This formed a proper new narrative on poverty and inequality, tackling the causes and not just the symptoms.

High and long-term unemployment is one of the issues at the heart of the poverty problem. That is why our pro-enterprise, pro-business approach – cutting businesses' taxes, giving investors new incentives to fund start-ups, encouraging unemployed people to quit the dole and start their own business – was so important to our poverty agenda.

The results were remarkable. Labour prophesied a million more unemployed. But for every job lost in the public sector, the private sector created five. By 2014, this country was creating more new jobs than the rest of the EU put together – 2.5 million during my time in office. The unemployment rate tumbled, hitting its lowest level since 1975.

In order to help people off welfare and into all these new jobs we launched the Work Programme, the biggest back-to-work scheme since the 1930s. Again, the scheme was recognisably Conservative: enabling businesses and particularly the voluntary sector to provide those services; paying them by their results; ploughing the savings we'd made on welfare back into the scheme; making sure the results were long-term by paying the provider once the employee had been in work for thirteen or twenty-six weeks, depending on their circumstances.

At one point, a thousand people a day were replacing their benefits cheque with a pay cheque.

Also important were new sanctions to make sure that people on out-of-work benefits were looking for jobs and taking available opportunities. These would be controversial: thousands of people would have their benefits docked because they were unable to prove that they were doing these things. And there were some mistaken decisions. But the results were undeniable. By March 2016 the number of workless households had fallen by three-quarters of a million, to the lowest proportion since records began.

It wasn't just unemployment that was holding people back, though. Too many hadn't a hope of escaping poverty because they were living with the terrible disease of addiction. However, unlike our success in helping people into jobs, helping others off drink and drugs was a tale of bureaucratic inertia, coalition mistrust – and failure.

Again, what had gone wrong was that the symptoms, not the causes, were being treated. Heroin addicts were given prescriptions for methadone, another opium-based drug. While this could help people lead less chaotic lives, it was still a highly addictive opiate; we were simply replacing one bad thing with a slightly less bad thing. Success was measured by whether you were engaged in a treatment programme – hooked on state-provided opium – not by whether you'd got clean and got your life on track.

Our aim was to change that by shifting the focus to a more progressive approach of residential courses and abstention and bringing in the voluntary sector, which I knew from friends who had suffered from addiction could do far more to help people escape the drugs' clutches for good.

But the resistance of the official machine to abstinence-based treatment, and the lack of protection for drug recovery spending, meant there was only a limited increase in non-residential rehab, and a decline in publicly funded residential rehab.

We never quite got the figures that would prove to the Treasury that rehab would pay for itself in reduced unemployment. I know this is one of Oliver's biggest regrets – and it is one of mine too.

Of course, welfare dependency and drug dependency might be the reasons an adult or their children lived in poverty. But the reason poor children so often turned out to be poor adults was inadequate schooling.

I would argue that all our education reforms – even those aimed at restoring proper discipline – were progressive, because they were aimed squarely at helping the poorest and making opportunity more equal. The actions we took to break open the state monopoly on education and tackle the dumbing down of standards are set out in the next chapter.

But one education policy particularly needs to be mentioned when considering the assault on poverty – and that is the Pupil Premium. This was a simple mechanism whereby schools received more money per pupil (around £600) for each child who came from a poor home. The idea had been developed over many years by a great centre-right thinker, and later my director of policy, James O'Shaughnessy, before it was incorporated into our 2010 manifesto.

By the time I left office, the Pupil Premium was set at £2.5 billion – a commitment from this country to its children that they will get the best chances regardless of where they're from.

Another factor that drove people into poverty and kept them there was family breakdown. I have always believed that the family is the original welfare state; it is the unit that brings up children, teaches values, supports people when they fall on hard times, inspires, loves and guides. Many people put my head start in life down to financial advantage. But what gave me a real edge, I believe, was the unconditional love and support of two amazing parents.

Successive governments hadn't thought enough about how important such an upbringing can be, and how vital it was for us to help families stay together if we possibly could. Indeed, they had taken policy decisions that actually made things worse – a benefits system that incentivised couples to live apart, a tax system that didn't reward commitment at all, and an adoption system that made it ridiculously hard for loving couples to adopt children.

In opposition, I decided that every policy had to pass a simple test: does it support families and actively promote them? A Marriage Allowance that would allow someone who didn't use up all their personal tax-free allowance to pass it on to their partner was part of our 2010 manifesto. But our coalition partners were against it, and so was George. At the time I was so keen to ensure our Budgets were a success that I sacrificed 'secondary' policies to get the big things through. I regret not pushing harder on this. Ironically, it would have to wait for a divorce – between the Tories and the Lib Dems – to become a reality.

Where we did make huge progress on strengthening families, however, was on adoption.

During my speech to the Conservative Party conference in October 2011, I revealed a fact about the state of adoption in Britain: 'Do you know how many children there are in care under the age of one? 3,660. And how many children under the age of one were adopted in our country last year? Sixty.'

There were all sorts of reasons for this, such as adoption agencies only matching children up with parents in their local borough. But the worst – the most shocking – was that some social workers were refusing to place black, mixed-race or Asian children with white adoptive parents.

In Ed Timpson, the winning candidate in the 2008 Crewe and Nantwich by-election, I had the perfect person to take on the challenge. Ed came from a family that had fostered ninety children, and he had two adopted brothers. A former family-law barrister specialising in care and adoption cases, he remained children's minister from 2012 until he sadly lost his seat at the 2017 election.

He and Michael Gove, who was himself adopted, fought hard for reforms that culminated in the Children and Families Act, together with £150 million of extra funding for local authorities to improve their adoption services. Information on councils' performance was published – an example of how transparency can help transform public services. Clear time limits were put in place for cases to get through the courts. Requirements to seek a perfect or partial ethnic match were removed. The number of children adopted increased by 72 per cent, the highest since records began in 1992. And 230 babies were adopted in 2016 – not enough, but moving in the right direction.

Each of these policies was aimed at a particular cause of poverty, but for some families, problems mounted one on top of another.

Local councils had been telling us for some time about 'troubled' families that were unable to function properly and that were often responsible for a vast amount of petty crime and truancy, stacking up court orders, police visits and social services inquiries.

Christian Guy – a young campaigner on social justice and a real visionary when it came to solving Britain's deepest, grittiest challenges – used to put it to me like this: there were plenty of people working *on* the family, but no one was really there working *in* the family. Working out why the relationships had broken down. Why the children never made it to school. Why the chaos had got so bad.

And so another flagship policy was the Troubled Families programme. It focused on families with the biggest problems, and attempted to lead them away from the things that fuelled the chaos. Jeremy Heywood helped me find the right person to lead it: the tenacious, no-nonsense Whitehall warrior Louise Casey.

I ended up in the run-down office of a London council, with Louise and a mum who had been through the programme, hearing about what a profound impact this intervention had. The woman had been in care as a child and trapped with an abusive partner in adulthood. Her three children, significantly neglected, living in a filthy flat, had been on the verge of being taken away until the whole family was taken in hand, with regular visits from a family intervention worker, a police officer and a psychologist. Together, they changed their fate. The pride with which this woman showed me pictures of her now spotless flat and told me of her son's award for school attendance was profoundly moving.

Their story was just one of 120,000 families whose lives had been turned around. It proved my point: that we should never accept that certain people are destined for chaotic, impoverished lives, or that they cannot escape them.

This, I would argue, was *real* fairness. Not the Labour fairness of giving someone a few extra pounds, or a school that was slightly better than terrible, or a heroin substitute instead of the real thing. This was giving people as good a chance as anyone else, through the means to support themselves, and the hope and self-worth that comes with it.

The result? Five hundred thousand fewer people in absolute poverty. A hundred thousand fewer children in absolute poverty. We even succeeded on Labour's measure: 300,000 fewer people in relative poverty.

This certainly wasn't the end of the line for poverty. But it was the end of the road for simplistic, flawed thinking on one of the gravest issues of our time. And when I returned to office in 2015, I planned to put the issue front and centre of everything we did.

Tackling poverty in innovative ways was one aspect of the Big Society. Another was promoting volunteering, which was embodied by our National Citizen Service.

The other aspect was opening up the delivery of public services. This could be – and was – lazily characterised as 'privatisation'. The reality was different. It was bringing all the expertise, ambition, competitive spirit, desire to do good, and yes, the imperative to succeed and make a profit, to bear on our biggest causes.

So we had social entrepreneurs – people who invest in charities – being backed by government. Social Impact Bonds, which pay investors according to results, having the potential to transform areas like children's services and rehabilitation. Big Society Capital, the world's first-ever social investment bank, funding scores of causes from dormant bank accounts. Communities taking over local assets through the Right to Bid. Public servants 'spinning out' services like leisure centres as 'mutuals', a type of cooperative. And, as I'll describe in the next chapter, public servants being put at the helm of their own services.

In many cases, our ideas on how to deliver reform and savings went hand in hand.

Much of the waste reduction and devolution was to take place in an eleven-storey glass office block in Victoria, London. This building itself, the home of Eric Pickles's new department – Communities and Local Government – was a monument to Labour's bloated, bureaucratic, big-state reign. Day after day, with the help of his scrupulous spad Sheridan Westlake, Eric unearthed absurd ways that Labour had found to waste taxpayers' money: £70,000 'peace pods'; £4,000 designer sofas; even an in-house pub, a leftover from when it was John Prescott's headquarters.

They also uncovered what was being done to create new layers of bureaucracy through a regional-level government, tying councils up in red tape and directing them from on high. So Eric's department led our mission to decentralise power and implement 'localism': devolution not to regional bureaucracies, but to local government and, where possible, to communities and people themselves.

By 2016, councils' revenues had fallen by 26 per cent (as well as central cuts, we'd given them all the chance to freeze council tax – many had taken it up and kept bills low for five successive years). But at the same time, satisfaction with these local authorities was maintained, demonstrating once again that spending does not equal success.

In the Cabinet Office Francis Maude played a key role in reducing the costs of central government. He sold off buildings we didn't need or couldn't afford to maintain. By the time I left office we had taken £3 billion out of central government costs, abolished or merged 290 quangos and reduced the civil service to its smallest size since the Second World War.

Behind these actions was a whole new way of running central government – introducing private-sector disciplines, digital technology and a smattering of behavioural science. Indeed, the digital agenda was where the principles of progressive conservatism – competition, transparency, market economics and devolution – would find fulfilment.

The words 'government IT' and 'passion' aren't often put together, but they were a combination I promoted with great fervour. I believed from the beginning that a digital revolution could save money, improve services, generate business opportunities, jobs and tax revenue, and even increase Britain's standing in the world.

It was no small task. Labour's thirteen years in power had coincided with the rise of the internet, yet their approach to this increasingly digital age was pretty analogue. Government websites offered a bewildering array of duplication, contradiction and frustration. In the second decade of the twenty-first century our public sector was still using pagers, fax machines and paper records.

Our aim was to make government 'digital by default'. Before long, 1,700 government-related websites were merged into one: gov.uk. Gov.uk made it easier to do everything from paying car tax to applying for lasting power of attorney. Technology came at the right time for this moment of economic belt-tightening: digital transactions cost around 20 pence each compared to £3 for a phone call or £7 for a letter.

Perhaps the most progressive element to this agenda was borrowed from behavioural science – a theory called 'nudge': the idea that you can change people's behaviour not through rules and regulations but by encouraging and incentivising them to make better choices.

I was convinced of its merits after reading Richard Thaler's book on the subject in 2009. We set up a Behavioural Insights Team – or Nudge Unit, as it became known – under the leadership of psychologist and civil servant David Halpern. Simply by changing the way people sign up for organ donation we could increase uptake; by redesigning the Inland Revenue's letters we ensured more people paid their taxes.

With the 'open markets' part of our mission complete, it was now time for 'open data'. Francis published a tidal wave of information online – Ordnance Survey maps, obesity data, local crime figures, real-time transport information, and all government spending over £25,000.

Again, this was true devolution: information is power. Not only did it help make us the most transparent government in the world – according to Tim Berners-Lee's World Wide Web Foundation – it became a vital part of our pro-enterprise, pro-work agenda. Almost every new public data set spawned new online businesses, as entrepreneurs used the raw information to create apps. Those like CityMapper, for example, were created out of the real-time public transport information we made available. Challenger banks like Monzo grew out of the reforms that made it easier for people to change their current accounts by forcing banks to share data.

By 2016 the UN ranked the UK as the most digital government in the world. Australia built its digital services based on our model. Barack Obama was keen for America to do the same, and sent a team over to learn from us in June 2015.

In an age when new superpowers are taking centre stage, our eminence in this area is crucial. We may not build the biggest bridges or make the most cars, but we build more websites and make more digital products than most countries on earth, leading in everything from fintech to e-commerce, coding in schools, big data and, thanks to our approach, digital public services. That gives us a new sort of soft power – software power, you could say – with which we can wield enormous global influence.

We led the way on the environment, too. That was not inevitable. With the black skies over our economy, I could quite easily have abandoned the green ambitions I'd set out at the start of my premiership. Instead, during my first visit to the Department for Energy and Climate Change I declared that this would be 'the greenest government *ever*'. As I've said: politics doesn't have to be either/or. I truly believed we could cut carbon emissions *and* cut the deficit – we didn't have to choose between them. Indeed, we needed the Conservative means of competition, efficiency and innovation to achieve that progressive end. That is how we could make the technology cheap enough to see the sort of mass deployment that would make a difference.

I was clear: everyone had to play their part.

Ministers had to play our part. At one of our first meetings I told my cabinet about a campaign which aimed to reduce Whitehall emissions by 10 per cent within 12 months, something Labour had deemed impossible. That 10 per cent cut was achieved, and exceeded, within just one year.

The finest minds in government had to play their part. Oliver Letwin unsurprisingly was at the heart of this, as was our energy minister Greg Barker.

Innovators needed to play their part – but they needed our help. The green technologies they were creating weren't going to be competitive from the start; subsidies were necessary to get them going. Our approach was to spread our bets to include onshore wind, offshore wind, biomass, carbon capture and storage, district heating systems and renewable heating systems. The best schemes rose to the top, and Britain became a global green-tech hub. I even launched the world's first Green Investment Bank, which encouraged investors to put their money into renewable schemes.

Indeed, the whole world needed to play its part, since the UK's carbon reductions were no more than a drop in the ocean. Deciding to use our aid budget to support green climate funds, so developing countries could cut their emissions, would be controversial but essential for making large-scale reductions – and giving us leverage in international agreements later on.

More and more solar panels began appearing on homes, offices and farms, soaking up the sun and eventually generating 4.9 per cent of Britain's electricity. I love the statistic that when I left office, 99 per cent of all such panels in Britain had been installed during my time as prime minister. Around a million households now generate their own energy from the sun. And we were able to achieve this because the cost of solar power more than halved.

The most remarkable transformation came from a resource more fitting to our climate: offshore wind. Critics said it would be ruinously expensive. But the cost of electricity generated by wind turbines at sea fell from around £150 per megawatt hour to £120, and is now pushing £50 – roughly the same as gas, and cheaper than nuclear. London Array, which I opened in 2013, became the largest wind farm in the world, cementing Britain's place as the largest offshore wind market in the world. By my final year in office, wind farms, solar panels and other renewable sources were generating more electricity than coal plants for the first time ever – 25 per cent of power was being generated by renewables.

UK carbon emissions have now hit their lowest level since 1894, and are still falling. After being the first major economy to pass a Climate Change Act (under Labour, supported by us in opposition), we met all our climate-change targets. As other countries cling to coal, Britain is decommissioning scores of coal-fired power stations, and has announced that no new ones will be built. We almost halved greenhouse-gas emissions between 1990 and 2018 – something that rapidly accelerated after 2010. In doing so, we smashed another orthodoxy by proving you can grow your economy *and* cut emissions.

So how were that historic building No. 10 Downing Street, and a famously reform-resistant civil service, adjusting to a fresh progressive conservative mission?

The answer is, remarkably well. What united us all – spad and civil servant, Conservative and Lib Dem – was not just the challenge of sorting out the mammoth deficit but attempting these never-done-befores, on public services, on poverty, on the part government played in people's lives in the twenty-first century. And that reforming spirit was to extend to universities, schools, healthcare and policing – with very differing degrees of success.

18

Success and Failure

Michael Gove was on fire. His denunciation of the people sitting around the cabinet table with us was witty, biting – and fair. The description of a politician as a 'sophistical rhetorician, inebriated by the exuberance of his own verbosity' was coined by Disraeli about Gladstone, but on this occasion it could equally have applied to Michael.

Who were these rogues who needed such a dressing-down? Step forward the head teachers or governors of Britain's leading private schools, including Eton, Harrow, Rugby and Wellington. The education secretary and I had invited them to Downing Street to discuss how the independent sector could help transform the state sector.

I told them that they had our support, including backing for the charitable status their institutions enjoyed. But in return we wanted them to do more to take failing schools under their wing or set up great new schools.

One – the headmaster of Rugby, I recall – seemed rather reluctant, and was explaining why. 'I'm afraid it's not really part of our skill set,' he concluded.

Michael exploded. 'Not part of your *skill set*? The head of one of the country's greatest schools, immortalised by *Tom Brown's School Days*, whose pupils painted half the globe pink, whose inventions and ideas have changed the world, spouting *management speak* when you should be forming *young minds*. I am struggling to recall anything as pathetic as the excuse for an argument which you have just offered . . .'

The diatribe continued. It was classic Michael: passionate, eloquent, theatrical. And it encapsulated our approach to reform: we wanted to deliver excellence in our public services, and we would do whatever it took to do so.

Before the economic crisis of 2008, the Blair and Brown governments had increased spending on key public services – and there had been some important advances. Levels of pay had risen, and the quality of some of the infrastructure had improved. Some process targets – for example on waiting times for hospital operations and on class sizes in our schools – had addressed important and legitimate concerns.

But there was still a sense that quality and excellence were in short supply – and that much of the investment had been wasted. Vast bureaucracies had been built up. The police were wrapped up in dozens of different targets and performance indicators. At one stage there were more people employed in the Department of Education (then the Department of Children, Schools and Families) than there were secondary-school head teachers. In the NHS the number of managers had grown five times faster than the number of nurses.

It was clear what was required. We needed to improve public services and make them affordable – not just because of the economic climate, but because if we could afford them we could protect them in the long term.

The answer wasn't privatisation, but some disciplines of the private sector were desirable. We wanted, where possible, to give people a choice; to ensure that providers of public services had to compete with each other; to open up the provision of services to new providers; to look at changing the structure of services so that new sources of revenue could be made available. Public services should be publicly funded, but not necessarily publicly delivered. That was our mantra.

Nowhere did it apply more than in our schools.

In opposition I had requested the shadow education job because all of those things we needed – a safe, open, opportunity-based society, and a dynamic, strong economy – relied on the schooling we give our children. Education is everything.

In Michael Gove I found a soulmate in this. The son of an Aberdeenshire fish merchant and a lab technician, he was educated at state school, got a scholarship to a private school and then went to Oxford before working as a local newspaper reporter. He brought a unique perspective to government – and a huge amount of inspiration and fun to my life. We were great friends.

It was over our lunches and dinners that we would shape the progressive conservative approach to education. We shared outrage at the system's decline. One in five of all children left primary school unable to read or write properly. Grades were inflated to the point where over a quarter of pupils got As at A-level. Universities began to complain that they were having to put on remedial classes for new students.

The worst thing was how this affected the poorest in our society. Just forty-five of the 80,000 sixteen-year-olds on free school meals in 2005 (a measure of disadvantage) went to Oxbridge in 2007. And the seeds of inequality were sown at a young age. Low-ability children from rich families overtook high-ability children from poor families as early as primary school (I remember Michael repeating that fact with venom). Far from making opportunity more equal, public services were cementing class divisions.

We discussed what had gone so wrong. The 1960s 'cultural revolution' in education – the dumbing down of standards, the all-must-have-prizes mentality, the idea that knowledge should be 'discovered' rather than taught. The fact that town hall bureaucrats had been calling the shots, while head teachers were unable to innovate or to sack bad staff.

We discussed, too, where our party had failed in its response. Our answer seemed to have been to try to get people out of the state sector. The Assisted Places Scheme set up by Margaret Thatcher had some incredible successes, but all it was doing was helping some children escape the state system rather than improving it for everyone. The grammar schools obsession, which I came up against in 2007, was similar. Again, these schools have helped bright children from poor homes, but only a very few, and they would only ever lead to an improvement to part of the state sector. Besides, I believed that separating children into 'sheep' and 'goats' – in entirely different buildings and environments – at the age of eleven was wrong. Wrong because children's brains haven't fully developed at that age, and wrong because you can't categorise children in such a binary way. Instead of having selection *between* schools, it is far better to promote selection *within* schools, through setting and streaming.

But my biggest complaint was this. A few private-school places here, or new grammars there, lacked sufficient reach. They were nutcrackers. We needed sledgehammers if we were going to pursue that ultimate progressive ambition – the thing that really excited me, my vision – which was to spread opportunity far and wide and do something many thought impossible: to make our comprehensive schools as world-class as our private schools.

I really believed it was achievable. I had *seen* that it was achievable. In

November 2007 I first went to Mossbourne Academy in Hackney. Here, in the poorest part of London, children from all backgrounds were learning Shakespeare and Austen; they were doing the most complex equations and excelling at sports. The headmaster, Michael Wilshaw – who would later become head of the schools inspectorate OFSTED – believed that every one of them had potential. Their results were proving that they had.

Burlington Danes, down the road from my house in North Kensington, represented the revolution in inner-city schooling that Labour had begun to undertake. It was in a deprived area, with nearly 30 per cent of pupils on free school meals. But given greater freedom, under the headship of the formidable Sally Coates it had increased its proportion of five A*–C GCSE grades from 31 to 75 per cent. In doing so, it was outpacing schools in my relatively wealthy West Oxfordshire constituency.

In opposition we studied schools like Mossbourne and Burlington, and we looked at what was going on around the world. That helped us develop the progressive conservative approach to education we would deploy in government.

First, many of these successful new schools were academies: independent schools run in the state sector.

This wasn't a totally new idea. Thatcher had partially introduced autonomy with grant-maintained schools and City Technology Colleges, both of which were free of local education authority control. Tony Blair started the academies programme – again, creating schools no longer answerable to their local town hall, but to parents. Allowing heads to steer their own ship worked wonders, because they could do the things they thought necessary to improve education, from lengthening the school day to introducing new extra-curricular activities.

But both Thatcher's and Blair's policies lacked the scale to make a real difference. We wanted all schools to be free to excel, giving every one of them the chance to become an academy. Even more radically, we wanted to enable groups of teachers, businesses, charities, faith groups and parents to set up their own institutions to increase choice and standards in their area. We called them Free Schools.

The evidence was there. In Britain, academies were more likely to be rated 'good' or 'outstanding' than other schools. And around the world, Free Schools or their equivalents were transforming education in the most deprived areas. We would boost academy numbers by rapidly expanding the numbers of *Sponsored* academies – those backed by philanthropists, businesses and charities. And we added a new type of *Converter* academy – schools that had reached key standards and were ready to take on new freedoms.

Second, Mossbourne and their ilk were employing the best graduates.

One of the problems we found was that teacher training was being controlled by lecturers, many of them on the far left of politics, who prioritised nebulous 'critical thinking' skills and 'creativity' over necessities like subject knowledge and how to manage behaviour. But then in 2002 the TeachFirst scheme was launched, enabling top graduates to train on the job, often in challenging inner-city schools. We would expand the scheme massively, raising the calibre of new teachers, and introduce a fundamental reform of teacher training through the School Direct programme, putting schools and the best head teachers (the real experts) in control of teacher-training courses and the selection of potential trainees. By 2015 half of all teachers were being trained through school-led programmes, and the number of Oxbridge graduates teaching in state schools had doubled.

We were convinced that raising standards rested on *who* taught children. But we also knew it relied on *what* pupils were taught. Therefore we would slim down the National Curriculum so the young focused on the fundamentals – mathematics, grammar, punctuation, spelling and phonics – and as they got older they studied more rigorous subjects.

The third thing we saw was that successful schools confronted failure, and mediocrity too. That meant insisting pupils had to sit and resit core subjects like English and maths until they passed. And by pass we meant C grade or above.

In addition, I knew that ring-fencing the schools budget, even in financially tough times, was important – specifically linking the number of students with the amount of funding. And that's what we did, effectively freezing (actually marginally increasing) the education budget.

So this was our progressive conservative, Big Society, open public services answer on education: rolling back the left-wing ideology that was wrecking state education; extending the school autonomy that Labour had started; and driving huge improvements in standards. The last of the 'prizes for all' had been given out.

That's not to say it all went according to plan.

When we came into office we had to deal with a Labour policy called 'Building Schools for the Future'. The scheme was poorly managed and wasteful, and Michael had to go to Parliament on 5 July 2010 to explain that 715 schools signed up to building projects under the scheme could no longer go ahead with the improvements they had been promised. This was our first big education announcement. It was a huge knock to many schools, and it was compounded by the repeated inability of Michael's department to give him the correct figures on the number of projects and schools – and by his own uncompromising tone.

Ever the decent man, Michael came to my office, his head bowed. 'I'll resign if you want me to,' he said. Of course I wouldn't let him go.

But that decision had an unintended consequence.

He told me he thought his department had been completely brainwashed by Labour and was working to thwart everything he did. He said he needed someone good to help him. 'I want you to let me have Jeremy Heywood,' he said. 'Hardly!' I replied. 'He's the permanent secretary at No. 10, for God's sake.' In a moment of weakness I let him rehire Dominic Cummings, his adviser in opposition whose departure I had made a condition of Michael's appointment in government.

I had seen how Rottweiler spads caused huge issues for the Labour government. Not only did Cummings seem to be set against the team that was running the party; the aggressive and abrasive edge he brought would have huge ramifications for his boss. Teachers often appreciated Michael's reforms, which gave them more powers. But many could not stand Michael himself, who, encouraged by Cummings, indulged in what was seen as unnecessary teacher bashing. This would store up problems for the future.

School reform dominated much of my thinking in opposition and my time in government. But the more prominent reforms during our earliest months in power were in higher education.

The policy will forever be remembered for the scenes of protesters smashing the windows of CCHQ in November 2010, hurling debris at riot police and storming the building. But it ended up with more people than ever going to university, and, crucially, more people from disadvantaged backgrounds. By 2017 students from low-income families were 30 per cent more likely to attend university than in Labour's final year in power.

How did we get there?

After the First World War, higher education became increasingly state-funded. It was only affordable because so few people attended. In 1970 just one in twenty school-leavers went to university. But as numbers rose – a third of school-leavers were seeking places by the mid-90s – a £1,000 yearly fee was introduced. When attendance approached 40 per cent in 2004, Labour increased the cap on fees to £3,000.

But it wasn't enough. With government funding and other sources of income not keeping up with the rapid expansion, institutions were finding it difficult to maintain their estates and their services. Even more seriously, young people were being turned away because of the limit on the number of places. Vice chancellors were pleading with the government to be able to increase fees and for the cap on the number of university places to be lifted.

The first question for me was whether an expansion of higher education was actually the right thing to do.

There were those in our party who said 'no': almost half of all young people going to university was already too many, and the numbers should fall. The respectable part to their argument was that there should be more technical training through apprenticeships, providing a genuine choice for school-leavers. But there was often an undertone that university should be reserved for an elite minority.

However, there were also some, like George Osborne, Michael Gove and David Willetts, by now devoting both brains to this issue as our universities minister, who said the proportion absolutely must increase.

There was the economic argument: that in a knowledge economy and a world of competition, the future belonged to highly trained graduates. There was also the opportunity argument: wanting your children to go to university is like wanting to own a home of your own – and as the party of aspiration, we should absolutely support this.

As someone who started off as a sceptic about the never-ending expansion of higher education, I began to warm to their point. Those opposing the further expansion of higher education had usually enjoyed a university education themselves – and hoped their children would do the same. Meanwhile, we had almost half of all young people going to university, but just one in five poor children doing so. This wasn't just about creating the workforce of the future – it was about smashing the barriers between rich and poor. And anyway, reducing the number of university places wouldn't solve the need for technical training. We had to do both.

Which forced the second question: how to pay for it?

Back in opposition, David Willetts had been in contact with the business secretary (who therefore covered universities), Peter Mandelson, as Labour planned a review of tuition fees. We agreed to the businessman and former BP chief executive John Browne leading it, and in our manifesto we committed to looking at its findings. We knew full well that Browne would recommend increasing fees after the election. David wrote to all our candidates ensuring that no one signed up to the campaign the National Union of Students was running to get MPs to pledge to vote against any rises.

Lord Browne, who was known for his fairness and his ability to advise across the political divide, wanted no limit on fees, and for universities to charge whatever they wanted. We vetoed this, but we did agree to students funding a much greater proportion of their tuition as the only way to achieve a highly educated, opportunity-driven society.

So, having promised to scrap tuition fees in our 2005 manifesto (which was opportunistic) and been ambivalent on the issue in the years since (which was lazy), we now had a proper policy on higher education: advocating an increase to a maximum of £9,000 a year.

On the face of it, this increase was dramatic. But the rationale was in the detail. Students would not have to pay any fees or charges in advance. In fact, they wouldn't have to start paying anything back until they earned £21,000 a year (up from £15,000 previously). They wouldn't start paying back in full until they earned £41,000. There would be generous bursaries and help for the poorest.

When it came to expanding higher education, it was always said that there is a simple choice about who pays – it's either taxpayers or it's students. Our answer was that it was only successful graduates who would pay. Indeed, under our plans, only a quarter were likely ever to pay off their fees in full. Taxpayers were still bearing a proportion of the cost, but now it was a manageable one. Because we had to consider fairness there, too. And there is surely no fairness in asking someone working on a checkout who left school at eighteen to subsidise a future lawyer.

Of course, making savings was part of the motivation. The Business Department was one of the highest-spending in government. Increasing fees was a way to make savings without reducing either the quantity or the quality of the service provided.

The system in my student days was hugely regressive. Yes, university was free, but it was only free because so few people – usually white, wealthy people like me – actually went. And that was the choice. Free university for the privileged few, or paid-for university for the many. We chose the many, in the only way that was sustainable.

But there was a problem: the Liberal Democrats. Nick Clegg had campaigned in the 2010 election on a promise not to raise tuition fees. He and his MPs had signed the NUS pledge – and he had proudly posed with a sign saying so. His manifesto included a pledge to 'scrap unfair university tuition fees', and the Coalition Programme for Government confirmed that his party wouldn't be forced to support the reform. With them abstaining, we'd just about have the votes needed.

I always held out hope that he'd change his party's policy, because I knew he had misgivings. He knew it was unaffordable and wouldn't enhance social mobility. Then, during one of our 'Quad' meetings around the cabinet table with George and Danny Alexander in October 2010, he declared, 'I'm going to support you on fees.'

But George did something surprising. 'Don't do it,' he told Nick. 'It would be a huge political mistake for you.'

George's concern for Nick was genuine. And he worried about the health of the coalition if one partner damaged itself like this.

I saw it differently. 'George makes a good point, but I want us to do things together,' I said. 'And this is the right thing to do.'

Nick was adamant: 'Our old policy was wrong; this is a good policy.'

It was one of the bravest steps I've ever seen a politician take.

The culmination of our reforms came during the 2013 Autumn Statement, when George removed the cap on university places completely. It was the best of all worlds: the universities had more money, the burden on taxpayers was reduced, around £7 billion in savings was made each year, and higher education became sustainable. We had – arguably – the best higher education in the world, and most importantly, the cap on aspiration had been lifted.

But when it came to the Lib Dems, George was right. It *was* political suicide.

The rage was initially directed at the Conservatives. It came to a head when I was at the G20 in South Korea. I sat on my hotel bed watching the television helplessly as protesters encircled party headquarters.

As hard as we had it physically on the issue, the Lib Dems got it worse politically. They were the party of students, university towns and academics. This was an attack on their core support at a time when they needed it particularly badly. Their tuition-fee reversal was also the worst type of broken promise: the type you break by actively going back on your word, rather than by failing to meet a target that you have set. In my view, that type is far more serious, and will be seen as such by voters. This is what I used to call 'Cameron's law of broken promises', and the Lib Dems are still feeling the effects of it today.

Police reform followed a similar pattern to university and school reform. The service needed improvement, and that improvement could also contribute to our much-needed savings.

We acted quickly to cut bureaucracy, make efficiencies and free up professionals to run their forces. We scrapped all the 'police performance indicators', and replaced them with just one: cut crime. We introduced 101, the non-emergency police number that prevented people calling 999 unnecessarily, thus saving police time. Pensions and pay were reformed. Instead of retiring after thirty years (meaning that some were receiving full pensions at just forty-eight), the police would now move towards retirement at sixty. Police funding by central government fell by 20 per cent (though as they're also funded by council tax, this was not a fall across the board).

The backlash was inevitable, and the Police Federation were as bad as anything Michael faced from the teachers' unions. But fortunately we had a long-serving, reform-minded minister to face them down and see it all through: Theresa May.

Lack of accountability was another problem. Who judged the performance of your local force? To whom was the chief of police in each area accountable? The answer was Police Authorities, but most people – including many politicians – had barely heard of them or knew who sat on them. An idea we developed in opposition was police and crime commissioners. They would be elected every four years to bring visibility and answerability to policing.

The first election for forty-one 'PCCs' took place on 15 November 2012. Turnout was depressingly low at just 15 per cent. But, seven years on, I believe that the idea of named individuals elected by the public to hold the police accountable is beginning to become well established.

And there were real results. The budget may have been cut, but crime actually fell by a quarter. Again we had crushed the old orthodoxy that equated levels of success with levels of spending. And we had made our streets safer while improving policing and repairing public finances.

Nigel Lawson was right when he said the National Health Service is the closest thing we have to a national religion. Its premise – universal healthcare, free at the point of use – is seen as sacred, its hard-working staff as saints, and the public is constantly on the lookout for any attempt to undermine it, which would be seen as sacrilege.

I share that devotion. I've been there, rushing into A&E at three in the morning with a gravely ill boy. I've spent the night sleeping at his bedside on a hospital floor. My family and I owe the NHS more than I can ever say.

So when I became leader I was determined to show that the modern, compassionate Conservative Party would be the party of the NHS.

Standing on the stage in Bournemouth at our party conference in 2006 I signalled the change. 'Tony Blair explained his priorities in three words: education, education, education. I can do it in three letters: N. H. S. We will always support the NHS with the funding it needs.'

Immediately, I scrapped Michael Howard's policy of Patient Passports, which would subsidise those who chose to go private. Like the Assisted Places scheme, it took people and public money out of the NHS, rather than trying to improve it for everybody.

These words and actions would, I hoped, do something to repair our unfounded reputation as cutters of the NHS (as I've said, the only cuts to the NHS in the last fifty years were in 1977, after reducing public spending was a condition of the IMF bailout – under a Labour government).

But our inheritance was mixed. The boost to NHS funding had raised standards, but a whole set of problems remained – and some had got worse. Cancer survival rates in the UK were among the worst in Europe. Mixed-sex wards were still a reality for many people. Infection rates and MRSA dominated the headlines. And the excessive target culture had serious consequences.

Stafford hospital, in the constituency where I had been a parliamentary candidate, was a small district hospital run by the Mid Staffs Trust. Just before we came to office, the shocking neglect there came to light. As families and staff who blew the whistle were ignored, possibly hundreds of people died between 2005 and 2009 as a result of poor care.

Health secretary Andrew Lansley announced a public inquiry during our first month in office. It would eventually confirm our suspicions about what had been going wrong: the priority in the hospital had been targets over patient care, not care itself; and with so many layers of bureaucracy, everything was assumed to be someone else's problem.

That was true of the health service more widely. This vast organisation – the world's fifth-largest employer – had suffered under Labour's top-down, centralising approach and constant reorganisations. Bureaucracy was rampant. Patient care was taking second place to meeting targets and Whitehall diktats.

The situation was only set to get worse. The population was growing, and life expectancy had risen – that meant more people to care for, and for longer. There were more people suffering from lifestyle-related diseases and chronic long-term conditions such as diabetes. Our most precious national institution was at a perilous moment – but so were our public finances. We simply weren't in a position to match the sort of increases Labour had made in the good years.

We said that we would ring-fence the NHS budget, and we did. In fact, we increased funding by £10 billion in real terms, leaving it almost 10 per cent higher.

Yet we knew we needed structural reforms to make the system work better. These would prove to be highly controversial. While I believe much of the controversy wasn't justified (a lot of it was fuelled by Labour and union claims of 'Tory privatisation', which it manifestly wasn't), I accept that mistakes were made, time and energy were wasted and a better approach could have been taken.

The reforms were *logical*: we were putting power in the hands of professionals, just as we had done in other public services. We gave control of budgets to the people who knew what patients need, generating greater choice and competition. Allowing clinicians and patients to choose from a variety of providers shifted the philosophy from 'get what you're given' to everyone getting the best treatment possible.

And they were *evolutionary* – following on from what had gone before. Margaret Thatcher had given some GPs the funds to commission services and had created the concept of Trust Hospitals, with greater freedom to manage (although, a bit like grant-maintained schools, the changes were too piecemeal). There was continuity with Tony Blair, who introduced semi-independent 'Foundation' Hospitals, and said that the NHS should pay for patients wherever the care was best, whether that was an NHS, private or third-sector provider.

But we were still battling against the preconception that Conservatives had sinister motives towards the NHS. People were told about 'coalition reforms', and all they heard was 'Tory cuts' and 'privatisation by the back door'. At a time when we were having to make savings, this assumption had particular potency. It didn't matter that we were increasing funding. Or that when we did encourage private or charitable providers, it was because they – people like Macmillan nurses – were excellent. Or that, in reality, under Labour 5 per cent of NHS services had been delivered by outside providers and under us that figure was just 6 per cent. Our reputation was set. Therefore we did not have the political space or the public permission to undertake such fundamental reform.

We were also battling against our own promises. I had repeatedly pledged 'no more pointless reorganisations of the NHS'. And the Coalition Programme for Government said, 'We will stop the top-down reorganisations of the NHS.'

We weren't lying; our reorganisation was never meant to be top-down. The local organisation of the NHS was based around bodies called 'Primary Care Trusts' (PCTs), which were bureaucratic organisations responsible for 'purchasing' care for their localities from the 'providers' of care, principally the district general and other large hospitals. Our view was that, as GP practices took on the purchaser role, PCTs would wither on the vine.

Yet when combining Lib Dem policy with ours, the Lib Dems insisted on abolishing the PCTs. It was neater – and we assented. This was a rare occasion when something turned from an evolution to a revolution because of the coalition. Not only was it now a top-down reorganisation, it became one of the biggest in the history of the NHS – so big you could see it from space, as NHS chief executive David Nicholson put it. More significantly, abolishing PCTs required legislation. And so the saga of the Health and Social Care Act began.

We were also hindered by the personalities involved. Andrew Lansley had shadowed the health portfolio since 2004 and would, I thought, make a sterling job of it in government. But he was *too* submerged in the detail. The jargon he'd use was baffling. I remember sitting in cabinet when he shared his reform white paper. It was like an artist unveiling a piece he'd spent years on, and everyone wondering what on earth it was.

The people who should have been supporting us, whose members we were giving more powers – the royal colleges and others – began falling away. While our school reforms had caused a ruckus with the old educa- tion establishment, which he had nicknamed 'the Blob', Michael Gove did manage to build a new coalition of support. The same could not be said of the health service. When Ken Clarke was health secretary he said that the BMA was the worst union he'd ever had to deal with, and it was now demonstrating its militancy once again by holding votes of no confidence in the government. One of the problems with the BMA is that you know it will oppose absolutely anything, whether it is a good idea or not. Back in 1948 it had opposed the creation of the NHS, for instance.

Looking back, I see the health reforms not as a triumph, nor as quite the disaster our critics predicted; but as something that took up a lot of energy and missed the biggest targets we should have been aiming for, which was the costs of an ageing population and reforming social care.

The controversial legislation all came to a head when the Lib Dem grassroots rediscovered their inner protest party on 13 March 2011 and voted against the reform bill at their annual conference. It was ironic, given the central role they had played on PCTs.

It seemed then that we wouldn't be able to get the Bill through Parliament. I feared – for the first time – that this might really be the moment the coalition would unravel, as many people were expecting it to. But on 20 March 2012 the much-amended Bill made it through both Houses – just.

But that wasn't the whole story. Look beyond the legislative behemoth and you can see that we achieved strides in healthcare I'd dreamed of since opposition. Waiting times for consultant-led treatment fell in our first two years. The number of people waiting more than a year for oper- ations halved. Hospital infections halved. Mixed-sex wards were all but abolished. There were almost 10,000 more doctors, over a thousand more midwives, and fewer managers. Patients finally had a choice in which GP they registered with. The NHS was ranked the best in the world by the Commonwealth Fund in 2014 and again in 2017.

On a few key issues, the picture on health changed markedly during my time in office. In these I was exceptionally well advised by my health spads, Paul Bate and then Nick Seddon, who told me to pick a small number of really big issues – perhaps those that others had failed at, or not even attempted – and try to effect change. I had four focuses – what I called my preoccupations: cancer, dementia, genomics and anti-microbial resistance.

It was before I was even party leader that my constituent Clive Stone brought the lottery of cancer drugs to my attention. People were dying, he said, because the treatments they needed were not being approved fast enough by the regulator, NICE.

Finally, in 2011, I could address this, introducing the Cancer Drugs Fund, which pays for treatments that NICE hasn't yet reviewed or has not deemed cost-effective. Sadly, Clive passed away in 2016, but because of his determination thousands of people who might not have survived are alive today. The fund's continuation is crucial.

My obsession with dementia was formed by that special vantage point you get being an MP and a PM.

As an MP I visited many care homes as part of my Friday constituency rounds. I noticed more and more new floors being built for dementia patients, and met many of those affected, including a sixty-year-old daughter and her eighty-year-old father. He had no idea who she was, or who he was. It was heartbreaking to witness.

As PM I surveyed the bigger picture too. I saw that, with an ageing population, many more people were slipping into this world of darkness. And I learned from experts that we could do little about the diseases, like Alzheimer's, that caused it. All the advances we'd made in cancer, heart disease and stroke – and there was, shockingly, nothing for dementia.

I saw the financial challenge very clearly, which is why I doubled funding for dementia research. But too many people thought dementia was a natural part of ageing, rather than a condition that is caused by diseases of the brain. As a result, breakthroughs had been held back. This was the communication challenge, and it was a huge barrier to advances. That is why I made sure my entire cabinet were trained as Dementia Friends, as part of a national initiative to help society better understand and accommodate the condition. And it's why I became president of Alzheimer's Research UK, putting the issue at the heart of my post-politics mission.

There is another focus I am continuing beyond politics, and it's about as personal as it gets: genomics.

When Ivan was born in 2002, genetic testing was in its infancy. We watched in agony as he had repeated blood and spinal-fluid tests, scans to record electrical activity in the brain, and any number of other interventions to try to find out what was wrong. And the 'genetic counselling' – to discuss whether this could affect any other children we might have – was little more than guesswork.

The intervening years, however, had brought us to the brink of a breakthrough. We now had technology that could provide not just the examination of individual genes, but whole-genome sequencing. Rather than testing for one disease at a time, this process could simultaneously test for all rare diseases (and, increasingly, for all forms of cancer too). The market was bringing down the cost of testing dramatically – and this could transform healthcare.

If some rare diseases are diagnosed early enough, the right medication can be administered so quickly that neurological damage can be prevented, or radically reduced. It brings me great hope when I read about babies born with Ohtahara Syndrome who have grown up with far fewer signs of developmental harm.

Britain has a great heritage in this area, but to keep us at the forefront – and to get ahead of these horrendous rare diseases – I set up Genomics England to run the '100,000 Genomes Project'. Over the next five years, 100,000 genomes from NHS patients with a rare disease or cancer had their genomes sequenced. I was proud to keep a copy of the first sequenced genome in a memory stick on my desk in No. 10, and I am proud to see the project expanding today.

The suffering caused by cancer, dementia and rare diseases was plain to see. But 'anti-microbial resistance' I'd never heard of until the chief medical officer, Professor Dame Sally Davies, came to my office in 2014. Many antibiotics were no longer working, she explained. Too many had been prescribed, and no new ones had been discovered since the 1980s. Diseases were evolving and becoming resistant to drugs.

I grasped how catastrophic this could be. Without action, we could return to a time when 40 per cent of deaths were caused by infections we'd been able to prevent for the best part of a century. And I grasped that it was an economic problem as well as a medical one. It was the lack of incentives to produce new antibiotics that was at the heart of the issue.

So that's why in July 2014 I launched the AMR review, led by the economist Jim O'Neill, and why I raised the issue at every global forum

I could. As PM, you get lots of people coming to you saying, 'This is the next Armageddon, Prime Minister.' Your job is to decide and prioritise. And this absolutely should be a global priority.

The recommendations of the report, which was funded by the UK government and the Wellcome Trust, are being followed today. The issue is on the global agenda. I am pleased that I listened to Dame Sally, and proud that it was Britain that kick-started this important endeavour.

When compared with the success of higher education, schools and police reforms, there are many lessons to learn from the failure of the flagship health reforms.

In the public's eyes, anything you do at a time of financial hardship will be seen as a cut. Especially if you're a Conservative. While university and police funding were cut and schools funding remained the same, the health budget actually rose. A government that was vilified for the cuts we made was, ironically, mauled most over the department in which we didn't cut at all.

We were never clear enough about the problem we were trying to fix. Compare it to schools. People knew there was dumbing down in education. They knew there was a lack of discipline. But they didn't sufficiently know that targets and bureaucracy were strangling our health service, or the benefits that would flow from a different approach.

In order to reform both standards and structures in schools we were doing tangible things, like banning calculators from exams, or building new, good schools. But we didn't explain how a GP's ability to commission services would help, say, cancer patients. Nor did we have an answer for the broader problems coming down the track, namely the social care crisis.

I had a relatively hands-off approach when it came to my cabinet. I trusted them to get on with it. But I trusted Andrew Lansley too much, and was blinded by his science. I could and should have stepped in earlier and got him to slow down, explain and modify his plans. Sadly, at the 2012 reshuffle I had to ask him to move to the job of leader of the House of Commons.

And finally, the holy grail: keep your word. We had pledged 'no top-down reorganisations of the NHS', and, albeit unintentionally, had done just that. When it came to Cameron's law of broken promises, I didn't follow my own advice.

But I didn't regret the pace or the philosophy, or the speed at which we saw results from our wider reforms. With every policy, including health, we started to see improvement (though our economic reforms would take a little longer).

Around the world we were being noticed. The front page of the *Economist* mocked me up with a red, white and blue Mohican, with the headline 'Radical Britain: The West's Most Daring Government'. I had it on the wall in the Downing Street flat, and it remains at home today.

We had a clear run at reform because of the political picture during those early years in office. Against all predictions, the coalition was holding together. Labour were looking less of a threat after electing as leader the more left-wing of the two Miliband brothers, Ed, in September 2010. Neither UKIP nor the Scottish National Party seemed a blot on the horizon. Indeed, we were on a long honeymoon as a coalition government.

It seemed the biggest battles I'd have to fight would be with my own party.

19

Party and Parliament

A leader of the Conservative Party can never for a minute forget one vital statistic: it only takes 15 per cent of your MPs (in my case that meant forty-six of them) to write to the chairman of the 1922 Committee and trigger a vote of no confidence in your leadership.

Not once during eleven years as Conservative leader did I feel secure for any length of time. You never do in a party that has been described by William Hague as 'an absolute monarchy, moderated by occasional bouts of violent regicide'.

Memories of Thatcher and IDS's dethronements at the hands of Tory MPs were still part of the current politics of the party. There were several times between 2010 and 2015 when I thought I'd be next. I never knew if I was just one angry backbencher's missive away from bust.

The good thing is that in this never-ending battle you are not alone. For a start, you have your parliamentary private secretary, who acts as your eyes and ears in the Commons. In my case, that loyal lieutenant was a major: Major Desmond Swayne.

'Dessie', instantly recognisable for his Panama hat and muttonchops, left politics for six months in 2003 to serve with the Territorial Army in Iraq. A dedicated Serpentine lake swimmer in Hyde Park, cyclist and jogger, he would have done a triathlon most days before the rest of us had even had breakfast.

A devout Christian, he was enormously popular with our MPs, and helped me, often via his notoriously colourful emails, with colleague management. 'It is not a question of being able to please everyone all of the time,' he would say. 'Frankly you are lucky if you can please anyone, ever.'

In one email he explained that Edward Leigh, the hardline Thatcherite MP for Gainsborough who repeatedly voted against the government, 'wants to come and tell you to your face that you are the Antichrist'.

When the time came we had a jokey and enjoyable meeting. And Edward's voting record continued pretty much as before.

Dessie also advised me on appointments, and seemed to have a nickname for everyone – ranging from 'Poshy Posh' to 'Mincehead'. I didn't always take his advice, and there were many ministers who remained in post despite Dessie's entreaties to sack them, including Theresa May, who he advised me to get rid of – twice. (Theresa, incidentally, was referred to by Dessie as 'Cruella', while her cabinet namesake Theresa Villiers was 'Morticia'.)

I spent whole days – whole parts of weeks – on party management. The Conservative Party really is the broadest of churches: a motley membership of centre-right moderates, hard-line Thatcherites and lite-libertarians. There are arch-Eurosceptics and ardent Europhiles. Social liberals and social conservatives.

It is also beset by groups and cliques – something I found baffling. I used to be a member of the One Nation dining club, a centrist group that met on a Monday night. There was also the Curry Club of 2010ers and the 301 Group of modernisers. But it was the right wing that had the serious factions: the 92 Group, the Free Enterprise Group, the No Turning Back Group, the European Research Group, the Cornerstone Group – it went on and on.

I once asked the Lincolnshire MP John Hayes – another eccentric who later became my parliamentary adviser – what Cornerstone was like. 'It's a mixture of high Tories, former regime loyalists and elements of the Christian right,' he said. 'It sounds like Beirut in the 1980s,' I replied. 'What have you done with the Druze militia and the Maronite Christians?' (Modernisers believed that following Cornerstone's advice would be the death of the party, and dubbed it the Tombstone Group.)

It is remarkable that this fractious, factional party could be a success at all. But it is: even despite our trials over implementing Brexit, it's the oldest, most successful political party in the world.

At its best, there is a magic combination of values in the party which embraces aspiration, patriotism, freedom and common sense. This allows the Conservative Party to transcend class, geography, gender, race and sexuality. Then there is the ability to move with the times. The country keeps changing socially, and the Conservative Party keeps changing with it.

What's more, the party has always had a desire to win. This instinct

Uncomplicated and unconditional love: my parents Ian and Mary with Alex, Tania, me and Clare outside the Old Rectory, Peasemore.

Playing fathers-and-sons cricket with Dad.

Eton, 1984.

Making my feelings on communism clear: the Berlin Wall, 1990.

With John Major, 1992.

Marrying Samantha, 1 June 1996.

At Norman Lamont's side, as he speaks on 'Black Wednesday', 16 September 1992.

Loving life as a candidate on the campaign trail, at Stafford Rangers FC, 1997.

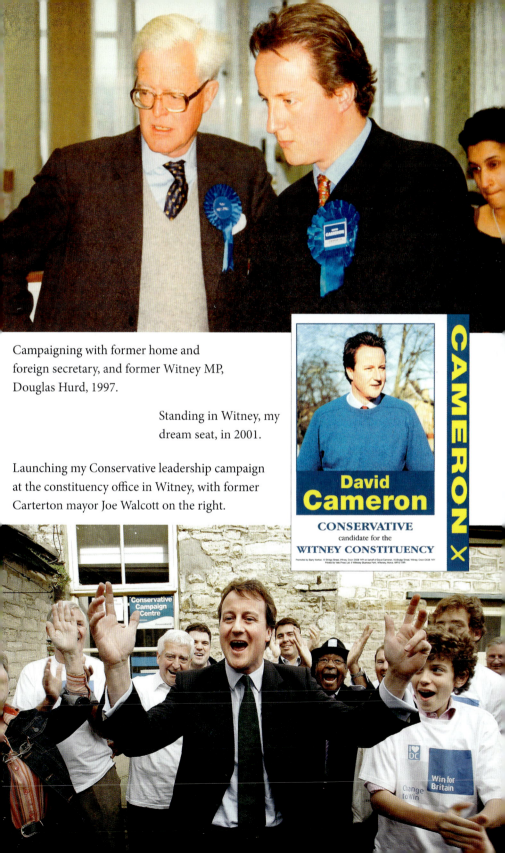

Campaigning with former home and foreign secretary, and former Witney MP, Douglas Hurd, 1997.

Standing in Witney, my dream seat, in 2001.

Launching my Conservative leadership campaign at the constituency office in Witney, with former Carterton mayor Joe Walcott on the right.

CAMERON X

David **Cameron**

CONSERVATIVE
candidate for the
WITNEY CONSTITUENCY

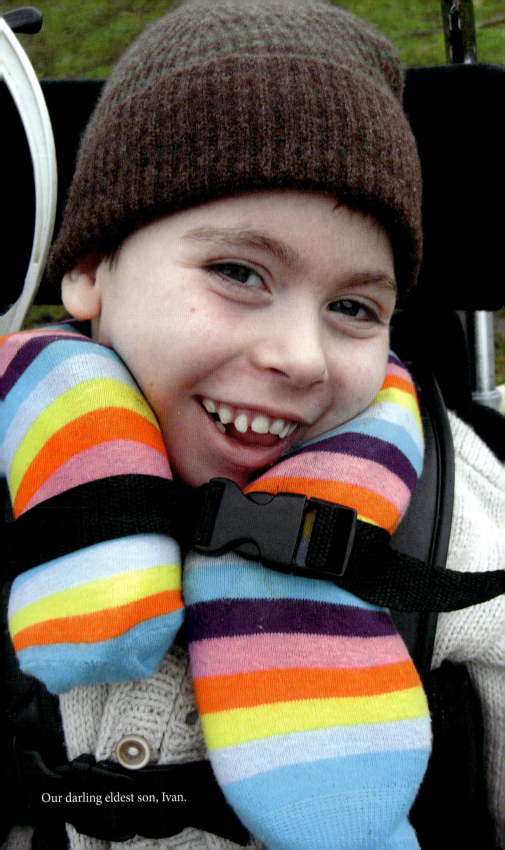

Our darling eldest son, Ivan.

In the garden at Dean with Nancy and Elwen.

Happy at home, with Samantha, Ivan, Elwen and Nancy.

The party conference speech, 2005. 'So let the message go out from this conference: a modern, compassionate conservatism is right for our times, right for our party and right for our country.'

After the announcement of the 2005 Conservative leadership second ballot results: a total of ninety votes to Davis's fifty-seven.

Celebrating in the moments after the leadership announcement at the Royal Academy of Arts, 6 December 2005.

Visiting a refugee camp in Sudan, November 2006.

A partnership. On the train with George Osborne, 2003.

Doing things differently: riding a dog sled in Norway, April 2006.

Campaigning hard, but mixing messages, 31 March 2010.

Working the tills under the supervision of Abdullah Rehman, who put me up for two days in Balsall Heath, Birmingham, May 2007.

A lighter moment with Boris Johnson at the Conservative Party Conference in Manchester, 4 October 2009.

The day after the 2010 election and hung Parliament. 'So I want to make a big, open and comprehensive offer to the Liberal Democrats.'

Watching from the Commons as Gordon Brown resigns, 11 May 2010 (left to right: David Cameron, Gabby Bertin, Ed Llewellyn, Kate Fall, Andy Coulson, Liz Sugg, Samantha, Oliver Dowden).

Our first night in Downing Street, with William Hague and George Osborne, Tuesday, 11 May 2010.

Like a wedding: the Rose Garden press conference, speaking with Nick Clegg on the new coalition, 12 May 2010.

Welcoming Mrs Thatcher
back to No. 10, 8 June 2010.

means that even though there is a home for those who like ideological purity, the party as a whole returns to pragmatism and moderation. (The definition of ideological purity does, however, change over time. At the turn of the last century, ideological purity meant committing to tariff reform to bind the Empire together. Later it meant free trade. Today, arguably, purity equals hard Brexit. At the time I became leader, many in the party thought purity meant a strict adherence to Thatcherism and a rejection of anything and everything that the Blair governments had done.)

The divide that began to emerge in the years of party modernisation and in the run-up to 2010 wasn't so much ideological as hierarchical: a growing gulf between the leadership and the backbenches. I remember during one meeting of the 1922 Committee executive, the veteran MP Nicholas Winterton railed at me, '*You* should be doing this' and '*You* should be doing that.' 'Nicholas,' I snapped, 'it's not "*you*", it's "*we*". We are one party.'

But when I moved my base out of Parliament and into Downing Street after the 2010 election, the inevitable gap between leadership and backbenches widened. Whitehall, the street that cuts through Westminster, physically divided MPs on one side and me and my cabinet on the other. It didn't matter that I took great pains to be accessible and inclusive. Being behind those black iron gates symbolised (and to a certain extent produced) separateness.

The truth is that a minority would simply never accept my leadership. There were those irritated that the party had 'skipped a generation', and that generation had been theirs. And there were those who disagreed with my progressive conservative views.

Kate Fall said, 'These people will never love you; they will tolerate you because you're good and you can win.' The complication was that I couldn't bring political success if I followed their advice, and I couldn't show clear leadership if I wasn't true to myself.

Our alliance with the Lib Dems gave them further reason to dislike me. It's not just that many in our party didn't feel part of that alliance, or ownership of our programme for government. It was that a significant proportion of our backbench MPs detested the fact that we were in coalition. They would have preferred a minority government, opposition – anything but an alliance with what they saw as the sandal-wearing, tofu-eating brigade.

If only we had been able to fight the election on more right-wing principles, some believed, we'd have won a majority. I thought this view was crackers. We didn't need less modernisation, we needed more. I still think that about the party today.

Trouble from the backbenches often came from the usual suspects on the Tory right who voted against the government regularly. But far more influential was the new intake – and almost half of all the MPs elected in 2010 were new to Parliament. The vast majority were supportive, crediting me with helping them to make it into the Commons.

However, rather than being a docile bunch who thought that filing through the right voting lobbies would lead to a ministerial job, they were independently minded and not afraid to prove it.

There were some amazing talents in the new intake who added great expertise and depth to our party. Their independence was in many ways an asset, because it garnered respect for Parliament and connected the party with voters. Yet it also made legislation and government harder to manage. The days when you could guarantee that your newest MPs would be the most loyal were over.

For some backbenchers, criticising the government was a way of ingratiating themselves with some of their most active party members. Coalition was part of that. When you rebel against a Conservative government, it can cause consternation in your local party. When you rebel against a coalition it's seen as sticking it to the Lib Dems.

While circumstances clearly made us prone to a more rebellious Parliament, I also admit that much of it was also down to my early failures at party management.

On 19 May 2010 I announced plans to – as I envisaged it – foster greater unity across the party by reforming the 1922 Committee. I wanted to foster the idea of us all working together to a common aim – nowhere was the 'one team' spirit more important than in our parliamentary party.

The 22 admitted all Tory MPs when we were in opposition. But now, in government, ministers – seventy-six of our 306 MPs – though able to attend, were excluded from voting. It was the us-and-them divide made flesh.

Even worse was that the make-up of the 22's executive committee was unrepresentative of the wider party in Parliament. With the divide between frontbenchers, who couldn't vote for the committee, and backbenchers, including those recently excluded from ministerial office, who could, this was likely to deteriorate further. Also, it met on a Wednesday, which made no sense since its purpose should have been to share information about upcoming business and political tactics for the week ahead.

Change was important. We had reformed trade unions all over Britain, but not the one at the heart of our own party. The vote to do so – crucially, to allow the whole parliamentary party to remain members when in government as in opposition – took place on 20 May 2010. We won it, but by such a small margin and against such opposition – some even threatened legal action – that we had to drop it. I had intended to hit the ground running on party management, but was left limping away, frustrated.

If I couldn't instil discipline via reform of the 22, what about via the whips' office?

The whips are, even more than your PPS, crucial to party management – your Praetorian Guard. I saw the chief whip, Patrick McLoughlin, most days. He was the most likeable and trustworthy colleague you could wish for. But we didn't see eye to eye on everything.

One thing Patrick wanted to do was to ditch our manifesto commitment to introduce Backbench Business Days, when a cross-party committee would decide on topics to debate every couple of weeks, rather than letting it simply be dictated by the government (as it was on most days) or the opposition frontbench (as it was on the twenty 'opposition days' held each year).

I had a feeling that all good prime ministers should do something to enhance the role of Parliament. Margaret Thatcher had established the Select Committee system in its modern form. Tony Blair had subjected the PM to questioning by the committees' chairmen. Gordon Brown had ensured that those chairmen were elected by MPs, not shuffled into place by the whips. I would – finally – give someone other than the government and the speaker a measure of control over at least part of the parliamentary timetable.

Dropping it would have made life easier for the government, but I wanted to keep our promise. Giving greater power to Parliament could play a part in helping restore faith in politics by making people feel their MPs were not mere lobby fodder. And for the same reason it would keep our party happy. But in truth it probably didn't do much of either of those things. Certainly I got little thanks or recognition for it.

* * *

While I was working hard to bring the Conservatives together, I was also trying to cement the coalition.

On 28 May 2010 I was on the train with Vince Cable to a manufacturer in Saltaire, West Yorkshire, when I received a call about the chief secretary to the Treasury, David Laws. He was about to be exposed by the *Daily Telegraph* for claiming expenses for the cost of his London accommodation which he was paying to his landlord, who was also his partner. This was against the rules, but he had let the lie continue for fear of his family finding out his partner was a man.

Suddenly, the Lib Dems' crisis became our crisis.

Andy Coulson, hardly a Lib Dem fan, hit the phones, trying to save David. The whole team, our spads and their spads, huddled around speakerphones, working out the lines to take, while I made a series of phone calls trying desperately to buy him some time.

I always thought Tony Blair was too brutal in pressing eject before an accused minister had got a fair hearing. Of course I knew you had to be tough and decisive – and when necessary, I was. For instance, when Patrick Mercer made racist remarks in 2007, I sacked him immediately from the shadow frontbench.

But when things were not so clear-cut, I didn't want to rush to ruin someone's career just to meet a newspaper deadline. So I attempted to convince David not to resign. The taxpayer wasn't worse off. If he and his partner had been open, they could have claimed for a mortgage as a couple. I was always sympathetic to gay people who hadn't felt ready to talk publicly about their relationships. But he was unpersuadable.

The coalition was working, even in crises, but our backbenchers, four months in, were generally feeling, in Dessie's terminology, a bit unloved. A chance to change this came during our party conference in 2010.

Over eleven years we had a pretty consistent record of using our annual gathering to galvanise and steer the mood of the party – even turn it around – through the policies we announced and the speech the leader gave at the end of it.

It was rarely me that came up with the conference slogan, but this time I was clear about the words I wanted emblazoned across the International Conference Centre in Birmingham and littered throughout everyone's speeches: 'Together in the National Interest'.

Party conferences are usually occasions for strident invective against the other parties to unite your supporters and put fire in their bellies. This was strikingly different. I recognised that Nick Clegg and the Liberal Democrats were proper partners, getting stuck in, making big decisions and working together with us.

If one side of my personality – pragmatic, laid-back, consensual – was suited to sustaining a coalition, the other side – tribal and fiercely competitive – was suited to what I had to do in Parliament. Nowhere did this come out more (sometimes more than I could help) than at Prime Minister's Questions.

As I've said, PMQs is a microcosm of the British parliamentary system: adversarial, noisy, partisan and unpredictable. It is important, and you have to do it well – to demonstrate that you're the leader of your pack, to engage your party, to take on the arguments being levelled against you.

I dreaded it all week. It is as intimidating, demanding, exhausting and downright terrifying as anything you do as a prime minister. I was always deeply relieved when it was over. You'd think that after hundreds of appearances that fear would fade. It didn't.

I developed a strict routine. It started with a very sketchy preliminary chat each Monday with Tristan Pedelty, head of No. 10's Research and Information Department and chief civil servant for briefings, and my principal private secretary, Chris Martin. A young star of the civil service, Chris was great with the ideas, and also brilliant at getting me to focus.

On Tuesday, Tristan would bring an official folder containing every subject under the sun, with a particular focus on the six or seven topics that were likely to come up. Later this duty was taken over by a wonderfully frank and robust civil servant called Hanna Johnson. At this meeting there would also be my briefings spad – and I was lucky to have some sharp and clever people throughout: first Olive Dowden, then Alex Dawson, Adam Atashzai, Meg Powell-Chandler and finally Ed de Minckwitz. There would also be the parliamentary questions expert, who during the whole time I was there was the civil service stalwart Nicholas Howard. We would discuss and commission further information that might be needed. By the evening they would bring me an updated – even bulkier – folder, which I'd take to bed with me.

On Wednesday morning I would do my box in the normal way from about 5.30 a.m., but as the minutes ticked by my mind would drift to

PMQs and the huge folder now on the kitchen table next to the rest of my work.

Downstairs, after a swift morning meeting I would meet with the PMQs team, plus a wider group including George Osborne, Michael Gove, Gabby Bertin, Kate Fall and frequently Danny Finkelstein, in my study. From time to time other MPs would join us, such as James Cartlidge. These meetings were often a riot of laughter as we tried to come up with the most topical jokes, put-downs and what we'd call 'zingers' – the comments that the House would howl at, the journalists would tweet and the broadcasters would clip on their evening bulletins.

Michael would turn up with a Pret sandwich or bowl of porridge which we'd have to watch him eat while he reeled off the material he'd diligently prepared – some brilliant, most unusable. He'd make up poems; he'd write raps. He'd link together two stories of the day, something from popular culture, something from the other side of the world, and then deliver it with *Carry On* campness.

I would either drive or walk over to Parliament at 10.30 a.m. and then sit in my Commons study, alone. On my desk was a smaller ring binder and multicoloured subject dividers, and I would remove the parts of the official briefing I wanted, writing out bits in longhand and assembling it in my own order. I'd even take out scissors and Sellotape to cut out quotes I liked and stick them on the inside of the folder. It sounds a bit *Blue Peter*, but I found that I only really absorbed the facts – and knew where to find them immediately – if I had arranged them myself.

At around 11.30 a.m. I would have a final session of 'net practice' with my PMQs spad, who would fire a set of six questions at me. Others from the private and political office would join in. Then at 11.55 I'd go into the Chamber, which was filling up during the tail end of questions to one of the smaller departments.

You're conscious that it's all going on around you. Behind: your backbenchers roaring. Above: the press and those in the public gallery focusing hard. In the air: TV cameras and dangling microphones. To your left: your PMQs team in the officials' box. In front of you: the opposition benches, jeering, muttering, braying, gesticulating – and that was just shadow chancellor Ed Balls, who I found really hard to block out. And all around you: MPs. Crouching in the aisles, spilling out into the corridors, sitting as close as lovers on a park bench.

Then it starts. 'Questions to the prime minister!' the speaker would shout.

'Question one, Mr Speaker!' the first questioner on the order paper would shout (there are twelve MPs chosen in a ballot – you know what their first questions are going to be, but not their all-important follow-ups). Typically, their first questions are all the same – 'What are the prime minister's engagements?' – but they then have complete flexibility to ask any follow-up question they choose.

Autopilot would go on. 'This morning I had meetings with ministerial colleagues and others. In addition to my duties in this House I shall have further such meetings later today,' I'd reply, as convention dictates.

Then autopilot came off and the turbo booster went on. Typically there would be a question from my own side, and once that was dealt with, six would then come at me from the leader of the opposition. Usually on the same topic, designed to expose and humiliate.

Both the prime minister and the leader of the opposition are under pressure to perform, but having been on both sides I can say – perversely – how much easier it is to be on the government side. Both of you have bricks to throw, but as PM you have a building to defend. And you always get the last word.

The Labour leaders I faced over eleven years were all strong in different ways, with one exception.

Blair was superb. He could turn from being statesman to party political jouster with consummate ease, and his sense of timing was excellent. He always claimed not to be a 'House of Commons man' but he played the Chamber like a music hall star at the London Palladium.

Brown was tough, but if you were good you could get round him and lob a grenade or two into the bunker. His big problem was that he often convinced himself of a defence that literally no one else would accept, for example when he bottled the general election in 2007. 'He's the first prime minister in history to flunk an election because he thought he was going to win it,' was my attack.

Harriet Harman was effective. She asked questions about subjects she was passionate about, like equal pay and justice for rape victims. And she had a strong but likeable character, with a good sense of humour, which meant she managed to be persistent without ever sounding strident.

Jeremy Corbyn was the exception. He seemed completely incapable of thinking on his feet. Every week as we did our preparation, my team would joke that Corbyn would ask the questions he should have asked the week before, like the Two Ronnies on *Mastermind*. For several months in a row this prediction was spot on.

Corbyn's tactic of reading out questions that came from members of the public actually made it easier, not harder, for the prime minister. While the rest of the House groaned, I thought, 'Great, I'm finally getting the chance to explain our policies to voters.' Any attempt to shout me down could be quelled by explaining that Mavis from Motherwell had asked a question and deserved a proper answer.

The public reaction to my new, more deliberative approach to PMQs was decidedly mixed. While some people claim not to like the Punch-and-Judy nature of the occasion, others tune in because they like their politics as a contact sport. One woman wrote to tell me that every Wednesday she met with her neighbours at twelve noon, opened a bottle of Chardonnay, switched on PMQs and looked forward to a good old punch-up.

Ed Miliband, who I faced for the longest period, was quick and annoyingly good at landing the class-themed blows on me that got his side of the House roaring.

One day, 18 March 2015, I had my revenge. He had recently done an 'at home with the Milibands'-type TV interview in his modest kitchen. Except that it turned out that this was the smaller of his *two* kitchens. It was the deception that got people. They don't really care if you have a massive kitchen – but they do care if you try to pretend you don't.

Eleven-year-old Nancy was coming to watch me at PMQs that day, and I told her at breakfast that morning in the Downing Street flat, 'Darling, I'm going to do something that I don't want you to copy. It won't be pleasant.'

I hammered Miliband as someone who 'literally does not know where his next meal is coming from', and then concluded with the inevitable: 'If he cannot stand the heat, he'd better get out of his second kitchen.' Nancy was in the gallery punching the air.

All that name-calling, backbiting, point-scoring and finger-pointing may seem strange when I was the one who said in 2005 that I was 'fed up with the Punch-and-Judy politics of Westminster' and would clean up PMQs. But I soon realised that there really was no alternative to playing the game. When you're under attack, you defend yourself, you fight back. When MPs come together for PMQs it's the entertainment of the week, and pretty soon you have to decide whether you will be the dinner or the diner. That is the reason I give for my somewhat aggressive delivery in those sessions – and for the fact that I used jokes to make my point.

Apart from my first one or two appearances, I rarely watched the TV coverage of PMQs, but I did read some of the newspaper sketches. Sam used to say that on Thursday mornings I was like an ageing actor reading my reviews over the breakfast table, complaining that the critics hadn't properly appreciated my performance.

But when you're on that spot at the despatch box, a primordial need to fight back kicks in. It's such a high-octane atmosphere that you feel you're suffering more from attacks than you actually are. The barbs get you. You fight back more than you need to. Blair would do well by soaking up the pressure over several questions and then hitting back with one powerful response, whereas all too frequently I found myself overreacting, or as George put it (usually with approval), 'winning the battle and then jumping into the trench and bayoneting the wounded'.

So yes, at their worst some put-downs were puerile and harsh, for example when I called Ed Balls a 'muttering idiot', which was how he was behaving, but it didn't need to be said. (It was a strange sensation watching him just a few years later doing his 'Gangnam Style' on *Strictly Come Dancing*, cheering him on and whooping with joy with the children as we all voted for him.)

For me, PMQs was also a real bonding experience with the parliamentary party. I think they felt respect, given how hard the performance is. As George would say, 'It's the only time the people behind you don't want your job.'

Afterwards, I would go to the Members' dining room, sit at the long table with Conservative MPs and chat about what had just happened, and listen to the parliamentary gossip. On Wednesdays lunch was roast beef and Yorkshire pudding, which I would wash down with a glass of red wine – and try to unwind after the adrenalin hit of PMQs. I may have had a reputation for being aloof from my MPs, but I am quite sure that this was a routine no other prime minister in living memory had kept up.

In the unique circumstances of the coalition, performing well in the House was never going to be enough. We had what I called at the time a 'head–body' problem. The head was made up of cabinet ministers or junior ministers delighted to be holding office after more than a decade in the wilderness. For most of them the coalition was an exciting challenge. The MPs in the body were either annoyed at not having achieved office or angry that the Liberal Democrats had taken their potential jobs and blocked some of our policies. For them the coalition was a source of frustration. To keep the head and the body connected, a charm offensive was required.

By August 2010 I had had every single Conservative MP for a drink in the Downing Street garden, almost every peer and every defeated candidate, funded by Tory coffers. I got much better at bringing people in and making sure everyone's voice was heard.

As well as the garden drinks, we had drinks in the flat and even bacon-butty breakfasts in the Downing Street dining room if MPs were having to sit on a Friday (to demonstrate how relaxed some were about rebelling, I even remember Andrew Rosindell turning up at one of these breakfasts, bold as brass, before going to vote against the motion I was backing).

The most valued invitation was to Chequers. These started off as being reserved for loyalists, but during the course of the Parliament more and more Conservative MPs visited. It was a physical way of sending the political message that we were in government together, ensuring that the leadership was seen as 'us' rather than 'them'.

Chequers is a sixteenth-century manor house near Aylesbury in Buckinghamshire that was gifted to the office of prime minister in the 1920s. Its owners, Tory MP Arthur Lee and his American wife Ruth, had been worried that future prime ministers would have neither the money, the time nor the inclination to appreciate country life. So they gave them the ideal country house in which to work and relax.

There are ten guest bedrooms. More than a thousand acres of land. An indoor pool. Tennis court. Two chefs. Plentiful staff. How can that possibly be justified? All I can say is that it makes the job more do-able, and frees the PM from the day-to-day fray so he or she can think and plan. The family and I would spend one weekend out of four there, and the rest at Dean.

It is a great setting in which to escape the noise and chaos of London and to be reminded of the higher purpose of politics. I would look out of my study there at the trees Churchill had planted along Victory Drive – saplings after the war, now a double row of great, towering beech trees. It helped me to forget about all the gossip and the intrigues and firefighting back in Westminster and think about the bigger picture, take the longer view, and think hard about the big decisions I was taking.

That is what Chequers allows you to do: work *and* relax. Everything is taken care of by the staff. Every Saturday and Sunday, much like every weekday, I would be working on my boxes at my desk from around 6 a.m., armed with a strong cup of coffee brought by one of the wonderful stewards, David and Harvey. But with no need to shop, cook, clear up or tidy, the day ahead gave me time both to work and to play with the children.

The rural retreat is also a great asset for diplomacy, formal occasions, charitable events, staff events and seminars. One day-long seminar was suggested by Angela Merkel, who said as we watched the Queen during one of her state visits to Germany: 'You know more about Islamist extremism than us, we know more about Russia than you. So let's have you present to us for half the day and we'll present to you the other half.' She suggested holding it at Chequers, because she had so enjoyed her first visit when we sat up drinking whisky in front of a roaring fire in the Long Gallery.

During that first visit I had taken her for a long country walk, and made a wrong turn at the top of nearby Coombe Hill. I ended up having to help her over a particularly treacherous fence. She laughed at the obvious joke about the Englishman, the German and the barbed-wire fence, and the weekend went extremely well. The farmer spotted our difficulties and put in a small stile, known from then on as the Merkel crossing.

Tony Blair came to visit to discuss the Palestine issue. The Queen and Prince Philip came for their first visit in nearly two decades, and Philip planted an English oak on the North Lawn, next to one planted by Her Majesty when she visited during John Major's premiership.

Boris Johnson came with his family one Sunday and there was a highly competitive game of football on the lawn, with Boris slide-tackling one of his children so vigorously they had to retire hurt.

At Chequers, Florence's cot was in the room next to where Lady Mary Grey, sister to 'the Nine Days Queen', was imprisoned on the floor above our bedrooms. Nancy would give guests guided tours, and proclaim, 'We won't be living here for long – it's only while Dad's prime minister.' She'd go into the house's history, talking about some of the figures who had a connection with it, like 'Oliver Crumble'.

The fact that I was known for spending time with my family and enjoying myself at a place like Chequers brought criticism. It fed into the idea that I was a 'chillaxing' prime minister. The accusation was irritating, as I was always on top of the detail, and made decisions promptly and with care. But it was also a source of some pride, because I knew that the ability to have a break and switch off made me much more effective at my job.

I remember Barack Obama telling me, 'You'll find that people will use your strengths as your weaknesses and vice versa.' He said that one thing we clearly shared was an ability to get on with the job effectively while managing to maintain an equilibrium and an element of normality in our lives. That was a strength, but it was frequently portrayed as a weakness.

The truth is, people want you to be normal, but the press criticise you for doing family things and taking time out.

One incident that does seem to stick in people's minds is the day in 2012 when we went for one of our regular weekend walks to our local pub, the Plough at Cadsden. On the day in question we had friends staying, and I remember bundling a big group of children into one car and watching as Sam did the same with another. Back at the house, as we were about to warm ourselves at the fireplace in the Great Hall, we both asked, 'Where's Nancy?' Then followed the usual Mum-and-Dad discussion – 'I thought you had her.' Then the inevitable panic.

Sam rushed back to the Plough, where eight-year-old Nancy was helping out behind the bar. All was well. (Months later Matt, the superb cartoonist from the *Telegraph*, sent us his drawing of Nancy sitting dejectedly at the bar of a pub, with the speech bubble saying she was worried about leaving her father to run the country. It is in her bedroom today.)

Really, I think being prime minister is a bit like being a parent. No one tells you exactly how to do it, you have great responsibility, you learn on the job, you make mistakes, but you slowly learn to master most of the different aspects.

And that's how I felt when it came to managing Parliament and my party. From the lows of 1922 Committee reform and the lack of direction to the whips, to the highs of conference, PMQs, coalition unity and, of course, the charm offensive with my own backbenchers, I was raising my game.

Looking back, I think that if I'd sat down more with my MPs, worked through the policies with them, explained, taken them with me, it would have fostered more loyalty and avoided some of the problems we faced.

A good leader has to be a good teacher, and in the beginning I wasn't. I was too busy learning myself.

20

Leveson

'I swear by Almighty God that the evidence I shall give . . . I give,'
I stumbled, 'shall be the truth, the whole truth, and nothing but the truth.'

You hear these words so often on TV. But when you have to say them
yourself before a judge, as I did in June 2012, the fear – of faltering, of
unintentionally misleading, of getting something, anything, wrong – is
tangible.

The phone-hacking scandal and Lord Justice Leveson's subsequent
inquiry into the culture, practices and ethics of British newspapers didn't
just put me, the prime minister, in the dock. It implicated the press, the
police and many other politicians. It dragged into the morass actors,
authors, musicians, sports stars, former ministers, members of the armed
forces, members of the royal family and, horrifically, the families of
murder victims (as well as those whose lives had been made miserable
by being falsely accused of such crimes).

It led to criminal trials, the imprisonment of one of my closest advisers, and the closure of Britain's best-selling Sunday newspaper. It brought
the coalition close to collapse – and my premiership to the precipice.

It changed the press forever, at a time when the media was already in
dramatic flux. And it added to the decline of trust in the current system.
Just as the financial crash had eroded faith in bankers, and the expenses
scandal had damaged faith in MPs, the hacking scandal brought shame
upon journalists.

When I became an MP in 2001, I had a pretty clear view on the media.
It was a symbiotic relationship: we needed them to communicate our
messages, they needed us for stories; we wanted them to publish our
successes, they wanted to scrutinise our failures. Inevitably, there is a
crossover between the two professions: there are friendships, poachers
frequently turn gamekeeper, and vice versa.

When you're running a party, all this becomes even more important. It's hard to win without the support of at least some of the media; it's hard to govern without a reasonably fair hearing; and when it comes to delivering change, it's hard to alter opinions without the media playing its part.

That said, when I became party leader in 2005, I started out trying to do things differently when it came to media relations. I never found the idea of politicians wining and dining newspaper editors and reporters particularly seedy or sinister; I just thought it was outdated. I hadn't relied on newspapers' backing to win the leadership (in fact much of the Tory press had been hostile). Why would I rely on them to win the election?

Instead, my strategy was to focus on broadcast media: getting onto the '6 and 10' (6 p.m. and 10 p.m. news bulletins) each day. I was also keen on using new media, hence my WebCameron vlogs and my regular column on the *Guardian*'s Comment is Free webpage.

But I didn't ignore the newspapers altogether. Of course, I'd meet with journalists, broadcasters and editors for on-the-record interviews and off-the-record briefings to help them understand more about my motivations, judgement and values. Those encounters ranged from dropping into the offices of the local newspaper where I was campaigning that day to meeting national newspaper journalists and their proprietors.

The biggest newspaper owner was the media mogul Rupert Murdoch, owner of *The Times*, the *Sunday Times*, the *Sun* and the *News of the World*. My first proper meeting with him was in January 2006 at a lunch at the headquarters of his company, News International. He was accompanied by the *Sun*'s political reporting team as well as Rebekah Wade, the paper's editor.

The meeting started badly. 'I saw that new film you've done, *Brokeback Mountain*. Are you expecting great things from it?' I asked Murdoch (I was referring to his company 20th Century Fox, not realising that the film was actually made by someone else). He looked at me, slightly bemused, and snapped back, 'That movie about two cowboys mooning at each other up a mountain? It won't play outside San Francisco.'

The next time we met, alone, for breakfast, I changed tack. 'Look, I'm just going to say what I think. I'm everything you don't like – upper-middle-class, a member of the establishment, a slightly left-leaning Conservative . . .' 'Well, you are,' he replied. With the problem defined, conversation flowed slightly more easily, but we never reached the point where he was enthusiastic about me.

The arm's-length approach to the press was fine in theory, but failed in practice. By 2007, as I've described, the Conservatives were losing momentum. The party was becoming restless. The grammar schools row had damaged us. Frankly, we were in trouble and drifting towards a fourth successive electoral defeat. George convinced me that we would struggle to win unless we galvanised all the elements of what he called 'the Conservative family' – and that included the Conservative-inclined print media.

He was right. But there was a problem. The Tory-leaning papers did not particularly lean towards modernisation – or towards me, for that matter. The *Mail* was unenthusiastic (its editor Paul Dacre was good friends with Gordon Brown). The *Telegraph* was also lukewarm ('Why is the *Telegraph* giving Mr Brown so much comfort?' the ConservativeHome website asked). The *Sun* was still supporting Labour, after switching allegiance to Blair in 1997. The *Times* had also supported Labour in the 2005 election, as had the *FT*.

Not only were they unsupportive; when you are the leader of a main political party at the centre of a crisis, you experience the media in full flow.

It was hard not to see the funny side. The Friday-night phone calls from my press team about some forthcoming revelation would lead to a conference call, often interrupted by gales of laughter. I remember having to deal with the allegations of Lord Laidlaw's bondage parties, and was soon joking in a speech to the parliamentary press gallery that while I had removed the whip from the peer, I should also have confiscated the handcuffs, fishnets and blindfold.

But the press was a force to be reckoned with, and I needed someone to grip it, a heavyweight media operator who knew how tabloids worked. George Eustice was very gracious, but said he couldn't do the job properly on his own, not least because when it came to dealing with some journalists, 'I just hate them too much.'

There were various other contenders from the broadcast world, some of whom I met but didn't click with. And there was a candidate from the tabloid press: Andy Coulson, an Essex-boy-done-good who shot to the top at the *News of the World* at a young age. It was he who coined the 'hug a hoodie' tag that stuck to me like glue. More recently he had dramatically resigned from the paper after its royal editor, Clive Goodman, and a private investigator, Glenn Mulcaire, had been convicted of hacking the voicemail messages of royal aides, including Princes William and Harry's private secretary. Andy always maintained that he knew nothing of this hacking, but he conceded that as editor the buck stopped with him, and he left.

His assurances had been enough for the existing press self-regulator, the Press Complaints Commission (PCC) and the police. They had investigated the allegations and concluded that they had found the criminals. The inquiries appeared to be over.

Andy reiterated his assurances during his interview with Ed Llewellyn and the then party chairman, Francis Maude, for the job of Conservative Party director of communications and planning. And he repeated them again to me during the course of our interviews in my House of Commons office.

I was satisfied with his replies, and believed that the authorities who had gone into the matter in some depth seemed to be satisfied too; Andy had done the decent thing and resigned for what had happened. I decided to give him a second chance.

His arrival helped put the wind back in our sails. He was hugely popular, and despite his frequent flare-ups with Steve Hilton, he injected camaraderie, energy and purpose into the operation.

Out went the green and blue tree on our logo, and in came a Union Flag-draped tree – an Andy innovation that remains to this day. In came more hard-hitting interventions from me on issues like schools, crime and jobs. And when Britain hit recession in 2009, Andy was superb at helping to deliver the message that we needed to live within our means.

Far from there ever being any hint of impropriety in his behaviour, he seemed to me someone of integrity: moral, upstanding, proper. There was a moment during the Caroline Spelman affair (the then party chairman had paid taxpayers' money to her children's nanny) when I said we must make sure the nanny knew what to say when the press got to her. Andy went ballistic. That was preparing a witness, he said. We would be breaking the law.

He also rebuilt my relationship with the press. People assume he was brought in to win the support of the Murdoch empire. But we worked just as hard establishing a relationship with the other main newspapers. I had previously met the *Daily Mail* owner, Lord Rothermere, as well as the editor, Paul Dacre. I had also met the various editors of the *Telegraph* and its secretive owners, the Barclay brothers, including flying with George and Ed to the Channel Island – *their* island – Brecqhou in September 2006. But with Andy in place, we stepped up our efforts.

Separately, I became good friends with Rebekah Wade. She started going out with my neighbour and old friend Charlie Brooks, who I would often play tennis and go riding with. Eventually she moved in to Charlie's house down the road from me, and they married in 2009.

Rebekah was clever, interesting, gossipy and easy to get to know. We remain friends. Of course it was part of her job to be close to political leaders – and part of mine to build relationships with key media figures. But our relationship felt as much social as political, not least because at the time it began the newspapers she was responsible for were still pretty staunch in their support for Labour.

But then on 30 September 2009, the morning after Gordon Brown's conference speech, the *Sun* switched that twelve-year allegiance to the Tories with the splash 'Labour's Lost It'. I was pleased with the endorsement, and I didn't feel it had come as a result of my efforts with the media so much as reflecting the national mood. But I was disappointed that it was framed as the paper having given up on Labour rather than taken up with the Conservatives.

We'd have to wait for the *Telegraph* to show its hand, which it did just two days before the election – again, more of a Labour and Lib Dem rejection than a Conservative testimonial. The *Mail* came out on election eve – also supportive, but not full-throated.

In the preceding months Andy had been in two minds about following us from opposition into government, but George and I persuaded him. And he thrived there.

At the same time as all of this, in June 2010, News Corporation, News International's parent company, launched its bid for the remaining 61 per cent of the satellite broadcaster BSkyB. I saw at once how significant this was, and recused myself from it. I told my office I didn't want to be involved at all, and didn't want to know when decisions were being made.

Privately, having been at Carlton and in direct competition with BSkyB when we established the rival pay-TV operator ONdigital, my view was that the media furore was actually overdone. BSkyB was already effectively controlled by News Corporation, and therefore by Rupert

Murdoch. Allowing it to buy out the remaining shareholders was a big deal in terms of scale, but not in terms of the impact on the market.

There was a legitimate argument that News Corporation had too dominant a position in the national newspaper market, where it had 36 per cent of circulation. And there was another legitimate argument that 'cross-media ownership' rules were needed to prevent companies having too strong a position in both print and television – although in my view Murdoch was already in that position.

I didn't buy into either of those arguments. While at Carlton I had made the case against ownership restrictions both within individual markets and between them. My view then and now was that the media were becoming more open and more competitive, with the blurring of distinctions between the different types of media and the entry of new giants, like Google, onto the scene.

Far from being frightened about strong media groups being established, we should recognise that it was bound to happen. Indeed, it would be good to have some British-based giants to compete with US titans like Disney and Time Warner.

The newspaper industry was in long-term decline, so consolidation was inevitable. Added to that, there was no need for concern about combining Sky News with a bunch of newspapers because a) they were already combined, and b) Sky News, while an excellent service, had a tiny audience compared with the news on BBC or ITV.

Politicians who love appearing on news channels tend to make the mistake of believing that everyone else likes watching them. They don't. I knew from my time in the industry, when I had studied the overnight ratings as soon as I got to my desk every morning, that even the most unpopular edition of the BBC's 10 o'clock news gets an audience five times the size of the most popular show on Sky News.

But none of this was my call to make. It was down to Vince Cable, who as business secretary had oversight of media deals, and he referred the bid to the media regulator Ofcom for investigation.

On 21 December 2010 the *Daily Telegraph* released a secret recording of Vince telling two reporters posing as voters at one of his constituency surgeries that he had 'declared war' on Rupert Murdoch and was planning to block the bid.

I pondered whether demanding his resignation might be appropriate, but the necessities of coalition and the weight Vince brought to the department prevailed, and he stayed put. The quasi-judicial responsibility for the bid, however, would have to go to another minister.

One allegation is that I chose Jeremy Hunt because he was sympathetic to Murdoch. Again, simply not true. He was not only the most logical person, as the secretary of state for culture, media and sport, but his appointment was made on the advice of the most senior civil servant in the government, cabinet secretary Gus O'Donnell, and was confirmed by the most senior solicitor in government, Paul Jenkins.

But it was phone-hacking that was to push media policy to the top of my prime-ministerial agenda.

Though the PCC had reviewed its evidence in 2009 and still found no evidence of wider hacking beyond Goodman and Mulcaire, the *Guardian* revealed that News International was making payouts – by that time totalling £1 million – to those claiming their voicemails had been hacked.

In February 2010 the Culture, Media and Sport Select Committee said that it was 'inconceivable' that more people didn't know about hacking, and continued to probe.

Then there was an essay in the *New York Times* that September revealing the extent of the practice and how exactly it was done: calling the target's phone to make it engaged, then simultaneously calling on another line and using the easily guessed manufacturer-set password to access their voicemail messages.

What had been claimed to be a couple of bad apples was now unfolding as an industry with a rotten secret at its core. Crucially, the article claimed that Andy had not just known about hacking, but encouraged it.

The charge against me has frequently been: why didn't you do something at this point, when the finger was being pointed directly at Andy? I was very much of the view that unless someone produced evidence that he knew about the hacking, then, like anyone, he was innocent until proven guilty. And while the claims about him were becoming more lurid, I still hadn't seen any evidence to countermand the assurances that he had given me.

But, looking back, of course my stubbornness was misplaced. Andy had been a capable and honest adviser to me, and had become part of my close team. Everything in our relationship to that point made me trust him. It wasn't only that I believed his assurances, it was that I very much wanted to believe them. And that always affects your judgement.

The Met Police then changed its position from the one I had relied upon initially. Operation Weeting later estimated that there were 829 likely victims of phone-hacking. The concurrent Operation Elveden began investigating payments by the press to the police for information. Accusations of corruption flew around: between politicians and the press, the press and the police, the entire establishment. Pandora's Box had been flung wide open.

On 21 January 2011 Andy came to me and said that his continued presence in government was too much of a distraction. He admitted that he had been wrong to have come into Downing Street when these claims – untrue as I thought they were – were still being levelled.

And then, that summer, the whole saga took a gut-churning turn when it was revealed that the phone of the murdered schoolgirl Milly Dowler had been hacked in 2002.

I was in Afghanistan, about to hold a press conference with President Karzai, at which I would be asked about the allegations: the *Guardian* had reported that voicemail messages on Milly's phone had been deleted by journalists, which caused her parents to believe she was still alive (in fact that allegation was later found to be inaccurate, though the hacking itself had indeed taken place). Overnight, the whole thing had moved from an issue that was unedifying and wrong to something that was despicable and wicked. The victims weren't just people in the public eye; they were grieving families who had already lived through hell.

Three days later my new director of communications, Craig Oliver, took me to one side to say that the *News of the World*, which had been on Britain's news stands for 168 years, was closing. We had an emergency meeting with Ed, Kate, George and others in the flat. I tried to make sense of this sorriest of sagas.

Three things had come to a head that hadn't been properly dealt with.

First, there were the poor practices by the newspaper industry. Reporters and editors were under such pressure – from twenty-four-hour rolling news, new media and declining sales – that they had chased new angles and exclusives, with some resorting to totally unacceptable practices. From rooting through bins (which the *Sunday Mirror* did to me) to hacking voicemails, the industry was in denial about its own activities.

Second, there was a chronic and persistent regulatory failure. Former media secretary David Mellor had famously told the press it was 'drinking in the last-chance saloon' two decades before, but if anything the situation had got worse. The self-regulatory body, the Press Complaints Commission, had proved itself toothless and ineffective. Its code of conduct was good, but it wasn't even close to being properly enforced. Nor were the police dealing with the elements that were clearly criminal.

It was difficult for anyone to get redress for falsehoods that had been published about them. And when people did successfully complain, an apology or correction would be printed so far back in the paper, with such little prominence, that it was hardly worth being printed at all.

And third, yes, there was a relationship between the press and politicians that was too close. And close in a particularly troubling way.

This is not a new thing. Unlike the BBC, there is no constitutional obligation for newspapers to keep their distance from politicians. But the desire of politicians to get good coverage, and the way society has changed, produced a relationship in which power was skewed. Because politicians were so reliant upon the press – and saw editors and reporters as colleagues and friends, but also feared them – few of them questioned how well the self-regulatory system was working or delved too deeply into press practices, even when the evidence was growing that the system was failing and poor behaviour was rife. That was the problem with the closeness. No one wanted to cross those on whom they relied, or risk angering those who could bring them down.

I'm sure this is why every time the issue of press intrusion had come up in the past, politicians stuck their heads firmly in the sand. For instance, in 2006 the information commissioner produced a report revealing that private detectives were selling people's personal data and information to newspapers. What was the political reaction? Did the Labour government condemn this theft of data? Was the opposition up in arms demanding action? Not a bit of it. There was close to total silence. (What had I done about the *Sunday Mirror* going through my bins? Made a fuss about it for a few days, and then dropped it.)

These issues would have to be dealt with, and dealt with properly. In terms of the most serious allegations – the hacking of phones – there was a strong case for launching a public inquiry. If a select committee couldn't find out the truth, if the PCC couldn't find out, if the Metropolitan Police couldn't find out, then who could? Parliament, the press and the police had failed. An inquiry was right and inevitable.

Nick Clegg and the Lib Dems wanted a fully independent, judge-led public inquiry with a wide remit. So did Ed Miliband and Labour. Doubtless they both believed that the Conservatives and I had the most to lose from broadening the terms to include the relationship between the press and politicians, but the case for doing so was pretty strong. And as there was a majority in Parliament for a comprehensive public inquiry, that was what we would have.

Indeed, on press regulation there was no Conservative–Lib Dem coalition and Labour opposition; there was, effectively, a Labour–Lib Dem coalition and a Conservative opposition. It was dangerous territory to be in. Either have an inquiry on our terms or have one forced on us by them.

The Labour position on the issue had fundamentally changed. The party under Blair and Brown had enjoyed an even closer relationship with newspapers and their proprietors than I ever had. But Miliband decided to go full-throttle on the issue, opposing the BSkyB takeover, criticising press practices, condemning the closeness of the media and politicians, even condemning his own party's two most recent leaders, and demanding that Murdoch's UK press empire be dismantled.

On 13 July, the day I announced in the Chamber the terms of a new inquiry into hacking – and the day that News Corp mercifully withdrew its bid for BSkyB – I was taken to task by Miliband over my friendship with Rebekah and my hiring of Andy.

I then went back to Downing Street and had the sobering and difficult task of meeting the Dowler family. They were charming and understanding, but passionately wanted something to be done. It was one of those days where you just can't believe what your day entails. As well as PMQs, the statement and the Dowlers, I had a meeting with a terminally ill teenager who was campaigning for more bone-marrow donors, a Eurozone crisis meeting with the chancellor and the governor of the Bank of England, interviews with the media, an audience with the Queen, and – perhaps proof that the symbiotic relationship never stops – a dinner with journalists from the BBC, NBC and ITV.

Even with the inquiry announced and the BSkyB bid stalled, the pressure on my position was building. Labour felt they had 'got us' over our links to Murdoch and employing this 'evil' guy in Downing Street. I could sense the pitchforks. I knew I hadn't done a deal with Murdoch nor offered policies for favours. I knew I hadn't known about Andy and hacking. But in such a dynamic, messy, high-emotion time, you just can't tell what will happen next – and you worry. That the pressure will build, that you'll lose the confidence of the party.

In August I was at Heathrow ready to depart for a long-planned trip to Africa to promote trade and British interests. Parliament was about to shut down for the summer and I had answered my last PMQs of the session. Surely, I figured, it was time to move on – Craig agreed. Not a bit of it. We were two hours into the flight when Craig went to the cockpit to take a call from Jeremy Heywood, who told him that the Met commissioner Sir Paul Stephenson had just resigned over hiring the *News of the World*'s former deputy editor, Neil Wallis, as a communications adviser. Wallis had been arrested days earlier on hacking charges.

On the same day, Rebekah Brooks was arrested. More and more questions were arising from the select committee hearings. Miliband was calling for Parliament to debate these developments.

I made it to Nigeria and South Africa. But I skipped Rwanda and Sudan, and returned on Tuesday evening before Parliament met the following day.

I answered a mammoth 136 consecutive questions, the most in parliamentary history. I got through the exchange, but I was vulnerable on the BSkyB bid (what more could I say than that it was a red herring?) and on Andy Coulson (what more could I say than that I had trusted him?).

I downloaded my feelings – a conflicted mix of contrition and self-justification – on tape a few days later: 'The worst thing I can be accused of is believing someone I shouldn't have believed. I didn't bring someone into government who then did terrible things in government, but I do feel . . . it shook me in that it's very personal because you feel your own integrity is being questioned, your decision-making, your judgement. The weird thing is, I'm not cross with myself for getting into this mess. I mean, with 20/20 hindsight I'd have made a different decision, but I can see why I made the decision: he was good, he was a nice guy, he did a good job for us, but it was a risk I suppose we shouldn't have taken, that's the truth.'

Less than a year later, the inquiry was under way. I'd sit in my study in Downing Street watching the latest witnesses giving evidence – from Hugh Grant to J.K. Rowling, Tony Blair to John Major, and of course Rupert Murdoch, Rebekah Brooks – and Andy Coulson.

It was before Rebekah's testimony in May 2012 that text messages between us were disclosed. Some of the coverage was mildly embarrassing but essentially harmless. Great fun was had over the fact that I routinely used 'LOL' to mean 'lots of love' when it actually means 'laugh out loud'. (Danny Finkelstein told me later that he realised I was getting it wrong when I sent him a text saying 'LOL' after the death of his father.)

Other texts caused a bigger political fuss.

One was Rebekah wishing me luck before a conference speech in 2009 and saying we were 'in this together'. A lot was read into this, but its origin was that I had complained about some press coverage, in the normal way politicians do. She was reminding me that whatever irritation I may have felt about an individual item, her paper was invested in our success.

Another text involved me talking about Charlie letting me ride Rebekah's horse. The newspapers obsessed about that horse. It was simply too marvellous an image: the racehorse trainer, the flame-haired ex-tabloid editor, the prime minister and the police horse called Raisa. It really wasn't that remarkable. Neighbours in the countryside arranging to meet up, a chief executive of a right-leaning newspaper group supporting a centre-right party leader.

But people read the chatty tone as conspiratorial, and the very fact of the exchanges as inappropriate. Everything was viewed differently through the prism of hacking, the failures of press regulation and a relationship between media and politics that had become too close.

At the same time as all this was being revealed, I was busily preparing my evidence and responses to the twenty-four questions the inquiry had sent to me. Soon I was swearing that oath in the Royal Courts of Justice, before five hours of questioning from Lord Justice Leveson and the inquiry's counsel, Robert Jay.

I felt caught between two contradictory conspiracy theories.

The first was – and still is – put about by Gordon Brown: that in order to win the support of the Murdoch press and prise them off their support for Labour, I had made specific policy promises. It was put to me in the questioning, for example, that our policy to 'trim back' the BBC was done in exchange for the support of News International and James Murdoch, who was particularly anti-BBC.

This was ridiculous. We had frozen the BBC licence fee at a time of national economic emergency, when most departments were having to make 19 per cent cuts. What's more, the BBC had totally overreached itself – a taxpayer-funded public broadcaster which was paying its executives eye-watering salaries and squeezing commercial competitors in areas like travel and publishing.

There was also the implication that I had put Jeremy Hunt in charge of reviewing the BSkyB bid in order to get a favourable outcome and the support of James Murdoch. Which must have meant that I rigged the Vince Cable faux pas.

At the same time as being accused of being in the pocket of the press, I was being accused by said press of launching this inquiry to spite them. That was the second conspiracy theory: Leveson was a dastardly attempt by me to settle scores with the papers that had given me so much trouble.

I can see why they might have thought this. Once I had decided we were having an inquiry, I did make a virtue out of necessity. I am certain the press's fury is what lay behind some of the bad headlines at the time.

The Leveson Report arrived on 28 November 2012. All the charges against me were knocked down, one by one. No evidence was found of wrongdoing or impropriety by me or by Jeremy Hunt. There was no evidence of News International offering support in exchange for policy favours.

The next hurdle was responding to Leveson's recommendations. He had proposed a new system for press regulation: a self-regulating body (an improved PCC) with a 'statutory underpinning', meaning that legislation would be put in place to make sure it remained independent and effective. Unlike the PCC, the board of this new body would not include any serving editors. The board, independent of government, would draw up a new standards code. Fines for breaching it would be substantial. Individuals would be able to complain without having to pay a fee, as before. Papers would have to publish more prominent corrections and apologies. There would be an arbitration service for a low-cost alternative to legal settlement of issues. This would only be open to publications that were members of the new self-regulatory body – those who weren't would face 'exemplary damages' in court.

Craig conceded that if we didn't back the Leveson findings we risked being on the wrong side of the argument, in the court of public opinion and in Parliament. But he was clear what would happen if we did implement them with statutory underpinning: 'You will be the prime minister who put in place a press Bill. That crosses the Rubicon – and it's not something which history is likely to treat kindly . . . None of us should be under any illusion that [the press] will bear a grudge, and they will fight dirty.'

I knew the PCC had to change, and that we had to take on board most of what Leveson was saying. But I absolutely did not want in any way to put Parliament or the government in charge of what the press could and couldn't do. Press licensing in England ended hundreds of years ago, and rightly so. Free speech and a free press were key components of our democracy. Politicians had to be able to speak to the press. The press had to be able to protect their sources. There was no way I was going to let that end.

Some might say – and some do – that broadcasters are regulated by a body established in statute by lawmakers, so why not the papers? The answer is that there is limited access to the broadcast spectrum, and that TV is a more powerful, potentially dangerous, medium.

But the idea of statutory regulation for the printed press is seen by many as a step too far – and I agree. Any statute introducing even an element of state-sponsored regulation could easily be expanded in future. And I have no doubt that a press law would end up being used for more than control of press behaviour of the hacking kind. It would be used to enforce control of content.

So I found myself stuck in the middle, trying to get something the press could live with that didn't damage democracy, that satisfied the victims and pressure groups like Hacked Off, and that stopped anything like the hacking scandal from happening again.

But my biggest challenge would be getting something that the Lib Dems and Labour could agree to. They were keen to stay as close to the Leveson blueprint as possible, and to legislate accordingly.

Nick Clegg was adamant that we should implement press regulation by legislation. As Oliver Letwin points out in his excellent memoir, it was strange to see a Lib Dem leader defending the conservative principle of personal privacy while the Conservative leader was standing for the liberal principle of keeping the press free from statutory regulation.

One Wednesday I invited Ed Miliband into my office after PMQs and told him how worried I was about statutory underpinning. He said there was no other way of doing it. In public he would portray my reluctance as a betrayal of the victims and evidence that I was in thrall to the press barons.

It all seemed impossible. But then Oliver stepped in with the ingenious idea of using a Royal Charter, rather than the law, to sanction the new press self-regulator. Many public bodies operate under Royal Charter, like universities, the BBC and the royal colleges. It was a neat solution: it meant there would be some oversight of how the regulator was established, but not of the type that would give MPs or government any control over the press.

However, while this was going on, Lib Dem and Labour MPs and peers were attaching Leveson-style amendments on the statutory regulation of the press to random Bills going through Parliament. I was exasperated. The contamination of government business left us having to delay urgently needed legislation and put the coalition in jeopardy.

On Tuesday, 12 March 2013, during a meeting with Oliver and me, Nick put it plainly: 'You have to realise that no piece of legislation matters as much to me as this, and I am prepared to fuck up all the legislation in order to get what I want on this.' It was the only time we nearly came to blows, and staff outside raised their eyebrows as they heard shouting inside my office.

I had one strong card left in my hand, which I decided to play. Two days later I pulled out of talks on the Royal Charter. There would be a backlash from Hacked Off. But there would also be praise from the press – which neither Labour nor the Lib Dems wanted. It brought them back to the table.

I got Nick to come to the flat at 3.30 p.m. on Sunday, 17 March. I told him we would accept his version of the Royal Charter, but with two changes: editors and serving journalists must be in the majority on the code-of-conduct writing committee, and the recognition body itself would not be a regulator.

From the Labour and Lib Dem point of view, this was close to their version of the Royal Charter. From my point of view it was a Royal Charter, not legislation. The Rubicon had not been crossed.

The press – now in full battle cry as guardians of free speech – would oppose virtually any proposals for meaningful change. I knew that this deal would look messy, but I judged it was better to make the compromise. We would avoid all those defeats, and anyway the coalition was splitting – indeed my own party was splitting, with twenty or thirty MPs threatening to vote with Labour.

But Ed Miliband couldn't make up his mind on the offer. That Sunday evening Oliver went over to Miliband's office in the Commons, to find him still sitting with Nick and members of Hacked Off. Oliver made a final concession, which involved changing the law on costs and damages in libel cases to give the newspapers an incentive to join an independent regulator which had been certified by a Royal Charter body. What the papers would hate about it was the free arbitration service for libel claimants and front-page apologies to those libelled.

Oliver called me at 3 a.m. and told me that he, Nick, Miliband and Hacked Off had agreed to a Royal Charter that would create a 'recognition panel', which would in turn verify a proper replacement for the Press Complaints Commission. It would sanction the new regulator, but it would not second-guess its work. It was very clearly not a second or state regulator.

We had lost the battle by giving in to some of their demands. But we had won the war – we got them off the dangerous idea of state regulation.

As for the response from editors, I spoke to several over the weekend. I knew some of them would blame the whole course of events on me, and they did. But these things had happened at their papers. They had to accept responsibility and accept change.

The Royal Charter on self-regulation of the press was granted on 30 October 2013, incorporating Leveson's recommendations. Independent self-regulatory bodies would be founded that would be recognised and audited by a Press Recognition Panel (PRP), which was established on 3 November 2014.

Most newspapers took a different path, signing up to a new Independent Press Standards Ombudsman (IPSO) regulator, which is not recognised and will not apply for recognition. Only a handful signed up to the Independent Monitor for the Press (IMPRESS), which is recognised. To date it has no national newspapers as members.

I had believed that we could come up with an answer that would provide closure on the entire issue – and I regret that we didn't. I see why the press rejected our compromise, but I can't help feeling that one day they may think that was mistaken. We had found a middle way between self-regulation and statutory regulation. If there is another crisis over press behaviour, and their regulator misses it, they will still be accused of acting as their own judge and jury.

On 24 June 2014, Andy was found guilty at the Old Bailey of conspiracy to hack phones. Rebekah, Charlie and a number of others were cleared. I gave a statement: 'I am extremely sorry that I employed him. It was the wrong decision.'

Can I really claim, after this story of messy muddling through, that the situation is any better than what I inherited? The Leveson arrangements for regulating the press were never put in place. The press and politicians are as close as ever. And most of the prosecutions failed.

I take a more positive view. The process forced the press into admitting that they couldn't go on as they had been doing. They set up a somewhat more effective self-regulating body, which does at least ensure more appropriate apologies and has a genuinely independent chair, and recently announced a low-cost alternative to legal settlement of issues. The law also now allows for exemplary damages to be awarded by the courts in cases of particularly egregious defamation. This never would have happened without the inquiry.

In terms of the relationship between politicians and the media, there are now unparalleled rules for transparency – ministers must declare every editor and proprietor they meet.

When it comes to prosecutions, decisions are made by the Crown Prosecution Service. Frankly, the decision to prosecute journalists for paying for information in the public interest was misdirected – it's a journalist's job to find out important things and print them. But there is no justification for public officials to take payments, and it is right that people who did so have been punished.

The strange thing is that, even just a few years on, talking about red-top scandals seems almost archaic. The quandary today is over the internet, to which Leveson devoted just a few pages in his two-thousand-page report. He dismissed internet regulation because the public supposedly take online content with more of a pinch of salt. As he put it, 'People will not assume that what they read on the internet is trustworthy or that it carries any particular assurance or accuracy.'

How wrong that was. In recent years we have had allegations of foreign governments pumping out fake news stories to sway election results; thousands of young people recruited to extremist causes by terrorist propaganda; hackers stealing and dumping private information online; and distrust in the 'mainstream media' compared with the supposed online truth-tellers. It's all the same issues – privacy, libel, harassment – but this time we are fighting in a virtual world and on a global scale. If anything, the established press and broadcast media are even more important in acting as a bulwark against fake news. We need trusted, responsible sources more than ever.

Besides, the printed press does still have a huge influence on public consciousness, as we would see during the debate over Britain's membership of the European Union.

But before that, the skills of politicians and the press would be put to the test as a totally unexpected upheaval swept across part of our world.

21

Libya and the Arab Spring

I didn't see it coming – I don't think anyone did. Of course, there were signs. The National Security Council papers I received told of mass unemployment and the inflation of food prices in north Africa. The diplomatic telegrams from ambassadors talked about the spread of social media and a rapidly expanding young population in the Middle East. But no one predicted that those things were kindling for revolts in fourteen countries across the region – or that our military would soon end up deeply involved in one of them.

The conventional wisdom is that the Arab Spring has been an embarrassing failure. Only one country caught up in the revolutions, Tunisia, has kept its new democracy. The legacy elsewhere is civil war, extremism, violence and worse.

As for Britain's enthusiasm for the uprisings, the theory goes that this was a mistake that should never be repeated, and that our intervention in Libya was misguided.

I don't agree. We made mistakes – and I will certainly admit to having made some of them. But I stand by my decisions. What we did was right.

It is simply too soon to tell the true outcome of the Arab Spring. Our own journey from autocracy to democracy took centuries and included bloodshed, extremism, civil war, and many false starts. Why should we expect modern transitions from dictatorship to democracy to be instant and painless?

There isn't room here to document every twist and turn of the Arab Spring in every country. But I will try to cover each of the phases of Britain's involvement, explaining why I took such a strong stance on the upheaval and why I decided to put British lives on the line by intervening.

*   *   *

Back in late 2010, a fruit-seller had set fire to himself in a Tunisian village, in protest at the heavy-handed authorities persecuting small traders.

Word of his death spread online, and rallies against the oppressive regime broke out. President Zine El Abidine Ben Ali, who for twenty-three years had rigged elections, censored the press, tortured opponents, imprisoned people without trial and plundered his country, was gone after just twenty-eight days. It was astonishing.

I received a detailed note from John Casson, my foreign affairs private secretary, in early January 2011 saying that these events were indicative of a wider problem across the region. It was in our long-term national interest that these countries moved towards the rule of law, not just 'the dead-end choice between dictatorship and extremism'.

John would later become our ambassador to Egypt. And it was Egypt where the spark of revolution would next catch light. The people there, inspired by what they saw in Tunisia – and they predominantly saw it online – took to the streets to call for 'bread, freedom and social justice' from their dictator of thirty years, Hosni Mubarak.

Much of the official advice was cautious, as Mubarak had been an ally in the fight against extremism and terror, and British companies had important economic interests in Egypt and the region. A note from the FCO on 27 January said that the situation in Egypt was '4 out of 10', with 10 being the collapse of the regime.

I thought it was right to align Britain decisively with those who wanted to see the spread of democracy, rights and freedom. If we, one of the oldest democracies in the world, couldn't support those with aspirations for the freedoms we enjoyed, who would?

I was, after all, influenced by earlier revolutions. When I was in my mid-twenties the Berlin Wall came down, and countries freed themselves from the shackles of communism. I had seen the 'before' and 'after' of a whole part of a continent, once repressed and now liberated. I had felt the surge of energy and hope. I understood; and I never took these things for granted myself.

What's more, I never bought the argument that some people cannot handle democracy. Yes, there are different challenges in different countries. But the desire for freedom burns inside us all, whether we're in Berlin or Benghazi.

More to the point, open and inclusive politics are necessary because

of the damage the alternatives do in the long run. Tunisia, Egypt, Syria
– all were in relative decline before 2011. They are even worse off now,
not because of the attempts to change the model, but because those
attempts failed. The 'strongman' leaders who so comforted Cold War
Washington, Moscow and London left their countries corrupt, embit-
tered and impoverished.

I felt, once the uprisings had started, that the status quo wasn't going
to be tenable. It simply wouldn't be possible, or desirable, to prop up
these regimes against their people. The damage to their countries and to
our long-term interests would have been great.

But of course I thought it worthwhile to try to stabilise the situation,
and to show these dictators a better way if we possibly could. I called
Mubarak on 29 January, begging him to show some flexibility if he
wanted to keep his job, and pleading with him not to carry out violent
repression. He could have survived if he had offered presidential elections. Instead, he cut off the internet and sent thugs onto the streets to
attack protesters. Almost a thousand people were killed.

On 2 February I stood outside 10 Downing Street alongside the UN
secretary general Ban Ki-moon, who was in London for a bilateral meeting. I made a statement supporting the Egyptian people's aspirations for
a more democratic future and greater rights, and condemning the
despicable violence meted out by their own government.

Meanwhile, William Hague was touring Jordan, Yemen, the UAE and
Bahrain. He was the first politician to visit Tunisia after Ben Ali was
toppled – and, it would turn out, the last Western foreign minister to
visit President Bashar al-Assad in Syria.

Britain was firmly on the front foot, in contrast to other Western
countries. France had already stumbled down the wrong path, as Sarkozy
initially backed Ben Ali. Germany had stepped back, with Merkel largely
absent from the debate. And America had a false start too. When I spoke
to Obama on the phone about the situation in Egypt, he told me that he
did want Mubarak to go. I had my iPad on my knee, and was reading the
latest BBC report from the region. 'But the special envoy you've sent
there has just said that Mubarak must be part of Egypt's future,' I said.
I rarely encountered Angry Obama. But when I did, it was because he
had been blindsided. He went quiet, spoke even slower than normal, and
was clearly fuming.

By 11 February, Mubarak was gone. I declared it a 'precious moment of opportunity', as inspired demonstrators took to the streets of Jordan, Syria, Yemen, Bahrain and, crucially, Libya, which was home to the tyrannical dictator of forty-two years, Colonel Muammar Gaddafi.

I was about to head off to the region myself, on a pre-planned visit to the Gulf which many in my team were now advising against. The situation was volatile. Questions would be raised about our relationships with the autocratic monarchs of the Gulf States – and about arms sales too. My view was simple: you can't influence events unless you are prepared to get stuck in. I was going.

I started the trip with a visit to Egypt on 21 February. I visited Tahrir Square, the epicentre of the revolution, the first foreign leader to do so. I spoke to some of the protesters, including an excited young boy with the Egyptian flag painted on his face. It really felt as if the place was on the cusp of change.

Our ambassador, Dominic Asquith, had arranged a series of meetings with pro-democracy activists and bloggers. We had agreed beforehand that these shouldn't include representatives of the Muslim Brotherhood or related Islamist organisations. Far from instigating the uprisings, the Islamists had been surprised by them, and were now trying to turn them to their advantage. We shouldn't assist them.

Indeed, when extremists tried to hijack the demonstrations they had been shouted down. The Egyptian people were rejecting both corrupt authoritarianism *and* extremism. The young, inspiring people I met there said as much. They knew we would be advised that the alternative to the strongman leader was extremism and chaos, but they believed that this time it could be different.

They told me they wanted jobs, a voice and the chance to choose their own future, free from corruption. That last word – corruption – was perhaps the most vital. The Mubarak regime was corrupt on an epic scale. Not simply taking some money for personal enrichment, but on a level that was holding the whole country back. Some of it was stashed in London, in properties and bank accounts, and I immediately set up a Whitehall taskforce and sent a Crown Prosecution Service prosecutor and team to the Cairo embassy for two years, freezing over £80 million of assets of the Mubarak family and their cronies.

The following day I was in Kuwait with John Major, marking twenty years since British forces had helped expel Saddam Hussein's invaders and restore the sovereignty of this small country. I was to address the

Kuwaiti National Assembly, where I would set out my stall on the current crisis.

I accept that I was optimistic about what the uprisings could lead to, but in my speech I made arguments about how we should encourage democracy and how we should fight extremism -- and the consequences of these things for our foreign policy. I am certain this remains the right approach today.

The choice leaders like me were told we faced – between the supposed stability of highly controlling, undemocratic regimes and the supposed uncertainty of freedom and democracy – was, I argued, a false one. Indeed, I calculated that supporting political change in some cases offered a lower risk of serious conflict and extremism. I was right. Look at what happened later in Syria: political change failed; the classic Middle Eastern strongman Assad ravaged cities like Homs; radicalisation and conflict proliferated.

There was criticism that my visit was hypocritical. How could I be standing there, in these oil-rich absolute monarchies with often terrible human rights records, while simultaneously condemning their secular autocratic neighbours?

We made pragmatic judgements about how best to promote our interests and values, and these differed depending on the domestic position of the leadership, the progress of democratic forces and the centrality of the relationship. If you want to promote British interests you can't only talk to democracies. If you did, you'd be having most of your conversations in western Europe. In meeting these governments you are neither giving up your principles nor endorsing their systems. And in foreign policy there is not just the promotion of democracy but a myriad of interests that includes security, prosperity, and the need to work with others in order to encourage them to change and to tackle common challenges, that must be taken into account.

The relationship between Britain and Saudi Arabia is a case in point. The fact that the regime in Riyadh uses cruel punishments, torture and excessive use of the death penalty is undeniable, and it horrifies me. Its journey towards equality, human rights and democracy is proceeding at a snail's pace.

I would argue strongly, however, that we are right not just to engage with Saudi Arabia, but to be its partner and ally. As the home to two of Islam's most holy places – Mecca and Medina – the country plays a special role in the Islamic world. As the only Arab member of the G20, it is the region's economic superpower. And, crucially, the Saudis are key allies when it comes to intelligence and security. Throughout my time as prime minister they held at bay a branch of al-Qaeda on the Arabian Peninsula, meaning that the UK, the US and allies could focus on areas where we didn't have capable partners.

Most notably, in October 2010 our security services informed me that Saudi Arabia had foiled the latest terror plot: two bombs smuggled in printer cartridges on flights from Yemen to the US. One had arrived at East Midlands Airport in Leicestershire, and could have blown up mid-flight. It was the Saudis who alerted us to the threat, and gave us details of the packages in which the explosives were hidden. When the planning, preparation and execution of an attack can take place in different countries, such security cooperation between governments is vital.

Excessive ideology in foreign policy is not helpful. Michael Gove had one outburst at cabinet about how the FCO had hugged dictators too close, and the Arab Spring was our moment to take the side of freedom. While I shared his view of freedom, I felt that William Hague got it right when he gave a dignified reply about it being slightly more complicated than that. The conversation made it into the papers the next day, something that often seemed to happen when Michael was involved . . . In any case, the response to the Arab Spring by the Arab monarchies was more effective – combining reform, cooperation with their neighbours, and repression, rather than just drawing the sword like the secular dictators.

At the same time as these theoretical debates, there was the urgent and practical matter of evacuating our citizens in those countries. We had already got people out of Tunisia on commercial flights, and had chartered planes to rescue expats in Egypt. But in Libya, the FCO was not quick enough to act. Here I learned an important lesson: government departments rarely react fast enough, and rarely work together, unless forced to do so by the centre. That meant me personally convening COBR, getting the key people into the same room, and metaphorically banging the table to make things happen.

The Ministry of Defence has extraordinary capabilities thanks to our armed forces and state-of-the-art equipment, but it can be the worst offender when it comes to working with others. There is a tendency to hang back and wait for things to go wrong before it steps in. There are also the financial constraints: the MoD and the FCO won't act until the

Treasury and DFID have committed to pay – another of the many lessons I would learn.

France, Germany and most other concerned countries had started to evacuate their civilians from Libya on 22 February. But here we were, on 24 February, being told there were still hundreds of Brits in Libya, some deep in the desert working in oilfields, and potentially in great danger.

I was at a grand dinner hosted by the Sultan of Oman, with the orchestra straining away, when palace staff brought in a note from Liz explaining the situation. Immediately I walked out and got on a conference call with William and Liam Fox. 'Just send the RAF in and do it now,' I shouted into my BlackBerry.

The military wanted an indemnity – in the form of a written instruction from No. 10 – before C130 Hercules transport planes left the runway. It was a nail-biting time. Thankfully, the Hercules and their passengers returned safely.

Initially, Libya was following a similar pattern to Egypt. But Libya wasn't Egypt. There was no American-backed national army with a special place in the nation's affections to step in and keep the peace. Libya hadn't functioned properly as a country for fifty years; it had been kept afloat by vast oil reserves.

Moreover, Libya's opposition wasn't Egypt's opposition. Its rebels, who quickly formed an anti-Gaddafi coalition which began to fight the regime, were organised and effective at political campaigning. It was the most credible opposition of any Arab Spring country. It had a presence in each of the key cities spanning the coast from west to east: Tripoli, Sirte, Misrata and Benghazi. It brought senior defectors from Gaddafi's government and long-term secular and Muslim Brotherhood oppositionists into a recognisable political body, the National Transitional Council (NTC). The FCO's view of its chairman, Mustafa Abdul Jalil, was that he was impressive, effective, had political legitimacy and was inclusive.

Most important of all, Gaddafi certainly wasn't Mubarak. He had made it clear that he was going to stay – and make his people pay. In his broadcast on 22 February the Colonel screamed in Arabic for a whole hour: 'We will cleanse Libya inch by inch, house by house, home by home, alley by alley, person by person, until the country is cleansed of dirt and scum.' He was defiant, deranged – and determined, promising to catch the demonstrators like 'rats' and 'cockroaches'. Watching it from a hotel room in Doha stirred something in Ed Llewellyn and me that bypassed our younger colleagues. To us, raised in the 1980s, he was 'Mad Dog' Gaddafi, a horrific figure in modern history who sold Semtex to the IRA, ordered the downing of Pan Am flight 103 over Lockerbie in Scotland, and admitted responsibility for the murder of PC Yvonne Fletcher in London. We knew what he was capable of.

To do nothing in these circumstances was not a neutral act. It was to facilitate murder. If Gaddafi had been allowed his way he would have murdered many, many thousands, and then consolidated his regime through repression. And even that wouldn't have stopped a breakdown, merely delayed it.

I was clear that our first response should be to try to isolate the regime. I joined forces with Sarkozy, speaking to him regularly on the telephone to enforce travel bans, arms embargos, asset freezes and, crucially, referring the perpetrators to the International Criminal Court. Our efforts resulted in UN Security Council Resolution 1970, which deployed our entire sanctions arsenal.

But it had little impact on Gaddafi's behaviour. On 28 February he intensified his assault against the rebels in Misrata, advancing eastwards, inch by inch, house by house, towards the key rebel-held second city of Benghazi.

At that moment, it was as if someone had turned over a sand timer. It was suddenly clear to me that we were in a race with the regime – them trying to get their hands on the rebel capital, us trying to coordinate an international political and military response. Whoever reached their goal first would either unleash bloodshed or avert it.

Of course, Iraq casts a shadow over all foreign policy – every intervention is seen through the prism of its failures. But it was Bosnia that was at the forefront of my mind as I discussed with Ed how to respond to the crisis in Libya. We knew too well what happens when the West drags its heels as an aggressor decides to 'cleanse' a country. Gaddafi had a long record of brutality and violence towards opponents of his regime, and this spoke volumes about his intentions. His army showed every sign of sticking with him – and having no compunction about firing on their own people. I was given information that Libyan soldiers who refused to shoot civilians were themselves shot.

On 28 February I told the House of Commons: 'We do not in any way rule out the use of military assets. We must not tolerate this regime using military force against its own people.'

The NTC leader was clear when I met him in London that what the rebels wanted was a 'no-fly zone' to stop Gaddafi mounting a war on his people from the air. I instructed the Ministry of Defence to draw up contingency plans. The MoD response was, 'What exactly are we trying to achieve?' The no-fly zone was not a solution in itself. Just stopping planes from bombing wouldn't prevent a ground war. But it would act as a deterrent, since sanctions hadn't worked, galvanising Western support and stopping Gaddafi on a broader front. I also believed it would be a foot in the door for the sort of action from the air that could stop him.

One thing on which we were all agreed, however, was that the situation did not warrant 'boots on the ground' – a level of intervention that would never get past Parliament.

Meanwhile, I was becoming increasingly concerned that the FCO was underpowered when it came to Libya. Its expertise is, of course, world-class. But I worried that, as with the MoD and MI6 too, 'traditional' thinking dominated. It wouldn't adapt to new ideas. And while it's good at analysis and process, it has lost a sense of 'can-do'. My response was therefore to centralise management of the intervention in my new NSC structure, with a new NSC(L) that dealt solely with Libya.

People assume the job of prime minister is simply to take the big decision, issue the decree, as it were, and then consider it done. But the Libya campaign brought home to me that so much of the job isn't just the deciding – it's the doing: cajoling, corralling, convening meetings, questioning official advice, offering up new solutions, being creative, seeking wider opinions, and keeping on and on at people until something happens.

The decision to ratchet up our response on Libya was, in many ways, the easy part, because I knew it was the right thing to do. What was tough was getting it done – and doing so against the clock.

On 10 March Gaddafi's forces were at Ras Lanuf, 220 miles by road from Benghazi. Meanwhile, we were dealing with our first obstacle: America.

It's a fact of modern conflict that we need other nations' help if we are to act effectively. We needed America's military might, from air power to intelligence. That meant convincing Obama to commit.

He had been elected on a pledge to disentangle the US from foreign conflicts, rather than start new ones – he had said in the *New York Times* that 'the best revolutions are completely organic'. I was about to see whether he applied this pledge to Libya. And when I found it hard to get a phone conversation with him, I feared that he did. I had the distinct feeling that the world's great superpower was dithering while Benghazi was about to burn. When I did speak to him, I was clear: 'Benghazi must not fall, or game over.'

On 11 March, Gaddafi's forces were approaching Brega, just 150 miles from Benghazi. I was at a special summit in Brussels facing our second obstacle: the EU.

I saw Libya as a key opportunity to create consensus, and like Sarkozy, I thought it was important to get the principle of a no-fly zone into the communiqué – a sort of statement of intent that everyone signs at the end of the event.

I found Europe in a peacenik mood. It felt as if the former Eastern Bloc countries were saying: 'Look, we did democracy with our revolutions, but these people don't understand democracy at all.' The southern countries were nervous because of their preoccupation with immigration from Africa. The Germans didn't really want to get involved. And Romania, which you'd have thought would sympathise with a country strangled by a dictator, tried to kibosh the whole thing.

I ended up in a corridor with Sarkozy and the Romanian president, Traian Basescu, shouting at one another. Back in the open forum with the other leaders, it was agreed that we would have some milder wording in the communiqué about protecting civilians. 'You don't have your no-fly zone,' said Traian defiantly. 'Yes, so people are being murdered from the air,' I snapped back in fury.

(A postscript to this shameful EU episode was at the next summit, following the successful UN resolution. The president of the EU Council, Herman Van Rompuy, was triumphalist: 'Europe led the way – we had our Special Council, we produced that communiqué, the UN then followed our lead. What we are doing in Libya is a great moment for promoting European values; it's a bit like the euro, we don't know exactly where it's going to lead but it was the right thing to do.' I nearly choked. Everyone in Europe apparently now thought it was a victory for the EU, when at the start the French and the British had been completely on our own.)

On 17 March Gaddafi's forces took Ajdabiya, a hundred miles from Benghazi, while simultaneously beginning the battle for Benghazi by bombing nearby Benina Airport. By this time we were facing our third obstacle: the support of the region.

While America seemed to dither and Europe deluded itself, William Hague reached a breakthrough in the Arab world. He persuaded the secretary general of the all-important Arab League, an Egyptian, to endorse action. It would guarantee the support of Lebanon, the only Arab country that was at that time on the UN Security Council. This was hugely important – and it's still important now. It demonstrates to those Arab countries that accuse us of breaking Libya that they were actively encouraging action.

Crises rarely happen one at a time. Just at this vital moment in the Libyan saga, Japan suffered an earthquake and tsunami that killed 15,000 people and triggered three nuclear meltdowns at the Fukushima power plant. It was impossible to comprehend the horror people suffered and the scale of the catastrophe. We held a succession of COBR meetings to coordinate our response. There was a real risk that radiation poisoning could spread to more populated areas, including Tokyo itself.

In another example of prime-ministerial micro-management I insisted that we had enough iodine tablets in the country for all 17,000 Brits in Japan. Prime ministers end up taking tactical decisions like this because there are always so many reasons for *not* doing something. 'They'll be stopped at customs.' 'It will be embarrassing that we haven't offered them all to the Japanese government.' On it went. I decided we wouldn't tell the Japanese, but would fly the tablets out in our diplomatic bag. And that's what happened.

As Gaddafi's ground forces advanced towards Benghazi, his son Saif al-Islam predicted that it would all be over in forty-eight hours. Meanwhile, we were facing obstacle four, the trickiest and most urgent of all. After Iraq, we knew it would be extremely difficult to do anything without a UN Security Council resolution.

Fortunately, we were seeing a welcome and crucial shift beyond the limited scope of a no-fly zone towards deeper and more effective involvement. We were now looking to achieve a resolution that included 'all necessary measures to protect civilian life'. This was what I'd wanted from the beginning. It was a genuine triumph for the FCO, which hasn't lost its ability to negotiate international statements. I think the Americans had come around because of the feeling that, if they were going to be involved at all, then any action ought to be decisive and effective.

Thus commenced the process of garnering support from the other countries on the Security Council. William was hitting the phones. Mark Lyall Grant, our ambassador to the UN, was flat out. I was speaking to my opposite numbers across the world.

Angela Merkel was sceptical. Why do this in Libya when we didn't in Iran, she was asking. Many thought Sarkozy was trying to make up for initially getting the Arab Spring wrong.

On the evening of 17 March, UN Resolution 1973 – pledging 'all necessary measures to protect . . . civilians and civilian populated areas' but prohibiting troops on the ground – passed by ten votes to none. Those who voted in favour were Lebanon, America, Britain, France, Colombia, Gabon, South Africa, Nigeria, Portugal – and, unsurprisingly, Bosnia-Herzegovina. Germany, Brazil, China, India and Russia abstained. It's a mystery why Russia did not vote against.

We were getting closer to action, but Gaddafi was getting closer to Benghazi, and we faced a fifth and final obstacle: ensuring the full support of the UK cabinet. Blair had lost four ministers over Iraq – and he had a landslide majority. We were in coalition with an anti-war party. Would I face my own walkouts?

But the legal advice was plain: 'This represents clear and unequivocal lawful authority to take action up to and including the use of force' and to protect civilians.

Nick had been a staunch supporter of the forward position on the Arab Spring right from the start, so the omens were good. Cabinet met on 18 March, and I made sure that on each minister's chair was a copy of the legal advice, the UN Security Council resolution and my forthcoming statement to the House of Commons.

Those four walls had heard so many arguments on the eve of war. Today they witnessed unity. One by one, the cabinet members spoke in support of the action.

That evening I finally spoke to Obama on the phone. He said the US would help in the first week – one week of heavy military support to take down air defences – but then we, Britain and France, would be on our own. He was unenthusiastic and matter-of-fact, but this was at least a clear and decisive response. It went beyond a no-fly zone and was militarily a more comprehensive solution. My foot-in-the-door approach had worked. But the tone of the exchange took some getting used to: I was so used to America, the leader of the free world, leading, that it was extraordinary to hear such reticence.

Nevertheless, I was relieved; Obama less so. He said he felt forced into it: 'You'll get our Tomahawks and Cruise missiles, then we'll take a less active approach.'

The next day I was on the Eurostar to the 'Paris Summit for the Support of the Libyan People'. Nick Clegg was in Westminster chairing COBR, taking advice from the military, who said Benghazi was under attack from Gaddafi's forces: we must act now. My permission was sought, and I gave it via a crackling mobile phone on the Eurostar that kept cutting out. Liam Fox rightly insisted on written instructions. After the Iraq War, and our armed forces being subjected to the European Court of Human Rights – another Blair legacy – the UK military was constantly pushing for paper trails.

The Élysée Palace event was carefully choreographed. Sarkozy and I had a brief conversation about the latest developments, sitting together on an ornate French sofa. Hillary Clinton arrived, and Sarko showed her to a chair opposite us. He seemed keen to emphasise that here were the Europeans – president and prime minister – and here was America, represented by a secretary of state – and there was space between us.

He surprised us both when he said that he had already issued orders for French jets to take off and launch attacks: 'They will be over the enemy in half an hour. What do you think?' I said, 'I think you're very brave, but Nicolas, we haven't yet taken out the air defences in Benghazi' – in other words, they were still in place to shoot down attacking aircraft. Nicolas looked at his military adviser and asked if there was a risk, almost as if it hadn't even been considered. His adviser reassured him that the mission was safe.

Clinton was completely unfazed by all of this. She invited an American military officer to give a short presentation. He stood up, opened a note-pad, and proceeded to reel off the awesome extent of US military power that would be brought to bear: 'At 1900 hours 113 Cruise missiles will be launched from ships around the Mediterranean . . .' followed by a list of all the things the Americans were going to do. It was clear who still wielded the real power.

I went back to London to tell journalists in a short speech that our planes would join in that evening. 'What we are doing is necessary, it is legal and it is right.' Translation: we are at war – but this is not Iraq.

And then it happened. On 20 March, American, British and French aircraft destroyed Gaddafi's tanks, armoured carriers and rocket-launchers, and his forces began to retreat. Benghazi was saved, and a Srebrenica-style slaughter averted. I've never known relief like it.

A couple of days later, Parliament would meet to vote on the action. I insisted on us having a substantive motion approving the action and its continuation. It was important to get Parliament's approval. In the end, 557 MPs voted in favour and just thirteen against.

We had to keep going. On the last two days of March, coalition forces flew 180 sorties, of which twenty-three were carried out by the UK. Tornadoes and Typhoons used Brimstone missiles to destroy Gaddafi's armoured vehicles, battle tanks and oil stores.

Soon I had a video call in Downing Street with Sarkozy and Obama. On one side of the split screen I saw Obama, clearly frustrated that he had been sucked in – blindsided by a bargain that asked for American support only in the initial phases, but that would now clearly require it in the long term. He said he would find it more difficult to trust us again. On the other side of the screen was an emotional Sarkozy. 'I can't believe you're doing this,' he said to Obama as he set out the limits of America's involvement.

I was trying to conciliate, though, like Sarko, I wanted America to remain engaged. I used a baseball analogy to convey the merits of staying: we just needed one more home run.

Obama was unmoved. Days later, NATO took over operations and focused on wider targets that would help the Libyan rebels. The symbolism of America's non-involvement was so important that Obama even made sure that NATO officers who happened to be American stepped back. A rather surprised Canadian lieutenant general in Naples found himself commanding the air war.

In the following months, Britain flew a fifth of all the strike sorties over Libya, while our frigates, destroyers, submarines and minesweepers enforced a maritime blockade. The scale of the operation was vast. Fifteen countries involved. Thirty-five ships on the Libyan coast at one time. On one day, 236 aircraft deployed.

The NSC(L), which met regularly under my chairmanship, had set out clear plans: immediate humanitarian support for the people of Libya; reconstruction, principally paid for by the oil-rich country itself and guided by Western help; and political transition, a Libyan-authored roadmap supported by us. International development secretary Andrew Mitchell was a key figure at those meetings – indeed, DFID took the international lead in developing these plans and corralling other states to support them. The aftermath was considered in every decision we took.

By May 2011 the war had sunk into stalemate, and needed a renewed focus. I agreed deals with France to commit Apache helicopters to help the rebels. I was on the phone to the leaders of the Gulf States to encourage their continued involvement, which turned out to be crucial.

In one of the most overlooked elements of the war effort, some advice from Alan Duncan, the minister at DFID and a former oil trader, was I believe decisive in breaking the deadlock. He sent me a note in April 2011 entitled 'The Libyan Oil War – and Why You are Losing it'. In it he told me bluntly that oil sanctions were having a perverse effect, cutting off oil supplies and revenue from the Benghazi side of Libya, which we wanted to support, while Gaddafi was still buying and selling everything he needed. If this were to continue, the anti-Gaddafi forces would run out of money and oil – and lose – and Gaddafi would keep going. Alan recommended the delisting of Libya's national oil company for eastern Libya, so that Benghazi could export crude oil, and the total disruption and blockade of the Gaddafi-controlled areas around Tripoli.

He put the NSC apparatus in touch with his close friend Ian Taylor, who ran the world's most successful commodity trader, Vitol. Their case was totally convincing, so I instructed the NSC to set up what became known as the Libyan Oil Cell.

Crucially, without any UK government support, and at massive commercial risk, Ian Taylor, a hidden hero of Britain's efforts, maintained supplies into Benghazi. Government action blocked Gaddafi; private enterprise supplied the rebels. The political and commercial worlds came together in the national interest, and helped turn the course of events.

There was another hidden hero from private enterprise. The UK banknote manufacturer De La Rue contacted the Treasury early on in the crisis, saying it had £900 million worth of Libyan dinars destined for the regime in Tripoli waiting in a warehouse in Newcastle. George asked them to delay sending it. We were desperate to get it to the official Libyan opposition – and frustrated when I was told it couldn't be done. Concerns were voiced about the legality, as the UN freeze covered all Libyan assets. By the time the NTC was increasingly recognised by countries across the world as the legitimate authority in Libya, we found a way. I'll never forget George turning up at the NSC and saying jubilantly, 'I've put all the fucking cash on a plane.'

There was another lever I was trying, too. Some say we never even tried to seek a political solution to ease the transition from Gaddafi to the NTC. Actually, we did. I pushed and pushed for a deal to be offered to Gaddafi – an 'exit with honour' – using political links built up historically between our countries.

All this was going on in the background. Andrew Mitchell sounded out the president of Equatorial Guinea on hosting Gaddafi in exile. At one point an ex-Spanish prime minister was lined up to make him an offer.

I also spoke to Tony Blair twice. He contacted me first to say that people around Gaddafi knew the game was up, and that he would take a way out if offered one. I said that if Gaddafi could leave he should leave, and we wouldn't stop him. However, Gaddafi then told Blair there was no violence in Libya, and an attempt to recolonise the country was being made. It was futile.

Despite all these efforts, summer brought stalemate. It became clear that we could stop Gaddafi's advances, but not get him to concede the need to do a deal. NATO alone couldn't put his forces under enough pressure. Oil and banknotes alone weren't enough.

We had people working on this issue round the clock. My team lived and breathed Libya. As did I. I spent vast amounts of time on the situation.

Hugh Powell was probing for equipment for the rebels – radios, Land Rovers, armour. I was commissioning work on military options – precision targeting, the mentoring of NTC commanders. I was constantly trying to keep the pressure up, and was furious after advice said the MoD was trying to dial down the number of frigates, helicopters, Typhoons and Tornadoes.

I was exasperated that too many parts of the government and military machines seemed more concerned about a future Libya war inquiry than about the war itself.

Something had to change. George, Nick and I mulled over the idea of shifting our focus from the east and Benghazi to the capital, Tripoli, in the west. This had been inspired by the UAE and Qatar, which had led on supporting the militias there.

Our military was cautious, but went ahead. With our allies France, Qatar and the UAE we ended up steering the ramshackle Libyan rebel army from a secret cell in Paris, providing weapons, support and intelligence to the rebels planning an assault on Tripoli. This quartet of countries – known internally as the Four Amigos – focused on training, equipping and mentoring effective militias in the West.

Though this was known to NATO and the US, once again we were operating outside the traditional structures – pulling new levers all the time. Had we not done so, I don't think there would have been the breakthrough that eventually came on 13 August, after NATO bombed Gaddafi's military in the Nafusa Mountains. The rebels were victorious. The end was in sight.

On 20 August the rebels took back Zawiyah, the city Gaddafi had devastated months earlier and which had become a symbol of his brutality. The following day, they pushed through to Tripoli, and a euphoric crowd gathered in Green Square. By 28 August they had taken the capital, and Green Square was renamed Martyrs' Square.

Within weeks the NTC was recognised as the new Libyan government, and on 15 September I visited Tripoli with Sarkozy. We had made a promise to each other that we would go together.

We went to Tripoli first. I'll never forget going to visit Gaddafi's victims in hospital (where we met an NHS doctor from Oswestry). It was a relief to see the hospital still standing. We hadn't destroyed the critical infrastructure. I thought to myself: this country is going to survive; it's going to work out.

We took a French military helicopter ride at fifty feet along the Benghazi waterfront. We spoke with a group of militia leaders with long beards and in three-piece Savile Row suits about the importance of compromise. We wove through the jubilant hordes to make our way to a stage in Freedom Square in Benghazi, and gave speeches as 10,000 people chanted 'Cam-er-on' and 'Sar-ko-zy'. Someone there had even named their baby Sarkozy.

Still, we had no idea where Gaddafi was, until 20 October when he was found and executed. My tone was sombre: this wasn't a time to celebrate, but to remember his victims, the people who rose up against him, and the many people from across the world who had risked their lives to help them succeed – so many of them British.

In the years following our intervention, elections were successfully held in Libya in 2012 and 2014. But the country then began to descend into factions, with the oil-rich east trying to break away from the west. Tragically, the extremist elements that were always there got a foothold, and the American ambassador, Christopher Stevens, was murdered in a September 2012 attack. There was a government misstep when it decided to bar all Gaddafi-era officials.

The UN failed to help reach a political solution, and for all the Western advice from our ambassadors and the heads of the UN support mission, little was heeded. This would enable the extremists of Islamic State to fill the vacuum in the important town of Sirte and continue to fight the moderates over Benghazi, and would also help to fuel the migration crisis. Eventually, in August 2016 Obama would end up committing more American resources to Libya in a bombing campaign against ISIS.

The problem, looking back, was that the UN-led political transition process focused too much on elections, and not enough on ensuring that the government had sufficient executive powers, critically over allocation of oil revenue and control of the militias.

Libya fast became a fragile state. Ministers lacked the authority to direct resources. International assistance had nothing to plug into. Worse, the transitional assembly gave the militias who had effectively defeated Gaddafi higher salaries than the government security forces – effectively bribing the militias not to attack them.

The militias became a tool for various factions to pursue their interests; time and again they were able to hold the government hostage to their demands, and they became the de facto police (or mafia) in 'their' bits of Tripoli and Benghazi. Which, in turn, left the field open for more radical elements to prosper.

With the UN process failing, I tried hard to find a political deal. I appointed Blair's former chief of staff Jonathan Powell a special representative to Libya. I also tried to use the G8 summit in 2013 to get politics and security back on track. But in the end, it wasn't the lack of leadership that was the problem – it was the lack of people following any lead.

There are those who say our error in Libya was, like Blair in Iraq, to support the removal of a dictatorship without any idea, beyond some vague notion of democracy, what might replace it. They have got it totally wrong. We *were* supporting a credible alternative government: the best-organised rebel group of the Arab Spring, led by someone respected in the West, which offered the best possible chance for Libya's future. This misconception is often accompanied by another: that we shouldn't have intervened at all – and by columnists like Matthew Parris stating that 'we would have done better' to leave Gaddafi advancing on Benghazi. I shudder when I read those words in print, and I will be relieved to my dying day that we chose to stop Gaddafi reaching that city.

Clearly, had we had troops engaged on the ground in the country, we would have had a better chance of guaranteeing success. But there were never going to be British boots on the ground. The Libyan rebels didn't want it, the world wouldn't have voted for it, and it was explicitly ruled out by the UN resolution. The British people would not have accepted it.

Libya still faces troubles. But in many ways it came out of its conflict with better prospects than some other post-conflict states. It had a functioning, 'moderating' political body in which various political, regional and even religious factions were already used to using a political process to settle differences; it had access to huge amounts of (Libyan) capital for rebuilding; and it had a full range of 'expert' on-the-ground assistance thanks to the preparatory efforts led by DFID. None of those things has gone away. Nor has the sense of hope, with the anniversary of those revolutionary protests celebrated every year with rallies and concerts.

So I take heart. As one commentator put it, the intervention succeeded in the short term, failed in the medium term – but in the long term Libya still has a chance. It never would have had that chance if I had listened to official advice and other countries' reluctance and decided not to take action, leaving Gaddafi in power. Just as I had hope in 1989 and in 2011, I still have hope for Libya.

22

Referendum and Riots

Should politicians hold referendums?

Margaret Thatcher didn't think so. She said, quoting Clement Attlee, that they were 'the device of dictators and demagogues', and pointed to Hitler, Mussolini and Napoleon III's use of plebiscites to confirm decisions they'd already taken.

Many people oppose the 'device' for less dramatic reasons. They believe that deciding upon specific issues is what we elect MPs to do.

My view is rather different – to the point where I put three country-defining issues to a vote (and a fourth on furthering Welsh devolution) during my premiership.

Of course, I believe in our *parliamentary* democracy – one in which we elect representatives to take decisions on our behalf – rather than a Swiss-style *direct* democracy, where issue after issue is decided through popular ballots.

However, I also believe that there are some issues – particularly where Parliament is giving up its powers, or fundamentally rewriting the rules of our democracy – where there is a strong justification for asking voters to take the decisions themselves; otherwise politicians can end up giving away powers that are not theirs to surrender. In other words, it is not for those who govern to fundamentally change the rules around governing.

Britain has a precedent for deciding constitutional issues in this way. We had the votes on national and regional devolution in the 1970s and 1990s. We had the vote on Britain's membership of the European Economic Community, or Common Market, in 1975 – until 2011 the only nationwide referendum ever held in this country. Countries around the world frequently determine such issues at the ballot box.

Not only are there past examples, there is a present compulsion for referendums. We live in an age when people fear that the rules are set by an elite from which they are excluded. A device that gives them control is more important than ever. As someone who believes in giving people as much power over their lives as possible, and who devolved power in many ways while in government, this is a natural instinct for me.

Of course, how these questions come to be settled in the first place often comes about through political necessity. The 2011 referendum on changing Britain's voting system is one such example.

Far from overriding parliamentary sovereignty, this referendum was a product of that system. It could only take place because it had been approved by MPs. More than that, it had been *demanded* by those MPs in the first place, in that it was a condition set by the Lib Dems for entering the coalition. Though I didn't want to replace the first-past-the-post system, and certainly not with the Alternative Vote system, I agreed that if the question was going to arise it should be put to the country. So that is how I ended up scheduling my first referendum for 5 May 2011.

The coalition programme had three legs: economic reform to deal with a broken economy; social and public service reform to deal with a broken society (though the Lib Dems would never use that term – they were too squeamish to admit that things were that bad); and political reform to mend our broken political system.

The pieces of political reform came together quickly.

In February 2011, MPs granted Conservatives our longed-for boundary review to reduce the size of the Commons and to make constituencies more equally sized. The average Conservative seat had 72,000 voters; the average Labour one 68,000 voters. But in some places it was far more uneven, with some seats having fewer than 65,000 constituents, and some over 85,000. Often, it took many more votes to elect a Conservative MP. 'Equal-sized seats' had been a cry for reform as far back as the Chartists in the 1840s. In the 2010 general election we'd have been just five short of an overall majority if there had been equal-sized seats. Rectifying that would increase fairness – and it would cut costs at the same time.

Another change was that petitions with over 100,000 signatures would automatically trigger a debate in Parliament. Then there was the power we gave to constituents to recall their MP if they were found guilty of wrongdoing (devised in the wake of the expenses scandal, this was criticised as timid and unlikely to be used; however, in the past three years it has already been triggered twice).

Then, in September 2011, the Commons passed the Fixed Term Parliaments Bill. General elections could no longer be called simply on the request of a prime minister: they would happen automatically every five years, unless two-thirds of the House agreed to hold one sooner. This was another long-standing Liberal Democrat policy, but the urgency now came from Lib Dem fears that the Conservatives might wait for an opportune moment, then collapse the coalition and seek a general election. The move made the coalition more stable. There was also the more profound advantage of making governments and prime ministers more focused on long-term decision-making, rather than trying to create a short-term set of circumstances that would help deliver a general-election victory.

I had always seen the force of the argument in favour of fixed-term Parliaments, and had said as long ago as 2006 that we should consider them. But it was the experience of politics between 2007 and 2010 that helped to convince me that the time was right for such a change. From the moment Gordon Brown cancelled his plans for an election in the autumn of 2007, I felt that he was trying to engineer favourable circumstances for a snap ballot.

The change meant removing from the prime minister the almost automatic power to call such a snap election (Theresa May would later effect a snap general election – but she had to ask Parliament to agree it).

The legislation for boundary reform also allowed for a nationwide referendum on whether Britain should replace its current voting system with the AV system. And electoral reform was the Lib Dems' holy grail. The existing first-past-the-post system – each voter puts a cross next to one candidate, and the candidate with the most votes wins – seems fair in theory. But it works better for bigger parties than smaller ones.

I would always remember the 1983 election. The Social Democrats, led by the 'gang of four' who broke away from a hard-left Labour Party, had burst onto the scene and, in alliance with the Liberal Party, secured over a quarter of all votes, just a couple of percentage points behind Labour. Yet this translated to just twenty-three seats for the new party, compared with 209 for Labour. Even in 2010, with 24 per cent of the vote, the Lib Dems won just 8.8 per cent of the seats.

AV wouldn't necessarily have rectified this. It wasn't *proportional* – it didn't directly link the number of votes cast to the number of seats shared out. It was *preferential*. Under the AV system, voters rank one, or some, or all of the candidates on the ballot paper. Those ballot papers are then put into piles according to which candidate is marked with a '1'. If no candidate gets more than half of all these first-preference votes, the candidate with the least first preferences is eliminated and their ballot papers' second preferences are added to the remaining candidates' piles. If *that* doesn't bring anyone up to the 50 per cent mark, the next least popular candidate is eliminated and *their* ballots are shared out according to their second preferences (or their third, if their second have already been counted). On and on it goes until – bingo – someone hits that 50 per cent mark. This is very similar to the system that is used in the London mayoral and some local elections.

The benefit of this system, its proponents argued, was that unlike most systems of proportional representation, every MP would have a constituency, and every constituency an MP. And all MPs would require at least some measure of support from at least half of their electorate.

AV would benefit Nick Clegg, because the centrist, fall-back space his party occupied in British politics would attract the second-choice support of many Labour and Conservative voters. Those second choices could stack up and tip them over the line in marginal seats.

But in some instances AV could be *less* proportional than our current system. For instance, it is estimated that had the 1997 election been carried out under AV, the Labour landslide would have been even larger. Clegg himself even once called the system a 'miserable little compromise'.

The reason for the Lib Dem switch to AV was Clegg's political calculation that there was greater cross-party support for it. Labour had officially backed a referendum on AV in their manifesto and it would have more chance of becoming law than any other system.

This would turn out to be a miscalculation. Labour was split down the middle on the reform, and there was a deep hatred of the Lib Dems for their decision to go into government with the Conservatives.

While the promise of the referendum was what brought our two parties together, the manner of the campaign brought the coalition close to collapse. The first big AV argument Nick and I had was over timing. Nick was determined to hold the referendum in May 2011, which was the first opportunity after the general election that coincided with local elections. Combining the polls would ensure a higher turnout and keep the cost of the vote down.

Backbench Conservative colleagues were strongly opposed to an early date for the referendum. Pointing out that the same Bill brought forward boundary reform got me nowhere with them. My party was determined that I shouldn't accede to Clegg's request for one simple reason: it was what Clegg wanted.

I got used to my party's hostility when I gave ground to the Lib Dems – and to the Lib Dems' hostility when I fought for the interests of my own party. It is the perpetual trap you're in as a coalition prime minister. If you anger your partners, you don't have a majority in Parliament. If you anger your party – especially when it's the Conservative Party – you might not have a job.

From other quarters I was accused of being transactional about the voting system. This was unfounded. At Oxford I had studied under one of the great experts and supporters of reform, Vernon Bogdanor, and had scrutinised the different systems. I was hostile to anything that eliminates the sacred principle of 'one person, one vote; one member, one seat', such as proportional systems.

The fact that every MP is personally accountable to the people who elect him or her – they meet them regularly, they raise their issues in Parliament, they are effectively hired or fired by them every few years – is vital to our democracy. That was reaffirmed by my own experience. I knew West Oxfordshire like the back of my hand; I loved it and fought for it from my first days as a newly elected backbencher to my last as a retiring prime minister.

I was hostile to preferential systems like AV, because they are inherently unfair. First past the post means every person's vote is equal. But AV makes some votes count more than others – literally, as supporters of unpopular parties end up having their ballots counted a number of times. We'd be giving all the Monster Raving Loony and British National Party crazies several bites at the electoral cherry, and letting them decide the outcome of elections by picking between the major parties, having already voted for their own.

It can also lead to the most unfair of outcomes. Someone who gets the most first preferences in the first round can be overtaken as more and more second and third preferences are taken into account. It could be, as I put it, a 'Parliament of second choices'. Ask David Miliband. He won the first three rounds of voting in the Labour leadership election, but then lost the final round to his brother Ed.

Both the proportional and the preferential systems would make our system less decisive. They would lead to more coalitions. This was perhaps an odd argument at a time when I was enthusiastically leading a coalition, but I did genuinely believe that the clarity and decisiveness of our system were good things.

From the very beginning we were clear that the Conservatives would campaign with vigour for the status quo. I do accept that we also gave the impression that the AV referendum wouldn't be allowed to hinder the coalition. In fact, we were all so keen to make it work that both Oliver Letwin and Michael Gove individually offered to me to campaign for AV, even though they were more inclined to my side of the argument. I told them it wasn't necessary: granting the referendum brought the Lib Dems into the tent; winning it would keep the Conservative Party from leaving it.

The truth was that I did crank up my involvement in the campaign to revive the flagging 'No2AV' campaign as the referendum approached.

Matthew Elliott from the Taxpayers' Alliance was doing a reasonable job as campaign director, particularly in signing up some big Labour figures and uniting Labour and Conservatives over the cause. Rodney Leach – a sort of Tory philosopher king, who had run Business for Sterling opposing British membership of the euro – was an effective chair.

But financially they were struggling. The Yes campaign had benefited from large donations, and we were seeing the impact in the polls. After No had initially been ahead, on 12 February 2011 a ComRes poll put Yes on 40 per cent and No on 30 per cent. Populus showed a strong Yes lead on 19 February. Panic – we might lose this – began to set in, especially after the 1922 Executive Committee paid me a visit and expressed their fears about a loss.

The fact was that we needed the big guns, and the big money – and I could do something to help that. What would be worse: damaging the coalition, or damaging democracy? I had to weigh it up. The latter was far worse.

I helped secure the hugely successful entrepreneur and Tory donor Peter Cruddas as treasurer, and I asked Stephen Gilbert to keep an eye on Matthew Elliott. I stepped up my own public involvement too, with visits, newspaper op-eds and speeches. The detail *was* important. Our polling was showing us that, when AV wasn't explained to people, it was more popular than first past the post; but the minute you told them how complicated and costly it was, they swung back to the status quo.

There was one more desperate measure to consider. Stephen Gilbert came into my morning meeting in Downing Street one day, and waited afterwards to talk to George and me alone. He said the No campaign wanted to add a third 'c' to the argument about how 'costly' and 'complicated' AV was: 'Clegg'. It could harness his dwindling popularity to show that AV would mean more coalitions, and more Cleggs playing king-maker in British politics.

The Tory tribe was gearing up for another of its uprisings. This time it probably wouldn't be regicide, but it would be another grassroots revolt that I could do without. So I said: 'Do it.' I didn't agree every word and every picture. But I did wince when I eventually saw the leaflets with a picture of Clegg holding that sign saying he wouldn't vote for tuition fees, and the words 'AV will lead to more broken promises'.

Politics is a brutal business. You have to campaign with all you've got. You have to put long-term interests above immediate concerns, and your own party and survival above other parties and leaders – however much you like and get on with them.

The weekend before the referendum was the first May bank holiday, and I stayed at Chequers with the family and some friends. In the early hours of Monday morning I received a call from a duty clerk saying that President Obama wanted to speak to me on the phone. When he came through at 3.15 a.m. he told me that Osama bin Laden, the head of al-Qaeda, had been killed by US forces in Pakistan.

I was delighted at the news. This was the man who had plotted 9/11 – the biggest terror attack in history, which had also caused the largest loss of British life in any such attack. He had inspired attacks like London 7/7 and others around the world. He was a mass murderer, and richly deserved his fate.

Before 7 a.m. I had filmed a statement for the TV cameras. Then it was back to Chequers for more meetings and phone calls, including with prime minister Yousaf Gilani of Pakistan – the country where bin Laden had been living, in the centre of a busy town, all along . . .

The next day at cabinet, Lib Dem energy secretary Chris Huhne confronted me about the AV leaflets. No was pulling ahead in the polls, and the Yes campaign was rattled. He tossed copies of the offending literature down on the cabinet table, asking if George had been behind them, and demanding I sack whoever approved them. I mumbled something about not personally approving every leaflet, and made the point that there was a Conservative campaign I *was* responsible for, and an all-party campaign that included other parties. Nick Clegg looked embarrassed. Everyone else was silent. Then George piped up: 'This is the British cabinet, not some sub-Jeremy Paxman interview on *Newsnight.*' It was just enough to deflate Huhne. I moved on to the next item on the agenda, and the meeting continued as if nothing had happened.

Then came polling day. That evening, as the results started to come in, I returned from dinner and watched the TV. The polls were good. There seemed to be nothing to worry about – and I slept soundly.

It wasn't until the morning that it was confirmed: No won by 68 per cent to 32 per cent. Turnout was 42 per cent.

I don't look upon the victory with much fondness. It was, in the coalition story, a miserable little episode. And things between our parties would never quite be the same.

That said, my relationship with Nick *did* recover. He came to Dean in August, just a few months after the result. We played tennis, had lunch and talked about how to get the coalition back on the road. It was a big deal after such a rocky patch in our relationship, and even my children were excited about his arrival. I remember Nancy saying, 'Dad, is NICK CLEGG really coming here to Dean? Wow!'

Between the result of the referendum and that lunch, something else happened. It helped to bring the coalition back together. But it showed that the Lib Dems were wrong on one thing. Our society *was* broken.

That summer I went with Sam and the children to Tuscany. We rented a large villa with friends, surrounded by olive groves and vineyards. It was blissful: playing tennis, reading by the pool, visiting churches and galleries, with young children running around everywhere.

Which makes holidaying as prime minister sound fairly peaceful. It isn't. No matter how remote your retreat, you're never completely alone. There are police with you constantly. Not only your own protection team, who are always nearby, but the host country's police as well – in Italy's case the regular Polizia, the military-style Carabinieri, and even the Forestry Corps.

There is always work. A small Downing Street team is permanently on hand in a nearby hotel or villa wherever you go. Every morning a small number of items were brought to me for signing, deciding and reading. And then there was my red box, full of 'summer reading' written by hard-working staff wanting to clear their desks before the break, analysing all sorts of complex issues at great length and presented to me just in time for my holiday.

And of course the press are never far away. When we arrived at Arezzo Cathedral to look at the frescos by Piero della Francesca there were about fifteen paparazzi waiting for us – tipped off, I'm sure, by the Italian police.

Usually the press would (sort of) stick to the much-mocked deal my press office struck with them. Our side of the bargain was one highly contrived snap of Sam and me – drinking coffee, walking along, or, for some reason – and this happened two years in a row – pointing at fish in a market. Theirs was to leave us alone for the rest of the trip.

On this particular holiday, the scene for our photo call was set in the town of Montevarchi. I wandered up to the counter of the carefully chosen bar and ordered tea for Sam and an espresso for me. I asked the waitress if she'd be bringing them over, but she said no, so I paid there, waited for the drinks to come and then took them to a table. Did I give a tip? No. It would have been like tipping when you leave Starbucks. When we left, the journalists interviewed the waitress and asked if I'd given her a tip . . . Now they had a picture *and* a story.

What had inevitably been named 'tip-gate' carried on when the poor waitress wrote to me a few days later saying her name was now mud in Montevarchi because she'd snubbed the British prime minister. She invited me back to the bar, and this time I went with seven-year-old Nancy. We ordered lemonade and a beer, and Nancy handed her a large tip. The waitress then gave me a bizarre-looking cocktail, which she had named the 'Cameron Tuscan Dream'. It was one part espresso, one part Vin Santo liqueur and one part cream. I drank it smiling happily – photographers had already assembled – and narrowly avoided giving them the even better story of the prime minister being violently sick on camera.

What I'll remember most about Tuscany was Elwen, then five, deciding he wanted to spend lots of time with me. Father–son relationships can be complex things. I am not sure I was the best dad when he was a toddler. Like many boys, Elwen was a human dynamo, requiring exercise and attention in vast amounts. Sam was much better at handling the oversupply of energy and the occasional tantrum. When I tried to help I often seemed to make the problem worse.

Looking back, the answer seems simple: the more things you do together, the easier it gets. Maybe we were both struggling a bit with memories of Ivan. During this summer the dynamics suddenly changed, and I had the time to make the most of it. We went for walks on our own, and seemed to make a new connection. Every morning Elwen would say, 'Where are we going to go for a walk today?' He just wanted to chat. It is a lovely thing when your young ones hit an age when they become your companion as well as your child.

Towards the end of the first week away I was having to spend more and more time on phone calls, with the problems in the Eurozone and America's reluctance to deal with its deficit creating a new storm in the markets. The steps by the front door of the house had the best mobile phone reception, and I would spend hours there, nursing a cup of coffee, taking calls from Mervyn King, then George, and then Angela Merkel and Nicolas Sarkozy. It wasn't panic stations for us. Britain was looking something of a safe haven: we had taken some of the difficult decisions, and weren't wrapped up in the disastrous euro project. But what happened on the continent would affect us, and we were trying to coordinate actions and messages that would deliver greater stability.

Every morning I would read the media summary that CCHQ's press office emailed to Conservative MPs, staff and activists. On Friday, 5 August I had seen the story about the shooting of Mark Duggan by police officers in Tottenham. The police said they were attempting to arrest him on suspicion that he had a firearm and was planning an attack, but the facts about what exactly happened weren't clear. People had begun to protest outside the local police station, and that evening the protests turned into an altercation with police.

On the Sunday morning I woke to find that local shops had been trashed and looted, and police officers had been injured. I was appalled, but had faith that the police would contain the situation. Yet that night the chaos – mainly vandalism and looting – spread across the capital. Police made a hundred arrests and charged sixteen people.

On Monday I left the villa at midnight, having said I wanted a COBR meeting first thing, Parliament recalled that week, and a range of visits planned. I was met by a small jet at Pisa, and spent the journey catching up on documents flown out in a red box. By 4 a.m. I was back in Downing Street and straight to bed.

It didn't take me long in the job to realise that 'If in doubt, get back' was the right motto. You are always better overreacting than underreacting to a crisis. This is not just about appearances: there is a moment when a situation is worsening and you know your intervention can be decisive. Time and again I would find that there are some things only a prime minister can do. We talk about the Whitehall machine as if the PM just presses a start button. The reality is that when things get rough it is the PM who must pull the levers and turn the cogs him or herself, day after day.

Two hours after my head had hit the pillow I was at the kitchen table studying the latest situation reports. The previous night had seen the worst rioting yet. Almost every London borough was affected. Much of the criminality was planned online, via messaging apps and social media.

There had been the most dreadful scenes. Masked gangs smashing shop windows. People brazenly walking down the street with arms full of TVs and trainers. Children kicking police officers. Historic buildings – people's livelihoods and homes – engulfed by flames. People jumping from burning windows. Others attacking firefighters as they tried to tackle the blaze.

I felt sickened that so many people were capable of such violence and criminality, and amazed that it could spread in this way. I felt embarrassed for Britain, too. Just a few months earlier, new technology had helped young people in the Arab world to fight for democracy – and here young Britons were, using it for theft and destruction.

Above all, I felt angry. Angry with the perpetrators and angry that the police hadn't contained the disturbances. I also felt angry with those trying to make political capital out of the whole thing. Some were saying the events were a response to the killing of Mark Duggan, and poor police and community relations. But I couldn't see what raiding Debenhams in Clapham Junction had to do with that tragic incident.

Others were saying it was about anger with politicians and cuts. But people weren't attacking Parliament, they were attacking private property. And anyway, the cuts hadn't kicked in yet.

It wasn't about race – this involved people of all backgrounds. And it wasn't about poverty. People weren't stealing food. They were stealing designer clothes and boasting about it on their smartphones. Some of the looters were from comfortable, middle-class backgrounds.

I was equally clear that the shortcomings in the response to the riots had nothing to do with cuts to the police budget. The problem was the numbers deployed on the streets, not the total numbers of police employed. The problem was also the approach they took to the disorder.

I asked to see Theresa May and the Acting Met Police commissioner, Tim Godwin, in my office. I was clear. What had happened the night before was unacceptable, and could not happen again. As I left, Tim Godwin said to me, slightly muffled, 'I'm very sorry, Prime Minister, about what's happened.' That, for me, was acknowledgement that the police had made a mistake. They didn't spot quickly enough that what had started as an attack on them had become an attack on private property. The number of officers deployed was too small, and they were slow to switch from dealing with a public order protest, where you protect life and not property, to criminality and looting, where you have to get physical and make arrests.

We sat around the COBR table: me, Theresa May, Hugh Orde from the Association of Chief Police Officers, the top officials from the Home Office, the Ministry of Justice, the Ministry of Defence, MI5 and GCHQ, and, on speakerphone, the chief constables of Greater Manchester and West Midlands Police, where unrest had also broken out.

The first job at these meetings is to try to get the facts straight. I challenged the police on how many officers had been on the streets the night before last, and how many there were going to be that night. There was an awful lot of 'Oh well, Prime Minister, there'll be forty PSUs [Police Support Units].'

So often in politics, jargon and acronyms prevent non-expert politicians from having a sensible conversation with experts. It frequently drove me mad – and never more so than now. 'Stop talking about bloody PSUs. I want to know how many actual officers will be on the streets.' I wanted him to be absolutely clear that there would be 16,000 – I'd been advised that this was the optimum number for dealing with such civil unrest. I wanted a figure, not jargon, and I wanted him to say it out loud so I could repeat it. That's the beauty of COBR: you nail people's feet to the floor.

The next task at these meetings is to make sure every avenue is being explored. I asked about contingency plans if the violence continued to escalate. Should there be baton rounds (plastic bullets)? Should there be water cannon? In what circumstances should we bring in the army?

The police were sniffy about water cannon: 'They've never been used on the mainland, Prime Minister. They're kept in Northern Ireland.' I said, 'Look, we don't know what's going to happen next. I don't want to come back to this meeting in two days' time and find we needed to use water cannon but we didn't have any contingency plans.' Hugh Orde said we could have two within twenty-four hours. So I said, 'Thank you, right, done,' and then, again, repeated it in public.

I went outside the door of No. 10 and, via the media, addressed those responsible for the criminality: 'You are not only wrecking the lives of others, you're not only wrecking your own communities; you are potentially wrecking your own life too.'

Then I began what I privately called my 'riots tour', starting in Croydon with the fantastic local Tory MP Gavin Barwell, who had been incredibly active. He'd convened a group of local people, and they all said the same thing: the police weren't there, they backed off, they didn't protect our property. Many were left fending off rioters with their bare hands.

One of them was eighty-year-old Maurice Reeves, who had watched the furniture business his family had built up over generations, House of Reeves, completely destroyed by fire. It had survived the Blitz, but not this. His ethic of work and duty, family and community, and his deep compassion for those who had suffered even more than him, made him a symbol of the personal cost of the riots.

That night there weren't 6,000 police officers on the streets of London, as there had been the night before – there were 16,000. Suddenly the rioting in the capital died down.

The following morning it was back to COBR. As well as receiving situation updates, I was driving the justice system to increase the capacity of our courts by introducing emergency night-time sittings. (I've always wanted faster justice, and was determined that our courts wouldn't be found wanting. Historically the British justice system has taken a very dim view of rioting – that such disorder is totally unacceptable – and it certainly demonstrated that tradition in the following weeks.)

But where was Boris? He'd rushed all the way back from a camping holiday in a remote part of Canada for this, and now he was fifteen minutes late – and missed the whole bit about London.

That evening the riots were effectively over. On Thursday I would have COBR, cabinet, then questions in the Commons from MPs recalled from their holidays.

First, though, I had to deal with Boris.

He had come out and tried to blame police cuts for the riots. I was furious, and called him straight away. 'Why the hell did you do that?' He said it was revenge for No. 10 saying this was his 'Hurricane Katrina moment', alluding to the fact that he had been away on holiday, and it had taken him several days to return to London. This had not come from my team, but from a *Guardian* article. He was being paranoid, and frankly at this stage of the proceedings a massive irritation.

And he was late again to the next key meeting.

While the mayor of London was veering all over the place, cabinet was pulling together as one. George Osborne and Theresa May had sought me out separately to check that I wasn't going to do a U-turn on police cuts. Even the Lib Dems were on board, and Nick Clegg appealed to me not to back down.

The speaker let every single MP who wanted to ask me a question do so. I was on my feet for two hours forty-five minutes, answering 160 questions – breaking my own record, set during the phone-hacking questioning three weeks earlier. It was hard work, but it showed that every question was answered and that I had a grip on the whole thing.

My statement was followed by a debate – and while that continued, I popped off to the tea room for a shepherd's pie and a glass of wine. Then I went back to the flat, switched on the cricket, and fell asleep on the sofa.

I returned to the Chamber for the wind-up speeches, during which Michael Gove delivered one of the best parliamentary orations I'd ever heard. I marvelled at how he had crafted such beauty from such an ugly episode – praising MPs from across the House, restating our shared British values, and championing those who stood up to the rioters.

And that for me is the most powerful image. I don't just look back at that summer and think of balaclavas and burning buildings. I think of the Londoners armed with brooms who came to clear up their streets. The Sikhs of Southall who didn't just defend their gurdwaras but local mosques too. Of Maurice Reeves, determined to see his store reopen. And the police, fire and ambulance crews who faced danger night after night.

Charlie Taylor, the government's school discipline adviser who had previously been head teacher for a school with many very damaged and disturbed children, came round that evening and we talked about what had happened. Yes, it was criminality. And no, the cuts weren't to blame. But there was a *background* to the behaviour – in terms of parenting and schooling and values – that we shouldn't ignore.

The following morning Hugh Orde came out and said that the police's tactics changed because of operational decisions, not the decisions of politicians. While this was irritating, I was clear that we needed to pour oil on troubled waters. Some politicians didn't agree. David Davis wanted to throw fuel on the fire – ringing me to say how important it was we won this argument with the police. Instead, Theresa May and I made emollient statements, saying that of course the police made the right decisions, but they had had the political backing of COBR.

That was the truth of it: the intervention of COBR gave the police the support to do what needed to be done. Whether or not they'd already decided to increase numbers on the streets, we had made damn sure the numbers were going to go right up, and would stay that way.

That evening I drove down to have dinner with my mother in her cottage in Peasemore. She was all on her own since my dad died, and I stayed the night. There was something very comforting about sitting with Mum and talking about what had happened over the past tumultuous few days. As a magistrate she was full of common sense about dealing with each crime on its own merits, and not overreacting. And, as ever, she was a patient listener as I sounded off about all the frustrations and difficulties of getting these things right. There was a small single bed in the spare room upstairs, no wi-fi and not much mobile phone signal. I slept like a baby.

On Saturday morning I got to Gatwick early to meet Sam and the children, and sat in Costa Coffee with my red box doing some work as I waited. More than usual, people were coming up to me and saying things like 'You've got to keep going.' That was the national mood: bleak, but firm.

I had garnered some very practical lessons for a prime minister.

I learned that if it's all kicking off at home, come home.

I learned that a comprehensive response to a crisis – chairing COBR, recalling Parliament, making visits, delivering speeches – does work. It stamps your authority on the issue.

I learned that a hard-line response is often right: despite Tottenham being intertwined with policing issues, Greater Manchester with organ-ised crime and the Midlands with inter-ethnic tensions, the riots were, by and large, simple criminality.

And I learned that if a national emergency is being lazily blamed on cuts, you can win the argument that the problem is not financial, it is social.

23

Better Together

I passionately believe in the Union between England, Wales, Scotland and Northern Ireland. It's in my make-up. I was born and raised in England. My mother's mother was a Llewellyn from Wales. My father's father was a Cameron from Scotland. Indeed, the motto of Clan Cameron is 'Unite'.

And I felt a special pride when I first saw those words, 'Prime Minister of the United Kingdom of Great Britain and Northern Ireland', during my early days in office, because I knew I would be leading a great and strong Union.

That's not to say it's a simple arrangement. Economically, culturally, even religiously very varied, our countries' differences have helped shape British politics for centuries.

The forty years before we took office had seen a steady rise of nationalism, and the main parties had struggled to keep up. In Scotland in particular, Labour sought to assuage nationalist feeling by becoming its champion. It had countered the idea of independence with the offer of devolved government, and after the election in 1997 it had put that plan in place.

Yet by the time I was in politics, I could see cracks emerging along our national borders. Each of the devolved administrations was controlled by a different party – the Scottish National Party (SNP) in Holyrood, Welsh Labour and Plaid Cymru in coalition in the Senedd, and the Democratic Unionist Party (DUP) and Sinn Féin in a power-sharing alliance in Stormont.

More profoundly, the appetite for self-government had been whetted, but not satisfied. The arrangements were incomplete and unstable. The devolution settlements seemed to have created a never-ending grievance culture, where no matter what the issue, Westminster took the blame.

Instead of politics in the devolved nations being about subjects such as housing, education or health – for which they were all now responsible – they were predominantly about the need for further devolution, more money and the shortcomings of the constitution.

This was taken even further by Labour, who embraced the argument that there were so few Tory MPs in Scotland that any Tory government wasn't legitimate. A narrative that was intended to damage the Tories north of the border ended up delegitimising the Westminster government, whoever was in power.

Those who pursued this course endangered the Union. The more that Scottish and Welsh people felt that they were still effectively ruled from afar, or not getting a fair deal – or, at least, the more their first ministers made them believe this was the case – the more the calls for full independence would gather pace.

An important part of my task in government, therefore, was to strengthen every part of the UK, democratically, socially and economically. The Conservative position has always been pro-Union. Our full name is the Conservative *and Unionist* Party, reflecting our alliance with Joseph Chamberlain's Liberal Unionists, who split from Gladstone in 1886 over the latter's plan for Irish Home Rule. The parties merged in 1912, and officially adopted the name.

We had – wrongly, in my view – opposed the devolution settlements in Scotland and Wales in the late 1990s, and had struggled ever since to find a constructive stance. Too many Tories thought that implacable opposition to devolution was the only way of protecting the Union. I totally disagreed. In every other area of policy we were the 'devolvers' – wanting to give people greater control of their lives. Only by giving people a real stake in their nation's affairs could we continue to justify the Union and retain support for it.

I also needed to bring the Tories up to date, paving the way for electoral breakthroughs. If an important part of modern, compassionate conservatism was about making the Conservative Party love modern Britain (and modern Britain love the Conservative Party), part of that meant making sure the party loved our reformed and devolved Union too.

My motivation throughout was my staunch commitment to the Union. My first foray into Northern Irish politics after I became leader was a demonstration of that – albeit ultimately an unsuccessful one.

My earliest experience of the issue had come many years earlier. In July 1991, when I was working in the Conservative Research Department, I had been keen to get my head around what was one of the most complex issues in British politics. I called up a journalist friend, the BBC's George Eykyn, who covered the Troubles at the time. 'I work for the Conservative and Unionist Party,' I told him, 'but I don't know much about Unionists.'

George invited me to Belfast to watch the Orangemen marching on 12 July. I lumbered around after him carrying the tripod as he filmed his package. It was fascinating: the bowler hats and bright orange sashes, the brass bands and the banners. It was also strange. These people were pledging deep allegiance to a country we shared, yet their traditions and their fervour seemed alien to me and to the UK I knew.

While the Conservatives' partnership with the Ulster Unionists had effectively dissolved in the 1970s, Conservative candidates were still able to stand in elections in Northern Ireland. But they were held back by lack of local organisation, a sense that they were somehow English imports, and by a fear among Unionist voters that a Conservative vote would split the pro-Union vote.

That left the Northern Irish political system cut off from the UK mainstream. People in Northern Ireland could get to the top of business, the armed forces or public services in the United Kingdom, but not national politics.

For many years during the Troubles and the subsequent peace process, changing that by re-establishing the old relationship with the Unionists hadn't really been possible or appropriate. But now, with the Good Friday Agreement in place, I saw a chance to merge the Conservatives with the old Ulster Unionist Party to form a new centre-right, non-sectarian political movement.

I announced my plan at the 2008 UUP conference to reunite under the rather *Star Wars*-sounding banner 'Ulster Conservatives and Unionists – New Force'. When I said that 'I will never be neutral when it comes to expressing my support for the Union', the room erupted in loud applause.

It would be good for them, as they had been eclipsed by the DUP and left with just one Member of Parliament. It would be good for us, as we would be the only party fielding candidates in every part of the UK, and if we could win seats, it could take us that vital extra step towards a majority.

But above all, it would be good for Northern Ireland, bringing it into the heart of decision-making at Westminster. And by attracting a diverse range of non-sectarian candidates it could break new and healthy ground in Northern Irish politics.

There was some initial success in the 2009 European Parliament election, when Jim Nicholson – a big, burly guy, as happy being labelled Conservative as Unionist – stood and was elected.

But alas, the force was not with us. The UUP was fixated on getting more money from CCHQ coffers, and its only MP, left-leaning Sylvia Hermon, quit the party entirely. In the 2010 general election we won just over 15 per cent of the vote in Northern Ireland. But no New Force MPs were elected.

While I had failed in my attempt to create party political unity in opposition, in government I had the chance to help heal one of the worst running sores in our modern history.

On 30 January 1972, twenty-eight people were shot and thirteen killed by British soldiers during a civil rights protest in Londonderry. After what became known as 'Bloody Sunday' a report by Lord Widgery said that the responsibility lay with those who had organised the march. This conclusion had not been accepted by the nationalist community. The famous image of Father Edward Daly holding a bloody white handkerchief as he tended to the dying had become a part of history, but the feeling of injustice was raw as ever. So Labour commissioned a fresh review in 1998 by Lord Saville.

The report was due to be published, finally, in 2010. I was clear, even before seeing it, about two things: I, the prime minister, should be the one to respond publicly; and if the findings were as bad as I expected, I, on behalf of the British government and the country, should be ready to give an unqualified apology.

At a No. 10 drinks party for the staff who had worked on the election, I went to one side with Jonathan Caine, special adviser in the Northern Ireland Office, who I had worked with many years earlier in the Conservative Research Department and who became the party's authority on everything to do with Northern Ireland and relations with the Republic. Anyone who disputes the controversial system of political advisers only needs to look at Jonathan. He is the best of spad-dom: a political brain, an authority in his field and a tireless behind-the-scenes presence. He was – and remains – passionate about securing peace and progress in Northern Ireland, and maintaining our United Kingdom. And on the Saville Inquiry, his instinct was the same as mine.

The 5,000-page, ten-volume Saville Report arrived at Downing Street on the afternoon of 14 June 2010. I had just returned from Afghanistan that day, and had to make one of the sudden subject-matter switches that you get used to when you're essentially the Minister for Everything. I was less than one month into my premiership, and about to handle something truly historic.

I sat at my desk in silence for an hour and a half, starting with the sixty-page report summary. None of the casualties shot had been armed with a firearm . . . Soldiers lost their self-control . . . One person was shot while crawling away from soldiers, another while tending to his injured son, another while lying mortally wounded . . . Nothing could justify any of the shooting . . .

The following day, as I took my place at the despatch box, I was conscious that my statement would simultaneously be appearing on a big screen in Derry, just half a mile from where the shootings had taken place. I thought of the victims' families, many now elderly, who had waited thirty-eight years for the truth. Of the reaction in the pubs that fly the Irish tricolour and the homes covered in murals of the fallen.

I thought, too, of the wider reaction – on the streets in Unionist areas where the kerbs are painted red, white and blue; on the roads where the Orangemen march; in the homes of the police officers, soldiers and Protestants who were murdered by the IRA and whose families would never see justice.

I explained how I approached an issue like this: 'I am deeply patriotic. I never want to believe anything bad about our country. I never want to call into question the behaviour of our soldiers and our army, which I believe to be the finest in the world.'

And that made my conclusion, set out shortly after the start of the statement, even more stark: 'There is no doubt, there is nothing equivocal, there are no ambiguities,' I said. 'What happened on Bloody Sunday was both unjustified and unjustifiable. It was wrong . . .' After detailing some of the most shocking findings, I concluded: 'On behalf of the government – indeed, on behalf of our country – I am deeply sorry.'

I heard later that David Davis had tried to whip up the ex-soldiers on our benches to complain about my statement, but he failed because everyone realised that it was not a reflection on the brave people who served in Northern Ireland. Indeed, Bob Stewart, the MP for Beckenham, who had led UK troops in Bosnia having previously served in Northern Ireland, was particularly blunt about it when I bumped into him in the tea room afterwards: 'I told Davis to fuck off.' I was absolutely clear in everything I said that the actions of that unit on that day in January 1972 did not define the British Army and the long years of Operation Banner in Northern Ireland.

In the days that followed, the press at home and abroad hailed my statement as a great act of reconciliation. But it was one piece that touched me most. It said: 'I am convinced that Saville's report, Cameron's speech and the welcome they received here will be forces for good in future relations between Ireland and Britain – a springboard for greater things and an increased generosity of spirit on all sides. It has been a blessing to live to see this day.' It was by a Catholic priest who later became the Bishop of Derry – Edward Daly.

The goodwill created by the handling of the Saville Report was something I didn't want to waste. Peace in Northern Ireland was one of the landmark achievements of the last twenty years, and here we had a chance to build on it.

Before the 2010 election we helped to fill in the missing piece of the devolution jigsaw. In February the Labour government, along with the Irish, had secured the Hillsborough Castle Agreement, which provided for the devolution of policing and justice.

This was still hugely controversial. It was effectively giving a party allied to the IRA, which had murdered many police officers, a say over policing. I knew how hard that would be for many Unionists to swallow. It required a vote in the Northern Ireland Assembly, and our partners in the Ulster Unionist Party were opposed to it. I came under huge pressure to bring them on board, including a phone call from George W. Bush asking me, 'How are we going to get that cat Empey [a reference to then UUP leader Reg, now Lord, Empey] over the line?'

In all my dealings with Bush, he demonstrated deep understanding of the issues in Northern Ireland and around the world. I always thought people got him wrong, and couldn't see past the Texan drawl – he was 'misunderestimated', as he might have put it in one of his trademark Bushisms. I was on the receiving end of one once. 'If ever you're in Texas, Laura and I would love to have you and Samantha to stay,' he said during one of our phone calls. 'We can offer hot sheets and a clean meal.' It's become a family expression ever since.

Eventually, when the vote came in March, we and the UUP simply agreed to differ: we backed the move and they voted against. It was passed in the Assembly, and policing and justice powers were devolved in April.

Now, in office, I stated that our intention was for a 'shared future, not a shared-out future', capturing how we wanted to move from a situation where the two parties in government in Northern Ireland – Sinn Féin and the DUP – would govern for the whole of the Province, rather than each merely trying to secure benefits for its own community.

While all this had been happening, economic trouble had been brewing south of the border. After the collapse of the housing bubble and a banking crisis followed by a deep recession, the Republic of Ireland's debt had risen from 25 per cent of GDP in 2007 to 90 per cent in 2010. I sent an adviser to Dublin, and he reported back that when he had asked for a three-point plan that the government was going to follow, one person had simply answered, 'I.M.F.'

I was determined that Britain should help its nearest neighbour, an idea that caused an expected but limited backlash. Arch complainers like Christopher Chope and the other usual suspects said Britain should be looking out for its own national interests. Well, yes, but I was clear that we were. Our economies, banking systems – not to mention our people and history – were intertwined.

The Loans to Ireland Act was quickly passed by Parliament on 15 December 2010. It couldn't have been more necessary or more urgent: Ireland's debt peaked at 120 per cent of GDP in 2012. Fortunately, a new Irish Taoiseach, Enda Kenny, acted fast to cut spending, raise taxes and address the banking problems. It was brave, it averted any Greek-style collapse, and it showed that if you take the tough decisions your economy can recover quickly.

Enda was prepared to do things differently with Britain, too. One of our most remarkable moments together was a visit to the Island of Ireland Peace Park and Tyne Cot Cemetery near Ypres, Belgium – the largest Commonwealth war cemetery anywhere in the world. The December wind whipped our hair and clothes as the rows of white headstones around us – most of them unnamed – lay still and silent. For decades there had been shamefully little recognition of the contribution of 200,000 Irish men to that bloody conflict. It didn't fit the Republican narrative to talk about us fighting side by side for the British crown. But here was an Irish PM, side by side with a British PM, forging a new narrative. (Enda and I got on well, and our relationship would culminate when he became my staunchest supporter during the European negotiations five years later.)

But our efforts were nothing compared to the brave gesture that was the Queen's breakthrough visit to the Republic in 2011 – the first time a British monarch had set foot in an independent Ireland. The trip culminated in a state banquet at Dublin Castle. It was incredibly moving to witness our monarch speak of forgiveness – not least as her own cousin, Lord Mountbatten, had been murdered by the IRA in 1979. Every carefully chosen word healed another wound of history. It was a lesson in reconciliation from the best. Typically, she downplayed her achievement, and I remember her saying to me something along the lines of 'All I did was decide it was time for a visit.'

At the banquet I was seated next to the great poet Seamus Heaney. On his other side was Prince Philip. Heaney had once written the lines 'Be advised my passport's green/No glass of ours was ever raised to toast the Queen.' I wondered how the two of them would get on. The Duke spent the entire dinner telling hilarious and often inappropriate stories, all to peals of laughter. Heaney turned to me at the end of dinner and simply said, 'Bejaysus, that man's a card.'

But back in Northern Ireland, the shared future agenda was beginning to hit the buffers.

In late 2012, Belfast City Council voted to stop flying the Union Flag over City Hall 365 days a year, which led to disgraceful attacks on the police by Loyalists, and one Orange parade caused serious rioting in 2013.

In late 2013 the Northern Ireland parties invited former US diplomat Richard Haass to chair talks aimed at resolving these issues, but by 2014 there was stalemate. By the middle of the year we were faced with an additional problem when Sinn Féin refused to accept the UK government's welfare reforms. Even though welfare was a devolved matter in Northern Ireland it typically mirrored the rest of the UK, and departing from that was putting enormous pressure on an already struggling Stormont budget.

In October 2014 I was left with no option but to ask Theresa Villiers, by now the secretary of state for Northern Ireland, to convene new talks. These were to last for eleven weeks. At a Stormont meeting in December, Enda's chief of staff gave me frank advice on how to handle the next phase of the talks: 'Prime Minister, just tell them they're not getting any more fecking money.'

Talks went on and on until 3 a.m., and resumed at 8 a.m. At one point a group of men turned up in the middle of the night, and when I asked who they were I was told, 'Ah, that'll be the IRA to give their view on the deal.' They blew in talking loudly, wearing football scarves and leather jackets, and disappeared upstairs. Everyone else seemed to be willing to accept this, and so was I if it meant peace and progress.

Due to our relentless focus on counter-terrorism, no longer did we need to worry that a breakdown in talks would automatically lead to a descent into violence, as had happened in decades past. The Police Service of Northern Ireland and our security services deserve a lot of credit for that.

And there had been a further shift. In 2010 we had consciously shut down Sinn Féin's direct channel to No. 10, requiring the party to work as an equal in the devolved administration. Though its leader Gerry Adams and number two Martin McGuinness didn't seem to accept this, we said that things had to be done out in the open, fully above board. They couldn't be marching up Downing Street every five minutes demanding further concessions.

Indeed, when Adams turned up at the talks on Thursday, 11 December for the first time and later declared it 'the most cack-handed process' he'd ever known, I replied: 'How would you know, Gerry? You haven't been here.' It was clear that no agreement was even close.

So I left. This was a shock to everyone, so used were they to Blair and Brown hanging around for days. But I wanted to signal that we were in different circumstances, and we would do things differently. Devolution and the preferential financial settlement for Northern Ireland gave them the wherewithal to be involved in active decision-making for the Province rather than the endless negotiation of the past.

Walking out worked. The negotiating team reached the Stormont House Agreement just before lunchtime on 23 December. It covered welfare, budgets, flags and parading, as well as agreeing far-reaching new bodies to help Northern Ireland address the legacy of its troubled past.

Implementation, however, stalled just before the 2015 general election, again over the refusal of Sinn Féin to accept welfare reforms. So after the election Theresa embarked on a further talks process. After ten weeks she secured the Fresh Start Agreement, which effectively resolved the welfare issue by Sinn Féin agreeing to accept legislation passed in Westminster rather than in Stormont.

The following months saw a period of relative stability in Stormont politics, with the DUP now led by Arlene Foster and Sinn Féin under the leadership of Martin McGuinness. Sadly, it was not to last. In late 2016 the Northern Ireland Executive became embroiled in a crisis over a renewable heating scheme which led to the effective collapse of power-sharing in January 2017.

At the time of writing, Northern Ireland has been without a government for over two years, despite the painstaking efforts of the UK and Irish governments. Talks are stalled over issues of language, identity and culture. That doesn't surprise me. Look at most issues in Northern Ireland – renewable heating scandals, welfare rows, whatever – and you'll usually find the old divisions lurking behind them. Sometimes it feels that every policy dispute is a proxy for the past.

Of course, Brexit has further complicated matters by upsetting part of the balance of the 1998 settlement and unsettling a large section of nationalism. But I believe that the Province can truly flourish. When I was PM I didn't just see the political Northern Ireland of Hillsborough Castle, I saw everyday, twenty-first-century Northern Ireland. I met entrepreneurs building a tech hub there, Church leaders building bridges there, film-makers using the province as their backdrop and tourists flocking to its beauty spots. I saw just as much passion about the future as there is about the past.

When it came to Wales, my instinct was similar to my view in Scotland: to go with the grain of devolution. The purpose was not to avert independence as such – the Welsh were far less secessionist. It was more about getting politics in Wales away from the constitutional question and onto the bread and butter issues that really mattered.

This, I admit, was linked to our party's prospects in Wales. We had last had a majority of Welsh MPs in Westminster in 1859. After that it was usually just a handful, and in 1997 and 2001, none at all. Yet I was strongly of the view that it was only when the Welsh devolution settlement was stable and accepted, and the Welsh Assembly had real powers and responsibilities, that the Conservatives could recover there.

In 2005, three new Conservative Welsh MPs were elected – David Jones, David T.C. Davies and Stephen Crabb – and when I was running for the party leadership I met with them. They tried to persuade me that no more powers should be ceded to the Assembly, but I wasn't swayed. Attitudes had changed since the devolution referendum, and more people were now supporting a stronger Assembly. What's more, I argued that we should be embracing Welsh national identity, the Welsh language and cultural events, moving into a space traditionally occupied by the left not only for our electoral outlook, but for the health of our Union.

The momentum was with us. We recovered strongly at the Welsh Assembly elections in 2007, the European elections in 2009 and at the 2010 general election, when eight Welsh Conservative MPs were elected, five of them new. But I wanted the recovery to go much further. My strategy was to show that Wales was about more than Cardiff, and more than devolutional wrangling. My visits always focused on businesses, infrastructure, investment and apprenticeships.

When it came to devolution I happily assented to the Welsh Assembly's vote to hold a referendum on devolving powers to itself in twenty areas, including education, health, environment and housing – powers promised in a 2006 law, but not activated. The referendum took place on 3 March 2011, and there was an overwhelming majority for this further devolution.

The problem was that this was still power without responsibility: the power to spend money, but not the responsibility to raise it.

As in Scotland, a spending competition was unlikely to settle the issue or to help a Conservative revival. It was always likely to favour those parties best able to stoke up national grievances and shout at Westminster, 'Give us more money.' So I was keen for fiscal devolution – to give Wales more powers over its own taxation and spending. I commissioned a review, which in turn led in 2015 to the St David's Day Agreement, which protected funding from central government to Wales, devolved more powers to Cardiff, and allowed for a referendum on income tax devolution.

Looking back, my principal regret is that we didn't get to this position faster. As is so often the case, I think that comes down to personnel. We started off with Cheryl Gillan as secretary of state for Wales, after she had shadowed the role for five years. Although she was born in Cardiff, I was keen to get an MP from a Welsh constituency into the post, and when the chance arose I chose David Jones, the Member for Clwyd West. He knew the issues well, and was well liked in north Wales. But unfortunately, like lots of north Walians, he didn't particularly like the devolution settlement. In many ways north Wales looks more to nearby Manchester and Liverpool than to Cardiff. So more powers to the national capital aren't exactly a priority.

It was only when I appointed Stephen Crabb, the young MP for Preseli Pembrokeshire, to the role in July 2014 that I found someone whose views had developed and who now had the same way of thinking as me. He was fantastic at turning the theory into reality. And we were to see the fruits of his efforts at the 2015 general election, when more Tory MPs were elected in Wales than at any time since 1987.

The majority of the time I spent on the devolved administrations would be devoted to Scotland.

Personally I spent a lot of time in Scotland, and I went on holiday every year to the island of Jura, in the Inner Hebrides. Twenty-seven miles long and five miles wide, Jura is home to just two hundred people. It is famous as the place George Orwell wrote *Nineteen Eighty-Four*, and as the home of one of Scotland's best whisky distilleries (a case of Jura whisky was my luxury when I went on *Desert Island Discs*).

Sam and I had been going there since the mid-1990s and it had become one of my favourite places on earth. There was no telephone and no road. You could only reach the house by boat, or by braving a three-and-a-half-hour walk from Jura's only road, which ran along the other coast. By the time I became prime minister the gas lamps had been replaced by electricity as a generator whirred away in a shed next to the house, but otherwise it was still the haven of peace that I loved.

But what forced Scotland to the top of the agenda when I came to office was the growing popularity of the Scottish National Party and its support for outright independence.

The SNP had been a steady, angry presence for decades, and had even survived Labour's landslide in 1997, doubling its number of MPs from three to six. It had also proved Labour's prediction that 'devolution will kill nationalism stone dead' to be completely wrong. Instead, devolution had exposed the fact that Labour had been taking Scotland for granted.

The main Unionist party – we had put ourselves out of the game by opposing devolution, among other things – was thus a weak opponent to the nationalists. This meant that until years later, when the Conservatives once more asserted themselves in Scotland, nationalism became a stronger rather than weaker force. The devolved Parliament provided an ideal platform for the SNP message of grudge and griev-ance. It allowed them to build the case for independence step by step.

Some Conservatives argued that this showed that devolution had been a terrible error, and that we should do our best to limit it, or even reverse it. In fact, what it showed was the importance of reviving conservatism in Scotland in order to protect the Union. And we couldn't do that while opposing devolution, which was now the status quo.

In early 2007, three hundred years since the Act of Union that brought England and Wales together with Scotland, the SNP emerged as Scotland's largest party, and governed as a one-party minority government.

By 2011 we were in power at Westminster, and the SNP was heading into another Scottish Parliament election with a manifesto commit-ment to an independence referendum. If it won a majority, it would claim a democratic mandate to hold that referendum. The UK govern-ment would therefore have a decision to make, since the power to hold a referendum was reserved, and so would need agreement from Westminster.

There was the option of making a referendum non-binding, or of denying it altogether. Many people wanted this. After all, a referendum was a massive gamble. The Union we loved could be broken in two. Some would do anything to avoid that – or, more realistically, to put it off until another prime minister's watch.

While I could understand the desire to avoid a referendum, I thought it would be a much bigger gamble to thwart it. The sense of grievance against a distant, out-of-touch Westminster government would only grow. That would be the fuel the SNP needed to turn an unlikely vote for independence – in 2012 only about 35 per cent of Scots were saying they would vote Yes – into a near-certainty.

On 5 May 2011, the SNP won that majority. Annabel Goldie had been a fearsome Conservative leader in the Scottish Parliament for several years. But, disappointed by the loss of two seats, she resigned.

At that election a thirty-two-year-old local BBC journalist, Ruth Davidson, was elected to Holyrood. A month later she would become leader of the Conservatives in Scotland. It would have been inconceivable ten years earlier that we could be led in Scotland by a young, gay, female Territorial Army signaller. Ruth and I were immediate political soulmates. I was so proud of how much our party had changed, and excited about what we could achieve with Ruth at the helm in Scotland.

I remember doing a photo opportunity with Ruth on my way to Jura one year with the Scottish Conservative MEP Ian Duncan. As the three of us stood together, it was clear that Ruth was so much shorter than us that the photo would look ridiculous. 'Hang on,' she said, going over to her car boot. She came back with a large box of jump leads and proceeded to stand on it. We got our shot – and I thought, whatever 'it' is, this woman has got it.

Ruth came to embody the pro-devolution, anti-independence, modern, compassionate Conservative Party. We might have failed to modernise ourselves along with the UUP in Northern Ireland, but Ruth, I believed, really would make us a New Force in Scotland.

In 2009 the Calman Commission had recommended further devolution, and Ruth and I were resolute about turning that into a reality. Again, the old guard made their unhappiness apparent. But there was more than an electoral mandate for it, and overwhelming public support. The Scotland Act 2012 would see the biggest transfer of financial powers from Westminster since the creation of the UK. With the passage of the Scotland Act 2016, half of all spending in Scotland would be raised north of the border.

Meanwhile, the issue of a referendum was unavoidable. People had voted for it; we would deliver it. We would fight to keep the UK together, and we would have to do it soon – within the Parliament. There was a feeling that the SNP wanted to put it off as long as possible in order to gain maximum support. I wanted to deny them that advantage. We would also have to make sure that the whole process was fair, legal and decisive. Any hint that it was not would be an open goal for the SNP to deny its validity. And above all, it would have to be a single question: Scotland in the UK or out of it.

We had a suspicion that the SNP didn't really want a proper independence referendum at all. They thought it might be lost, and were actually happy to carry on governing and to keep the grievance. What they might do was insist upon a second question on the ballot, offering far greater devolution – an attractive option to those nervous about independence. A vote for it would mean the nationalists could chalk up the result as a victory, as progress, and use it to pave the way for a referendum on total independence – which would be easier to win – at a later date.

I always believed we would win. We had the right argument. The polls were in our favour early on, and however alluring the *Braveheart* imagery from the SNP, the reality of Union – a shared currency, an open border, ease of business, shared welfare system and armed forces – was surely more appealing than going it alone. Hold the referendum, do it in a way we wanted, and win, and we could turn the tables on the nationalists and show that independence wasn't inevitable. So on 8 January 2012, during an interview on *The Andrew Marr Show*, I announced that we would be holding a referendum.

There would be a lot to negotiate to even get to that point. I hired two brilliant characters to take charge. Ramsay Jones would be my Scotland media spad, while Andrew Dunlop would be my Scotland policy spad. They were the perfect duo: one a former journalist with a thick Scottish accent, a golf-loving, politics-loving wheeler-dealer; the other a meticulous, clever and universally respected adviser who I'd known back when I was working for Norman Lamont.

I said to them: 'Alex Salmond will be thinking about the Scottish referendum 24/7, but I'm PM of the UK and I can't. So I need you to be thinking about Scotland every hour of every day, and bringing me in when I need to take a decision.' They did just that.

A few days after the announcement, Gordon Brown came to see me in the prime minister's office in the House of Commons – a dark, wood-panelled, lead-windowed room overlooking Parliament Square. I expected it to be strange. This was a man who always found it fantastically difficult to have any frank or open conversation with a political opponent, and here he was, in his old office, with his successor.

But this issue was so much bigger than that. He made two points. One was that we must have independently verified information about what independence would mean. The second was that if we just presented an argument about the head and not the heart, we – and by 'we' he meant the new pro-Union team that was going to fight this thing – would lose.

The argument about the heart, he explained, must be put in the following way: 'It's not just that you can be proud of being Scottish and British, it's that for many people being Scottish is more important than being British. But the crucial point is, even if that is how you feel, you can still want to be part of the United Kingdom.'

He was absolutely right. Of course the reality of the Union was the main selling point, but without acknowledging those deep emotional ties, we would be sunk. And I would make this argument in speeches and interviews for the next year and a half.

So the ball was rolling. George began speaking to Alistair Darling about leading the anti-independence campaign. The next big step was negotiations about the referendum itself: the what, when, where and how. The SNP wanted the question phrased in a way that people would be voting positively – with a Yes – to independence. They wanted the referendum in autumn 2014. They wanted sixteen- and seventeen-year-olds to be able to vote. And – though they didn't admit it at the time – they wanted more than one question on the ballot paper.

In February 2012 I went up to Edinburgh for my first negotiation meeting with the slipperiest of characters, Alex Salmond. No matter how well you prepared for any telephone call or meeting, he would always raise a completely unexpected curveball issue to try to wrongfoot you. You felt constantly that he was trying to extract anything that could be used against you. I always used to say you had to count your fingers on the way out of a meeting with Salmond.

I still felt we had the advantage at this point. We had the machine – the whole of government, the majority of political parties. We had the money – business would want to fund a campaign to keep the UK together. And we had the message – the truest message there ever was: that our four nations were stronger together and weaker apart.

The talks with slippery Salmond continued. That summer, during my drinks at No. 10 for the lobby, I said to some Scottish reporters that I was beginning to wonder if there was no agreement because Salmond had no intention of signing up for a single-question referendum. They agreed. At that moment it really seemed it wasn't going to happen, and that Salmond's plan was to use the failure to get a deal as yet another argument for independence.

So I decided to force the pace, and apply some pressure on the Scottish government. During a visit to Glasgow at the end of July I told the Scottish media that an agreement could be reached by October. Nothing focuses a negotiation like creating a deadline. Salmond got the message. A few weeks later he reshuffled his deputy Nicola Sturgeon from her

Health portfolio to head up the referendum preparations. This was the first indication that the Scottish government was prepared seriously to move matters forward.

And by October 2012 the referendum was back on. During a short meeting in Salmond's office in St Andrew's House he and I signed the Edinburgh Agreement. It paved the way for an autumn 2014 referendum, including sixteen- and seventeen-year-olds in the franchise, with Yes being a vote for independence – but, critically, just one question on the ballot paper.

People would argue that I gave too much away to get that single question. That with votes for the young, a long campaign and the positivity of a Yes vote, it was 3–1 to Salmond. But there was more to my concessions. I was determined to win the battle of perceptions as to whether the referendum was genuinely 'made in Scotland'. I knew that if there was the merest hint of unfairness or sense that it was a Westminster stitch-up, Salmond would exploit it mercilessly. We needed it to have a force and legitimacy that could be used in future arguments against calls for a Scottish referendum.

I was happy. As I put it on tape shortly after the agreement was signed: 'I think we genuinely have outmanoeuvred Alex Salmond and he is now in a position where he has signed up to a referendum he won't win . . . So the negotiation was successfully done, I went up to Scotland, signed on the dotted line, and I hope that in 2015 we'll be able to look back and say we saved the United Kingdom.'

If only it had proved that easy.

24

Treaties and Treadmills

Sitting around the huge dining table with twenty-six other European leaders, I looked at the time on my phone: 2 a.m. We had been at an 'informal dinner' discussing a new treaty to deal with the Eurozone crisis since 8 p.m., and it still wasn't my turn to speak.

Herman Van Rompuy, the former prime minister of Belgium and now, thanks to the Lisbon Treaty, the first-ever formally appointed president of the European Council, was doing his characteristic 'talk and talk until everyone collapses in agreement'. My team and I referred to it as 'the Belgian method'.

I sat under the bright lights, surrounded by uncleared plates, with Ireland's Enda Kenny on one side and the Spanish prime minister, José Zapatero, on the other. I waited patiently. My moment eventually came at 2.30 a.m. on 9 December 2011. I pushed the activate button on the microphone in front of me and launched what I privately called 'my Exocet'. Without safeguards for Britain's financial services and for the single market, I said, we would not be signing the new EU treaty they were proposing. And without our signature, there could be no treaty.

There were three big European moments in 2011, and this was the third.

It wasn't what Nick Clegg and I had expected as we formed the coalition and looked ahead to what we thought would be five relatively Europe-free years.

In 2010 a new treaty (a significant change to European law, that usually sees power flow from member states to Brussels and has to be agreed by all those member states) looked unlikely, because the Lisbon Treaty had just been ratified. The process had exhausted Europe, and in December 2007 leaders committed to making no more changes 'in the foreseeable future'. It was as though the volcano had erupted, and was now sure to lie dormant for some time.

In addition, we were in coalition. Not emphasising European policy – where we sharply disagreed – was a price we both agreed to pay to make the government function properly. Nick's party was avowedly in favour of European integration, to the extent of still hankering after British membership of the euro. Mine was still the broad church of the Major years, but the balance had shifted markedly. There were still enthusiasts like Ken Clarke, Alistair Burt and Nicholas Soames, but the heart of the party was Eurosceptic. William Hague had summed up the mood with the slogan 'In Europe, but not run by Europe'. Some were zealously, obsessively opposed to everything the EU did – and they happened to be members of the rebellious right-wing usual suspects. But then there was a large part of the new intake – modern, compassionate Conservatives – who were more stridently Eurosceptic than I had expected at the time and than people realise now.

It was still possible to have agreed Conservative positions on Europe – and we did. Our election manifesto set out support for the single market and enlargement, combined with the return of certain powers from Brussels to Britain, and a 'referendum lock' – a law that stated there could be no future transfer of powers from Britain to Brussels without a referendum.

The word 'Europe' wasn't even mentioned in the 'Coalition Agreement', and in the 'Coalition Programme for Government' there was just one European policy, this 'referendum lock'. After our aborted referendum pledge, caused by the Lisbon Treaty's full implementation, it was the next-best thing.

Getting the lock into the 'Programme for Government' was made slightly easier by the fact that the Lib Dems' election manifesto included a rather complicated referendum pledge of their own. In the event of another EU treaty, the Liberal Democrats had proposed that it should be accompanied by a full in/out referendum.

All this makes it sound rather as if I was not bothered by the EU's failings. Far from it. I saw myself as a pragmatic Eurosceptic: I wanted to be in the EU – it would affect Britain regardless, so we should be at the table, fighting for our interests – but I also wanted to change it to make it work better for us.

I felt strongly that the Lisbon Treaty – the proposed 2005 European

Constitution in all but name – had been agreed over the heads of the British people, without the referendum they had been promised by the Blair government. I disliked its substance, not just its form. It was yet another treaty that reduced national vetoes and centralised power in Brussels. And I wasn't just trying to stop more powers going to Brussels. I genuinely wanted to get powers back.

I had supported the Conservative opposition to Blair's decision to grant Poles, Hungarians, Czechs and others immediate access to our labour market when the so-called 'A8' countries joined the EU in 2004. And as leader of the opposition I had pushed for transitional immigration controls to be put in place before Romania and Bulgaria joined the EU in 2007.

In 2009 I took our party out of the EPP grouping because of its commitment to federalism and 'ever closer union'.

In 2010 we promised in our manifesto to get out of the Social Chapter, which added labour and employment legislation to the rules of the single market. John Major had sensibly opted out of the Social Chapter at Maastricht in 1992; Tony Blair had foolishly surrendered this opt-out in 1997, without getting anything in return.

And even as prime minister, with all the coalition constraints, I had not been entirely silent on the Europe issue. I said in a *Spectator* interview in July 2011 that now we were in government, there might prove to be opportunities for the repatriation of powers.

So, in an ideal world, yes, we should recalibrate Britain's relationship with Europe. But this wasn't an ideal world; it was a world whose economy was in crisis and where we were in coalition. 'Control the issue, rather than allowing it to control us' became my mantra, for now at least.

And yet in an interconnected global economy, you're never fully in control.

So far I've used the word 'Europe' as a proxy for the European Union of twenty-seven (later twenty-eight) countries out of fifty in the whole continent. But there was also the club *within* that club – the seventeen countries that used the euro as their currency. And it was this currency union that was to drive Europe to the top of the global agenda when I became PM in 2010.

If I was sceptical about the EU, I had always been *deeply* sceptical about the euro. My problem wasn't just with the fact that the euro was an avowedly political project to unite Europe and override national governments. It was also that the currency union was established within a flawed framework. The euro was introduced on the basis that monetary policy – interest rates and the amount of money in circulation – was set by the European Central Bank (ECB).

Yet economic and fiscal policy – taxing and spending – continued to be set by national governments. Yes, there were rules about deficits, and procedures for monitoring Eurozone economies – the so-called 'Stability and Growth Pact'. But they were incomplete and regularly broken (indeed, Germany and France were among the first countries to break them).

Monetary union without economic and fiscal union doesn't work, not in the long term and particularly not if the countries it covers diverge in terms of productivity and economic performance. And it certainly won't work if the rules are incomplete, fudged and frequently ignored.

Can a single currency spanning different nations ever work? Of course. The pound sterling functions effectively as a single currency across the four nations of the United Kingdom because we have common monetary, economic and fiscal policies. When one nation, or even one part, of the UK has a difficult year, it pays less in tax and receives more in public spending. These so-called 'fiscal offsets' are vital in making the system work. And because we share a common British identity and feel solidarity towards one another, we don't (generally) complain that London is 'subsidising' Liverpool, or Glasgow is 'bailing out' Cardiff – we just let the system and the offsets work.

A single currency that crosses different countries, with widely differing economies, without such fiscal offsets – and without such solidarity – is, in my view, doomed to eventual failure.

My strong instincts about this issue flowed directly from Britain's experience inside the Exchange Rate Mechanism in the 1990s, when the high interest rates across Europe left Britain with a tighter monetary policy than our economy needed. It was a disaster.

And disaster also struck the Eurozone. In this case it wasn't high interest rates that initially caused the problem, but low ones.

After the euro was first established on 1 January 1999, these low rates enabled countries that had previously struggled to borrow at a low cost to go on a borrowing and spending spree. This created dangerous credit and housing bubbles, together with debts and deficits that the performance of these economies couldn't justify. When the economic crisis struck in 2008, markets started to question the creditworthiness of these countries and the safety of their banks, and the interest rates they faced sky-rocketed.

First were the Greeks, who had manipulated their economic statistics to join the euro, subsidised an inefficient public sector, and failed to collect taxes. Portugal, Spain and Ireland were also vulnerable.

This was terrible, of course, but what had it got to do with Britain? We weren't even in the single currency.

The fact was, if our neighbours were in trouble, so were we. The Eurozone was one of the largest markets in the world – and our biggest trading partner. 'Contagion' became the watchword, hence the document waiting for me on my first day in office, warning, among other things, that creditors might want to restrict their investments in other high-deficit countries – including ours.

And then there were the political consequences. I soon came to see that our European neighbours, predominantly Germany, would want to change the EU treaties to deal with the crisis. If they were going to have to pay to fix the crisis and save the euro, they would be looking to strengthen the 'Stability and Growth Pact' in a number of different ways. That was something we would all have to get through our respective Parliaments.

I hoped that no powers would be ceded from Britain to the EU, so the referendum lock wouldn't apply and no referendum would be triggered. But the situation would need watching very carefully.

And you didn't need a crystal ball to see that some Eurosceptic MPs, still smarting over Lisbon, would ignore whether or not the referendum lock applied, and just block any such treaty out of frustration with the whole EU system. What's more, they would see the door open to their longed-for idea of treaty change and say: 'If *they* can change treaties to sort out the Eurozone, why can't *we* change treaties to get out of the Social Chapter, get rid of "ever closer union", or get some of our national vetoes back?'

Moreover, as a Eurosceptic I thought it was a completely fair question. But such changes would be hard to achieve in coalition, and even harder in the midst of a global financial crisis.

So I had to manage a delicate balancing act to help unblock a currency crisis that was hampering Britain's economic growth, while not standing in the way of any necessary treaty change that might help resolve the Eurozone crisis.

And so it began.

On 25 March came the first big European moment of 2011. I went to the European Council and agreed a small amendment to the Treaty on the Functioning of the European Union that would allow the creation of a European Stability Mechanism (ESM), worth €500 billion. This was basically a new bailout fund. It was for the Eurozone only. Britain would neither pay in, nor take out. But it was treaty change. The door was now open.

I saw straight away that a process was beginning that might allow us to make positive changes, changes I believed that Britain *did* urgently need. I also appreciated that without such changes it would be hard to get Parliament to go along with the moves Europe wanted to make. This wasn't just about managing the parliamentary party (although that is essential in a parliamentary democracy). It was about responding to a growing public mood that couldn't just be ignored.

I refer to the ESM as a 'new' bailout fund because the European Commission was already providing bailouts through an existing mechanism, the European *Financial* Stability Mechanism, or EFSM. This was the Commission at its most brazen: employing an article of the Lisbon Treaty which was meant to offer financial assistance in the case of natural disasters or exceptional occurrences. In other words, they were using something intended for the aftermath of earthquakes to prop up the euro.

What was wrong with that? Well, because this article was an *EU* mechanism, not a *Eurozone* one, the UK was expected to contribute. We could not block it, either; decisions were made by qualified majority voting, which allocated votes according to a country's population. If we opposed, we'd lose.

On the Sunday in the middle of the coalition negotiations, with Alistair Darling still chancellor, he phoned George from the meeting of European financial ministers and consulted him. George told him he shouldn't commit the government to a bailout. Our position was simple: while we didn't rule out bilateral loans, or assistance through the IMF (both of which we did agree to), Britain wasn't in the Eurozone and so shouldn't be part of Eurozone bailouts. Even so, Labour did end up signing us up to the first Greek bailout by the EU.

On taking office, my team debated what sort of positive change this open door might allow us to get through. A note William Hague wrote to me said: 'We have to propose something significant enough to satisfy UK public opinion, but not so unpalatable to our own EU partners that we end up being isolated and blamed for blocking the German proposals to underpin the Eurozone.'

I saw this as a moment to get one thing sorted. I told our fellow member states that our ratification of the treaty change was conditional on permanently stopping the Lisbon Treaty being used in this way; what I legitimately described as 'getting Britain out of the euro bailout mechanism'.

It worked. It was a win for Britain – a major achievement. I had repatriated one power from Brussels to Britain.

But – and there was a but – it was not a legally binding change to a treaty. Rather, I got an explanatory note added to the text of the treaty. In a manner that was repeated over the next year, EU lawyers were mobilised to explain that it was not legally possible to make the changes I wanted.

This may have seemed a mere technicality, but it mattered. The Eurozone got the treaty change they wanted; I didn't. It was a sign of things to come. Whenever there was pressure to transfer powers *to* Brussels, the lawyers always found a way, but when I wanted to take powers back, those same lawyers always opposed it.

Why couldn't I get something so small and simple? The truth was, they didn't like UK exceptionalism – *'extrawurst'*, or 'extra sausage', the Germans called it – as in the Brits always wanting a bit more than everyone else. They hated the fact that we would not join the euro. I would say to them: 'Sterling is my currency, and the euro is yours. There are "ins" and "outs".' They would reply: 'You are a member of the European Union, and the euro is the currency of the European Union.'

For them, there were no 'outs' – there was simply Britain being an irritant by keeping the pound. Our opt-out got us out of using the euro, but not of propping it up, because for them the single currency was at the very heart of the EU. It was almost theological – and our position thus wasn't simply a disagreement with the others, it was a heresy against the scripture.

None of this was new. From Thatcher's rebate to Major's opt-outs, Britain had always been the awkward one in the club – something that was celebrated by the British press and Conservative backbenchers, and derided by Europeans. And my EU counterparts didn't like the fact that

– as they saw it – I was pushing for more on the back of emergency measures to deal with the crisis in the Eurozone.

However, this time Britain's intransigence was a footnote, not the main story. As often happened in the EU, the big division wasn't Britain versus the rest. It was between north and south, between those in the Eurozone who wanted greater fiscal discipline and a rigorously independent central bank – the Germans, Dutch, Luxembourgers and later the Finns and Estonians – versus those who wanted greater solidarity, like the Italians, Spanish, Portuguese and Greeks.

I saw this play out during my earliest European Council meetings in smoke-filled rooms (literally – I once counted eight leaders all puffing away).

The drill was always the same: arrive at the clinical Justus Lipsius Building in Brussels and make a brief statement to the banks of television cameras outside. Attend the main meeting around the gigantic hexagonal table to hear a report from the president of the European Parliament. Then pose for a 'family photo' with your fellow leaders before returning to the hexagonal table for the first leaders-only working session.

These started with often lengthy reports from the president of the Council and the president of the Commission, often followed by the president of the European Central Bank and the president of the Eurogroup. The fact that you'd often heard from five different EU 'presidents' before the lowly leader of a sovereign state got to say anything spoke volumes about the way the organisation worked – or didn't.

At 8 p.m. or thereabouts the proceedings would break for everyone to reassemble on the eighth floor for a working dinner. These rarely started at the allotted time (nothing in Brussels ever does), and leaders would take time to deal with other matters in their delegation rooms – or in some cases assemble in the bar opposite the dining room for a drink and a gossip. When these dinners finally got under way they could – and often did – stretch on into late nights and early mornings. I would spend any downtime in the UK delegation room on the sixth floor.

The 'artwork' throughout the building had a strong Belgian, surrealist feel. A large neon sign outside our office always made me laugh when I passed it: 'Order Emerges Again and Again from the Magma of Chaos'.

The contrast between the vital subjects under discussion and the grandiose but dysfunctional way this organisation addressed them had a contradictory yet consistent effect on me. I became more disillusioned about the way the EU worked, yet more convinced that, for the sake of our national interest, we had to remain a member of it.

I could see the power of cooperation with allies. I knew that decisions would be made without us even if we weren't there. I came to understand how difficult it could prove to work bilaterally with each country if we refused to accept the way they wanted to work as an integrated unit. All this while my frustrations grew about both the procedure and the intransigence.

At both meetings and dinners it was 'leaders only'. Just a handful of EU officials were allowed in the room, and officials from member states were excluded. Therefore it was essential that you were there, completely on top of your brief, using every opportunity to push your positive agendas and watching like a hawk for anything that went against your national interest.

I used technology to keep my team abreast of what was happening. A BlackBerry Messenger group for all my key officials was set up before any meeting started. If new documents were tabled I could get messages back to the team. Other delegations would wander up to the British team outside the meetings because they were usually the best-informed about what was happening inside the leaders' meeting.

I'm not sure everyone else was as engaged. Sometimes I would sit there and count how many people around the table were asleep. Sarkozy would often be reading the French papers. Once Merkel had to be told to turn the football down on her laptop because she'd forgotten to plug her headphones in.

At the early euro crisis meetings I had some sympathy with the chain-smoking southerners in the Eurozone – Zapatero of Spain, José Sócrates of Portugal – and (the decidedly anti-smoking) George Papandreou of Greece. Yes, they had overdone it in the so-called good years of early Eurozone membership, but they were now being throttled by the conditions that the Commission (effectively Germany) had imposed on them. Their interest rates were going through the roof, their economies were tipping towards another recession, and their political positions were being threatened by emerging populist parties.

Theoretically, there were two approaches that could have been taken to ease the crisis. You could jettison the weaker members of the Eurozone, allowing them to devalue their currencies and trade their way out of trouble, leaving the remaining members to form what economists would call an 'optimal currency zone' of like-minded, mostly northern European countries. Or you could accept that far greater coordination of fiscal and monetary policies would be needed to make the Eurozone work, starting with the ECB letting it be known that it would do 'whatever it took' to defend the euro.

Yet Germany, the dominant player, with so much at stake, couldn't seem to make a decision. When I argued with Angela Merkel that it might be better to let Greece leave the euro, she said she was undecided whether this would be 'throwing off ballast to allow the rest in the balloon to rise' or 'knocking over a domino' that would lead to a collapse of other weak member states. The 'ballast versus dominoes' argument raged inside German political circles, and the portmanteau word 'Grexit' – Greek exit from the Eurozone – began to creep into news commentary. But nothing happened. Meanwhile, in the weaker economies interest rates soared and unemployment rose.

Germany's indecision over what to do about the Eurozone was, frankly, brutal. In October 2011 I decided to opt for some 'public diplomacy' – in other words, telling the Europeans, via the press, to get on with it. The course we advocated, via a *Financial Times* interview, had three parts.

One: the Eurozone shouldn't rule out ejecting the weaker members, to set them free from these impossible constraints. If it did, it should follow the 'remorseless logic' of greater fiscal union and burden-sharing. What it mustn't do was continue to fail to choose one course or the other.

Two: it should recapitalise the Eurozone's troubled banks and stress-test them.

And three: the ECB should buy up bonds of Eurozone countries. This wouldn't just act as 'quantitative easing' – putting central-bank cash into the system – but through buying bonds and pushing down yields it would also demonstrate to the markets that these governments should be able to borrow money at affordable interest rates.

That third part I likened to the 'bazooka' approach once advocated by former US Treasury secretary Hank Paulson: if you want to send the markets a message about defending the integrity of your currency and financial system, don't use a popgun, instead wheel out the biggest weapon you can.

The response from Eurozone countries was distinctly mixed. Those in real difficulty welcomed anything that pressured the ECB to act, but disliked any speculation that a country might have to leave the Eurozone.

While the French always wanted a more activist ECB, they didn't want to hear about any plan from the authors of what they saw as the 'Anglo-Saxon capitalist model', which they still wrongly blamed for causing the crisis in the first place. Unsurprisingly, Sarkozy was particularly frosty about this interference. Ahead of the G20 in November that year I got the Mexicans, Canadians, Australians, Indonesians and South Koreans to join Britain in writing to him saying what we needed to achieve at the Cannes summit. It was billed as 'six weeks to sort things out'. He was furious.

Privately, Angela Merkel initially seemed to welcome our interventions, accepting that Britain had a huge stake in this, and acknowledging our expertise in financial services. But all German politicians – right, left and centre – are allergic to activist monetary policies and interventionist central banks. Memories of printing money and the hyperinflation in Weimar Germany in the 1920s linger on.

And so the crisis limped from one emergency European Council to another, with partial solutions and endless prevarication.

For mainstream parties there was a political price. The Eurozone suffering cannot be separated from the rise of far-left parties such as Podemos in Spain and Syriza in Greece, and the outrage at the bailouts from the rise of populists on the right, like AfD in Germany and PVV in the Netherlands, and the continued growth of the Front National in France. But there was an economic and human cost too – huge unemployment rates, vast cuts to public services, and uncertain prospects for many young people.

The episode confirmed to me that the Eurozone was indeed a potential disaster zone. But Germany and the European Commission drew the opposite conclusion – that because the single currency survived, the project had been vindicated.

This drama had an important domestic element too. 'Some problems are just insoluble because they're insoluble,' is how I put it on tape. 'The Tory Party and Europe is just one of those things that, like the Mississippi, keeps on rolling.'

In October 2011, the second big European moment hit me. I was facing the biggest Conservative parliamentary rebellion of my premiership so far, on a vote about whether to hold a referendum on the issue of Europe. With Backbench Business Days ceding some control of the parliamentary timetable from the government to backbenchers, and an EU referendum petition attracting massive public support, the stage was set. It became inevitable that MPs would schedule a debate and a vote on it.

The Backbench Business Committee had to decide exactly how this would be handled. It was left to Bill Cash to help pen the motion – and he chose one that would cause maximum discomfort for the government. The motion proposed a multiple-choice referendum that asked whether voters wanted (a) to stay in the EU on current terms, (b) leave the EU, or (c) renegotiate the terms of membership to 'create a new relationship based on trade and cooperation.'

The proposal was incoherent, especially without knowing if 'c' was even possible. But coherence wasn't its point. The aim was to ensure that as many MPs supported a referendum as possible.

My aim was different. I didn't believe it was the right time for a referendum. That would only come if the lock kicked in, and since we were so focused on fighting the economic crisis, I hoped it wouldn't. It also seemed redundant to ask the public whether they wanted to renegotiate better terms for Britain in Europe when I was right there, in Europe, pushing for better terms. If proper treaty change was coming, surely we should wait for that opportunity rather than jump the gun.

Fortunately, a backbench vote isn't binding. Such motions can be treated as 'unofficial', and therefore be dismissed by the executive as 'expressions of interest'. But I decided to ignore that approach. I reasoned that if we did nothing, Labour would abstain and we'd be left with 120-odd Tories voting for this referendum. That would probably be sufficient to carry the motion. Not only would we look divided, but I'd be asked at every PMQs: 'This House – and your party – has voted for a referendum on Europe. Why aren't you having one?'

A referendum was not government policy. We had the power to stop a defeat by whipping the vote, and that's what we did. We moved it to a Monday, 24 October 2011, so William Hague and I could attend. In the end, we won – but eighty-one Conservative MPs rebelled.

I wasn't angry. As I confided at the time, 'It wasn't bad blood, it wasn't like Maastricht, it wasn't bastard talk.' The rebels were a mixture of people who were absolutely committed to an in/out referendum, and a crucial and larger group, those who had campaigned for referendums on the Lisbon Treaty and felt that the words they had used then bound them to support other reasonable-sounding proposals for a referendum.

Lisbon didn't just rile the usual suspects. It also had an impact on the younger, liberal, what some might call 'Cameroon' wing of the party – reinforced by the fact that MPs were keen to reflect their constituents' views. The bright and loyal new MP Stuart Andrew came up to me and said that he backed my leadership, and knew he wouldn't be there if it wasn't for me, but because of the boundary review he would potentially be going for the same seat as anti-EU MP Philip Davies, and he had to rebel. I didn't like it, but I understood.

The ringleaders were rapturous. They had brought about the biggest parliamentary rebellion on the issue of the EU for years. And they could argue that I was being stubborn in doggedly defending the status quo. Of course, I didn't see it that way. My tactics may have been cack-handed on this occasion, but I believed that preventing the government from being undermined by a badly timed and dangerously drafted motion was in the national interest.

In the event there was little doubt that the 'PM versus party' narrative was as bad as, if not worse than, a landslide defeat on an insignificant Thursday motion. As Commons victories go, it was as pyrrhic as they get. And it showed the extent to which the ground was moving beneath us.

While the focus seemed to be about *form* – how I dealt with the debate – the argument of *substance* – how I should deal with the issue – began to matter more and more towards the end of 2011. The Eurozone crisis was growing. The pressure for more treaty change – from European leaders and from our own backbenches – was growing. And, crucially, our economy *wasn't* growing, stifled by the lack of progress in the Eurozone. All of this was creating a storm that was very difficult to navigate.

And then, having envisaged no dramatic changes to European law for five years, I was eighteen months in and facing the prospect of a second treaty change.

The Germans had come to the conclusion that they would support more bailouts and monetary activism by the ECB, but only if there were tougher rules on Eurozone countries and their ability to run up excessive deficits. Merkel came to the view that treaty change was the only way to achieve this.

The French didn't really like it, but if it meant the ECB pumping money into the system, they would go along with it. It was, as I called it at the time, 'the traditional Franco-German stitch-up'. This Fiscal Treaty was so harsh we privately named it 'the German torture chamber'. It would cap at 0.5 per cent of GDP the ailing countries' annual structural deficits, bring in 'automatic consequences' for those whose public deficit exceeded 3 per cent of GDP, and enshrine this in their constitutions

I could see the case for treaty change to help the euro, but we were back to the difficulties it could pose for the UK. While any treaty would principally concern the Eurozone and its members, it represented more powers for part of the EU, and a further change to the nature of the organisation. It also posed a political difficulty: in Britain people would be asking, if the Eurozone could tinker with treaties, why couldn't we?

While I thought I could handle both these difficulties, there was a bigger problem at hand. What was on the table was a new treaty that would allow the seventeen Eurozone countries to shut out the ten other EU countries while deciding things that would affect us. Anything they did on financial services or single market regulations, for instance, would have an impact. Financial services was our biggest sector – a huge source of GDP and jobs. The single market was in some ways our creation, and with such high exports in services, it had the potential to benefit us more than many others.

I knew I'd have a fight on my hands to get the safeguards we needed. But isn't that what we were in this thing for? Not to roll over and do what Germany wanted every time a decision arose, but to fight for our interests and ensure that, if the Eurozone was going to caucus, Britain's interests were protected. That, after all, was central to the case for membership.

The showdown was scheduled for the next European Council and the Eurozone meeting three days afterwards. I insisted on attending the latter – who knew what they'd cook up without us?

But before that, I arranged a video call with Barack Obama, Sarkozy and Merkel. I agreed with Obama beforehand that I would go in quite hard on the need for a proper deal and genuine bank recapitalisation, reiterating my 'bazooka' intervention.

Obama played it brilliantly: he was the good cop, setting out the scale of the problem, sounding very reasonable. I came in hard. Merkel and Sarkozy were both frosty.

When, at the end of the call, they said goodbye to Obama but not to me, I realised that the stock of Brownie points I'd built up with France and Germany over the past year and a half had been run down in one go. But it was in a good cause, because prevarication on the Eurozone was stultifying Britain's economy, and that of many other countries.

Then came the emergency Eurozone European Council meeting on Sunday, 23 October 2011. Sarkozy was bad-tempered: 'You're not in the euro, you hate the euro, you're always attacking it, and now you want to come to our meetings' – the accusations flew out of him like rounds from a machine gun. I remained calm. I said we must have some safeguards for those countries that were in the single market but not in the single currency.

Jean-Claude Juncker, the prime minister of Luxembourg, was particularly dismissive of British concerns. As a finance minister, he'd been there at Maastricht when the journey to monetary union began. He'd been there when we refused to join the euro. He always pushed and pushed for more powers for the EU. Often the smallest nations get the biggest hearing, and here was the proof: the leader of a tiny country, with a population the size of Manchester's, trying to sideline the interests of the biggest financial services exporter in the world.

Then José Manuel Barroso, the president of the European Commission, chipped in. He was meant to be the keeper of the EU of twenty-seven. Yet when the Germans proposed a treaty change that would benefit the seventeen and potentially disadvantage the others, he didn't defend the twenty-seven, he attacked one: Britain. I was furious, and I called him afterwards to tell him exactly that.

On Wednesday, 26 October I was back for Round Two. Sarkozy was still angry at my presence, though Swedish PM Fredrik Reinfeldt supported my decision to attend. When the leaders met, I made a polite intervention about how we wanted to help; we wanted to endorse the plan for greater Eurozone coordination when it was announced; we were a member of the IMF, indeed a leading shareholder in that vital organisation, so there might be ways that we could help. I was met by complete silence. Nobody was interested. I sent a message to my team outside: 'It's like karaoke night in a Trappist monastery here.'

Merkel came up to me afterwards and said that I had lectured the

European Council like a schoolteacher. 'Angela,' I said, 'you have come here asking for a treaty change that none of us want and we all have to go "Yes, yes, yes, of course" – so to get that from you is bloody rich.' To be called a schoolteacher by Europe's schoolmarm-in-chief was exasperating. Still, we said these things as you'd say them to a friend. And that was good, because I knew she was the key to nailing down Britain's demands.

The following month she and I sat in her office in the Chancellery, a modern building in Berlin colloquially known as 'the washing machine' because of its square shape and large round window. Ed Llewellyn and John Casson joined us, as did Jon Cunliffe, the cerebral Treasury man I'd brought in as our permanent representative to the EU – someone with a sceptical eye to scan Europe both for changes that could disadvantage Britain and for opportunities to make progress.

I looked Merkel in the eye and said I had a choice. 'I can try and galvanise all those countries that don't want treaty change – and there are quite a lot of them: the Swedes, the Dutch, the Danes – and try and lead all that lot to try and stop what you're doing. Or we can reach an agreement that you get your treaty change. But in return I *do* need some safeguards, or something for Britain, to help me take a treaty through Parliament.'

Perfectly reasonable. Merkel was conciliatory: 'David, I understand. You will ask for more than I can give, I will propose less than you need; but in the end we will find the way.'

We followed the meeting with two alternative proposals: safeguards around financial services or safeguards around the Social Chapter. That was then expanded into five key asks drafted by the Treasury: one about safeguarding the single market, two about financial services, and two about the Social Chapter. Merkel had agreed that she'd come back with some counter-proposals, and would make sure there was what she called a 'decent overlap' to enable me to get any treaty change through Parliament.

But no counter-proposals were forthcoming. So I would have to prioritise my demands. I weighed it up – what our backbenchers really wanted, versus what would actually help Britain's long-term economic strength – and concluded that the safeguards for our financial services were far more important than the Social Chapter. They were also more straightforward to implement, and easier to explain to other EU countries as directly related to the new Eurozone arrangements.

The European Council meeting to decide on the Eurozone plan was to take place on Thursday, 8 December. This was it: my third big European moment of 2011.

I started that day at the DHL logistics factory in Feltham near Heathrow, where a by-election was to take place the following week. I skipped across London to watch five-year-old Elwen play a mouse in his school nativity play. He had two lines, and I was poised to dart out the moment he'd delivered the second – 'It's snowing so badly I can't see' – straight into the car and onto the plane at Northolt, bound for Brussels.

The pre-meeting with Sarkozy and Merkel was so frosty *I* could barely see a way through. Merkel wanted a treaty. Sarkozy said he wanted a treaty (in fact, I think his real desire was for an intergovernmental arrangement). And I wanted safeguards for Britain's financial system – otherwise there would be no treaty.

Still, Sarkozy went off like a rocket. It was ridiculous, he said. We were trying to carve out space for Anglo-Saxon capitalism, trying to opt out of the single market in financial services. He went on and on. But it was Merkel's reaction that was key. She said, cold as ice, that we could have *assurances*, but we could not have treaty changes. This, even though she was seeking treaty changes herself. She knew it wasn't enough. I knew I couldn't get her treaty through Parliament with nothing more than a few polite assurances.

Normally everything is sewn up at the pre-meetings before the dinner, but not this time. I took my seat feeling that no one was talking to us and no one was listening to us. In the loos I bumped into the Dutch prime minister, Mark Rutte. He was one of my greatest allies – so often the Netherlands and the UK were on the same page. But he said to me, 'You are asking for too much on financial services.'

Van Rompuy said we should discuss the substance of what was required before we moved onto the method – in other words, whether or not it was necessary to have treaty changes at all. That's why I waited until 2.30 a.m. to say my main piece, because I thought it better to play according to the rules as set out by the president.

That was a mistake. Van Rompuy's strategy was to wear everyone down through lengthy arguments, while at the same time marginalising British concerns. It didn't work, but had I deliberately pushed pause on the proceedings right at the start, I might have increased the chance of British concerns getting a fair hearing.

In the end, my refusal to agree an EU treaty was swiftly met by a proposal to create an equivalent measure that was formally outside the European Union structure. It was what Jon Cunliffe had predicted: they'd make a treaty without us, and just call it something else.

I had one final card to play, which was to say that without Britain signed up to this thing – what they were now calling a 'compact' rather than a treaty – they couldn't legally use EU institutions to staff it. When I said that, there was an audible sigh in the room.

It was now 3 a.m., and they called the head of the European Council's legal service, Hubert Legal (his actual name), to clarify whether I was right. Could they use the institutions as they had the previous year? Or couldn't they? All those expectant eyes on this unassuming, greying, bespectacled man, waiting for his judgement.

I knew that, at a pre-meeting the day before, Mr Legal had been categorical about the constraints on forming a pact outside the EU treaties that used the EU institutions. He had said that it was full of difficulties and potentially illegal. However, he now proclaimed from on high – holding some sacred-looking EU text – that their use probably *was* in order. Anyone who says that the EU is an organisation based on law and not politics has never seen it act under pressure. We were powerless to stop them and to get what we needed.

I said to the team afterwards that it reminded me of that scene in *Monty Python's Life of Brian*, when the title character has endlessly insisted he is 'not the messiah', only to be told by a character reading from an ancient-looking scroll, 'Only the true messiah denies his divinity.' 'Well, what sort of chance does that give me?' Brian says, exasperated. At that moment I knew exactly how he felt. The sort of miracle that can only happen within the walls of the Justus Lipsius Building.

At 5 a.m. the meeting broke up, and I held a press conference at which I explained why I had vetoed an EU treaty. I genuinely believed that I had behaved with reason and decency, as well as determination. And I was genuinely angry that this organisation – supposedly governed by the Council of Ministers acting unanimously when it came to issues of treaty change – could behave in this way. I went to bed for an hour at the UK ambassador's residence before coming back to Justus Lipsius to sign – you couldn't make it up – *another* treaty.

This time, however, I was happy to put pen to paper, because this was the treaty that would admit Croatia to the EU. It felt momentous:

a country that just over two decades earlier had been a communist dictatorship, and that was then ravaged by civil war, had come in from the cold, and was ready to trade and cooperate with its European neighbours.

We had always pushed for an enlarged EU. Not only was it good for peace and prosperity, creating a wider trade area for Britain and a greater bulwark against aggression from the east; it was a counterweight to those who wanted a tighter, federalised Europe – the 'German horses and French coachman' that de Gaulle had originally envisaged.

Derided in Europe, I was now lionised at home by the press and parliamentary party. Never had the 1922 welcomed me with louder table-banging or heartier hugs. I didn't share their elation. The volcano had erupted, and all I could see ahead of me was the magma of chaos.

I later heard that Sarkozy and Merkel had met and decided that the risk of a British veto was preferable to my proposed treaty changes, since they could fall back on an intergovernmental treaty outside the European Union structures.

Looking back, I think this was a very important moment and choice. I had made it perfectly clear that Britain was willing to support a more integrated Eurozone, even to encourage it, but that we needed our status outside the Eurozone but *inside* the EU to be recognised. They had decided that uniformity was more important than accommodating our differences. I think this was a historic error for the EU, and quite unnecessary.

Some argue that my demands were unreasonable. But I had dropped the Social Chapter asks. Indeed, William Hague told me I was being brave in asking for so little.

Perhaps a bigger proof point came from Nick Clegg – lifelong Europhile, College of Europe alumnus, former MEP, speaker of five languages, husband of a Spanish wife, Mr Continental. He urged me throughout to oppose any treaty change, as he was convinced that it was unnecessary, that others opposed it, and that there were other ways of doing what the Germans wanted. He also agreed the negotiating strategy which unambiguously involved insisting that if there were treaty changes for others there would have to be safeguards for financial services and the single market.

Another criticism was that my demands were irrelevant to a treaty that was about the Eurozone, pure and simple. By even trying to amend it with safeguards we would be making the job of getting it through the British Parliament harder. I don't agree. If the Eurozone members were going to have their own treaty and their own meetings, using all the machinery of the EU – including its legal base – we needed to make sure that caucusing by them didn't result in damage to key British interests.

The third theory is that I mishandled the negotiation. That I waited too long to set out what Britain wanted, both before the meeting and on the night itself. People were convinced we were asking for something unreasonable before I'd even opened my mouth.

There is some truth in this, certainly about the timing. But anyone trying to negotiate safeguards from the EU is always in a dilemma. Hold your key goals close until talks are properly under way, and you are accused of trying to bounce your colleagues. Set out what you want in full and in advance, and when you inevitably get less than you asked for, the mission is branded a failure.

The real story behind the veto is more complex, and more deeply rooted in the European dynamic.

Germany's unfailing ability to get what it wants in the end is matched only by its devotion to the euro. I wasn't wrong to attempt a deal with Merkel to get the British safeguards – after all, two years later we would, together, drive through a deal to cut the EU budget that very few other countries wanted. But I was wrong to think that, at a time of crisis, she would expend German political capital to help Britain out. Another pathway – a treaty outside the EU – was open to her. She hated it, but she took it.

My battles over bailouts, the referendum motion and the fiscal compact had two other clear lessons for me. On Europe, the Conservative Party (reflecting broader opinion in the country) was becoming increasingly ungovernable. And Britain's current status in Europe was becoming increasingly unsustainable, as the whole project continued to mutate into something so different from what we signed up to all those years ago.

By January 2012, my feelings on the issue had developed since May 2010. As I explained on tape: 'My long-term view is that Europe is changing and Britain is changing in its relation to Europe because of the creation of the euro and a multi-speed Europe.' The consequences for our Europe policy were potentially profound. As I said: 'At some stage, altering Britain's relationship with the European Union in some regards and then putting it to a referendum I think would be good Conservative policy for the next Parliament.'

So there it is. As early as January 2012 I was thinking that this would be the only way to change things. Long before the UKIP threat. Long before immigration started to rise. Two years before the European elections, and three before the general election. Anyone who claims we were bounced, or didn't give the strategy enough forethought, has their riposte right here.

Even then, I believed our goals should be limited and specific. As I put it: 'I wouldn't alter [the relationship between Britain and the EU] as much as some of my colleagues. I think some of them are dishonest, in that they endlessly object to things that are actually part of the single market whilst at the same time saying that the single market is the key thing we want in Europe.' The tape concluded with the observation: 'There is a lot of dishonesty about the debate.'

Popularity in politics is a fickle thing. At any one time you can be fêted in one quarter and pilloried in another. And so it was with Europe and America: at the same time as achieving *persona non grata* status in Brussels, I was treated as a king – almost literally – when I went to Washington, DC for what was probably the closest thing a prime minister could get to a state visit.

That was thanks in part to the relationship I had forged with Barack Obama over several years. But the person I really had to thank was the Queen. With the exception of Lyndon Johnson she has met every one of the US presidents who have served during her reign – a quarter of all the presidents there have ever been. Yet only two had the privilege of a full state visit to the UK: George W. Bush and Barack Obama. When Barack and Michelle came in May 2011, they loved it, and I knew how much that was down to the relationship they struck up with our head of state. The warmth of my visit to Washington in March 2012 was, I felt, largely due to the success of their London trip.

After we landed at Andrews Air Force Base in Maryland, Samantha and I stepped off the plane and onto the red carpet, blinking in the sunshine. The US Air Force Band played both national anthems and we stood like statues. There were cameras everywhere.

As we were being driven off the tarmac I stared out of the window, looking ahead to the meetings with Obama, the ball at the White House, and everything else the trip had to offer. But when I turned around I could see tears in Sam's eyes. 'I can't do this,' she said, panic-stricken. While this was my idea of heaven, it was her version of hell. I was often guilty of forgetting that while I was a volunteer, she was a conscript. The pressure on political spouses can be huge – especially if this just isn't their world. All eyes were on what she did and said and wore.

I said in a rather hopeless way, 'Don't worry. You'll be great.' Ever practical, Sam replied that if I wanted to do something helpful I could make sure she could have a vodka and some painkillers.

We got to the president's guesthouse – which, amusingly, is named Blair House – just across the road from the White House, and the door was opened by the splendidly named Randy Bumgardner, the charming head of protocol who ran the place. The vodka and painkillers did the trick, and Michelle Obama's presence during the school visit they made together also helped to calm Sam's nerves.

There would be several firsts on the trip. I was the first foreign PM to hitch a ride from Washington on Air Force One. And when we landed in Dayton, Ohio, I would see my first-ever basketball game. Obama spent much of it explaining the rules to me, and I spent most of it pretending to understand.

Travelling back on Air Force One, I was beginning to flag. It was about 3 a.m. UK time. 'Why don't you use my bed?' Obama asked. He opened a door at the front of the plane, to reveal a double bed in its nose. As I leaned back on it, he proceeded to tuck me in with a blanket emblazoned with the White House crest. 'I bet Roosevelt never did this for Churchill,' he said.

The next morning Sam and I turned up at the White House in a US government limo. There was a crowd of thousands, a full military line-up, a nineteen-gun salute and, once again, our national anthems. I stood alongside Obama on the stage in front of the White House that had been set for our press conference, and looked out over the Washington Monument. The sky was blue, the blossom was out, there wasn't a breath of wind in the air. I wanted to press pause on my life right then, walk away from where I was and have a good look at what I was doing, because frankly I couldn't take it all in.

That evening we had drinks – the Obamas, the Camerons, the Bidens, William Hague, Hillary Clinton and George Osborne – on the balcony of the president's private apartment in the White House.

Later that night I had a reminder of the different natures of US and British politics. As Samantha and I stood in line with Michelle and Barack to greet the guests for the grand White House dinner, a whole series of Democrat-supporting gay couples came up to me, wanting to shake my hand. 'We're right behind your stance on gay marriage,' one man said, adding, as he pointed at the president, 'When are you going to convince this guy?' 'I know,' I said. 'It's odd. You're asking *this* Conservative to persuade *that* liberal to back gay marriage!'

The Obamas had suggested that we both ask a favourite musician or group to play after dinner. They chose John Legend; we chose the British folk band Mumford & Sons, who at that stage were only beginning to reach stardom. They played their hearts out, and were fabulous. I had Barack on one side of me and Michelle on the other, and to her side was George Clooney. Needless to say, I didn't really get a word in there.

Over the next few months my bond with Barack grew stronger, including at the G8 summit he hosted that May at his presidential country retreat, Camp David.

It was interesting how the EU leaders reacted to America's involvement in the Eurozone crisis. Before the summit I had a video conference with Merkel, France's new president François Hollande, Italy's Mario Monti, Barroso and Van Rompuy, who were discussing why they didn't want to talk about it in front of the Americans. There was an element of delusion about the severity of the situation. Van Rompuy even uttered the phrase: 'Were it not for the situation in Greece, actually everything else in the Eurozone is going quite well.'

At Camp David each leader stayed in a beautiful log cabin, and everyone travelled around the wooded hills of Maryland by golf cart. When I arrived at the opening drinks, Obama took me to one side and said he wanted to meet with Merkel, Monti and Hollande, alone. His plan was to get them to see sense on action on the Eurozone. He didn't want the Commission and the Council presidents there – or me. I was put out. We were six times more exposed to the Eurozone than the US, I told him. But he said the Europeans wouldn't open up in front of me after what I'd said in the *FT* urging the 'bazooka' approach. He invited me instead to come to the gym with him first thing the next morning, so he could download the contents of his meeting.

Early the next day there was a knock on the door of my cabin.

I opened it to find Obama standing there – I hadn't realised he was coming to pick me up. I pulled on my trainers, and was soon being driven by the president in a golf cart, trailed by a convoy of secret service people in their own golf carts.

This secret 'treadmill bilateral' brought about a welcome change in the way Obama conducts meetings. Normally he talks at great length, in great detail. Exercising forced him to make his points more concisely. I said that the EU was just putting off the problem. He agreed. In fact, he seemed to think that an ongoing Eurozone crisis threatened his chances of re-election later that year, and was desperate for Grexit not to happen during the election. We agreed that a contingency plan was absolutely necessary.

Some people question the purpose of bilateral visits and summits like those I attended in America in 2012. What was the point? Why meet in person?

The answer is that during those few days in Washington and at the G8 in Camp David, I got more one-on-one time with Obama than I'd ever had. In the car, in Marine One, in Air Force One, during the basketball game, at the various political events, at the dinner, and yes, on the treadmill, we had an enormous amount of time to talk. And I don't just mean shooting the breeze, or talking about politics in general. We were covering all sorts of issues. Drawdown in Afghanistan. The growing crisis in Syria.

Some of these were uncomfortable topics, like the – frankly perverse – US decisions that were preventing us from selling munitions and other arms to Gulf allies. They claimed that this was to prevent vital US technology contained in UK-made weapons from falling into the wrong hands. Yet at the same time they were happily selling their own munitions, containing similar technology, to the same countries. America may be the land of the free, but it is not always the land of free trade.

The advantage of meeting like this is you get inside your fellow leader's head, and really get to understand what they're thinking. And it was on these trips that I got to understand Obama's position on Iran, and agreed a strategy with him which I believe was to prove crucial.

For several years, war with Iran had been considered imminent. If Iran got close to developing a nuclear bomb, it was clear that Israel would strike pre-emptively in self-defence. By now Iran was probably less than a year from having enough fissile material to produce one bomb. That meant strikes were a growing possibility, to hit Iran's nuclear-bomb-making sites before they could be hidden deep underground.

Obama and I agreed that such an attack would risk major escalation, but that the alternative of Iran having a nuclear bomb would be even more calamitous. An arms race. Regional instability. Major conflict. Even world war. The intended policy that we had inherited from our predecessors was sanctions so severe that Iran would uproot its entire nuclear infrastructure. Most of them hadn't even been introduced, and were never going to work in time anyway.

It was time, therefore, for a new strategy. The priority was to make sure Israel didn't act unilaterally. The country's prime minister, Benjamin Netanyahu, was constantly talking in apocalyptic terms. The Israeli Defence Forces were rehearsing attacks. Obama knew that he had to convince Netanyahu that the US would do the job a lot better than the Israelis ever could, if all other measures failed and Tehran got too close to the red line. I said I would support his efforts to persuade Israel.

The second element was keeping Iran far enough from having a bomb, while also maintaining a realistic negotiating strategy. A carrot was needed as well as a stick. Iran should not have the capacity to produce too much fissile material before our joint intelligence could spot it and US-led military action could stop it – but in return it should be allowed a substantial civilian nuclear industrial capacity.

It would be a political hard sell for Obama domestically, and a big shift for his intelligence, security and diplomatic services. Nor would it be easy to contain a hyperbolic Israel and an inflamed Iran. I said that Britain would row in behind the US, supporting it publicly, garnering EU support, and squaring our sceptical Gulf allies.

The third step was introducing the sanctions we'd spent so long talking about. I had approved the next wave in November 2011 – going beyond what the UN had done – by banning UK financial institutions from doing business with their Iranian counterparts, including the country's central bank. In response, a mob attacked the British embassy in Tehran.

In January 2012 the EU, led by Britain, ratcheted up the sanctions and imposed an oil embargo. Iranian oil exports soon nearly halved – hugely important, since they formed 50 per cent of the Iranian government's revenue.

The fourth step was intelligence. With the best intelligence agencies in the world, the UK's covert programme could play a big part in stopping Iran getting a nuclear weapon. I received frequent reports on this work, and it was key to keeping Israel from leaping into action.

At the heart of this policy – and therefore key to preventing war – was keeping alive the credible threat of force. I was serious when I told Obama that I supported his approach. I said I felt that force was indeed the last resort, but that we had to prepare for it, and make it known that we were doing so.

But introducing the threat made this strategy hard for the Foreign Office to stomach. So for months Hugh Powell on my own staff led secret contacts with Obama's NSC over contingency plans if Israel started a war or Iran crossed our red line.

At one point in the summer of 2012, Iran got within a few months of being able to produce a bomb's worth of fissile material, and there were increasing signs of imminent Israeli strikes. It seemed we were planning for when, not if, a war would begin. To this day I'm not quite sure why Netanyahu didn't act.

But by that time we had begun pursuing this new negotiating strategy. It had to be a largely American effort from Obama and his new secretary of state, John Kerry. Our role was to help keep the UN Security Council's other permanent members (France, China, Russia), Germany and the EU on board. The EU's high representative for foreign affairs at the time was Britain's Cathy Ashton (we never accepted that the EU had a 'foreign minister', so didn't use that title as other countries did). She would be crucial to delivering the Joint Plan of Action that built on the central US–Iran talks.

In July 2015 a historic deal was reached with Iran. In return for lifting sanctions, both Iran's stockpile of nuclear material and its number of centrifuges would be radically reduced, so it could not rapidly produce a bomb. International inspectors would have much greater access, making it far more difficult for Iran to cheat. Iran could still in theory build a bomb. But, crucially, we had bought time; if it did violate the agreement, the US and its allies could take action – action the Iranians were now convinced would happen.

The following month, foreign secretary Philip Hammond visited Britain's reopened embassy in Tehran and declared that our two countries could, after so many years of hostility, 'draw a line and move on'.

I look back at the Iran deal and see many common threads with my wider approach to foreign policy. Getting a good-enough deal is a lot better than keeping an unachievable ideal alive. Being prepared to be a 'hawk' on military action is what allows you to be a 'dove' in diplomatic action. Weighing the value of a policy by whether it stops something worse happening. Challenging, and even bypassing, the UK government machine to get it to change direction. Using our place in multilateral bodies to steer them in the right direction.

And I look at the deal with pride, too. Yes, it left Iran able to develop its missiles and to make trouble in Iraq, Syria, Yemen and elsewhere. But the most important thing is that we stopped a war, and did so safely. Of course, the problem that is Iran was not 'solved', but it was made less dangerous. The neo-con alternative of aiming for regime change to solve everything looked then, and still looks now, a lot more dangerous.

25

Omnishambles

'When sorrows come, they come not single spies, but in battalions,' wrote Shakespeare. He was, of course, talking about the state of Denmark. But he could have been talking about British politics in 2012.

In the space of just one week, I faced a whole brigade of woes. A Budget that went so badly wrong that a word used to describe it – 'Omnishambles' – entered the *Oxford English Dictionary*. A recently appointed Conservative Party treasurer allegedly promising 'premier league' access to me in exchange for money. And a senior minister advising people, before a potential fuel strike, to stockpile petrol.

In just a few short days we managed to alienate pensioners, charities, churchgoers, caravan owners and, most famously, pasty eaters. It was, to borrow another term from *The Thick of It* and *In the Loop*, which coined 'Omnishambles', a total 'clusterfuck'. It made the government more unpopular, our reputation weaker and the remainder of our parliamentary term tougher. It helped to reverse our poll lead, lose us seats at the local elections and put us on the back foot.

Yet – and this is what made the whole saga so infuriating – the Omnishambolic *appearance* of things masked a *reality* that was in many ways sensible, substantial and frankly unshambolic.

We were better prepared for a fuel strike than at any time in our history. One donor's blathering overshadowed a system of party finance that was more robust than ever. And a few, broadly rational, tweaks to VAT totally drowned out some vital tax reforms that would help to kick-start our economy and give millions of low-paid people a pay rise.

I am not saying we didn't make mistakes. No British politician will try to tax baked goods again in a hurry. But the truth was that we needed to make some bold decisions to get the economy growing, without abandoning our work to reduce the deficit. Politically this was the critical

time to do so, since we had three years for the policies to make a difference and three years for any arguments to fade.

Still, there were mistakes, and these began in the formulation of the
Budget.

Getting a bold, gear-shifting Budget right means keeping it tight.
Draft Budgets are notorious for finding their way onto newspaper front
pages before the chancellor utters a word of them from the despatch box,
which blunts their political and even economic impact.

The conversations about the actual content of the Budget in 2012 were
initially kept strictly between the Quad, plus James Bowler, who was now
the Treasury's Budget director.

Nick Clegg and Danny Alexander had been living their own
Shakespearean tragedy – hammered over tuition fees and their failure
to stop the European veto. They wanted a significant increase in the
personal tax-free allowance, their flagship policy, to win back support
and fulfil their manifesto.

We couldn't believe our luck – after all the years listening to Lib Dems
wanting spending increases, they were now actually asking for a tax cut.
It was a hugely necessary one. While employment was beginning to grow
– we had created more private-sector jobs in two years than Labour did
in ten – wages weren't keeping up with inflation, and people were feeling
the pinch.

But such tax cuts would have to be accompanied by spending reductions or other tax increases. The Eurozone crisis, coupled with the rise in
oil prices, had stifled growth forecasts, and this in turn meant the deficit
wasn't falling as fast as the Office of Budget Responsibility originally
forecast. We had always defined 'Plan A' as the spending plans we had set
out rather than the deficit target. However, it would be wrong for us to
add to the deficit through our own discretionary measures.

Meanwhile, our appeal to the world as a great, global trading nation
was being undermined by the fact that we had higher corporate taxes
than Germany, and higher top-rate income taxes than France. One idea
was to cut the top rate of income tax, currently 50 per cent for earnings
over £150,000.

I had always thought that the increase from 40 per cent to 50 per
cent that Gordon Brown introduced in the dying days of his government was a mistake. But in the age of so-called 'motivation politics',
where you are judged as much on what people believe to be your

motives as on what you actually achieve, cutting it now could be deeply harmful for the government. Whatever the economic case, we would be seen as the party of the rich, cutting taxes for the well-off at a time of austerity.

Yet George was persuasive. He described the increase in the top rate from 40p to 50p in April 2009 as a classic Gordon Brown bomb. Brown knew that it wouldn't raise extra revenue. He knew that it could deter people from investing in Britain, or push them to find tax loopholes. But he also knew it made him look good because it was a tax on the rich. And he knew that whoever opposed it or felt it right to reduce it to encourage investment – us – would pay for it politically.

George accepted all of that, but he knew we couldn't win the argument that Britain was the home of enterprise if we kept this tokenistic and damaging tax rate.

I came around to his thinking. In politics you often come across what I call 'the last piece of the jigsaw' problems. You identify an issue. You marshal the arguments about the need to tackle it. You introduce a series of measures to do so. But no one takes you seriously because there is something – usually something controversial or unpopular – that you are holding back. You often get nowhere with either the experts or the press or the public unless you are prepared to put in place the most difficult piece of the jigsaw.

A good example would be early efforts to try to solve the housing crisis without radically reforming planning. Another would be initial attempts to tackle obesity without taxing sugar or fizzy drinks. In both cases I started cautiously, and then persuaded myself to get on and complete the picture.

We could have all the pro-business schemes you wanted – from visas for foreign entrepreneurs to interest-free start-up loans for domestic ones (both of which we did) – but they would never complete the friend-of-enterprise picture without the most vital piece in the jigsaw. We needed a top rate of tax that said 'Go for it.'

For the moment, George and I didn't debate how much the cut should be, because there was another question to be settled first: how to get this through the coalition.

Here, George pulled off something extraordinary. In a succession of Quad meetings, the arguments had flowed to and fro. The Lib Dems wanted their increase in the personal allowance, we wanted to go ahead with a cut in the top rate of tax. They wouldn't agree our policy, and so (even though we supported it) we told them that while they held us back, we wouldn't find the revenue for theirs. We didn't see these as policy alternatives – they fitted together – but at the same time we had to get the Lib Dems to shift.

We had a plan for wearing them down. We produced idea after idea about how we could pay for the cut in the top rate by taxing the richest in other ways, through targeting aggressive tax avoidance, closing down loopholes and looking at how we taxed the most expensive properties in the country. These made political and policy sense anyway.

George would then dangle in front of them just what a boost to incomes a big increase to the personal allowance could provide. And they could also see that our motivations for cutting the top rate were sincere. We really believed that the current rate was holding back our economy – the issue on which we would all ultimately be judged.

And so, on the evening of Sunday, 4 March, the four of us sat on the two yellow sofas in the kitchen of the No. 10 flat, discussing the issue once again. And on this occasion George not only persuaded Nick and Danny of the need to cut the top rate; he convinced them – a left-leaning, not particularly business-loving party – that we should cut the top rate not to 45 per cent, which was one of the options, but all the way back to 40 per cent.

Contrary to some reports, it wasn't the cautious Lib Dems who reined in a gung-ho chancellor. It was me. To cut the top rate to 45 per cent was reasonable. It would still send out that important pro-investment message. It would cost much less than a 40 per cent rate in terms of lost revenue in the short term. Most authorities, including HMRC, agreed that 45p was the revenue-maximising rate. And crucially, the new rate was *higher* than the 40 per cent Labour had stuck with throughout all but one month of their thirteen years in power.

All this, however controversial, was the easy bit. We were reducing revenue by raising the personal allowance and cutting the top rate of income tax, which though eventually neutral would cost us in the beginning. Yet we were determined not to add to the already deteriorating deficit figures. So we needed to raise revenue. And this, ultimately, is what brought the Budget down.

It was essential that a good portion of the revenue raised came from the richest in society. Which is where the vexed issue of a 'mansion tax' came in. The idea had been circulating in British politics for some time.

The Lib Dems wanted a mansion tax – it was in their manifesto. The chancellor wanted a version of one, having become convinced by the Treasury that we taxed income too much and property too little. After many lengthy discussions, the leading option became having two additional council tax bands with a percentage rate of the value of the house: one for homes over £2.5 million and one for homes over £5 million.

The idea has merit. Excessive taxes on income create disincentives and can have a bad economic impact, as people can change their behaviour or even move abroad to avoid them. Unlike people, properties can't move – and there is a strong argument that in the UK they are undertaxed. Council tax bands treat houses worth £500,000 and £5 million the same – and they hadn't been reviewed for years.

But the original mansion-tax idea – a blanket annual percentage levy on the most expensive homes – was effectively a wealth tax. Someone with a £2 million house would pay £20,000 every year. This was something I found unpalatable. Taxing people as they accumulate wealth, or when they sell their home or pass it on, is one thing. If you tax a property at the time people buy it, then it's clear they can afford it. Taxing someone's home year in, year out on a percentage basis seemed to me to be quite another.

I was also thinking of the forthcoming London mayoral election. The overwhelming majority of houses over £2 million were in the capital. Yes, many of them were mansions whose owners could afford to pay more tax. But a good number belonged to people who couldn't afford a £20,000 or more annual bill, paid out of their hard-earned income or pension. Frankly, I didn't think we could win the city if we pursued it.

So I'm glad we avoided a wealth tax. Sweden has recently got rid of theirs; France did the same in 2017. We will need to return to the idea of higher council tax bands at some stage in the near future, but disconnected entirely from the idea of a mansion tax.

Instead, we chose to hike stamp duty on the most expensive homes – from 5 per cent to 7 per cent on those worth more than £2 million – a far better measure, since it taxes people at the time of purchase.

My mistake was that I should have simply said 'No' to a mansion tax at the start. It meant we spent too much time killing the idea rather than looking more carefully at the other revenue-raisers. Because if the money wasn't going to come from 'mansions', it would have to come from somewhere else.

The Treasury produced a series of anomalies in the tax system that needed sorting out and would raise revenue at the same time.

First was the so-called 'age-related allowance'. Pensioners had an extra tax allowance, over and above the personal allowance, for money they could earn before having to pay income tax. As raising the personal allowance was the centrepiece of the Budget – and would eventually overtake the old personal allowance and the age-related allowance combined – it made sense to freeze the latter. The basic state pension was soon to rise by over £5 a week, all the pensioners' benefits had been protected, and there would be no cash losers as it was a freeze, not a cut. What could possibly go wrong?

Next were anomalies in the VAT system. Some of them are hard to explain, and even verge on the comic. We charge VAT on moving caravans, but not static ones. We charge VAT on chocolate-covered biscuits, but not cakes. (Jaffa Cakes, which I am sure fans like me would count as a *biscuit*, once baked a giant version to prove it was really a *cake*, and thus exempt from VAT. They won.)

I vetoed the cakes, but agreed to what seemed a sensible change on caravans.

We also tried to correct the anomaly that VAT is charged on building repairs, but not when the building is listed. Most listed buildings are houses owned by rich people, so this seemed another anomaly that would both raise money and ensure that the wealthiest paid their share. It was ridiculous that ordinary families should pay VAT when they have a kitchen extension, but a duke pays no VAT on building a ballroom. Again, another disaster – this time involving churches – lurked in the fine print.

Then came food. Nigel Lawson had extended VAT to hot takeaway food in the 1980s. It was unpopular, but there was a logic to his change. He was trying to support restaurants and pubs, which already paid VAT, in the face of competition from takeaways. And he was raising money at the same time.

But it had created a new anomaly. If you bought hot food from a takeaway you paid VAT on it; if you bought it from a supermarket you didn't. Our plan was, quite simply, to put VAT on it all. This to me seemed sensible and fair. More to the point, it seemed unambiguously in favour of the little guy. The chip shop or kebab shop owner, who had to charge VAT, was being undercut by supermarkets, which didn't.

We thought we had already averted disaster when the Lib Dems proposed levelling the playing field between eat-in and takeaway food – which would have meant taxing all sandwiches. The political pitfall of targeting people's lunch screamed out at George and me. Of course it would have been a big revenue raiser, but we'd be the modern-day 'milk snatchers' – the sandwich taxers.

So far from thinking we were running into a hail of bullets, we believed we had dodged the worst of them. A big mistake.

The only easy part of the Budget was the fact that because of the faster than expected drawdown from Afghanistan, we had a £2 billion windfall that would help the numbers to add up.

So this Budget seemed to stand firm. Its contents were known only to the Quad. It added up financially. It was fair – stamp duty more than paid for the cut in the top rate. In fact, the rich were already paying more income tax than ever before: 1 per cent of people paid 27 per cent of income tax. It could be the masterstroke that eased the cost of living *and* lifted our economy out of permanent stagnation.

But then, the weekend before Budget day, details started appearing in newspapers and on television. Most damaging of all, the *Guardian* revealed the reduction in the 50p rate, and ITV and BBC News reported the rise in the personal allowance.

On Budget Day itself, 21 March, the advance headlines confirmed the leak. 'Tax Bills to Fall for 20 Million' trumpeted the *Telegraph*. George was furious. With the good news out there, Budget day and the following day's coverage would be dominated by the revenue-raising decisions.

The following day the whole thing started to come apart. The top story was the phasing out of the age-related allowance: 'Five Million Pensioners Robbed in the Budget.' No one was interested in the fact that millions of low-paid people would be paying less in tax – that was yesterday's news. No one cared that the age-related allowance freeze would soon be offset by the personal allowance increase – that would get in the way of a good narrative, which was the Tories hammering the elderly with the 'Granny Tax'.

What made it all worse was that it was juxtaposed with the cut in the top rate of tax – 'Rich Win, OAPs Lose' said another splash. Again, no one was interested in the fact that it was higher than Labour's top rate, or that the rich would be paying more through stamp duty.

It wasn't just that the leaks destroyed the vital news balance between tax cuts and tax increases. It was that with the lost revenue from the personal allowance we had set ourselves too high a hurdle to jump in terms of raising fresh cash. And we had done this in a difficult environment economically and thus politically, and at a fraught moment in our relationship with the media.

It was clear there was to be no respite at the weekend when Craig Oliver rang me to say that the *Sunday Times* had a big story about our new party treasurer, the businessman Peter Cruddas.

I always erred towards the benefit of the doubt, but when I heard what Peter had said, I knew that this wasn't survivable. He had been recorded telling undercover reporters that a donation of £250,000 would give them 'premier league' levels of access to me and other senior Conservatives.

Peter is a self-made man. He had helped out over the AV referendum and done great charitable work. I was sure he hadn't meant to overstep the mark. But although he urged us all to wait for the full transcript of his conversation with the undercover journalists, my mind was made up. There was no context that could mitigate his comments. We didn't sell access or honours or policies or anything else, and to fall into a trap that made it look as if we did was deeply damaging. He had to go.

But there was no time to dwell on this. I had signed up, together with Samantha, Nancy and Elwen, for a charity fun run – the Sport Relief Mile – on the Sunday. I ended up giving my response to the scandal standing in a field in Oxfordshire wearing a Sport Relief sweatshirt. It was the same day that the first post-Budget polls were published: they showed that, having been neck and neck with Labour, we were now seven points behind.

On the Monday I thought, 'New week, new start.' It was a big day for me, as I was visiting the National Neurology Hospital and the Dementia Research Centre to launch something incredibly close to my heart.

But during this month – this mad March – any attempt to seize the agenda was futile. Within hours the agenda once again seized us, as fuel-tanker drivers from the Unite union held a ballot on a walkout scheduled for a week's time.

Fuel strikes have the power to bring an entire country to a halt. They can also be disastrous for governments and their reputation for competence. The crisis of 2000 was the one moment in his first administration that Tony Blair's grip on power looked in danger and his poll ratings fell below the Conservatives'.

Mindful of this, we tried to get ahead of the game in our early years in government. Oliver Letwin was meticulous in finding out how many fuel-tanker drivers the army had, how long it would take to train more, how much fuel could be stored on our military bases and elsewhere.

I knew that what we communicated was almost as important as what we were doing behind the scenes. If there are rumours of a fuel strike, the sensible thing to do is to advise drivers to keep their tanks topped up, because there is more space in the fuel tanks of thirty million cars than in all the country's petrol stations. But on the other hand, if at any time this tips into panic buying, the petrol stations' supplies rapidly empty and you help to *cause* the crisis you are trying to avert.

By Wednesday I was trying once again to shift to a more positive, proactive agenda, as the International Olympic Committee were visiting London to prepare for that year's Games. But as I scanned the newspapers it was clear the shaky Budget was about to collapse further. 'Let Them Eat Cold Pasty' blared the front page of the *Sun*, a reference to the new VAT rates on hot takeaway food.

Pasties. I hadn't even thought about pasties. Spring rolls, saveloys, chips and chicken wings – we'd discussed the lot as we stress-tested the VAT rise. But no one had mentioned pasties – and, importantly, no one in the Treasury had raised the problem with the Quad.

They should have done. Pasties were in a sort of VAT no-man's-land before the Budget. They were hot food – heated up, not cooked – and not necessarily sold for instant consumption. So in the past they hadn't always been subject to VAT, but now they would be.

Just like 'Grannies versus top-rate taxpayers', the significance was the symbolism. We were targeting a snack so ubiquitous at motorway services, in food vans and supermarkets that it was associated with ordinary working people. I knew it was bad when George came into my morning meeting and said he had just been followed around Whitehall by someone from the *Sun* dressed as Marie Antoinette holding a tray full of pasties.

Cabinet Office minister Francis Maude then poured fuel on the firestorm. In an impromptu BBC interview he reassured the public that no fuel strike was imminent, but added that it might be best to keep a full tank of petrol, and maybe a little bit more 'in the garage, in a jerry can'.

This was absurd. We'd told people to fill up their tanks, not to decant flammable liquid into twenty-litre containers and take them home. Two days later a woman in York was badly burned while pouring petrol into a can in her kitchen. Francis was absolutely mortified.

At the press conference with the IOC I decided to tackle the whole thing head-on. When I was asked a question on fuel, I said that I thought a strike would be 'completely irresponsible'. The *Sun*, of course, asked a question about the so-called pasty tax. I was able to say what I'd been longing to: that the small pasty seller was already paying VAT; this was just making it fair to them by ensuring that the big chains did too.

But I made a spectacular error when I was asked the inevitable question: where had I bought my last pasty? It shouldn't have been difficult. I love pasties. As a child I had sat shivering on Cornish beaches devouring piping-hot steak-and-potato ones. These days I often popped into the West Cornwall Pasty Company and scoffed one on the train. Not having a photographic memory of every hot snack I have consumed, I took a punt on the last place I did that: 'Leeds station,' I said.

It not only turned out that that particular shop had shut five years before, but that that chain of shops didn't charge VAT on its pasties anyway. So it was literally the wrong pasty at the wrong station at the wrong rate of tax.

By the following month, whole chunks of the Budget were just falling off. Interest groups and newspapers went apoplectic over the series of VAT tweaks that could be characterised as the so-called 'church tax' and 'caravan tax'.

If all that wasn't enough, there was the 'charity tax'. On charities, the Budget stipulated that anyone seeking tax relief of more than £50,000 would have a cap set at 25 per cent of their income. There's a good argument that there should be a limit to the tax relief you can claim. And there was a caveat to ensure that this measure would not impact significantly on charities that depended on large donations.

The motivation was that a lot of people were abusing charity relief to minimise their tax bill by giving to slightly dubious charities that benefited their families in various ways. The Quad were shown examples of rich British families having paid virtually no income tax since the 1960s because they supposedly gave all their annual income away, or set it against family expenses and employees. Our change was meant to be a tycoon tax, but very little in this Budget ended up being received as intended. Encouraging philanthropy was part of my credo, but the policy appeared to fly in the face of my Big Society philosophy. Far from completing the jigsaw, I seemed to be taking a piece out of it.

But it was the 'caravan tax' that would lead to the largest Conservative backbench rebellion on any aspect of the Budget. The caravan lobby in our party turned out to be bigger than those for churches or charities. 'Lots of colleagues have discovered that they have rather more holiday parks than they had realised,' my parliamentary aide Desmond Swayne wrote to me.

By May, Labour had notched up a ten-point lead. The press dubbed the Budget an 'Omnishambles'.

Many people – friends, journalists, colleagues, Dessie – urged me to sack George. I didn't consider it for a second. Yes, there had been oversights, but many of us were guilty of that. Plus, George and I were in it together.

I was briefly concerned about him. He wasn't his usual waspish self during those weeks. The following Friday night I rang him from Chequers to try to cheer him up. But by the end of the conversation he was telling *me* not to worry. The old George was back: optimistic and self-confident. I was clear, however, that there would have to be U-turns.

The public would deliver its verdict on the Budget on Thursday, 3 May at the local elections, the London mayoral contest and referendums on city mayors. The comparatively calm sailing from the early days of coalition was over. We lost twelve councils and four hundred councillors. Only one city out of ten, Bristol, decided that it wanted a mayor, which was disappointing after all the work we had done on localism.

As for the London mayoral election, there was a widespread theory that I didn't want Boris to win a second term – that we were such great rivals that I would rather see the capital go to Labour than my old rival triumph. It really wasn't like that. Boris was the one who was full of jealousies and paranoias, which so often influenced his behaviour. I wanted him to win. He had been a great mayor, but I felt City Hall was dysfunctional, and that he could do the job a lot better. I had been advising him to pick winnable fights with the government on things he controlled and that we could try to accommodate – like transport – rather than economic policy, which wouldn't change and which would leave us both coming off badly. But he chose to pick fights over everything.

In one meeting I had with him before his campaign, I asked him his top line, and he replied, 'Driverless trains.' This made perfect sense on brand-new networks, like the metro system in Dubai, but sounded hopelessly vague – and probably even dangerous – for a complex, older system like the London Underground. And anyway, Boris saying 'driverless trains' had the potential to frighten the life out of people. 'I have a nine-point plan,' he then said. But he couldn't even remember the points.

On the eve of the election I sent him a text saying, 'I've given you a great headline' after an *Evening Standard* interview in which I'd said, 'I want a Boris in every city.' He replied, 'Well, I hope I can survive your endorsement.'

For all our mini-run-ins, I liked Boris and he made me laugh. But I didn't always trust him.

Thankfully, he won back the mayoralty – but by just 60,000 votes. There was no victorious hand-holding between us this time. Instead, we had a private chat in his office in City Hall. He came in – late – and said, 'I'm going to do this job, and that's me done with public life, I'm leaving public life after this. People say I want to be an MP. I don't. I'm not going to do that.' I took it with a huge pinch of salt.

At the end of May came the U-turns. George clarified that food would no longer be liable to VAT if it was 'cooling down' after being removed from an oven (though food on a warming plate would remain subject to VAT). He also said that VAT would only be charged on static caravans at a rate of 5 per cent, rather than the proposed 20 per cent. The tax-relief limit on charities was also restricted. Churches were to be compensated for the VAT on their repair bills by a grant scheme. An amazing programme of church and cathedral repairs followed, and vicars and archdeacons still point out their renovations to me, reminding me that our government paid for it. The age-related allowance, however, remained.

Looking back, should we have had a more simple, steady-as-she-goes Budget? After recounting the pain of the Omnishambles Budget, I can't answer anything but yes.

But I don't regret the 45p income-tax top rate for a second. It was a fight worth having and a reputational risk worth taking. It sent out that pro-business message, and fired up what John Maynard Keynes called the 'animal spirits' of enterprise. The figures justified our approach: when the new tax rate was implemented in 2013–14 there was an £8 billion increase in revenue from additional-rate taxpayers.

I also learned some big lessons.

The first two were fundamentally contradictory. Leaked Budgets are disastrous for positive coverage. Yet the more eyes on a controversial policy, the better.

The rest were more obvious. Don't go against your own politics – if you say you back the Big Society, don't undermine it with a tax on charitable giving. And frankly, never touch VAT definitions. It's not worth the trouble. Beware symbolism. 20p on a pasty can cost you more than £1,100 off income tax. And remember that your motivations are as important to people as outcomes.

There were other forces at work too. The newspapers were gearing up for Leveson. Many of our changes to taxes and benefits were just coming in. We were in the mid-term, and the weariness that comes with that is unavoidable. And we were still in the middle of the Eurozone crisis.

The 2012 Budget also tested my partnership with George, and since I stuck by him throughout, it brought us closer together. Still, this tragi-comedy of Shakespearean proportions was largely avoidable. It had been a terrible spring. We were determined to have a better summer. I didn't want to get to the end of the year and find anyone channelling *Richard III*: 'Now is the winter of our discontent'.

26

Coalition and Other Blues

When it seems things can't get any worse in politics, they frequently do. If you were writing about the year 2012 in the style of a *Daily Mail* editorial, you could say that the government lost the public over the Budget, the press over Leveson, coalition unity over Lords reform and, potentially, the next general election over the failure to redraw constituency boundaries. These last two traumas brought the coalition the closest it ever came to collapse.

For both of the coalition partners these issues were seen as existentially important. For the Conservatives, the unfairness of the size of constituency boundaries had to be addressed. For the Liberal Democrats, it was unthinkable to be in government without at least attempting not just electoral reform but Lords reform too.

The AV referendum had been difficult enough, and I was relieved when the tensions of the campaign had begun to fade. But then came the Lords.

The red leather benches in the cavernous gilt-and-oak chamber were occupied by nearly eight hundred people, the overwhelming majority of whom were there because they had, at one time or another, been appointed by the prime minister. They included experts from various professions, former MPs, party supporters and advisers, Church of England bishops and those who were simply born into it.

I could see the point of the Lords. It could be effective at amending legislation. A fully elected second chamber might either simply pass everything the Commons passed or reject everything – a rubber stamp or a roadblock. However, a revising chamber with limited powers and a suitable range of expertise could be – and often has been – useful. That said, our second chamber is an anachronism. In 2010 it had no proper arrangements for retirement, for discipline, or even for throwing out convicted criminals.

All prime ministers are stuck, as I was, with a dilemma. If you add peers, you are accused of patronage. But if you don't, your programme is frustrated by a previous government's appointments. I was accused of making too many appointments in my attempt to rebalance the huge anti-government majority in the Lords. The result of these long-term dilemmas is a chamber now second in size only to China's National People's Congress.

I had a permanent reminder of the idiosyncrasy of the current arrangements because Samantha's stepfather, William Astor, was one of the remaining hereditary peers. Labour had capped the number of inherited seats at ninety-two; some 660 had been appointed for life by governments past and present, and twenty-six were from the Church.

William has always been more than a stepfather-in-law to me. We are good friends, and share a love of the countryside as well as of political intrigue. He has a notoriously mischievous sense of humour. When Blair's government was simultaneously threatening Lords reform and the abolition of hunting, he said to me, 'What's a man to do? I'll have nowhere to go during the week, and nothing to do at the weekend.' He was only half joking.

A great source of information about what was going on in the Lords, he played a vital role for me of explaining how the Upper House functioned and what its grandees were thinking. He – quite rightly – observed that most MPs are joyously ignorant of how the place works. But just as he would remind me of this, so I would frequently remind him of the reason he had an automatic lifelong seat at the heart of our democracy: his family had bought a hereditary title from Liberal prime minister David Lloyd George in 1917.

Yet despite my strong feeling that such quaint historical quirks as hereditary peerages had outlived their usefulness, I might have decided to leave the issue for another day. A second term perhaps.

The Lib Dems, however, didn't see things that way. For them, constitutional reform was crucial, a matter of principle and – they believed – central to their long-term success. And now they had their first, potentially their only, chance to make it happen. Plus, they had lost 336 council seats in the local elections and a third of their membership in the last two years. They needed a win. And given that the Conservative Party's policy set out in 2010 was to move towards a wholly or mainly elected Lords, it would have been bizarre for us to block what they wanted.

As for my own view, it hadn't changed much since I was a politics student in the 1980s, when as part of my homework I read a proposal by the former Conservative prime minister Alec Douglas-Home, who argued for a mixed House of Lords with an elected element, an appointed element and the steady phasing out of hereditary peers. This, I believed, was the key to sorting out the House of Lords: not some blinding flash of reform, suddenly going from a bicameral system to one chamber, as Labour wanted during the 1980s, or going for an instantly fully elected senate, as Liberals tended to argue for.

I never understood the argument that you couldn't have two classes of peer, one elected and one not. We already had different types of peer, and other countries seemed to manage with it.

Nor did I buy the argument that we needed to reform the role of the Lords before deciding on its composition. It was a successful revising chamber that improved legislation without directly challenging the authority of the elected government. It should carry on that way.

I watched Nick Clegg consulting, listening and compromising as he forged a White Paper and turned it into a Bill. I thought the process brought out the very best in him: collegiate, measured and meticulous. And when he came to report back in June 2012, what he produced was evolutionary and incremental. It would start with 120 elected peers, rising to 240 in a second phase and 360 in a third. The number of appointed peers and bishops would be steadily reduced.

Having accepted that we were going to move on Lords reform now rather than wait, I was strongly in favour of it.

While the 'Programme for Government' included a large package of constitutional reforms, two of the headline changes – a referendum on AV and the redrawing of constituency boundaries – were clearly linked. They came together on page 27 of the document, and they came together in law as part of the Referendum Bill. They were two sides of the same coin: we would decide whether to reform the voting system, and we would reform constituency sizes. More to the point, they were negotiated together, as a quid pro quo.

But on that point Nick began to show the worst of himself. He said that if we failed to achieve Lords reform, his party would sabotage boundary reforms. His argument was that all of the reforms were linked – that we shouldn't expect the Liberal Democrats to deliver boundary reform if we couldn't push ahead with Lords reform.

And because of a foolish mistake, he had us by the short hairs. We hadn't ensured that boundary reform would go through automatically. The Referendum Act established the boundary review, but the recommendations of that review would ultimately go to straight Yes or No votes in both Houses of Parliament. Therefore the Lib Dems, combined with Labour, could make the passage of boundary reform dependent on Lords reform. They could make a false tit-for-tat link between Lords and boundaries. Which is exactly what they did.

Crunch time was scheduled for 10 July 2012, when the Commons would vote on the second reading of the Lords Reform Bill and, most importantly, the 'programme motion'. This was essentially a timetable to get the Bill through all the necessary stages. Without it, MPs could talk it out of existence or amend it to death (talking and amending being an MP's deadliest weapons). Labour supported Lords reform, but they would find the opportunity to defeat the government and cause tension in the coalition irresistible. So it would come down to whether our own MPs would let the programme motion pass.

While I genuinely wanted Lords reform, and was willing to put up with rebellions in my own party from those who hated it, I was not prepared to put the whole of our legislative programme in jeopardy by having the Bill stuck in the Commons for weeks or even months on end.

Because this Bill was a constitutional reform, the 'committee stage' of scrutiny would have to take place on the floor of the House of Commons, and not upstairs in a committee room. I had seen how Bills like this (the most famous being the one that implemented the Maastricht Treaty) had brought the Major government to a state of combined paralysis and civil war. I was determined not to let that happen to my government.

On 2 July, Nick wrote to me confirming that if my party rebelled and the programme motion was defeated, the Lords Bill would be dead – and so would boundary reform. We haggled and argued, but to no effect. He rejected the proposal I made of having a referendum on Lords reform, something I knew would change many of my MPs' minds. He suggested delaying the implementation of both Lords reform and boundaries until after the election. That was unacceptable: it made it unlikely that either would ever happen.

I felt cheated by him. Here was this reasonable, decent person I had worked with for over two years being disingenuous and – frankly – dishonourable.

But while I was fuming, I also knew that I had to get over it. What was the alternative? Collapse the government and hold an election, which we would probably lose? We were bound together – and we both knew it. The considerable political pressure Nick's party was under had created a dynamic of its own. He was desperate.

If a desperate deputy prime minister wasn't going to budge, I thought there was a chance my hard-to-manage MPs might. But for some reason – and I've never quite got my head around why – many had started to rally around the cause of preserving the House of Lords as an issue of high principle. And their devotion to the status quo gathered pace in the days leading up to the vote. I had intelligent MPs like Jesse Norman getting on their constitutional high horse and firing off all sorts of specious arguments about why we should protect the House of Lords in its current state. One of his arguments was that it was so diverse. I thought, What planet is he on? There were two dukes, twenty-six earls and three marquesses – and the average age of all peers was sixty-nine.

On 3 July, John Major of all people came out and said that this was 'not the time', and the reforms were 'wrong in principle'. I was surprised at this unhelpful intervention from a former prime minister who had always been so supportive, and amazed that this was the issue he'd picked. A few days later came a letter to all Conservative MPs attacking the proposals from former cabinet ministers including Geoffrey Howe and Norman Lamont.

It would have been less puzzling if it was just Tory grandees standing in the way of progressive modernisers, but the new intake was proving just as stubborn. Some who, as far as I was aware, had never expressed a view on the Lords were now speaking about it like it was some great moral crusade. Charlotte Leslie, the modernising, boxing MP for Bristol North West, sent me a note declaring that she could 'never live with herself' if she voted for Lords reform. Penny Mordaunt and Mike Freer – sane and sensible new MPs – felt the same.

Were people really that passionate about keeping the Lords as they were? Or were they just anti-change, anti-coalition, or anti-me? Whichever it was, the whole thing had become a bandwagon – one that was about to knock me down.

Looking back at the lead-up to the vote, I can see my mistakes so clearly. The first was that I thought the position we had developed in opposition, going back to Iain Duncan Smith's leadership, was a moderate and reasonable one, and therefore one that our MPs had got behind. The truth was that a lot of Conservative MPs had always disliked the idea of an elected house, and had only allowed it to become part of the programme because they never expected us to do it.

My second mistake was that I never sufficiently articulated the facts that the reform we envisaged was not just moderate and reasonable, but evolutionary. It would be done in stages. And if no one wanted to go beyond stage one, we could leave it there. One hundred and twenty elected peers out of five hundred – that was all they had to vote for, and it would mean we would secure boundary reform. Was that really the end of the world?

I was personally convinced that Lords reform would never get beyond stage one. As soon as the first intake of elected peers had one dinner in the Peers' Dining Room they would, I was sure, be seduced by the luxury of the place and the new status quo would assert itself in the corridors of power. I would get the mixed House, with elected and appointed elements, that I wanted.

But I didn't convey that message with enough force to enough people. Great leaders have to be great educators, I know that. But this was one issue among so many during what I'd call the 'year of horrors'. I simply didn't have the time to guide my colleagues through every twist and turn, and I'm not sure they had the inclination to listen or comply.

The third crucial mistake was that I never conveyed how important boundary reform was. Not only did Tory MPs feel little ownership of the 'Coalition Agreement' that had been thrashed out in 70 Whitehall; they didn't feel part of the bargains within it. They regarded boundary (and therefore Lords) reform as my issue, not theirs. It was the head–body problem all over again. Not enough people understood that boundary reform was hugely important to an overall Conservative victory at the next election.

And of course when it came to individual constituencies, the truth was more mixed. If boundaries changed, our present MPs might gain a more Tory-voting area (good), acquire a more Labour-leaning area (bad), or see their seat eliminated altogether (very bad). Some otherwise perfectly sensible Tory MPs saw the boundary changes as threatening the very identities of their constituencies. For instance, there was outrage from the new Conservative MPs in Cornwall that one of their seats would have to cross the county border and include a part of Devon. The horror. I had once joked before an interview with the veteran presenter of ITV West Country, Bob Constantine, that a constituency crossing the River Tamar couldn't be that serious a problem. 'It's the Tamar, not the Amazon for Heaven's sake,' was how I rather indelicately put it. They couldn't resist broadcasting my comment.

Fourth, I never showed how passionate I was about the whole thing. There were times I let my reasonableness spill over into being too accommodating and conciliatory. My colleagues couldn't see the belief, the passion: on this issue I didn't give enough of a lead.

People tend to assume that prime ministers surround themselves with yes-men and women. Not true. The chief whip, Patrick McLoughlin, warned in the days before the vote, 'If you're going around telling people that it doesn't really matter, don't be surprised if we fucking lose.'

On 9 July, Nick Clegg introduced in the House of Commons the second reading of the Bill and the programme motion. The second reading debate was due to last for two days. We had allowed a further ten days for debate in 'committee of the whole House', which was plenty. Labour, as expected, scented blood and wanted to inflict a defeat on the government by voting against the programme motion.

Even before Nick rose to his feet, backbencher Jacob Rees-Mogg stood up with a spurious point of order. Though he was from the new intake and still in his early forties, Jacob was a caricature of an old-fashioned Tory: double-breasted suits, cut-glass accent and socially conservative views. He was also a prolific parliamentary rebel. He tried to argue that this should be a hybrid Bill, which would take much longer to be passed. The speaker ruled it shouldn't.

Others were constantly rising to interrupt Nick's opening speech. It was maddening to see my own party sabotaging this – self-sabotaging, given the ramifications for the next election.

I sat down with Nick the next morning and made one last attempt to save the situation. I proposed that we ditch the programme motion, carry on putting the second reading to a vote the following day, and then try to get Conservative MPs on board – through a bit of retrospective education I should have done earlier – before we tried to take it to a third reading. If they agreed not to talk it out, we wouldn't need the programme motion. But I knew that if the rebellion on the second reading was big, that would be it for the reform. We would never be able to win enough MPs round, and the Bill would be killed with verbosity.

Later that day George Young, the leader of the House, withdrew the programme motion. At 10 o'clock that evening the vote on the second reading took place. It passed, but ninety-one Conservatives rebelled against the three-line whip.

I wasn't angry with the ninety-one. I knew the party was split on this issue. But I was angry with one MP who had misled his colleagues in order to manipulate the vote. Jesse Norman had texted fellow Conservative MPs saying, 'The PM desperately needs the Bill to be knocked off. There is no manifesto commitment now that the programme motion has been withdrawn . . .' He said that rebelling 'will help the PM . . . the government won't need your vote'.

With these words he had given the impression that I had sanctioned a vote against the Bill. My even temper left me right there, and when I spotted him sauntering through the Lobby I went up to him and said it wasn't acceptable behaviour or the action of an honourable Member, and that he should send a text round saying that it wasn't my view. He was deeply apologetic, and did clarify his remarks. But it was too late.

Oliver Letwin did his best over the summer to try to find a way of winning round enough of the ninety-one to make the proposal, or at least a version of it, fly again. But he couldn't make it happen. Lords reform was dead. So was boundary reform – and potentially our chances of a majority in three years' time.

What about the coalition? Was that dead?

The experience left us more hard-bitten, but we still wanted to make the coalition work. So these issues that brought us to the brink of divorce ended up in a renewal of our vows.

Oliver Letwin and I had been discussing for some time the idea of a formal stock-take of the Parliament so far, and an official look ahead to the second half. It developed into a plan to reboot the coalition by launching the 'Mid-Term Review' and 'Programme for Government Update'.

Our future plans included a new Single-Tier Pension which would reduce the unfair means test. Rather than getting £100 per week plus however much of the £40 pension credit you were deemed entitled to, you'd get the whole £140 regardless. There would be more provisions for childcare for working parents, balanced by the 2012 Budget's measures to taper child benefit for those earning over £50,000 and stop it for those over £60,000. We did, however, fail on one of the biggest long-term problems – capping the cost of social care for individuals. That is one of my greatest regrets.

On the evening of 10 January, three days after our relaunch (no Rose Garden this time, just the wood-panelled State Dining Room), Clegg came up to the flat for a drink in a really good mood. As I opened a bottle of wine, he told me his thinking. He was enjoying being in government; he felt he'd got the levers, and he now knew how to pull them. He had a stronger team around him, and despite the pressure from his party he wanted the coalition to go all the way to 2015.

His priorities for the year ahead would be nailing down all the main policies and shoring up his MPs to help each of them defend their seats based on incumbency. There was a slight worry that he'd lose his own seat in Sheffield Hallam. But he said that if the next election resulted in another hung Parliament, and it was possible to have another coalition with us, he would want to do that.

This wasn't the first conversation we'd had about this, but it was perhaps the most explicit. Usually the only time I'd air such thoughts was when occasionally the Quad met for dinner in the Downing Street flat. George and I would choose our language very carefully. We'd mention historical parallels, mainly the National Governments of the 1930s, and 'would not rule anything out'. Nick and Danny would respond by not ruling anything out either.

Whether that meant not fielding candidates against each other in some seats, or running on some kind of joint ticket, we never discussed in any great detail. But we knew that the coalition's task on the economy, on politics and on society was unlikely to be a one-term job, and that there would have been merits to extending our alliance.

'Would the Tories stomach it?' Nick wondered during that conversation in January 2013.

'I think it's difficult,' I said. 'I think we'd have to do it in a different way. We'd have to give Conservative MPs the chance to vote on the deal. Though it would of course,' I added, 'be a vote on whether they wanted to remain in government or go into opposition.' I said this with a smile, because I knew such a choice would inevitably lead to such a deal's approval.

I regret that we didn't find a way to deliver this, on their side or ours, and I often wonder what might have happened had I been able to do so.

27

Wedding Rings, Olympic Rings

It was a boiling-hot July morning in 1981 when my brother Alex, sister Tania and I packed away our sleeping bags after a night on the hard ground opposite Buckingham Palace.

We were thirsty and exhausted; none of us had slept a wink. But the atmosphere on the Mall was amazing. People from all over the world had come together for this strange summer sleepover, and now we were chatting, cheering and craning our necks to get a glimpse of the soon-to-be-weds going past in their royal carriages on the way to St Paul's Cathedral.

It was called 'the fairy-tale wedding', and for fourteen-year-old me the marriage of Prince Charles and Diana Spencer *was* like a fairy tale. I thought I had the best view on earth as I watched the beautiful young bride and her new husband whisk past us before appearing on the balcony of Buckingham Palace. Despite all that would happen in their marriage, it still shines out to me as a special day. And never could I have dreamed that, thirty years later, I would have an even better view of their son at his own wedding in Westminster Abbey.

I have always been a passionate monarchist, but never able to explain precisely why. A person's future should be determined by their talent and hard work, not by the accident of their birth – my whole political life has been dedicated to that meritocratic ideal. And yet here I am, ardently supporting an institution that is founded on hereditary privilege.

For me, it comes down to two things: patriotism and practicality. Steeped in our history, the royal family is a focal point for the nation. It embodies the most British of values – duty, tradition, stoicism – and projects those values around the world. It provides our country with a head of state, and gives us stability as a result. The boost to tourism and diplomacy is immeasurable, and it also brings us the gift of national occasions we can all share. It's no wonder so many republics envy us.

The three decades between these two royal weddings – Charles and Diana in 1981; Kate and William in 2011 – were a turbulent time for the monarchy, with multiple divorces, the Queen's 'annus horribilis', and of course the tragic death of that beautiful princess, who was still so young.

But my time as prime minister coincided with a time of recovery for the royals. There was the engagement and marriage of a future heir to the throne, and two births. There were also national milestones – the Olympic and Paralympic Games, the Commonwealth Games, the centenary of the outbreak of the First World War and the seventieth anniversaries of D-Day, VE Day and VJ Day – in which they played a key role. The Queen celebrated sixty years on the throne, and shortly afterwards became the longest-serving monarch in our history. In this age of change we value more than ever the continuity our constitutional sovereign brings. Far from there being uncertainty, the monarchy is here to stay.

I wanted to do what I could as prime minister to cement this position. Early on, George and I decided we needed to address their funding. Until 1760 the monarch met all official expenses from hereditary revenues, which included the profits of the Crown Estate (the property portfolio which today includes everything from London streets to coastal paths). George III agreed to surrender these hereditary revenues in return for an annual grant set by government, called the 'Civil List'.

While this was a sensible and enduring arrangement, in recent years it had meant a painful annual discussion on every aspect of royal expenditure, accompanied by a tabloid-led debate about whether individual members of the family were 'good value for money'.

All that meant that it was almost impossible to increase the money that went to the royals, even though some of their vital infrastructure was crumbling – and even though in the twenty years up to 2012 the royal financing had seen a cut in its budget of 50 per cent in real terms.

This was compounded by the fact that, for various complicated reasons, the Civil List had given them too much money in the early 1990s, which had then been saved during the course of the decade, and spent over the following decade. This meant that, come 2010, to coin a phrase, there was no money left. So in 2012, to put the royal finances on a more secure and sustainable footing, we scrapped the Civil List and replaced it with a 15 per cent share of profits from the Crown Estate. In the event, it turned out to be a generous settlement.

Soon we had the good news that the Queen's eldest grandson, Prince William, was to marry his long-time girlfriend Catherine Middleton. I had already got to know William and Kate quite well. They were a warm, charming couple, who genuinely loved each other and wanted to build a life together. I was thrilled for them.

The prime minister has an odd role in such national events. You're not directly responsible for running things – but if things go wrong, it's your fault. I got stuck in, overseeing everything from street parties (slapping down spoilsport councils that tried to stop the parties by preventing road closures) to security (chairing COBR meetings to make sure enough police would be on the streets).

Friday, 29 April 2011 was another hot and dry day – and we had made it a national holiday, just as William's parents' wedding had been thirty years before. Samantha looked stunning in a turquoise dress, and our car to Westminster Abbey followed the Household Cavalry along the gravel-covered Horse Guards Parade in a cloud of dust, their metal breastplates glinting in the morning sunshine.

We watched the ceremony from the quire stalls alongside Ed Miliband, Nick and Miriam. It was magical. The princes in their military uniforms. The twenty-foot trees lining the aisle. The feeling that you were watching something that could have happened in any one of the last ten centuries, and wouldn't have looked that different.

On that day, the country – indeed the world – came together. The thousands of street parties, the million people lining the route, the near-billion viewers watching the vows, the flypast and the kiss on the balcony on their TV screens. Our own street party outside Nos 10 and 11, which included local schoolchildren and residents of local care homes, is a particularly happy memory.

The palace, where we went after the ceremony, was familiar territory for me. Every Wednesday evening while Parliament was sitting I followed the same routine: arriving at the palace, walking down the corridor, trying not to step on the sleeping corgis, and being escorted to the Queen's study, where the small yet sturdy woman is standing to greet you. 'The prime minister, Your Majesty,' says the equerry, and you both bow. He then leaves, you walk forward and bow again as you shake the Queen's hand and say, 'Good evening, Your Majesty.' She points you to the usual chair. And from then on it's 'Ma'am' – as it says in the films – rhyming with 'jam'.

There were two essential bits of preparation for the famous 'audience with the Queen'.

One: always check the BBC headlines, in case you've missed something (I usually turned up just after the 6 o'clock news, and in any event, she is phenomenally well-informed).

Two: always check what's going on in the horse-racing world. A quick call to Tom Goff, my racing expert friend, would bring me up to speed on whether one of the Queen's horses had won that week, or another had recently had a foal. Her knowledge of the turf is prodigious. During a separate conversation, the week after my father died, the Queen said how sorry she was, and asked if his horse was running at Windsor that evening. It was. I had absolutely no idea about it, and was completely lost for words.

There is never anyone else in the meeting, just the two of you. She was better-informed about foreign affairs than many politicians, not least because she would see most British ambassadors before they left the country, and all foreign ones shortly after they arrived.

After the allotted hour, the conversation would come to an end. I often found myself leaving with greater clarity about the issues of the day and a greater resolve to tackle them.

And no matter how miserable the political scene or the news that day, I never left without something of a spring in my step. The Queen has that effect on you. Not surprising really – you've just spent an hour with one of the world's greatest public servants. Although I admit that a little edge was added to proceedings by the fact that, when things were going wrong, you remember that she started with Winston Churchill, I was her twelfth prime minister, and she had – quite literally – heard it all before. Whatever her political views – and she genuinely never gave anything away – it felt as if we were a team. I was a much better prime minister than I would have been without those weekly doses of wisdom.

It was at Balmoral, though, that I saw the Queen at home, informal and relaxed. How *can* it be informal and relaxed when you're in a castle in the Scottish Highlands? In early September Samantha and I would arrive for a weekend stay in the castle. I was often tempted, but we never took the children – the pressure would have been too great. Florence was still a baby when she first met the Queen, at a No. 10 lunch to celebrate Prince Philip's ninetieth birthday. She grabbed the Queen's pearl necklace and briefly refused to let go.

As the rhythm of our years was quite similar, the Queen and I would bond over our summer exhaustion. The difference was, I had been on that schedule for six years – she had done it for over sixty.

The visit to Balmoral always included a formal audience. We would meet in the Queen's study, full of pictures, letters, memorabilia and dog paraphernalia. I lapped up the Balmoral experience. Throughout the place there is the stamp of Queen Victoria, VRI ('Victoria Regina Imperatrix') – even on the wallpaper – and in the park and up into the hills are monuments to her children, to Prince Albert and to her dear friend John Brown.

Every year I was asked whether I'd like to fish for salmon, shoot grouse, ride one of the Queen's Highland ponies or go red-deer-stalking. I love doing all those things, but I had cut down on country sports after becoming Conservative leader in 2005. I had enough problems dealing with the 'posh' accusation without being photographed with a gun in one hand and a dead bird in the other. Still, I managed all the sports at Balmoral except the deer-stalking.

Alongside her racehorses the Queen has one of the best – if not *the* best – collections of Highland ponies. These hardy creatures can carry dead stags off the hill on their backs, and can work in all conditions. They are also fun to ride, and can pick a safe path up the steepest hills. The only problem is their broad backs. After two hours exploring the hills and glens around Balmoral I would be walking like John Wayne for a week.

One year Sam and I chose to go for a walk, and were given a map and a suggested route. The Queen is keen that you should have complete solitude, and Balmoral is the one place your close protection team is left behind as you head up into the hills. It was bliss to be genuinely alone for a few hours, and we wandered for miles, forgetting to consult the map. Eventually we saw a Land Rover and flagged it down to ask the way. The head of the Duke appeared out the window and he snorted, 'You're completely lost, aren't you Prime Minister?'

The pinnacle of the visit was Prince Philip's Bothy Barbecue. The Queen drives you at breakneck speed across the moor to a bothy – a stone hut originally built for deer-stalkers and shepherds. The Duke of Edinburgh is outside, tongs in hand, smoke rising from a row of sizzling grouse. And then the two of them, and whichever other Windsors are around at the time, cook and serve you dinner. Literally, the Queen of the United Kingdom and Commonwealth Realms topping up your drinks, clearing up your plates and washing up.

Trips to Balmoral were usually accompanied by visits to Birkhall, the beautiful house on the river left by the Queen Mother to Prince Charles. While I would see the heir to the throne on many other occasions, and started having regular audiences with him towards the end of my time at No. 10, it was at Birkhall that we had time for some longer conversations. His appreciation for art and design and his love of nature are everywhere to see. The house and gardens are immaculate, and a door to the garden is always left open so the red squirrels can come in and feed off a variety of titbits he has left for them.

While I was PM the controversy about the letters Charles wrote to ministers and prime ministers exploded, after the *Guardian* had used a series of Freedom of Information requests to get hold of them. I received a few of these letters – always in his own hand and always detailed and thorough – and had no sympathy with those who complained about them. Why shouldn't the heir to the throne write to ministers with suggestions, questions and ideas? And why should those letters be made public? In my view, he had a perfect right to ask questions, and to do so privately. Ministers of course had the right to reject his suggestions if they thought they overstepped the mark.

In my own experience, in both letters and conversations Charles restricted his enquiries and entreaties to subjects in which he was an expert. And when I say 'expert', I really mean it: he has a detailed knowledge, reads deeply and consults with authorities from all over the world.

I also have no truck with people who think that his strong opinions will make it harder for Charles to stay inside the correct boundaries for a constitutional monarch. He understands perfectly that the role of a monarch is different to that of an heir to the throne – and said so specifically to me on more than one occasion. His interest in these subjects shows how much he cares about his country and his people.

Did I act on his suggestions? In some areas – protecting our oceans from environmental degradation and acting against wildlife poaching in Africa – he did encourage greater attention and action. In others – like planning – he encouraged me to think more about the issue. And on some matters – complementary medicine, for instance – I would listen politely and say clearly where I agreed or disagreed. When the Supreme

Court ordered the release of the letters he'd written in 2004–05, many people were disappointed to discover that they were as I've just implied: uncontroversial and occasionally esoteric. After the election I sought cross-party support to protect confidential letters.

As an ardent monarchist I had been looking forward to June 2012, when Her Majesty would mark sixty years on the throne. It would precede another great national event, the Olympic and Paralympic Games just two months later. After the Omnishambles Budget, and with the slow economic recovery, a fractious coalition and restless backbenches, we – *I* – needed a golden summer.

Once again, we decided that it should be a national holiday. As I explained the idea to Her Majesty, I detected a modest reluctance, although of course she agreed that it was a matter for Parliament.

As ever, she was probably right: the Office for National Statistics is meticulous in marking down these days as reductions in GDP, and in this case the extra bank holiday possibly tipped the figures for the second quarter of the year into the negative.

The highlight of the Jubilee came on Sunday, 3 June, when a celebratory flotilla of a thousand vessels sailed down the Thames. Barges, cutters, sailing boats, gondolas, canoes, dinghies, even the Dunkirk little ships were led by the royal barge, *The Spirit of Chartwell*. And it poured with rain. Incredibly, the Queen and Prince Philip, aged eighty-six and ninety, stood on deck for four whole hours. Duty, tradition, stoicism – their profoundly British behaviour matched the profoundly British weather. I was at the end of the route at Tower Bridge to greet the shivering royals. The Duchess of Cornwall made a beeline for the cups of tea, and told me she had thought she would expire out there. No one wanted to go against the Queen's example, so everyone spent the proper amount of time on deck.

It was fitting that the Jubilee should fall in the year Britain was to host the greatest event on earth – the Olympics – the first time we'd done so since 1948.

John Major had set up the National Lottery and guaranteed a sustainable source of funding for UK sport, and from that acorn grew the mightiest team of competitors in our history. Tony Blair helped bring together the bid team, and did a great deal to help win. Gordon Brown helped to make sure the infrastructure was in place. Former banker Paul Deighton and construction boss John Armitt delivered the stunning

Olympic Park. And Seb Coe, who led the bid team and then the London Organising Committee of the Olympic and Paralympic Games (LOCOG), deserves more credit than anyone else for realising Britain's Olympic dream.

Yet as the Games grew closer, there was a big task left for me. The PM's role was – as ever – one of oversight and course-correction. On my first day in office I was given a briefing that left me in no doubt about where the buck stopped: 'The government carries the ultimate financial and reputational risk and has to deliver significant services to ensure the successful operation of the Olympic and Paralympic Games in 2012.'

I appointed a young civil servant, Simon Case, to lead the Delivery Team in Whitehall (and marked him out as a brilliant future principal private secretary – a job he would do for both me and Theresa May). And I made Jeremy Hunt secretary of state for culture, media, Olympics and sport, because I knew how good he was at gripping issues. With a successful business career behind him, Jeremy brought an easy manner and a great brain to politics. He was the most collegiate of all my cabinet ministers, a team player to his fingertips.

I buried myself in the details, receiving regular updates from Jeremy on every aspect of the planning – immigration queues, ticketing, road closures, everything. We held monthly stock-take meetings, and come 2012 turned this into a proper cabinet committee. When I was given a note by a private secretary tipping me off about Theresa May and the Home Office finding Jeremy's interference aggravating, I replied, 'Hunt is right to be aggravating.' I totally had his back.

I was determined to do more than just get through the Games, tapping the ball into the net after the goal had been set up. Our party had promised in its manifesto to bring about lasting benefits for the country as a whole, and I wanted to wring every conceivable advantage out of this huge opportunity.

Transforming east London, where the Games were being held. Igniting the spirit of volunteering. Bringing more trade and investment into the country. Boosting 'brand Britain' around the world. Nurturing the medal-winners of the future. Demonstrating that Britain could deliver big, transformational projects. And getting more people, particularly young people, into sport.

But in the lead-up, security was my biggest worry. The murder of members of the Israeli team at the Munich Games in 1972 has hung over every subsequent Olympics. We couldn't take a single chance with safety, which meant deploying 15,000 police officers on peak days. It meant having the helicopter carrier and assault ship HMS *Ocean* berthed in the Thames. I even went to visit the Puma and Lynx helicopter pilots and their teams who had the responsibility of shooting down any terrorist light-aircraft attack. We had snipers on the roofs of buildings and anti-aircraft missiles on the top of high-rise blocks of flats.

G4S had been contracted to provide guards for things like bag searches and door security. The original contract had been for 2,000 staff, but they had agreed to increase this to 10,000 when it became clear that many more would be needed. But on 11 July 2012, two weeks before the Games were meant to begin, they revealed they couldn't provide anywhere near that number.

Total panic. We were hosting the world's biggest event, and we hadn't got enough security guards. Thank God there were troops on standby, but there were huge disagreements about how many of them should be deployed. Listening to all sides, and hearing the clock ticking down to the opening ceremony, I had one of my bang-the-table, cut-the-nonsense moments. I insisted that 17,000 troops would be deployed overall. I didn't care what it would cost or how 'sending in the army' looked. Security was non-negotiable. As ever, our service personnel were amazing, and the public loved their presence, which added to the patriotic atmosphere.

Finally, the day of the opening ceremony came. Two years earlier the great film-maker Danny Boyle, top director and producer Stephen Daldry and Seb Coe had come to Downing Street to present their creative vision. We couldn't outdo Beijing on budget, pyrotechnics or scale, but we could wow the world with something even more special: culture, charm and creativity. They showed me the mood boards, they played me videos, we even went to visit their studios. I loved it – this magical musical tour of Britain so ambitious it would make a West End production look like am dram. I said I thought there needed to be more national pride – more Churchill and Battle of Britain. And I immediately vetoed one of the more bonkers ideas – a section featuring Gerry Adams and Martin McGuinness.

They told me about one brilliant, top-secret sketch idea that might be a tall order. I told Her Majesty I really thought she should do it. She and James Bond are two of the best things about Britain, and bringing them together would be fantastic. It wasn't tacky, it wouldn't dumb down the monarchy, it would be really brilliant . . .

The Olympics opening ceremony was an occasion so daring it surprised everyone. The centre of the stadium was covered in grass, with children playing, actual livestock grazing and a choir singing 'Jerusalem'. Suddenly, the bucolic scene was swept away as great chimneystacks pushed up through the ground to the sound of a thousand drummers, heralding the arrival of the Industrial Revolution. Kenneth Branagh played Brunel and read from Shakespeare. Steelworkers forged giant Olympic rings.

Parades and performers poured onto the stage: Sgt Pepper-era Beatles, *Windrush* immigrants, First World War Tommies, Suffragettes, Chelsea Pensioners, colliery bands. As the Red Arrows flew overhead and the London Symphony Orchestra blared out Elgar's 'Nimrod', I thought it doesn't get madder or more British than this.

Then there was an act devoted to the health service, featuring children, nurses and doctors from Great Ormond Street Hospital. The lights dimmed and the illuminated beds spelled out the initials 'GOSH', then 'NHS'. It was particularly moving for Samantha and me, as we had spent so many days and nights in that extraordinary hospital with Ivan.

And then it happened: a short film was played on the big screens, featuring Daniel Craig as James Bond. He wanders into Buckingham Palace, along a corridor, avoiding the corgis, and meets a person, her back to the camera, who you first think must be an actor playing the monarch. 'Good evening, Mr Bond,' she – the actual Queen – says, before they depart the palace and head towards a helicopter. Suddenly there is an actual helicopter above the stadium, and two people looking remarkably Bond- and Queen-like parachute in as the 007 theme tune plays. Moments later, the Queen and Prince Philip appear next to Sam and me in the VIP box. Absolutely brilliant.

The ceremony then took us through the decades, via a celebration of British film and pop music. I wondered what the Queen was making of it all. Was she a fan of Led Zeppelin? Did she like the Eurythmics? The Happy Mondays? Dizzee Rascal, who came on to sing 'Bonkers'?

During the ceremony I felt as if all my problems had melted away. But something always turns up. In the following days it was the fact that there were vast numbers of empty seats at the events. How could that be? People had entered ballots, paying top whack, praying they could be a part of this. Who the hell had a ticket and wasn't turning up?

Well, the rules state that a certain number of tickets must be allocated to the media, athletes and the all-powerful IOC, together with large numbers allocated to national Olympic organisations. So many officials and dignitaries from the IOC show up that it's like an invasion by a foreign country. The problem was that many of these dignitaries could not be bothered to attend the early rounds of Olympic events, while many national organisations had not sold their full complement of tickets.

The Olympic rules also stipulate that host countries must do things like seal off entire traffic lanes for VIPs, and change local bylaws so as to prevent protests. It's absurd, but you have to play along with it. But I wasn't prepared to tolerate this seat-allocation rule. It wasn't a nuisance, like the roads, it was a scandal. I made sure – another cut-the-nonsense decree – that any empty seat would be offered to a schoolchild or a soldier.

I wanted to be there in the stands too, cheering on Team GB. I thought it was my duty, although I confess it wasn't one that I performed reluctantly.

But it didn't start well.

First up, I watched the men's cycling road race on the Mall. Britain's Mark Cavendish was tipped for gold, but sadly he finished twenty-ninth.

Then I went to the Aquatics Centre to see our gold hopes Tom Daley and Peter Waterfield in the synchronised diving – and they came fourth. As if to rub it in, François Hollande watched his team win a surprise gold in the pool.

I watched the judo with Vladimir Putin (he knew the sport so well that he told me a move would be overruled before the referee even declared it so). Russia's best competitor won gold in his weight division, ours lost in the final to the USA in hers.

The press had a new narrative: this was the Curse of Cameron. If I was in the stands, Team GB bombed. It was superstitious nonsense, of course, but weirdly, on the Thursday we did find ourselves actually debating whether I should go to the velodrome for the cycling men's team sprint and risk fuelling the 'curse' charge. I sat with Princes William and Harry, and felt doubly nervous. I didn't need to worry. We flew out of our seats as Chris Hoy, Philip Hindes and Jason Kenny pedalled to victory.

Then the golds started rolling in. On the Saturday – Super Saturday, as it became known – I watched as Team GB won three gold medals within a single hour: Jessica Ennis in the heptathlon, Mo Farah in the 10,000 metres and Greg Rutherford in the long jump. I thought to myself: there is nowhere better on earth than this place, at this time. Whatever else happens in my life, this is a magic memory.

I was living the Olympics day and night. With the beach volleyball being held in Horse Guards, which backs onto Downing Street, Sam and I could hear the whole thing from our bedroom. We would fall asleep each evening to the sounds of cheers, gasps and the steady thud of the sound system. 'Can you hear us, Prime Minister?' came an announcement over the Tannoy one night. 'Yes, I bloody well can,' I said aloud.

On the final day, I was invited to speak to all of our medal-winners in what was known as Team GB House. They were superstars. Third in the medals table, ahead of huge sporting powers like Russia, France, Germany. Compare that with 1996 in Atlanta, when we came thirty-sixth.

The previous summer, images of London ablaze had beamed around the globe during the riots. This summer, the world had seen the city lit by the Olympic flame, and our home team – many the same age as the rioters, many from similar backgrounds – showed what hard work, patience and discipline could achieve, and what being British was really about.

I wanted them to know what power they held, beyond their excellence in their own field. 'Every school you visit, every playing field you go to, every school assembly you address . . . you can change children's lives,' I told them. I was given a Team GB kitbag as a memento. It became my travel holdall, and for the rest of my time as Prime Minister I jetted around the world with this piece of London 2012.

In the 120-year history of the modern Olympics, ours can make a claim to being the best ever. But there is no doubt that our Paralympics *were* the best the world had ever seen. From the beginning I wanted to make sure that the event, which took place two weeks after the Olympics ended, were treated with the same enthusiasm, credibility and respect as the other Games.

I needn't have worried. Ticket sales broke records. In the nation's eyes Sarah Storey, Jonnie Peacock and Ellie Simmonds were every bit as heroic as Mo, Jess and Greg. Indeed, my highlight of the entire summer was presenting the seventeen-year-old Ellie with her second gold medal.

If anything, the Paralympics felt even more special than the Olympics. We were watching people who had probably been told from a young age all the things they *couldn't* do showing exactly what they *could* do.

Samantha and I felt that awe very personally, and we were particularly captivated by the wheelchair basketball. I think we both noticed how much changed that summer in attitudes towards disability. As I would later put it, when I used to wheel Ivan around in his wheelchair, I had always thought that some people saw the wheelchair, not the boy. Because of what happened in London in 2012, I think that today more people would see the boy and not the wheelchair. Perhaps that is the greatest legacy of all.

And it was a wide-ranging legacy. In the run-up I had thought, if we can get the world to unite over sport like this, we can get it to unite over the next big challenge in aid policy: global malnutrition. With so many big names in town, I convened a Hunger Summit at Downing Street, co-hosted with the vice president of Brazil, Michel Temer, whose country had made huge strides in tackling hunger, and would host the next Games in 2016. The stars of the event would be the Brazilian footballing legend Pelé and the British track legend Mo Farah. Because the traffic was so bad, even with VIP lanes, I had to order the police to get Mo there by motorcade.

Another legacy was the rise in volunteering. So many people were inspired by the 70,000 Games Makers who directed traffic, collected tickets, offered first aid and marshalled athletes, all for free. Forty per cent were volunteering for the first time ever. It was the biggest peace-time volunteering recruitment drive in modern history.

There was a massive economic legacy, too. There were parties at Downing Street for investors every night, conferences every day, receptions, investment forums, even an economic summit at Lancaster House on the eve of the Games. Commercial diplomacy in practice.

We had talked for a long time about wrapping up the Games and the events surrounding them into one overarching campaign. Steve Hilton had been key to this, and we'd signed up a creative agency, Mother, to help. The idea was a campaign called 'Britain is GREAT' – a single brand that could be used all over the world to push what we had to offer, capitalising on the Olympics. 'Countryside is GREAT', 'Fashion is GREAT', 'Sport is GREAT'. It went everywhere, and reaped £1.2 billion for our economy. I still see it around the world today.

And there was the sporting legacy. I was determined not to be like previous host nations, which had plummeted down the medals table at subsequent Olympics after the money and the interest had departed. So on the final day of London 2012, we announced that we would continue elite sport funding right up to Rio 2016. The results would be seen four years later, when we came second in the gold medals table, beating our tally in London.

One other thing I'm proud of. While you see many tragic photos of former Olympic sites falling into disrepair, I was determined that the story in east London would be quite different. I spent almost as much time in the run-up to the Games focusing on the legacy as on the event itself.

Once among the most deprived parts of the capital, the Olympic Park is now a go-to destination. Its digital cluster complements nearby Tech City. It houses a hub of art and design. The adjacent Westfield is Europe's largest shopping centre. The Victoria and Albert Museum is planning to build a new site there, as is University College London.

The good cheer of 2012 didn't end there. In December we heard the fantastic news that Kate and William were expecting a baby. The rules of primogeniture dictated that if that baby was a boy, he would be king one day. If it was a girl, she would not be queen, unless she had no brothers. Which was sexist and outdated. But we were one step ahead. Two months earlier I had attended the Commonwealth Heads of Government meeting in Australia. I arrived in the middle of the night, hardly knowing what day it was. It was one of those trips on which we spent more time in the air than on the ground. Luckily, though, the verdict – from the Solomon Islands to Papua New Guinea – was unanimous: the law of succession would change. The Cambridges' baby would reign over Britain and the Commonwealth, whatever his or her gender. It also ended the ban – which was, amazingly, still in existence – on a Catholic becoming the British monarch.

It may appear strange that my favourite four weeks as prime minister came during such a difficult time for the government. But the Games weren't just an antidote to my own malaise. They seemed to be an antidote to so much that was wrong in our country. To the social breakdown we'd seen in the riots, proof that young people were a positive force. To the bleakness in the economy, proof that we had a brilliant brand the world wanted to buy into. To fears about integration and national identity, proof that we could all come together as one nation, and get behind the most diverse Team GB in history. Race by race, medal by medal, it felt as though Britain was stepping up onto the global podium.

Perhaps most important of all, they showed that we, the UK, could still do the big, bold, transformational things. We could still wow the world. We had no shortage of ambition. One of the core ideas of my politics, that our best days are ahead of us and not behind us, found its emblem in the Olympics. That belief hasn't faded. Indeed, I don't think Brexit should alter it. Having a strong relationship with Europe outside the European Union doesn't diminish or damage the determination, pride, openness and warmth we displayed in that glorious summer of 2012. Indeed, it is through those things that we will flourish in the future.

28

Resignations and Reshuffles

So often when a government, or even an opposition, hits a bad patch, a reshuffle is suggested as a kind of miracle cure. Becalmed in the polls? Unpopular leader? Strategy not working? Economy not growing? It doesn't matter what the problem is, the cry goes up: 'Shuffle the pack!' Not least because the press – and, to be fair, most MPs – love the drama of who is up or down, in or out.

I was determined to do things differently. Whether in opposition or government, I kept shadow and cabinet ministers in post for proper periods of time. The same chancellor and home secretary for six years. The same foreign secretary for four years. Many of my ministers went straight from the roles they held in opposition – often for the full five years – into the same role in government.

In my view, the only arguments for a reshuffle were if it improved the way you governed, helped you manage the parliamentary party, or if it gave new MPs the experience they would need to hold senior office later.

However, I did accept that in politics if you do have to make changes, it is better to do them all at once. Anything else can lead to a sense – in the press to begin with, but more widely as well – of permanent instability. But I didn't rate reshuffles as a way of relaunching a government or solving political or presentational problems. The central point in a reshuffle, as far as I was concerned, was getting round pegs in round holes – people who would do the job well in the right positions.

The mix mattered too. You need the right spread of male and female, young and old, old hand and newcomer, left and right, north and south, Eurosceptic and Europhile. This isn't just a matter of presentation: diversity makes government better. The cabinet also needs to reflect the parliamentary party, some of whom are only satisfied if they have their man or woman representing their strand of conservatism at the top table.

There had been some shuffling already that year – though not from the Conservative side. For some time Chris Huhne had been facing allegations that he had made his wife take the points for a speeding offence when he had in fact been driving. Nick and I discussed what should happen, and decided that he should remain in post unless he was charged. In February 2012 he was charged with perverting the course of justice, and agreed to leave cabinet while he faced trial. His fellow Lib Dem Ed Davey took his place.

There had been some shuffling in my team, too.

As in opposition, Steve Hilton's ideas continued to be one part brilliant to several parts bonkers, the latter of which included cutting the civil service by 70 per cent, scrapping maternity pay, closing all Jobcentres and – no joke – introducing cloud-bursting technology that would, he claimed, stop it raining over Britain. True blue-sky thinking.

However, his relationship with people in government wasn't working. He was no longer excused as a free spirit when he was late for meetings – he was seen as someone who had disregard for others. His antagonistic style was no longer helping him advance his cause in Whitehall – it had started to hurt it. And because we were friends, people felt that when he hit an obstacle he would go directly to me, and undermine their legitimate concerns.

And the relationship between the two of us became strained, too. Steve is a real ideologue in a way I'm not; I'm ideological *and* practical. This important difference was exposed now that we were running the country, not just talking about running the country. He thought I was losing my radical zeal and falling for the trappings of prime minister. But I knew that to be a successful radical you have to play the game. And he wasn't interested in playing the game, just tipping it over and throwing the pieces all over the floor.

He had told me a while earlier that he needed a break, and now he'd found the ideal moment: his wife, Rachel Whetstone, was working in California for Google, and he thought he would be able to join her there, lecture at nearby Stanford University, and pursue some other business and political interests. He came to tell me his plans one weekend at Dean. We sat on the floor in front of the fire and talked it through. I was sad to see him go – he was a creative thinker, and his energy had helped to get a lot of things done. But disruptive forces like Steve have their pluses and minuses – for every initiative he boosted with his zeal there was an idea that misfired, or a relationship that subsequently needed repairing.

He had given me incredible loyalty and service for well over seven years, and was starting to think about his own political future. He was a strong believer in the city mayors we were creating, and he wanted to be one himself. Maybe in his home town of Brighton, or perhaps even, in the future, in London. It was announced on 2 March 2012 that he would be leaving in a couple of months' time to go on a 'sabbatical'; but there was little expectation of a full return.

On 2 May Steve departed for America. He left me a long note – a 'valedictory telegram', I called it – detailing his frustration with what he saw as the lack of progress and the limited scale of our ambition. I objected to his implication that we were not pursuing a radical agenda. Indeed, the full reshuffle I was planning was all about putting the right people in place to deliver that agenda.

The reshuffle, which took place in September 2012, was, I believe, probably the longest and best-prepared in living memory. The last time a prime minister had gone that long without shuffling the deck was Thatcher in 1981. Its purpose was to sharpen our focus, and the new line-up would do that in four key ways.

It would support our economic strategy, strengthen weak flanks, get the party into better shape and demonstrate that we were bringing talent through the ranks. If I wanted to end up with secretaries of state from the 2010 intake by the end of the Parliament, I couldn't promote them straight from the backbenches to the cabinet; they needed time in junior ministerial positions. That meant shifting lots of people up a level – and that in turn meant shipping out some of those near the top. You can only promote people if you fire others.

Lots of ministers, including the chancellor, home secretary and foreign secretary, would all remain in post. I had already consulted Patrick McLoughlin about a move during the summer. He had been loyal and hard-working, but I wondered if a change might help with the management of the party. More support for the coalition, rather than just tolerance of it. More backing for my brand of conservatism, and less prominence in Parliament for its opponents. But I didn't want an unhappy former chief whip – and Patrick deserved a promotion. I had just the role for him. I offered him secretary of state for transport, and he was elated. We get so used to egos in politics, and you can't blame people: they've had to blow their own trumpets to get ahead. But here was a man with more ministerial experience than most of us, behaving with more grace and modesty than anyone.

He said, 'You may notice that I don't ever give you handwritten notes even though I'm chief whip. The reason I don't do that is I'm not sure I'd spell everything right. I left school at sixteen; I'm entirely self-taught. I know about the Whips' Office, and I know what I think about Transport . . .'

Not only was he happy with the move, he was happy with who I had in mind as my new chief whip. I had asked Andrew Mitchell into my office before the summer holidays, and told him my thinking: 'You've done a brilliant job as international development secretary – and now I'd like you to be my chief whip.'

While Andrew had run David Davis's campaign in the 2005 leadership election, he had become a good friend and one of my most committed secretaries of state.

There was more than a touch of Marmite about Andrew: colleagues either loved or hated him. He had rubbed some people up the wrong way, and been abrasive in the past (his nickname was 'Thrasher', which said it all). But I knew he would be effective – and, I believed, respected. After two major defeats on Europe and the Lords, the Whips' Office needed a Thrasher. Andrew had played 'good cop' in DFID for years – it was time for him to resurrect his inner 'bad cop' (this would soon prove a particularly bad analogy).

I thought he would be delighted with the idea. But he was very committed to his current DFID projects, and was in two minds. 'I'd love to be your chief whip,' he said, 'but I'd want to do it in six months' time.' Which wasn't really the deal. Indeed, some key parts of the reshuffle were resting on this appointment. I said I'd cook him dinner, take him through the whole of the reshuffle, and he could help me shape it.

That was the clincher: the chance of being in control. Even so, when we sat down again he confessed he was still reluctant. 'You've got to realise that I was a whip in the past, and that appealed to the dark side of my character. Then you made me shadow international development secretary in opposition, and that made me a better person. It changed my life and changed the way I think about things, and I think I've done it really well.'

He had done it well. Brilliantly, in fact.

'It's changed me,' he said, 'and now you want me to go back to the dark side.'

'Yes,' I said. 'I'm afraid I do.'

I wasn't just clear about the appointments I wanted to make. I was clear about how I wanted to play it. In some previous reshuffles, people had endured the humiliation of walking up Downing Street waving to the cameras, then shuffling out of the back door half an hour later after being sacked. I decided to do the decent thing and hold the difficult conversations in my Commons office on the Monday evening before the reshuffle was properly under way and then make the appointments in the Cabinet Room on the Tuesday morning.

First in the queue for my Commons 'bloodbath' was Ken Clarke, an effective lord chancellor and justice secretary for the past couple of years, and a great help in shaping our economic programme. But I just couldn't see him driving through the innovative reforms needed to alleviate the overcrowding in our prisons. It would also become an overcrowded cabinet unless I shifted things around.

He sat down, and I offered him a whisky. With Ken you know he is going to talk for ages, so you might as well turn the whole thing into a proper chat. I said I wanted him to remain in cabinet, but without a department, to continue to lend weight to our economic reforms. I thought he would stay – once Ken is committed to something he is loyal and dependable – and I was delighted when he said yes. He explained – using almost exactly the same words as when he had agreed to join the shadow cabinet years earlier – that he liked me and George, he thought we were doing a good job, and it was an interesting time to be in politics. He trundled off cheerily. An excellent start.

Then Caroline Spelman, my secretary of state for environment, food and rural affairs, came in. I launched straight into my spiel. 'We've worked together a long time, but I have to make changes because I've got to promote new talent. So I'm going to ask you to step down from the cabinet.'

She said, very legitimately, that there had been no performance appraisal – no one ever says if you're doing a good job or a bad job. And the thing is, she hadn't done a bad job. It would have been unfair to lay all the blame for the forestry fiasco at her feet, and she'd done good work brokering a vital international deal on biodiversity. But I felt I needed someone with stronger political support, both in Parliament and outside, and I thought a stronger champion for farming and rural affairs would help the government through what would be a difficult patch.

Then came Cheryl Gillan. As I've mentioned, once we had some Tory MPs sitting for Welsh constituencies it would make sense to have one of them as secretary of state. Cheryl represented a Buckinghamshire constituency, and so was at a clear disadvantage when dealing with Welsh MPs and the Welsh Assembly. I thought she would understand that. I wasn't prepared for her tirade. 'You're saying I just don't fit in, do I?' she began. 'I'm old, I'm a woman, you're not offering me anything else. This is appalling.' I didn't know what to say. It was awful.

George Young, the leader of the House, lived up to his nickname 'Gentleman George' when I asked him to step down. He simply said, 'I've loved working here, and I've enjoyed the job. You promised me two years, and I've had more than two years. I quite understand.' He had given phenomenal service. Indeed, at seventy-seven he is still serving as a minister in the House of Lords today.

One extra river I wanted to cross was moving IDS out of Welfare and into the now vacated Justice Department. I thought he would be good at prison reform because he understood the issues and cared about improving our prison system. Plus, his obsession with Universal Credit and his consequent failure to implement other welfare reforms was becoming a bit of a roadblock. But he still provided important balance to the cabinet.

I thought the best way to get him to accept was, rather than offering him the prisons role and nothing else – and risk him becoming a disgruntled troublemaker on the backbenches – to give him the choice: stay or move. I talked up the new role. 'It's the most senior job in the government, and it's the other half of the whole broken society agenda that you've pioneered so successfully,' I said. He seemed tempted. I thought: the fish is halfway in on the line.

Then one of those strange flukes of politics intervened. The idea of IDS moving from Welfare to Justice was floated by Danny Finkelstein on *Newsnight*. Danny had no idea about the plan, and was simply speculating. But because everyone knew he was a close friend of both George and me, it looked like a Cameron–Osborne plot to lever IDS out of Welfare – and the plan was scuppered. Iain said he wanted to remain at the DWP, and I had to decide whether to keep him there or risk letting him go altogether. I didn't take that risk. On reflection, that was a mistake.

Finally there was Sayeeda Warsi, the party chairman. I thought she would be all right, because she knew we needed an MP in that role. She was always at a disadvantage doing the job as a peer. A bit like Cheryl, that was a simple fact that wasn't her fault.

I said: 'I want you to have a really good career in politics, and I think the right thing for you to do now is to be in a department. I think the best department for you is the Foreign Office. There's a great job to do there which involves the Commonwealth and Central Asia. I also want you to stay in the cabinet, and I'd like you to do the Women and Equalities job, because that needs to be done by a cabinet minister.'

It was an amazing role, and I could see her thriving in it. I had been looking forward to a constructive conversation and a happy future together. Instead, she objected to every part of the role and was very difficult to deal with.

She left without us resolving what would happen next, and went into a rather strange overnight phone negotiation with Ed. He found a neat solution: she would sit in two departments – the Foreign Office and the Department for Communities and Local Government. At the Foreign Office she would be responsible for Central Asia, but not the Common-wealth. She would carry out the cohesion job – focusing on faith, where she would excel – rather than the Equalities job.

But even with all these roles, the position wasn't resolved until the morning, when Ed's final offer made her turn her car around and come back (a politician making an actual U-turn). She was given a seat on the NSC and the title 'senior minister of state', which although it hadn't existed on the ministerial ladder (parliamentary private secretary, parliamentary undersecretary, minister of state, secretary of state) until that moment, illustrated that she was number two in the FCO and the DCLG.

Tuesday was a much happier experience. In they came, one by one. Chris Grayling was delighted at being justice secretary. Theresa Villiers became secretary of state for Northern Ireland. Maria Miller, who had done a great job as disabilities minister, went to the Department for Digital, Culture, Media and Sport – she would carry through my equal-marriage plans. Owen Paterson, who was passionate about the countryside, went to Environment, bringing more of a right-wing balance to the cabinet.

And Jeremy Hunt, one of my smartest minds and safest pairs of hands, went to Health. Grant Shapps became chairman. He was loyal, energetic, and really wanted it.

The more earnest Michael Fallon was also considered for the chairmanship – as deputy chairman he was regularly sent out onto TV to fight any fire going, and was very effective at it. But I wasn't yet sure he would be 100 per cent on my side in a crisis. Instead he went to the Business Department as the new business and enterprise minister.

Because of Patrick's move to Transport – and I was really keen to get a Midlands MP into that job – there would have to be a sideways move for Justine Greening. There had been some talk in the press about her objection to a third runway at Heathrow Airport (it affected her constituency of Putney) making her position at Transport untenable. But that wasn't my reason for her move. The airport decision had been delayed anyway, with the inquiry reporting at the beginning of the next Parliament. My motivation was simpler. It was not so much that I wanted her out, but that I wanted Patrick in.

I said my piece to Justine across the cabinet table. The role on offer would be good for her career development, and was certainly not a demotion. DFID was a massive foreign policy job, with a protected budget and a cause close to my heart. She was a modernising Tory who believed in the aid programme and had taken part in Project Umubano in Rwanda.

But a completely unexpected reaction followed. She said I was making a huge mistake, her volume increasing as she spoke. But eventually she assented. When I saw her at cabinet the next day she was completely calm, and said, 'I think I owe you a note.' 'Of course you don't,' I said. 'It's all right; forgotten and forgiven.' But the truth is that you never completely forget a reshuffle experience like that.

My cabinet was finally complete. Now it was time to appoint forty-two junior ministers. But once again, before I could hire, I had to fire. That's why I began the following morning back in the Commons, giving out bad news again. I saw everyone I let go, except the agriculture minister Jim Paice, who I had to speak to on the phone. He was reason personified.

The whole process of identifying – and then seeing – ministers for the chop is soul-destroyingly awful. There are one or two who clearly aren't cutting it, but on the whole you are dispensing with people who work hard, enjoy what they do and expect to go further.

And I made some bad mistakes.

Nick Gibb had enthusiastically backed my leadership campaign, and he'd been a brilliant junior education minister under Michael Gove. He was fixated on returning schools to using the phonics method of teaching children to read, and I had backed his work at every turn. And we turned out to be right. Studies show that children learning by the phonics method are two years ahead of their peers.

But while I admired Nick's work, I had bought into the theory that if you are a minister and you are not destined for cabinet, at some stage you have to go. Good rule, but terrible example. I should never have moved him, and I made good my mistake twenty-two months later when I gave him his old job back.

The consoling thought during all this bloodshed was that people go down, but they also come back up. Also, many who had missed out because of the coalition were now being given a chance. Experienced MPs – like Andrew Murrison, a former doctor in the Royal Navy who had come into the Commons at the same time as me and had served on the front bench in opposition – now had the chance to serve in government. He had sat patiently on the backbenches, and was given the chance to be a very effective defence minister, leading on the First World War centenary commemorations.

Another upside was more round pegs in round holes. Nick Boles, who had written and spoken convincingly about the need to build more houses, would do Planning. Matt Hancock, a former Bank of England official who had done much to help policy-making in opposition, would sort out Apprenticeships. Mark Harper, who had demonstrated his effectiveness at getting things done and managing tricky politics when working as Nick Clegg's deputy, went to Immigration.

As with every reshuffle, not everything went like clockwork. There are so many moving parts that you always end up with a colleague without a job, or a department with too many ministers.

Next, we heard a rumour that Home Office minister Nick Herbert was so cross he wasn't being promoted that he had cleared his desk and was going to walk out of the government altogether. I asked Ed Llewellyn to ring him up and find out if this was true, because if it was, it would solve a problem for me. It was true. Nick had done a great job as prisons minister and policing minister during the police and crime commissioner elections, and would later work incredibly hard on the EU referendum campaign. He was also a friend and supporter. But for me at this moment during a tricky reshuffle, he was a round peg in a round hole marked 'exit'.

'Brutal, bloody, horrible, but completed . . . and completed in pretty good order,' is how I described the reshuffle at the time. It may not have been fun, but the job was done.

You don't just lose ministers through reshuffles, you lose them through resignations too. I would face several. In every case the resignation was preceded by a press furore, and followed by endless debates about whether the saga was well handled. But nothing hooked the newspapers like the saga of Andrew Mitchell.

For all his talent, before Andrew had even started as chief whip he was pulling rank. But on 20 September 2012 I found out it hadn't only been political staff he'd come to blows with. When Craig forwarded me an email from the press office saying there had been an altercation at the Downing Street gate, I wasn't entirely surprised to find that it was Andrew Mitchell who had displayed a bad temper to the police officers.

The following day, the *Sun* said it had the full story. It alleged that when a police officer refused to open the main car gate for Andrew, who was on his bike, he had said, 'Best you learn your fucking place,' and called the officers 'fucking plebs'.

Ed, Craig and Jeremy Heywood checked the CCTV footage, and saw that there clearly *had* been an altercation, after which Andrew had been ushered out of the pedestrian gate and had cycled off from there. I summoned Andrew to my office, and he admitted that he had lost his temper and used the f-word, but he swore he didn't say the word 'plebs'. I thought right from the start that swearing at police officers is bad enough, but if he had said 'plebs' it would have been truly appalling.

The police logbook from the gates was then leaked to the *Telegraph*, and its contents published four days later. It alleged that Andrew indeed swore and said 'plebs'.

He maintained that he had said something like 'You're meant to be fucking helpful.' I suspected the truth lay somewhere in between. I was sure he had said more than just that one sentence, but I didn't believe he had used the word 'plebs'.

I phoned Andrew from the flat and said, 'Look, I want a public apology, and a grovelling one at that. "I'm incredibly sorry, it should never have happened, it will never happen again, but I'm going to get on with my job."'

When the moment came, he apologised for showing a lack of respect. He said he'd had a difficult day, but that he hadn't said the words that were attributed to him. It was one of the worst political apologies in the history of political apologies.

Over the following weeks the rather limited inquiry I had asked Jeremy Heywood to undertake was inconclusive. I didn't want to sack Andrew – I'd just appointed him, and he had (sort of) apologised for the swearing, and denied using the loaded word 'pleb'. And I believed him.

The most peculiar incident of the whole saga was the arrival in my office of an email purporting to be from a member of the public. This person – a constituent of John Randall, who happened to be deputy chief whip, and to whom the email was sent – claimed to have witnessed the entire episode, and said that he and a crowd of people had heard the word 'pleb' being used.

John Randall was despatched to find out who this person was, and he reported back that the man was a genuine, if rather uncooperative, constituent. And this is where I really failed: we should have taken this opportunity to launch a proper full-on inquiry into what was going on. Andrew was absolutely convinced that he had been stitched up: by the police, who wanted revenge for cuts, and by the papers, which wanted blood over Leveson. It didn't help that he had long been going in and out of the gate, running late for cabinet, demanding that the officers open it for him.

The toll of all this on Andrew was immense – he lost a stone in weight. Meanwhile, the story wasn't melting away; members of the Police Federation were even wearing 'PC Pleb and Proud' T-shirts at demonstrations against funding cuts. Andrew was losing the support of Conservative MPs, and because a chief whip should avoid the press and cameras, he was severely constrained as to what he could do, including attending the party conference. This was an unsustainable situation.

So, in the end, it became clear what had to happen. Andrew came to Chequers on 18 October and handed in his resignation. It was amicable.

I know I should have had some sort of proper inquiry at the time. But trying to demonstrate that I had a sense of proportion was an important part of my approach. Given the importance both of the chief whip and of respect for the police, however, that was a misjudgement.

After Abrasive Andrew it was time, I thought, for Gentleman George. So I rang George Young and offered him the job of chief whip (proof that anyone can make a comeback, at any time of life). Typical George, he was out on a Friday night speaking for a parliamentary colleague. He accepted the job with delight and to my relief.

But that was not the end of it for Andrew, or for me. The week before he left, he produced a recording of a meeting he'd had with the Police Federation which contradicted its public statement on their discussions.

After that, he was truly on the rampage. I said to him in one meeting, 'You can either just accept what's happened, go away for a bit, work hard in your constituency, and I'll try and find some way of bringing you back' – I mentioned maybe being an envoy to Somalia, a big interest of his and mine, as a precursor for a ministerial role. 'Or,' I said, 'you're perfectly entitled to say "I'm not satisfied with that, I want to find out what the hell happened and pursue this."' I tried to steer him towards the former. Sometimes, in politics and life, you just have to let things go.

But I understood why he didn't let it go. Two months later an officer was arrested in connection with the incident. Then the email from the member of the public claiming to have witnessed the altercation turned out actually to have been from a police officer. The whole thing unravelled in a Channel 4 *Dispatches* documentary in February 2013. I watched it in the flat, and felt delighted for Andrew but sick to my stomach about my failure to get to the bottom of it all. He *had* been completely stitched up.

Andrew sued the *Sun* for libel. Then the police officer at the centre of the affair sued *him* for libel. Over the following months, four police officers were sacked for gross misconduct. A text message in which an officer boasted that she could 'topple the Tory government' was exposed. Leaks from police to the *Sun* came out. The police officer who had posed as a member of the public was sentenced to a year in prison. Andrew was vindicated. Justice had been done.

In the cruellest postscript, however, he lost his High Court libel action against the *Sun*. The judge, Mr Justice Mitting, said he was satisfied Andrew *did* say the word 'pleb'.

Andrew had been rude to people for years, he had sworn at police, he had played much of this affair badly. But none of that warranted what happened: a legal bill stretching into millions, and a man in pieces. I know he feels that he was badly treated by No. 10 and by me. And I concede that we should have held a full investigation immediately. But I held out for him for as long as I thought was right, and I genuinely had a plan for his return. He was content to return as a minister of state in either Defence or the Foreign Office, and I planned to bring him back in a 2016 reshuffle. It was not to be.

In April 2013 another surprise followed the Mitchell saga. Chris Huhne, who had protested his innocence on the saga of the speeding ticket throughout, changed his plea to guilty and was sentenced to eight months in prison.

Resignations aren't just difficult because of the effect they have on individuals and their families. They are bad because they destabilise the government, take up time, distract from your agenda and reduce politics to daily instalments of a drama. By the autumn of 2012 we'd had the soap opera of 'plebgate', the crime thriller of Chris Huhne, and the Liam Fox mystery – not to mention the Omnishambles comedy of errors. We desperately needed to grab the narrative back – and the party conference would be the ideal moment to do so.

The lead-up was similar every year. After a few speech-prep meetings in Downing Street in July, my core team would reconvene at Chequers in September for our annual conference speech 'sleepover'. As ever, Ed, Kate and Craig were there. Ameet Gill, Steve's replacement as strategy director, came too, as well as Michael Gove.

But the vast majority of the detailed work – constructing the paragraphs, crafting the killer lines, choosing the 'moments' that would make people gasp or laugh or connect emotionally – would be the work of the speech team. There was no one more up to the task than my long-time wordsmith Clare Foges. After weeks hammering away at a laptop, producing draft after draft and watching me go over and over them, first on paper or on my laptop, and then in autocue sessions in the conference hotel in Birmingham, this was her finest yet.

It summed up what we'd achieved in just two and a half years: a cancer drugs fund; life-saving vaccinations for 130,000 children in developing countries; a European treaty vetoed; council tax frozen; two million of the lowest-paid taken out of income tax; the deficit down by a quarter; a cap on benefits; a million new jobs created; exports up; business creation at its fastest-ever rate; 2,000 more academies; seventy-nine new Free Schools; and the greatest Olympics and Paralympics in history.

Most importantly, it framed one of the big, long-term issues of our times: how to respond to the geopolitical power shift from west to east. The countries on the rise – China, India, Brazil, Nigeria, Indonesia – I said, were lean, fit, obsessed with enterprise, spending money on the future: on education, incredible infrastructure and technology. Whereas those on the slide – which would include us if we weren't careful – were fat, sclerotic, over-regulated, spending money on unaffordable welfare systems, huge pension bills and unreformed public services. 'We are in a global race today,' I said. 'And that means an hour of reckoning for countries like ours. Sink or swim. Do or decline.' It wasn't about clambering up the global league tables; it was about lifting up lives. Those forward-thinking countries delivered the things people needed to get ahead and to prosper, and we could do so too if we built an 'aspiration nation' that realised the potential of all its people.

Conference, as I've said, was one of those rare moments when we had the nation's ear. At this moment I felt we were saying exactly what needed to be said, and had finally mastered the art of what this government was for. It had been a difficult year, but Britain was back on track – and so was I.

29

Bloomberg

How are the biggest decisions made? They are usually rooted in convictions and beliefs. They tend to be contemplated for a long time, but are often expedited by circumstances. They are frequently influenced by other people's views, and events that have taken place over many years.

One of the biggest decisions I would ever take – to renegotiate Britain's relationship with the European Union and then hold a referendum on our membership – was an example of all those things.

Months, even years, of thinking, arguing, listening and planning brought me to this moment: sitting in a traffic jam at 7.45 a.m. on Wednesday, 23 January 2013, my blood pressure rising as I contemplated the difficulties of getting everything done that day and doing it well, and as I thought about the consequences beyond that.

I was travelling to the media and finance firm Bloomberg, where I'd make a speech announcing my policy of renegotiation and referendum. Time was tight, and the police protection officer driving me put on a few blips of the blue lights – just to 'make some progress, Prime Minister', as they put it.

I'd be facing Prime Minister's Questions at midday, before flying to Switzerland to speak at the World Economic Forum in Davos. But before all that, at a time when I'd normally be at my kitchen table preparing for the most nerve-inducing event of the week, PMQs, I was on my way to make probably the biggest speech of my career.

'There are days when you think, "Why on earth have I created such an appalling set of hurdles for myself?"' I mused later when talking about the schedule. But critics could ask the same thing of the announcement itself. Why, when so many of my predecessors had avoided it, had I chosen to head straight towards this towering obstacle?

There are plenty of theories.

The most often repeated are that the commitment was a tactical measure devoid of strategy, or that it was only made because of the fractious state of the Conservative Party at that moment. Another is that it was driven predominantly by electoral considerations, and that I was spooked by the rise of the UK Independence Party. Or that the referendum pledge, far from being cast in iron, was a gimmick: I didn't expect the Conservatives to win outright in 2015, and so didn't believe I would have to deliver on my promise. And finally, that even if we did win the general election, my approach to the pledge was essentially flippant, because I assumed that Britain was certain to vote to remain in the EU.

There is a whole series of more detailed questions that I must answer. Was it the right deal? The right time? The right campaign? I will address all of those questions – and I will not say 'yes' to all of them.

But the speculation over my *motivations* behind the decision – why I was walking to the podium at Bloomberg, clutching fifty-six pages that could change my country and continent forever – I reject almost entirely. I want to answer that speculation in some detail.

The accusation that I decided to make the referendum pledge because I didn't believe I'd actually have to hold one is absolutely false. One of the reasons I had thought so long and hard about whether promising a referendum was right and necessary was that I knew that once I made the pledge there was no going back. I understood from the beginning that Conservatives (not just MPs, but the press and voters) would not allow and sustain a future coalition that was created by abandoning the referendum pledge.

So I made it very clear from the outset, in public and in private, that if I didn't get a majority, this was a red line: this policy was so important, so unlike any other, that I would not become prime minister after the next election in a government that was not committed to do it. In other words, it could not be traded away in some future set of coalition negotiations.

Actually, I didn't think I'd need to insist on this red line, because I didn't believe the Liberal Democrats would block it. After all, holding an in/out referendum if there was to be a 'fundamental change' in Britain's relationship with the EU had been their policy in 2010. People often forget that the Lib Dems had theatrically walked out of the Commons in 2008 when the speaker refused to call their amendment asking for a referendum – and not just on Lisbon, but on Britain's wider membership of the EU. And who was it that called for the 'in/out referendum that the British people really want' in the Commons in March 2008? Well, that would be Nick Clegg.

The idea that the referendum pledge was frivolous, because I complacently assumed that 'remain' would win, is simply not true. I knew from the start that this was deadly serious, and I thought it very possible Britain would vote to leave. I said this repeatedly on tape from the moment I decided to call the vote. In October 2012, for example, I confessed: 'This is very, very difficult and very dangerous. We could end up not being able to achieve enough to be able to put into a referendum and therefore effectively leaving the EU if we're not careful.'

I had thought more about the pledge I made in the Bloomberg speech to hold a referendum than about any other decision I was to make as prime minister. From the moment I said those words into the camera lens – 'It is time for the British people to have their say. It is time for us to settle this question about Britain and Europe' – I worried about the outcome.

But I made the pledge because I genuinely believed it was the right thing for Britain. It was right to try to get a better deal for us in Europe. It was right for us to have our say on that deal, and on our membership more broadly. It was right, if we got a better deal, to remain in the EU. I believed this strategy was the most likely way to keep Britain in the EU. Indeed, that's how I put it to my fellow European leaders: this is my strategy for keeping Britain in Europe.

The strategy failed. I failed. And that failure has had some serious consequences for the UK and Europe. But it all flowed from an attempt to do the right thing.

I had reached the conclusion that the risks of failing to act were greater than the risks of acting, and that we would have ended up on extremely weak ground if we had failed to reconfigure our status within the EU and put it to the people to decide.

The euro, and the British decision to stay out of it, had produced a new constitutional relationship that was simply not stable or tenable. The instability of this situation became particularly apparent during the Eurozone crisis and the EU's response to it. The simple truth was that the EU was changing – and the Eurozone crisis was driving that change quite rapidly. Britain was in danger of being left in an organisation whose future was being dictated to by an integrationist core, those for whom the euro was their currency.

We were beginning to see decisions being made that threatened our national interest, the attempt to take vital trading in euros out of the City of London being a case in point. The crucial moment for me was, as I've said, when we vetoed a treaty because it failed to give legitimate safeguards to the UK, only to see that treaty implemented, using the institutions of the EU in full. And somehow it went through – miraculously – without the need for the unanimous agreement of member states. Such unanimity had been required for every other treaty change since the Union had been founded.

The choice as far as I was concerned was between sitting back and letting events take their course, accepting that the case for leaving would almost certainly get stronger, or having a proactive plan that dealt with the problem and secured our national interests, *within* a reformed EU.

I was increasingly convinced that an in/out referendum would be held in the not too distant future, quite probably by a successor Conservative government that might well recommend leaving. To my mind, holding a referendum was not only necessary to achieve the changes we required to secure Britain's interests *within the EU*; it was needed to settle this issue *within the UK*. Better, I thought, to drive the process, rather than be driven by it.

All of this raises a bigger question: why was it that Britain seemed to have such a problem with the concept of a more united Europe, when other countries seemed, by and large, to be able to go along with it?

To explain this in detail, you have to go right back to the 1940s, when France, Germany and others formed the core of the new European Coal and Steel Community. This became the European Economic Community in the Treaty of Rome in 1957 – one which committed its members to 'ever closer union'. It was this economic community that Edward Heath took us into in 1973, and which the public voted to remain part of in Harold Wilson's 1975 referendum. But the vast majority of Britons were never going to love it like our European neighbours. We hadn't founded it, and we hadn't shaped it. Nor did most British people want 'ever closer union'.

That difference of perspective, like so many other things, goes back to the Second World War. Many political leaders and thinkers in western Europe saw the nation state as having been the cause of fascism and conflict. Yet in Britain we saw it as a solution: in 1940 we stood alone, as a nation state, and by doing so helped to save European civilisation. So we were always going to be wary of proposals for political union.

Yet increasing political union is what happened. The economic organisation we were told we were joining gradually mutated into a political one. To be fair to the other countries in the EU – and to those who opposed our membership from the beginning – the concept of a political union was there in embryo right from the start.

The European Economic Community, or EEC as it was then called – and soon simply the EC – was never just a 'common market'. This was the point that Enoch Powell and Tony Benn made over and over again in powerful speeches at the time. Powell had called joining the EEC potentially the 'biggest event' in Britain's entire political history.

So the tensions between the economic benefits we wanted and the political union we opposed were there at the start. But it would take more than a decade before they started to reach breaking point.

In the 1980s, Margaret Thatcher helped deliver the economic benefits of a genuinely common market by securing the treaty that delivered the European single market, and was rightly cheered at the time. But this came at the price of accelerating the political union that she claimed to oppose.

The single market was more than a free-trade area, and it could only be completed by wave after wave of EC regulations. The European Commission is the only body in the union that can propose new legislation. It has – to use the jargon – the 'sole right of initiative'. And these regulations would be passed by a qualified majority, and not require unanimity. So the single market in effect required the ending of national vetoes in area after area. It represented a significant transfer of sovereignty from the nation state to the European Community.

Margaret Thatcher recognised the dangers of this – in particular the danger of onerous regulations being imposed on the UK in order to protect other countries from competition – and from 1988 started to develop the 'thus-far-and-no-further' approach to the EC of her late premiership.

I remember, as one of her researchers, nodding in agreement at the stand she took in her Bruges speech that year: 'We have not successfully rolled back the frontiers of the state in Britain, only to see them re-imposed at a European level.' I also agreed with what she said in the Commons in 1990 during one of her last appearances at the despatch box as prime minister: 'The president of the Commission . . . said he wanted the European Parliament to be the democratic body of the community, he wanted the Commission to be the executive, and he wanted the Council of Ministers to be the senate. No. No. No.'

But just as Britain was pulling away, the European Community was pulling together. The end of communism raised the prospect of a reunited Germany, and the fear that the tensions that had characterised the first incarnation of German unification might re-emerge. For many European leaders, the answer was more European integration – to bind powerful, united Germany within the European institutions and to press ahead with the already suggested common currency, the euro.

With 1992's Maastricht Treaty, the pace of advance towards a more federal Europe quickened further. It eroded more national vetoes over legislation, created the European Central Bank, the euro and the Social Chapter, and established whole new areas – economic and social policy – over which the organisation could legislate. In a genuine negotiating triumph, John Major and Norman Lamont secured opt-outs from both the single currency and the Social Chapter, solidifying what became a special status for Britain within the newly named European Union.

I thought that accepting Maastricht was – just – a price worth paying. We stayed in the decision-making body that governed the single market, but were out of the parts that most threatened our sovereignty: the single currency and the Social Chapter.

The runaway Britain-to-Brussels train had thus far been driven by those with a foot on the brake. But in 1997 there was a new driver, Tony Blair, and he was intent on using the throttle. He gave away our opt-out over the Social Chapter, and announced that the euro was 'our destiny'. Further treaties – Amsterdam and Nice – swiftly followed. Each time, more national vetoes were eroded. And each time, despite vigorous campaigns by the Conservative Party, no referendum was offered to the British people.

Then came the big one: the grandly titled 'European Constitutional Treaty', in 2004. It was a detailed and ambitious attempt to provide a unified constitution for the EU, amalgamating all the previous treaties into one text and further driving political union and federalisation. Tony Blair was in favour, but he at least acknowledged – under political pressure from the Conservative Party – that this time the British people should have their say in a referendum.

I remember watching as he announced the referendum from the despatch box, declaring: 'Let the battle be joined.' I was absolutely convinced that the Conservative Party would be almost completely united in opposing the European Constitution – and that we would win.

So the three main political parties went into the 2005 general election promising a referendum on the proposed European Constitutional Treaty. But it was never to be. The derided European Constitution was rejected, via referendums, by several countries, then modestly amended and rebranded as the Lisbon Treaty, retaining all the worst elements, including a shift of power from the European Council of national leaders towards the European Parliament of unknown MEPs and powerful party groupings, and then railroaded through. Tony Blair took the opportunity to ditch his promise of a referendum. The Conservative Party renewed its calls for one, but the treaty was passed through all of the member states' Parliaments, including our own, and the moment was missed. And so the democratic deficit between Britain and the EU grew greater still.

While never a believer that Britain should leave the EU, I was genuinely angry about the way our country was being treated. I used the first Prime Minister's Questions in autumn 2007, after Gordon Brown had bottled the election, to raise the issue. 'The prime minister said that the Labour manifesto was an issue of trust,' I said. 'That manifesto promised a referendum on the European Constitution. Do you understand how not holding that referendum damages your credibility?' I would keep up the pressure all the way through to the moment Brown himself signed the Lisbon Treaty and passed the legislation through the UK Parliament. As described in Chapter 10, from that moment a referendum on the treaty alone, as I'd hoped for, was no longer a viable option.

When it came to our own manifesto for the 2010 election, we did think seriously about including a pledge for a referendum. Oliver Letwin particularly pushed for one. But because Lisbon had been ratified, it would run the risk of turning into an in/out vote. The EU could legitimately argue that every Parliament, including our own, had ratified Lisbon, and so a referendum to reject it was effectively a call to leave.

George argued powerfully that this wasn't the time: 'We are going into an election after the biggest financial crisis in our history, and people will say, "That's your answer – a referendum on Europe."' It would be 2001 all over again: a Conservative 'Keep the Pound' campaign in an election about the NHS. So instead, the referendum lock came in its place.

In a stable and unchanging EU, that probably would have been sufficient. The referendum lock – no powers could be passed from Westminster to Brussels without the express agreement of the British people in a referendum – was the ultimate expression of the 'thus-far-and-no-further' policy. The British people would have an absolute lock on any proposed changes. But a stable EU wasn't on offer, particularly as the Eurozone crisis took hold.

There was the Greek crisis, where we had to fight to stay out of bailing out a country that was suffering dreadfully, in part from its membership of the euro, even though we weren't part of the currency bloc.

And there were the treaty changes that came out of Brussels and had to be ratified by us (with the aim of easing the Eurozone crisis) even though, after Lisbon, the EU had promised that there would be no further constitutional changes.

Ivan Rogers, my adviser on Europe from 2011 to 2013 and later our top diplomat in Brussels, has written about how the decision to hold a referendum stemmed from the Eurozone crisis and centred on whether our place inside the EU but outside the Eurozone could be made sustainable in the light of the changes that were having to be made. The biggest single moment on the journey was the veto and its aftermath described above: clear evidence for the proposition that Britain's place in the EU needed fixing.

All the while, the public concern about being caught up in it all – through bailouts and the like – strengthened support in the country for a referendum.

The depth of the Eurozone crisis made it likely – at times it seemed almost inevitable – that more fundamental treaty change would soon be contemplated. The southern countries wanted more solidarity; the northern ones tougher rules on delivery of structural reforms to improve competitiveness. Major changes couldn't be fixed within the current framework. So surely it was also the right moment to find a settlement for the UK, and confirm it through a popular vote.

Indeed, that is how I explained it to officials. I was bringing forward something that would have happened regardless. When the inevitable new treaty came, the lock would have kicked in and meant we'd have to hold a referendum – and the pressure for an in/out one would be huge. Our national interests meant we'd want to push for reform of the EU and renegotiate our place within it.

But the referendum pledge was also about what had gone before. One after the other, these changes led to a moment of reckoning for the UK and EU. Indeed, you can join the dots from Maastricht to Lisbon to the euro crisis and my veto – and there is a relatively straight line towards renegotiation and referendum.

That was intensified by something else. Because while there was a *practical* problem building in the relationship between Britain and the EU, I accept that there was also major *party political* pressure over the issue at home. My parliamentary party was becoming increasingly restive about the failure to hold a referendum in the past and the need for one in future.

I make no apology for paying close attention to the views of Conservative MPs. That is how our political system works. A prime minister serves while he or she retains the confidence of Parliament. Parliament represents the people.

One of the arguments made against having a referendum is that it was all about managing the parliamentary party. Yet the very same people who make that argument also complain that we shouldn't have referendums because we are a parliamentary democracy. They do not appear to realise that these points are directly contradictory. In a parliamentary democracy, the leader of a parliamentary group has to pay attention to the views of that group. And that group was pressing for a referendum. So holding the EU referendum was not a rejection of parliamentary democracy, it arose out of it.

And what about the argument that the MPs were cut off from their constituents who did not want a referendum? In other words, that the vote came from obsessional Tories out of touch with the public – it arose from Parliament, but from a Parliament that had lost the plot.

I have one simple response to this. People voted to leave. They voted in unprecedented numbers. You simply cannot deny that there was democratic pressure for this vote. You might be able to argue that the AV referendum was a bit of political management that didn't engage the public. You can't say that about the EU referendum.

The political pressure to hold a referendum was strong and growing. There was even a risk that Labour would use the turmoil in Europe to leapfrog us and protect its relatively weak flank on Europe by proposing one themselves.

And of course there was UKIP. UKIP's role in the formulation of the referendum pledge tends to be overstated. As I mentioned in Chapter 24,

I started seriously mulling over the possibility of a referendum in January 2012. At that point UKIP was still a small force. Polling showed it regularly on less than 5 per cent, and it was less popular than the Lib Dems.

Yet I did see the attraction of UKIP to Tory voters as an indication of public feeling. And I also thought that if UKIP came first in the May 2014 EU parliamentary elections (when they would receive a higher profile and were more likely to perform well), the pressure to pledge a referendum would be too great to resist – rather as I saw the SNP victory as a mandate for the referendum it was calling for. That was undoubtedly an additional incentive to move before being forced to do so.

The truth is that all of these issues – the potential rise of UKIP, the difficulties of managing the Conservative Party, the need for a compelling and popular policy offer on Europe at the next election – were concerns of mine. But they were not the determining force that led to the referendum pledge.

If it was only about managing the party, I could have come up with a formulation for a different sort of referendum, rather than the full in/out version. A nationwide plebiscite asking for a fundamental change in Britain's relationship with Europe – a so-called 'mandate referendum' – was popular at the time with some of the party's leading Eurosceptics. Many of my MPs would have been happy with any pledge as long as it involved a nationwide referendum on the issue of Europe. If it was all about not being outflanked by the opposition, I could have waited for Labour to make a move and then neutralised it with a pledge of my own.

And it is worth remembering that all these party-political considerations were brought about by a simple fact. The British public were becoming increasingly concerned about the direction that the EU was taking. Trust in the EU fell across the continent after the euro crisis began, but nowhere more than in Britain, where it plummeted from -13 per cent to -49 per cent in just five years. This reflected a broader trend: according to the British Attitudes Survey, after 1997 the proportion of the British people favouring a reduction of the EU's powers or an exit altogether was *always* above 50 per cent. So I wasn't responding to Conservative Euroscepticism so much as to British Euroscepticism.

And I thought it was important to respond. The proposition that we needed to settle Britain's relationship with Europe was bolstered by the argument that when it came to consent for what was happening, politicians had kicked this can down the road for far too long. The broken promises on referendums. The changes to the powers of the EU without the explicit backing of the British people. The fact that with the euro the organisation that we were part of was changing before our eyes.

Those who say now that it was wrong to hold the referendum are arguing, in effect, that it would have been better for the country to have been forced to continue down a road it didn't want to take, and which we couldn't turn back from, having never been asked to make the judgement.

So I am not apologetic about having been the prime minister who promised a referendum and delivered on the promise. I couldn't foresee a possible future for the UK *without* a referendum – and I thought it right to hold one and try to win the argument. I deeply regret my failure to do so and the consequences that have followed. But that's not the same thing as believing that the whole attempt was misguided. I strongly believe that it wasn't.

I also don't accept that I was appeasing populism. In fact, I was confronting it head-on. And far from holding the views of Eurosceptics in disdain, I shared a lot of their dissatisfaction. The EU *had* become too big and too bossy. Our current position was becoming unsustainable. The democratic deficit was increasingly intolerable. I knew we were better staying in: what happened in the EU would still affect us, whether we were in or out; better that we were at the table, having our say. The veto incident had made me even more convinced of that, but had also sold to me the pressing need for reform.

Ultimately, I thought the choice about whether or not to stay in a reformed Europe was bigger than one politician or indeed one Parliament. It was for the people to decide. That's what brought me to recording on tape in January 2012 that I was now seriously considering this strategy.

So what about getting my team on board?

Oliver Letwin was still strongly in favour. In fact, in May 2012 he sent me a note saying it was time to discuss the matter of a referendum seriously. As Europe split into the Euro-ins and Euro-outs, he said, we should be rallying the Euro-outs. I agreed with him that it was crunch time, but I thought this was going to be far from a team effort. 'Every time Europe has a tremor, the calls for "more Europe" come from the familiar quarters and we seem to be the only ones objecting,' I wrote back.

William had become more pro-EU since becoming foreign secretary, but he was still of similar thinking to me. It's been said that he and I decided on the referendum over pizzas at Chicago Airport after the NATO summit in May 2012. Not true. We did discuss the subject at a pizza restaurant on that trip, but it was a discussion, not a decision. There would be many, many discussions over the months. We certainly wouldn't have finalised something like that without George being there.

It's been widely reported that George was opposed to the strategy. That's true, and he still says to me now, 'I told you not to do that fucking referendum.' But it's more complicated than that. He was one of the first to raise the need for a proper discussion about it. He was worried that we couldn't keep our party together until the next election without offering one; and he specifically raised the concern that we might end up as the only major party at the election *not* offering a referendum.

Ultimately, his fears over the business community's reaction and the timing, and about not being able to negotiate a good deal, led him to caution against the referendum pledge. But he said he would support my decision, and like me, he believed that ultimately a referendum was probably inevitable anyway.

Michael Gove, one of the more ardent Eurosceptics in my team, was surprisingly opposed to promising a referendum, principally on the grounds that other priorities were more important.

Boris Johnson, who was Eurosceptic but had never argued for leaving the EU, was now echoing the call for a referendum. He seemed to have done almost no thinking about what sort of referendum, when it should be held, or what the government's view should be. Nevertheless, as a popular figure, out of Parliament but in power as mayor of London, his support for a referendum – however inchoate – was a potentially dangerous development if we decided against holding one. And his apparent support, however muddled, for an immediate vote was an irritation.

I invited him and his family to Chequers for lunch later in 2012, and the subject was discussed over the table. 'Let's hold an in/out referendum now,' he said. I explained my developing thinking – a referendum was now becoming close to inevitable, and we should make it part of a strategy for reforming Britain's place in the EU. Boris's wife Marina rather effectively shouted him down, saying, 'Dave's thought it through. I'm not sure you have. Why don't you let the prime minister just get on with it?' – or words to that effect.

In June 2012 there was a European Council that in my mind further vindicated the strategy. It was additional proof that the EU was changing, and that I should – and could – negotiate to protect Britain's interests.

The change that was being made on this occasion was to create a banking union, and there was pressure on Britain to join in. Effectively, if another country's banks went down, we would have to support them financially: we would have to stand behind Greek or Portuguese banks, and allow our banks to be regulated by the ECB, not the Bank of England. This was completely unacceptable. Thankfully, George and I managed to make sure that it would be agreed by the Eurozone seventeen rather than the EU twenty-seven, and led by the ECB, as part of the single market. But again we had come close to more power passing to Brussels without the British people having a say.

At my press conference afterwards I said that Europe was changing rapidly, that this would bring opportunities for us, and that we needed to think how to make the most of that – with the backing of our people. However, the newspapers missed this hint completely, and the stories the next morning said, 'David Cameron has ruled out any referendum that could see Britain leave the EU'. So the following day, on the helicopter flight to Plymouth for Armed Forces Day, I typed out an article for the *Telegraph* on my iPad, setting the record straight.

In it I gave a further hint of a renegotiation and a referendum: 'The fact is the British people are not happy with what they have, and neither am I. That's why I said on Friday that the problem with an in/out referendum is that it offers a single choice, whereas what I want – and what I believe the vast majority of the British people want – is to make changes to our relationship.' I then pointed to the sorts of things I would be seeking: 'Far from there being too little Europe, there is too much of it. Too much cost; too much bureaucracy; too much meddling in issues that belong to nation states or civic society or individuals. Whole swathes of legislation covering social issues, working time and home affairs should, in my view, be scrapped.' I made it clear that I would clarify my position on Europe in due course, and that 'For me the two words "Europe" and "referendum" can go together . . . but let us give the people a real choice first.'

I wanted to get all the autumn European Councils out of the way before I gave the promised speech. That took us to the end of 2012, which meant the speech would actually have to take place in 2013. But even then we struggled to pin down a date. We settled on 18 January in Amsterdam. Ed drafted my speech, with the help of John Casson, the long-serving and energetic speechwriter Tim Kiddell, and Helen Bower, my impressive foreign affairs press chief, who I would later make the prime minister's official spokesman. Clare Foges wrote the opening few pages: a beautifully crafted homage to British–European history.

But then, two days before I was due to deliver the speech, I was told that al-Qaeda-linked terrorists had seized a gas facility at In Amenas in Algeria. Several British citizens were among the 132 foreign nationals held hostage there, eight hundred miles from the capital Algiers and one of the most remote places on earth.

COBR sprang into action, coordinating with BP and the Norwegians and Japanese, who also had nationals working out there. The situation was desperate: we were threatened with the biggest loss of life of British people in a terrorist incident overseas since 9/11. After a frantic few days the Algerians finally recaptured the compound. Forty hostages had lost their lives, including six British citizens and one Colombian who lived in Britain.

The tragic episode meant that the speech had to be postponed – and the only suitable day was the already packed Wednesday on which I had PMQs and would then be flying to Davos.

As the fact of the referendum had already been trailed, the news on the day would be its timing. It would take place by the middle of the next Parliament – in other words before the end of 2017. William and the Foreign Office were worried that we were boxing ourselves in. But I felt that if I started pushing it too far back, I would lose credibility. We knew from voter research that people welcomed the prospect of a referendum but were, naturally enough, suspicious that we would promise one and not hold it. A date was essential.

The speech went well. I set out the three major problems: the changes in Europe, the continent's lack of effectiveness in the global race, and, above all, the democratic deficit. My five major principles came in response. Europe needed competitiveness. It needed flexibility. It needed to be able to give powers back to member states rather than just taking them away. It needed democratic accountability. And it needed fairness – to make sure that any changes didn't disadvantage those outside the Eurozone.

The speech was tricky to land with so many different audiences. It had to explain to European leaders that we weren't just being awkward neighbours, and make them understand why our attitude to the EU was different to theirs – more hard-headed, less emotional. It had to keep business leaders on board; they would likely fear the uncertainty that would hang over them. It had to demonstrate to the British people that I understood and shared their concerns, and was doing something about them.

Three years later, immigration would be cited as the reason many people voted to leave the EU. So why did the issue of free movement between member states – the 'fourth pillar' of the EU – not warrant a single mention in the 5,000-word Bloomberg speech? It should have done, but there was a reason for its absence. At that moment it was immigration from *outside* the EU that was driving the numbers. I thought – wrongly, as it turned out – that that picture would continue. It was a bad mistake that I would later try to correct. But I shouldn't have made it in the first place.

The context for this was that I agreed that immigration to Britain was too high, and had been for many years. In opposition I had developed our policy on the basis that it was *net* migration that mattered: not so much the number of people coming into our country, but whether that number outweighed the number of people leaving, therefore putting pressure on public services and rapidly changing our communities.

Until 1998, immigration had remained broadly in balance with emigration – there were as many people leaving Britain as there were people arriving. But ever since then, immigration from outside the EU had risen steadily. Then, in 2004, with the accession of seven former Eastern Bloc countries to the EU, including Poland – and, crucially, Labour's failure to activate the seven-year transitional controls that all other countries except Ireland and Sweden imposed – immigration from within the EU shot up: from 66,000 to 195,000 per year in four years.

However, by the time we took office in 2010, it was once again non-EU immigration that was on the steeper trajectory. We managed to get a handle on it, and thanks to the steps we took, the numbers started to fall sharply. But at exactly the same time – as the Eurozone crisis deepened and our economic growth far outstripped our neighbours' – EU immigration rose again. We should have focused earlier on the importance of controlling not just one type of immigration, but all of them.

What's more, I didn't see how deep the disillusion was – in some cases outrage – about our inability to control immigration. That was the EU trade-off: access to the single market in exchange for free movement of people, and therefore unlimited immigration from inside the EU. For many people the very idea of lack of control, coupled with changing communities and strained services, was hugely important.

However, the immigration theme didn't quite start to bite until after I'd taken my referendum decision. Ivan Rogers argues very powerfully that 'The decision to offer an in/out referendum long predates the point at which that issue became central . . . [I]t was the domestic and cross-Channel political tensions unleashed by the Eurozone crisis which put us on the tramlines to Brexit.'

Another communication failure in the Bloomberg speech was not pointing out how the caucusing of euro countries would really affect us. I spoke about the need for fairness between Eurozone and non-Eurozone countries, but I wish I had said more. When people complained about Europe it was about the cost, the border control, the influence it had over our laws – not about the threat it posed to our economy. I should have explained that issue more frankly.

One reason for this failure is that I was speaking to more than one audience. I wanted to pitch this to Europe as the opening of a negotiation, as well as to British voters. I spoke to European leaders on the phone afterwards, and to Barack Obama too. Angela Merkel warned against arguing for British exceptionalism from the first. She said I should argue about reform for the whole of Europe, and then 'let's see what we can achieve'. If we couldn't achieve some things, she said she'd be prepared to consider British opt-outs, but we shouldn't start there. Dutch prime minister Mark Rutte said there were reforms in Europe that we'd all like to see, but that he'd like the twenty-seven to make them together.

That night I ate at the Buffalo Grill in Davos with my team, ahead of another packed schedule. We were in good spirits. The speech had gone down well with European leaders, businesses, the press, parliamentarians and the public. 'Mr Cameron has not caused a problem, but elucidated one,' wrote *The Times*. I felt that captured it well.

I knew it was high-wire stuff. But I believed that what would have been more dangerous was not having a plan. In that case we would have been gently heading towards the EU exit. As I put it on tape at the time: 'The risks of playing with fire are now safer than sitting and watching the fire burn.'

The following month I proved that things could change in Europe, and that Britain could lead that change.

The level of the EU budget for the next seven years was coming up for renewal. I was adamant that there should be a cut. Member states like us had been cutting our own spending after the crash; indeed, the EU had imposed stringent cuts on many Eurozone countries. This wasn't just Britain being difficult – as the third-biggest contributor, we had a big stake. And it was more than possible: the last budget had been very generous, and there was an appetite for a cut (besides which, they were hardly living on bread and dripping in Brussels).

As ever, I was faced with achieving something between what Parliament wanted, what I thought was right, and what Europe would agree to. The sort of dramatic cuts those at home might have wanted were impossible. The EU had expanded, and we had obligations to the new member states. So it was extremely unhelpful when the Treasury published a projected figure of 886 billion euros, which I knew was far less than the other EU leaders would agree to.

Then in October 2012 we suffered a parliamentary defeat when fifty-three Conservative rebels, joined by Labour, voted for a motion demanding a cut 'in real terms' to the EU budget, which I thought was raising expectations we would find it hard to meet.

I knew there would be two factions in Europe. There would be those I called 'the Alliance of Fiscal Discipline' – the Danes, Dutch, Swedes and Finns – who would want to work with us: we were in regular contact with one another by text. Then there would be the Poles, French and Spanish, who didn't want a cut. Sitting in the middle – as ever, the casting vote and kingmaker – would be the Germans.

So I began to work on Angela Merkel. I invited her to Downing Street for dinner and gave her a presentation, half of it in German. I barely speak a word of the language, but Ed helped translate bits, and I made it as entertaining and understandable as I could.

My aim was simple: to show that what she and I were arguing for was not that far apart. The fiscally prudent countries weren't headbangers, I told her; what we were asking for was sensible. But we would have to veto anything that wasn't in our interest. She wasn't sold, but she was interested. And although we didn't get an agreement at the following European Council in November 2012, we did get the total budget down from a trillion euros to 973 billion.

My ambition, and that of my mini-alliance, was around 900. A little over if necessary. The last seven-year budget had a higher ceiling of 943 billion euros. We judged that around 908 would keep the budget fixed below the 2011 level, so there was my red line.

I realised that if we got the three big countries to agree, then we could bring everyone else along with us. Merkel and I agreed this approach in Davos in January 2013. At the following European Council that February, she and I arranged a meeting with François Hollande. We sat there with his translator, waiting and waiting, but he didn't show. In the meantime, Merkel and I summoned José Manuel Barroso, the president of the European Commission, and told him what we were looking for. Around 907.

Then, in the middle of the night, Hollande and Merkel appeared in my office. She said she *thought* they had a deal on 908.5. I wasn't happy, because that was above the average of the 2011 and 2012 budgets. It was absurd: we were arguing at 2 a.m. over £150 million – a huge sum, of course, but less so when you consider that it is spread between twenty-eight countries, over seven years, and just how much it is paying for. Eventually we settled on 908.4 (I thought I could round that down), and shook on it. But Hollande wasn't happy – he had said nothing below 913 in public.

After they left I slept for an hour in my chair. Then I heard that Hollande was trying to get the figure up again. I said to Merkel, 'You're going to stick with me, aren't you?' She said, 'Yes. François gave his word and is now going back on it.'

Eventually, 908.4 was agreed at the full Council. A 3 per cent cut on the previous seven-year budget, the first-ever real-terms cut in the EU budget, thirty-five billion euros lower than the deal agreed by the last government – and a win for me. Good for Britain and good for Europe, I declared it when I returned home. And, I thought to myself, very good for my strategy to improve Britain's place in the EU – and keep us there.

If there was one person who might have sympathised with my trials in Europe, it was Margaret Thatcher. Now eighty-seven, she was increasingly frail, living with dementia and recuperating from a serious operation.

When you're in politics, you have to be rather more prepared for deaths than you would otherwise be. The BBC frequently updates its obituaries, so you can find yourself filming a moving, past-tense tribute about someone you've just had lunch with. Not only were the Thatcher tributes in the can; Clare had my speeches polished and perfected, and the funeral plans were also in place.

Indeed, Margaret Thatcher's funeral had loomed large ever since I became leader of the opposition. There were meetings about the planning – 'Operation True Blue' it was named – and I fought hard to make sure it was as close to a state funeral as possible.

The only prime minister since the war to have been given a state funeral – where the costs are met by the government, there is a lying in state and full military honours – was Winston Churchill. Why did I think Thatcher should get something similar?

Whatever your political view, you had to acknowledge that she was our first – and at this point only – woman prime minister. She had won three elections in a row, and served for a longer continuous period than any prime minister in over 150 years. While I recognised that opinions throughout the country were hugely divided, I believed she didn't just change Britain, she saved Britain.

On 8 April 2013 I was in Spain, having lunch with prime minister Mariano Rajoy during a tour to promote my Europe speech and strategy, when I heard that this great figure of history had died. I flew home and recalled Parliament, and there was a day of very moving speeches from all sides of the House.

My position was slightly more nuanced than many of the uncritical devotees who spoke. I felt that my premiership owed a huge debt to Margaret Thatcher, but also marked a break from her. Some people would say that I had broken the spell: since 1990, all Tory leaders before me – indeed, all my fellow contenders in the 2005 leadership election – had been very closely linked to her and her ideology.

By contrast, I had always felt myself more of a Thatcherist than a Thatcherite. I wasn't always convinced by her approach, and thought some of the rough edges needed to come off. But on the big things – trade union reform, rejecting unilateral nuclear disarmament, our alliance with Ronald Reagan's America, privatisation, Europe – she was absolutely right.

In my speech, I talked about the turnaround she produced, and how the scale of it was all the more astonishing when you remembered the defeatism of 1970s Britain. Her credo – 'sound money; strong defence; liberty under the rule of law. You shouldn't spend what you haven't earned. Governments don't create wealth, businesses do' – not only saved our country, it now seemed to be the consensus in politics.

The night before her funeral I held a dinner at No. 10 for many of the dignitaries who had come to Britain for the occasion, including former leaders of Australia and Canada, Reagan's advisers, and many of those who had worked for her here. Their passion for her, which was evident in the many brilliant stories they told, was matched by that of the leaders from eastern Europe who had also come for the funeral. They were very clear about what she meant to them. Her hatred of communism helped tear down the Iron Curtain and free them from tyranny. We often fail to appreciate in this country that Margaret Thatcher is celebrated from Gdansk to Bucharest as a great defender of liberty.

At the service in St Paul's Cathedral, I spoke from the Gospel of St John: 'I am the way, the truth and the life.' I later received a letter from Mark Worthington, her longest-serving aide, who was with her until the end, saying that if it wasn't for me she wouldn't have had the funeral she deserved. I was very touched. But it was the reaction that made the ceremony special. Because while protests had been predicted, and thousands of police lined the streets, the odd placard and chant were totally eclipsed by the spontaneous applause that followed her coffin along its route to St Paul's. Then, as she left the cathedral for the final time, the sunlight flooding in, those of us inside could hear the applause ring out louder than ever, and the crowd erupting into a hearty three cheers. It was magical.

And there was something magical too about this person, who had come from a grocer's shop in Lincolnshire to change the course of our history. I was lucky to have known her, to live in the Britain she had built, and to have the enormous privilege of being able to, hopefully, build upon her legacy – including in one crucial aspect. She was never afraid to take on the biggest issues, and she recognised that acting was far better than apathy. I was determined to do the same.

30

The Gravest Threat

The train journey between Brussels and Paris takes about an hour and twenty minutes – enough time for me and François Hollande to catch up between the European Council meeting we'd just attended and my visit to France on 22 May 2013.

I liked Hollande. He was free from the hauteur and grandeur of many French politicians, and was warm, amusing and down to earth. Though we had had recent disagreements over the EU budget, he seemed genuinely as keen as I was on defence, security and intelligence cooperation between Britain and France.

But our discussion was soon interrupted by events. Indeed, by the time we reached the Gare du Nord, the world would feel different to the one we had left at Bruxelles Midi. More dangerous. More brutal.

Ed came over and told me that a man, possibly a soldier, had been hit by a car and attacked with knives near the Royal Artillery Barracks in Woolwich, south-east London. 'PM' – I remember him making this plain to me – 'this wasn't just a stabbing; they tried to behead the victim.'

We got the live news coverage up on a stuttering smartphone. In the most extraordinary and disturbing footage, one of the attackers, wielding a knife and a meat cleaver, his hands soaked in blood, explained his actions to someone filming him on a phone. It wouldn't be the last time I would watch a murderous lunatic address me personally as he carried out the most gruesome of deeds.

As we sped through the French countryside, I thought of the as yet unnamed victim and his mother. Imagine being her; imagine being told her son had died – and then finding out *how* and *why*.

I thought, too, of the new reality we had entered. No longer were we at risk only from complex attacks plotted by well-organised terror cells.

Now it was also 'lone wolves' armed with nothing more than a car or a knife. I was touched by Hollande's genuine concern. There was a sense that this attack in London was an attack on us all. I took huge comfort in that.

By the time I stepped onto the platform in Paris, I had taken some early decisions. Theresa May would chair COBR in London. I would stay and carry out my plans for that evening, then go home. Abandoning my entire programme – giving off signs of panic – was what the terrorists would have wanted.

Yet at the same time, at moments like this the public needs to be reassured. Responses need to be coordinated. A prime minister needs to take charge. And, of course, I felt very strongly that I had to go to Woolwich – and that I had to go to MI5 too.

I have a huge sense of pride in our security services. As PM, every night my red box was full of vital intelligence reports. And every week I would receive a comprehensive update from my deputy national security adviser, Paddy McGuinness. Bald and burly, with ears that looked as if they had seen close combat on the rugby field, he was kind, intelligent and tough. A public servant to his fingertips, his notes always made fascinating – and chilling – reading.

Because of this, I was one of the few people in the country who had a window onto an underworld of malice and hatred. In my three years as prime minister so far, MI5, MI6 and GCHQ had foiled plots to blow up transatlantic cargo planes, the London Stock Exchange, British army bases and the London Olympics. And those were just the ones that had been made public; there were many more. Every single year I was PM, at least one attack the size of 7/7 was averted. It is impossible to know how many lives had been saved by these quiet heroes. But this brutal attack in Woolwich was the first Islamist plot that had slipped through the net in eight years. I knew how the intelligence officers would be feeling.

By the next morning I had been told the identity of the victim. Fusilier Lee Rigby of the 2nd Battalion of the Royal Regiment of Fusiliers came from Middleton, Greater Manchester and had served in Cyprus, Germany and Afghanistan. He was twenty-five and had a two-year-old son.

After a COBR meeting and a statement outside Downing Street, I travelled to Woolwich, where I met the local MP and former Labour minister Nick Raynsford, and members of the community. Some of the people who had guarded the soldier's body at the scene were dubbed the Angels of Woolwich. The story of one of them, Ingrid Loyau-Kennett, touched me deeply. She had leapt off a bus to help in what she thought was the aftermath of a car accident. When she realised it was an attack, she bravely engaged one of the terrorists in conversation in order to distract him. He said he wanted a war. She said he'd lose. 'It's only you versus many,' she told him.

I went to Fusilier Rigby's barracks and met his comrades. They were shocked but stoic; they had served in Afghanistan, and understood more than most the worldwide phenomenon of Islamist extremism, the poisoning of young minds and where it could lead. And they were fiercely proud of the job they did and the uniform they wore.

The day before, Philip Hammond, the defence secretary, had decreed that following the attack, soldiers should not wear uniform in public. Unusually for him, this was a stupid response: we shouldn't be asking our armed forces to hide away. To do so would be a victory for the terrorists. Fusilier Rigby hadn't been wearing uniform anyway; he was wearing a Help for Heroes hoodie. So I overturned the decree immediately.

Back in central London, I walked through the ornate archway of Thames House, the home of MI5. As you wander around the operations rooms, each with large screens showing live pictures, you realise how much surveillance is going on at any one time. The sea of faces at the computer screens was a microcosm of our nation: black, white, male, female, young, old – but mainly young. You could tell from the juice cartons and sandwich packets that these people worked night and day.

I stood in front of one of the big screens and spoke to the assembled staff. By this point it was clear that the two attackers had been on MI5's radar. I knew they would come in for criticism, and that there would have to be an investigation. But I also knew something else: there is such a large number of people who potentially want to do us harm that you simply cannot follow all of them all the time. The job of MI5 is to prioritise, but they are not omniscient. I wanted them to understand that I understood that. I told them they had my full support.

I had spent a long time working out what I thought about the rise of Islamist extremism. It created the defining events of my early years in Parliament – 9/11 happened during my first year as an MP; 7/7 the year I became party leader.

First and foremost, I thought the poisonous ideology that linked the atrocities of the attacks on New York on 11 September 2001, the Bali bombings of 2002, Madrid in 2004, London in 2005, Mumbai in 2008 and more was, and remains, the gravest threat we face. I also believed that, so far, the challenge had been met with muddled and messy thinking.

The criticism from many on the left was that it had nothing at all to do with Islam. It was instead the result of grievances against Western actions, and issues such as poverty, in Muslim countries. Meanwhile, there were those on the right who believed that the problem was entirely with Islam itself, that there was an inevitable 'clash of civilisations' between Muslims and Christians.

I certainly didn't believe the problem was the religion of Islam itself – over a billion people follow it peacefully, and the vast majority of them completely reject extreme Islamist ideology.

But nor was extremism simply the result of Western actions abroad. 9/11 happened before we intervened in Afghanistan and Iraq. Indeed, British armed forces and aid workers were defending and saving Muslim lives around the world. And while the poverty of parts of the Muslim world is undeniable, many of the terrorists came from middle-class families and were well-educated.

The problem was, predominantly, the ideology of Islamist extremism. Hate preachers had twisted and perverted the religion of Islam and turned it into a dogma of hatred, division and ultimately violence.

I became obsessed with the barriers that prevented us from dealing with Islamist extremism. Since 2005 the UK had been trying to deport the radical preacher Abu Qatada. He was wanted in Jordan on terrorism charges, but the European Court of Human Rights and Court of Appeal repeatedly blocked his deportation because he might face an unfair trial there.

Another radical preacher, Abu Hamza, was wanted in the US, but we were caught in a legal wrangle, and unable to extradite him. I once talked about 'human wrongs under the banner of human rights', and there was no greater example than Britain being stuck with these dangerous men because the law considered the principle of their individual comfort as more important than the reality of our national security.

A key part of dealing with the ideology was how to respond to it as a society. Early in my premiership I decided to make a speech setting out my stall on the subject, and the Munich Security Conference in February 2011 was the ideal moment. As so often, it was during the speechwriting process, this time led by Ameet Gill, that the policy approach was properly forged.

'Frankly, we need a lot less of the passive tolerance of recent years and a much more active, muscular liberalism,' I said. 'A passively tolerant society says to its citizens, as long as you obey the law we will just leave you alone. It stands neutral between different values. But I believe a genuinely liberal country does much more: it believes in certain values and actively promotes them. Freedom of speech, freedom of worship, democracy, the rule of law, equal rights regardless of race, sex or sexuality. It says to its citizens, this is what defines us as a society: to belong here is to believe in these things.'

When Lee Rigby was murdered in May 2013, so much of what we knew to be wrong, so much of what I had railed against in my Munich speech, had been present in the murderers' journey to jihad.

The hate preachers were there – a key feature in the radicalisation of both of these young men. One had even been filmed at a rally with Anjem Choudary, a notorious British cleric who peddled the sort of extremism I had talked about in Munich: a brand of bile – anti-British, pro-Sharia – that just kept on the right side of the law while inspiring volatile, vulnerable youths to push it to its logical, violent conclusion.

The online radicalisation was there. Much of the hate preachers' material had been accessed by the pair on the internet – indeed, this is where the two of them had been communicating and concocting their plan.

The broken society was there – a mixture of educational failure, petty crime and gang membership. Both attackers had done spells in prison. And, as I had said in Munich, while the lack of integration was a factor, poverty didn't really come into it: both of them had benefited from good upbringings. It was almost as if Islamist extremism was the ultimate gang to belong to, the most sadistic and uncompromising of them all.

The identity issue was clearly there, but it was complex. The two young men were born in Britain to Nigerian Christian parents. Yet they found no sense of belonging in any of that hinterland, instead forging an allegiance to a global '*Ummah*', or community, of Muslims. When one of the attackers said 'our lands' on that phone footage, he didn't mean the country that had given him every opportunity in life; he meant a region of the world he'd never even visited.

Yet in spite of setting out my thinking early on, our response to Lee Rigby's murder was not quite right. It was a little too gentle.

Yes, 'the perversion of a peaceful religion' that I had spoken about was clearly there. Indeed, the fact that such a high proportion of Islamist extremists are not born Muslims suggests that jihadists tend to have a poor grasp of the faith. Because – the gentler argument goes – if they actually understood the religion, they would know that there was nothing in it that justified this.

But is that really true? Is there *really* nothing in Islam to justify such dreadful acts? That was what I said in my speech in Downing Street after the murder of Fusilier Rigby. It was what I said again in Parliament when I gave a statement the following week. But I look back now, and think my words were not entirely consistent with the clear and strong argument in the Munich speech. There clearly *is* something in Islam, otherwise these extremists across the Islamist spectrum wouldn't quote its scriptures and honour its God.

That doesn't mean they were right to do so, but they *were* finding *something* in it to twist and manipulate and distort into a widely held world view (as is true of sects in many religions, including Christianity). This is not a war between Islam and the West, but a war *within* Islam, between the moderates and the extremists. Our job is to help the moderates in that conflict – and denying that any of this has anything to do with Islam doesn't help. The right answer is to state clearly that Islam must, and can, and often does assert itself as a religion of peace, but also to acknowledge the direct line the extremists draw from Islam to Islamist extremism and to violence.

I talked on tape a couple of days after the attack about how I wanted to see government policy reflect this position. I said: 'The sort of argument that's been made by the Lib Dems, former governments and a lot of Whitehall is there is a difference between violent extremism and extremism and you must concentrate on the violent extremists. That's crap. We've got to do more to go after the radical preachers, the radical mosques, the radicalisation on campus. The swamp in which these people swim has got to be drained more effectively.'

Pretty quickly I decided against holding a public inquiry into the murder of Fusilier Rigby. We needed answers urgently; there simply wasn't time for a judicial procedure to drag on and on. Instead, on 3 June 2013 I appointed the House of Commons Intelligence and Security

Committee (ISC), led by former foreign secretary Sir Malcolm Rifkind, to conduct a special report into events.

I also launched the first-ever Extremism Task Force.

Whenever there was even a hint of a specific type of threat, the intelligence machine would kick into action and try to work out how to prevent it from happening, and how to handle it if it did. We were, for example, hugely well prepared for a multi-gun attack as happened in Mumbai in November 2008.

In dozens of meetings of COBR I ordered a thorough testing of all our systems, more armed police, more and better body armour for the emergency services, and any number of technical improvements. All of this was critical to national resilience when five terrible attacks slipped through the net in 2017.

But after this lone-wolf attack in Woolwich, it was clear that we needed to look again at whether we were doing everything we possibly could to prevent not just terrorism but extremism in every form.

The good thing about the task force was that it drove policy from the centre, with me as chair, and covered all types of extremism, including the far right. Two years earlier, a far-right nationalist had murdered dozens of people, most of them teenagers, in Oslo and at a youth camp in Norway. Just a month before the Woolwich killing, an eighty-two-year-old Muslim grandfather from Birmingham was stabbed to death by a Ukrainian who was plotting to blow up several mosques around the Midlands. There was also the English Defence League (EDL) manipulating events and inciting hatred against people of all races and backgrounds. As far as I was concerned, you were an enemy of Britain whether you were Anjem Choudary or Tommy Robinson, the EDL's leader.

A number of measures were proposed by the task force when it reported at the end of 2013. One bold move was to agree a definition of Islamist extremism. Were you an extremist if you said women were inferior to men? Or did you have to promote violence against women? What about believing in loyalty to the *Ummah* over your own nation state? Or saying that, while terrorism is wrong, the terrorists do have a point?

There were all sorts of cabinet arguments over the wording. Theresa May wanted a narrower definition; she was cautious about criminalising people simply for holding unpopular views. Michael Gove advocated a wider definition that included anyone at odds with British values. Their row over this issue was to rear its head several years later. We managed to come up with an agreed form of words. It summed up Islamist extremism as an ideology which seeks to impose a global Islamic state based on Sharia law, rejects liberal values such as democracy, the rule of law and equality, and is distinct from the faith of Islam.

Another bold move was shifting the responsibility for dealing with extremism from government alone to society more widely. The onus was on everyone – schools, universities, the Charity Commission, prisons, councils – to actively intervene in stopping this (now defined) Islamist extremist ideology from flourishing in their institutions.

During one meeting in March 2015 with Nick Clegg, he and I clashed over the implications for universities. He saw them vetting their speakers as an affront to free speech. I saw it as a sensible and reasonable requirement to prevent young minds being brainwashed by speakers who had been given the credibility of a campus platform.

Yet there were other bold proposals that should have been acted upon more quickly. One excellent idea was for an Extremism Analysis Unit that would monitor which bodies and individuals the public sector should and should not engage with. During the Cold War, we would never have caught a communist spy without having a picture of all the different far-left groups out there. The unit was announced in March 2015.

This was far too slow, and I should have pushed harder. The power to prevent public bodies being infiltrated by extremists was needed urgently – as we were to see in some Birmingham schools later on. Also too slow was the investigation into the application of Sharia law in England and Wales, which was announced as part of the Extremism Task Force. The Home Office sat on it, and failed to launch it until May 2016.

A year after the task force wound up I was given the ISC report. It found that the intelligence agencies were not in a position to prevent the murder of Fusilier Rigby. While it did identify processes that needed to be improved – for example, there were delays in granting surveillance of one of the attackers – there was no evidence that this would have provided advance warning of the attack.

However, it did identify one thing that could have been decisive. Months before the attack, the killers were speaking on social media, and one wrote about his desire to kill a soldier. Though the internet company automatically shut down the accounts he used on the grounds of terrorism, it did not notify the authorities that it had done so. The report said that 'This is the single issue which – had it been known at the time – might have enabled MI5 to prevent the attack.'

Technological transformation was a big feature of my period in office. For decades our intelligence services had fought threats to our national security on two fronts – international and domestic. But now there was a third domain, a vast, ungoverned space: the internet. And it seemed to me there were three types of adversaries using that tool to do us harm in various different ways.

The first was extremists using it to communicate with one another, indoctrinate others and plan atrocities. It was clear we needed to go further and faster on interception and potentially blocking such communications. Simple telephone data was no longer enough: the internet led to the proliferation of different ways for people to contact each other. So our first big move was to compel companies to retain email and mobile phone contact data for twelve months.

But that alone wasn't enough. We needed these companies to keep records of each user's internet browsing activity, email correspondence, voice calls, internet gaming and mobile phone messaging services. We developed this into the draft Communications Data Bill.

That was where we ran into problems. The civil liberties lobby stood ready to resist anything that could be remotely spun as an invasion of individuals' privacy. The political wing of that lobby, the Liberal Democrat Party, catchily branded the Bill a 'snoopers' charter'. Without parliamentary support, it was dead in the water. Our security services would have to wait until after the election to get the powers they needed to keep us safe.

We also had to counter the extremists' propaganda. So we supercharged the Home Office's existing Research, Information and Communications Unit (RICU).

The second way the internet was being misused was the depraved criminals and paedophiles using it to share images of child abuse. I was sickened to learn that searching online for child pornography was neither blocked nor automatically reported. On 22 July 2013 I went public with my dissatisfaction, saying in a speech that while Google, Microsoft and Yahoo were going some way to address the problem – blocking images of abuse when they were reported – they were being purely reactive. Left unreported, those images – those crime scenes – were still out there, and nothing was being done to stop people getting to them. There needed to be a blacklist of terms that would offer no returns on search engines. It was their moral duty.

A meeting I'd had the previous week made me even more determined. Tia Sharp and April Jones were two little girls who had been abused and murdered by paedophiles. It was discovered in both cases that online images of child abuse had played a part in influencing their killers. I sat with their brave and grieving families, all of us incredulous that anyone could type this sort of thing into a computer and get away with it.

The tech companies were still arguing that they were just platforms, not content providers. So I said that I was willing to carry out one of these searches publicly, from his platform, to show how bad matters were.

By November there was a breakthrough. Google and Microsoft, which between them owned 95 per cent of the search-engine market, agreed that up to 100,000 search terms would return no results, except a warning that child abuse imagery is illegal. It was a huge success – and it was only the beginning. An organisation called WePROTECT was set up to broaden this work. It is currently bringing together eighty-five countries to tackle the scourge of child abuse online.

The third threat from the internet was hostile states and other actors determined to hack our systems, steal state secrets, leak classified information, disseminate propaganda and even launch cyberattacks on hospitals, airports or financial infrastructure that could bring society to a standstill and endanger lives.

I'll never forget visiting GCHQ and seeing a world map on the wall that showed real-time red lines of attack striking out from Russia and China and into the UK – and being told that our ability to deter such risks was sorely lacking.

That's why I said we'd spend an extra £1 billion on cyber-security defence, even in tough times. It's why I gave the go-ahead to a National Cyber Security Centre, which opened shortly after I left office. And it's why I was uncompromising when it came to those acting to subvert our state and security services.

In 2010, 250,000 stolen US diplomatic documents were published by the organisation WikiLeaks. Then in 2013 a contractor for America's National Security Agency, Edward Snowden, stole an enormous cache of confidential data and gave it to various outlets for publication. Experts compared the magnitude of the breach to that inflicted by the Cambridge spy ring in the 1950s.

When I read several files of the Snowden material I realised it wasn't so much the individual pieces of information that were damaging – though of course they could be – but that, put together, they formed a bigger picture of our security services' methods and techniques. Yes, I understood the need to scrutinise what the security services did in our name, but this data was indiscriminately dumped online – there was no scrutiny in that. I supported whistleblowing when there was a clear public interest, but there was no public interest here, only public injury.

The *Guardian* was one of the outlets that were given access to the data. Civil servants advised me to issue an injunction to prevent publication. I advocated a different approach. I knew that the paper's editor, Alan Rusbridger, and I had different politics, but I also knew he wouldn't want to do anything that jeopardised people's safety. So I asked Jeremy Heywood to go and talk to him about how dangerous this stuff was, and how susceptible to Chinese and Russian hacking. Although the *Guardian* would go on to publish stories about some of the individual revelations, Rusbridger allowed the hardware to be smashed up with GCHQ officials' oversight.

Then in April the European Court of Justice declared the rules we had used around retaining communications data since 2009 were invalid. I was warned that the public would be significantly more at risk as a result of this decision.

It was essential to get that power back. Even though we were essentially retaining the status quo, the opposition was huge. I sent Paddy McGuinness and security officials to brief some Lib Dem MPs on the realities, and they were won over. Jeremy Heywood worked wonders in talking to Labour, culminating in a meeting in my office that included Ed Miliband and Theresa May, at which we explained that the law we were shoring up with this emergency fix had been passed by Labour. So in July 2014 we passed the Data Retention and Investigatory Powers Bill. It was made permanent in 2015 with the groundbreaking Investigatory Powers Bill, which brought together all the UK state's intrusive powers while subjecting them to new and more stringent oversight.

None of this is to demonise internet companies. But I was always clear that the market should only be free to flourish if it behaves acceptably.

'You are not separate from our society, you are part of our society, and you must play a responsible role in it,' I said in my speech on child protection. It was a warning that could have applied to them on many sectors. Anyone who says we refused to call out companies' bad behaviour – the monopolising, the tax avoiding, the blind eye to criminal activity – that has led to the backlash against capitalism has a selective memory.

The temptation to be reactive rather than proactive is a danger faced by all of us in public life, not least in politics. It is a sad fact of our world that it often takes a tragedy to induce change. A stop sign is erected at a junction after a fatal accident. A drug is banned only once it has claimed lives. This was certainly the case with technology. I was determined that from now on we should be one step ahead of the criminals, the hostile states and hackers, and of course the extremists.

On that front, the situation was more desperate than ever. Indeed, just two years after that fateful train journey, François Hollande and I would find ourselves side by side in Paris once again, commemorating yet another attack in Europe, this time the deadliest to date.

31

Sticking to 'Plan A'

I've often said that I felt as if I was prime minister twice: once when Britain's economy was growing, once when it wasn't.

When your GDP is on the up, your power rises with it. Your global stature increases, public confidence grows, your party's fortunes rise, and your economy's success sparks the interest of investors. Growth begets growth.

But when GDP is stagnating or shrinking (or at least when you are told it is – the provisional figures don't always turn out to be true), you're in a permanent state of precariousness.

From the start of 2012 until mid-2013, that was the position I was in. I felt exposed, one of the few leaders of a major power arguing that a radical strategy of monetary activism, supply-side reform and fiscal responsibility – a form of what has been labelled 'austerity' – was the right way to get our economy growing.

Rescuing the economy was my biggest test – it was the thing upon which everything else rested. Yet, three years into office, it still wasn't growing as we'd forecast. Indeed, figures showed in April 2012 that it had tumbled back into recession, and therefore into a dreaded 'double-dip'.

As the seasons rolled by, even when we were plunged into what felt like an eternal economic winter, I was determined to stick to the plan. After all, politics isn't just about having the right philosophy, the right policies and the right people in the right positions. It's about perseverance: knowing that you are doing the right thing, and giving it time to bear fruit, even when the critics are circling and evidence of progress is hard to point to.

I believe it was that perseverance – and a handful of key people – that made the difference in the end.

From the beginning of our time in office, growth had been sluggish – and there had been one quarter of negative growth at the end of 2010 (two consecutive quarters means you're in recession). As I've said, we were doing everything we could to fire up the economy, from inward investment to infrastructure, education to entrepreneurs. There was a hike in the oil price, which had a big negative impact. But it was the Eurozone crisis, which reached its peak between mid-2011 and mid-2013, that really stopped us from seeing the fruits of our policy earlier in the Parliament.

There were obvious political consequences. While the OBR would say in 2014 that – as we had argued – the timing and depth of the slowdown was much better explained by the Eurozone crisis than by our domestic policy, our opponents at the time found it easy to blame us.

I spent many hours in meetings on the Eurozone chaos: whole nights under the harsh lighting of Justus Lipsius at emergency European summits. We were caught between the caution of Angela Merkel and the urgency of the Eurozone problems, as the member countries kicked the can down a road that stretched from Lisbon to Athens.

It wasn't until July 2012 – that magical Olympic summer – that we saw a glimmer of progress. On the eve of the Games I held that economic summit at Lancaster House to drum up investment from our overseas visitors. But Mario Draghi, the president of the European Central Bank, turned up with something even more valuable, our longed-for 'big bazooka'. In carefully chosen words, he made the statement that 'The ECB is ready to do whatever it takes to preserve the euro.'

Whatever it takes. With those three words he achieved what we couldn't during four European summits that year: calming the markets and restoring confidence.

A former Goldman Sachs banker, Mario had the intellectual self-confidence to take on the Germanic view that monetary activism was a dangerous heresy that would lead to inflation. In doing so, he was to prove one of the key people in saving Britain's economy.

This new approach, making clear that the European Central Bank would stand behind the euro and pledging to buy bonds across the zone if necessary, had a rapid and profound effect. Interest rates across Europe came down; investment and growth could begin once again. The contradictions of the euro hadn't been solved, but the immediate danger had passed.

I felt enormous relief. We badly needed that decisiveness. Now the euro was off death row, Britain's economy could start to grow properly.

Yet as that Olympic summer faded, global growth remained on a go-slow, credit continued to be rationed, and the crash was reassessed as being worse than originally thought.

Also, on 25 July 2012 the second-quarter estimate for GDP growth was published at -0.7 per cent. At the time this appeared to be the third consecutive quarter of negative growth. The row over our economic strategy – was it the right approach? Was it time to change tack? – was raging. It also drove the OBR to dramatically downgrade its annual growth forecast in the 2012 Autumn Statement to -0.1 per cent for 2012 and 1.2 per cent for 2013.

As a result, that Statement would have to contain an embarrassing admission. We wouldn't be able to get our debt falling as a share of GDP by the end of the Parliament as we had pledged. We would have to delay it by a year.

The wrangles with the Lib Dems over the Statement were fraught. They wanted another big increase in the personal allowance – effectively a tax cut. We wanted bigger welfare reductions to get government spending under control.

Was this a case of good cops versus bad cops? No: it was idealists versus realists. Of course we wanted to cut taxes. We just believed that any cuts had to be paid for; with an ever-widening deficit they simply weren't sustainable. And we knew that unless we got the deficit under control we'd never inspire the confidence that would lead to a genuine, long-term recovery.

We also understood that structural changes to spending had to be made while voters still appreciated the importance of the problem and would support difficult decisions.

Twice a year, the media would indulge in a frenzy of speculation about what rabbits George would pull out of his red box. But few saw the sorcery he performed in the lead-up to Budget Day.

What he conjured this time was little short of genius: the support of a wary Treasury, the backing of an increasingly lily-livered Lib Dem leadership, and a Statement that would get the rich paying more taxes, government expenditure falling, the deficit dropping, and a series of inventive measures to help stimulate business and the economy.

It's important not to forget the position he was in. This was a chancellor who was serving up more bad economic news, who had not long delivered the 'Omnishambles' Budget, and who had been booed by the public when he handed out medals at the Paralympics that summer.

Despite all this, he gave another forceful speech at the despatch box. We might not get debt falling until 2016, he said, but we had cut the deficit by a quarter already. Unemployment wouldn't peak as high as expected – it was only 8 per cent, compared with 25 per cent in parts of Europe. And to put it all into context – to show what a vast difference our tough measures were making – we had already saved £33 billion on lower debt-interest payments. That was almost as much as the UK's entire defence budget.

Economists talk about stress-testing banks, and this was the ultimate stress test for George. He passed. Indeed, this display of confidence was one of many masterstrokes that would make him an absolutely key person in rescuing Britain from the brink.

But we would have to wait a little longer before the results came through. Indeed, on the day after the Bloomberg speech in January 2013, at the dinner in Davos, I was given early sight of the next day's GDP figures: Britain's economy apparently shrank in the last quarter of 2012. Suddenly the double-dip recession was nothing: we might be heading towards a triple-dip.

This seemed hard to believe. And Rupert Harrison didn't believe it. As well as being a talented economist, Rupert had an acute political brain. In February 2013, at one of the many agonised meetings in my office discussing the forthcoming Budget, he made a daring prediction. He talked the team through the impact the Funding for Lending policy and rising business confidence were having on the economy. Low interest rates were finally being passed on to households through falling mortgage rates. Businesses were restarting projects that had been put on hold due to the chronic uncertainty over the future of the Eurozone. People were starting to feel more optimistic. This was the shot in the arm we needed, he said, adding: 'The economy will be going gangbusters by July.' Everyone laughed. We made him write down his prediction in Ed's notebook. Ed then stuck it on the wall outside my office. I hoped to God Rupert was right.

His confidence wasn't shared by other experts. Support for our strategy from an external coalition of institutions, publications, economists and credit-ratings agencies began to fall away.

First, the IMF's chief economist Olivier Blanchard said it was time for the UK to 'take stock', and released a report that hinted at the need for a different approach.

Moody's then downgraded our credit rating from AAA to AA1. Our recovery, it said, was slower than after the recessions of the 1970s, early 1980s and early 1990s. The *Economist*, which had been steadfast in support of our approach, also began to voice doubts, calling for more government investment spending.

Allies of Vince Cable briefed that he was ready to walk, and he wrote an article in the *New Statesman* that came very close to openly criticising the government's economic policy.

Boris had little to add to the rescue effort except a metaphor, this time complaining about our 'hair shirt, Stafford Cripps agenda' after Labour's dour post-war austerity chancellor.

Day after day, the airwaves were thick with hyperbole from interest groups. We were cutting just £1 in every £100 spent, but you'd think we had reinstated the workhouse. The gap between the much-discussed 'pain' and the delayed gain showed up in the polls: 67 per cent of people thought George was doing a bad job as chancellor, and only 7 per cent thought our economic strategy was working. That strategy was more and more frequently being referred to as 'Plan A', because there was talk of 'Plan B': a higher-borrowing, more Keynesian approach.

But none of these factors convinced me to change tack. Of course our recovery was going to be slower than in the 70s, 80s and 90s – the 00s financial crash was the worst in modern history. Of course the IMF was turning on us: the pro-stimulus US government was using it to fight the pro-austerity Republicans in Congress.

You'd never please Vince. Boris was Boris. The polls were unfortunate, but as Margaret Thatcher had said to me when I told her we were nine points behind, 'That's terrible – Labour's lead should be *far* higher than that during the middle of your term.'

And as spring sprang in 2013, there were signs that we were on the right track. A quarter of a million new businesses had been started. Unemployment was down. We had exported more cars than we'd imported for the first time in nearly forty years. But still there was no breakthrough – and certainly no gangbusters.

I was confident that if we stayed the course, we would be successful in the end. The one thing that would be fatal was if the public lost a sense of the problem we were attempting to resolve. It was time to set out all over again not only what we were doing and why, but also why it was taking time.

Standing in front of the hulking machinery at a manufacturing plant near Keighley, West Yorkshire, I delivered a speech reminding people of all the things that had gone so badly wrong in our economy under Labour. Excessive government spending. Reckless bank lending. High household debt. Over-regulated businesses. A housing price boom. Uncontrolled migration rather than proper training for UK workers. More and more public-sector jobs plugging the fall in private-sector jobs. Five million functionally illiterate adults. Half of all teenagers failing to get at least a C in GCSE English and maths.

Understanding what Labour did wrong was key to understanding what the coalition was doing to put it right. At the heart of it all was the deficit reduction programme, vital to restoring domestic and international confidence in UK PLC. A plan that risked losing international confidence, or that threatened higher inflation and higher interest rates, was no plan at all. Inflation wasn't just a government statistic: a 1 per cent increase in interest rates would mean an extra £1,000 on an average family's mortgage repayments.

Britain's fiscal credibility on the global stage really mattered. We had a mountain of both public- and private-sector debt. A sharp rise in interest rates, as had happened in countries that had lost the world's confidence, would cause enormous economic damage, with companies going bust and families losing their homes. Low interest rates helped to steady international investors, who might otherwise have been ready to jump ship. Exports and foreign investment weren't just good for businesses, they translated into jobs. Indeed, the factory I was speaking at was French-owned, exported 90 per cent of its goods overseas, *and* had hired more staff during the recession.

Our approach was undeniably about people, about their lives and livelihoods. Still, Ed Miliband continued with the lazy attack that deficit reduction hurt ordinary people, and our plans for growth were all about helping the rich. But I suspected the British people would understand the simple logic of our position, given that it mirrored the tough choices they were having to make in their own businesses or household budgets. I concluded therefore that there was 'no alternative' to what we were doing – deliberately echoing the line used by Margaret

Thatcher in 1980, when things were getting worse before they got better.

That speech set the tone for the Budget. This was the moment of maximum danger; the growth forecast for 2013 had been halved. The following month we would find out if there had indeed been a triple-dip. The pressure on George was enormous, but he stuck to his guns.

He deftly defused the bombs of fuel and beer 'duty escalators' (automatic tax increases, set out in law) fixed by the last Labour government, freezing fuel for the fourth year in a row. One of the biggest packages in history aimed at tackling tax avoidance and evasion was unveiled. That would pay for making the first £10,000 of everyone's earnings tax-free – at a stroke increasing the incomes of twenty-four million people and taking 2.7 million more out of paying any income tax at all.

Corporation tax falling to the lowest level in the G20 was meant to be the big business-friendly story, but the policy that turned out to be most popular was one that was thought up by David Gauke, the Treasury minister: exempting every business from the first £2,000 of National Insurance contributions. This wasn't worth much to a vast supermarket chain, but it made a big difference to a small store or start-up business.

And then, finally, summer arrived, and with it a moment we had been waiting for. It wasn't just positive growth forecasts for the future; it was more the past that made the headlines. On 27 June 2013 the ONS revised its economic data for 2012. Not only had we averted a triple-dip, there hadn't even been a double-dip in the first place. In fact, the middle quarter of the three negative quarters (the first quarter of 2012) was revised up from -0.2 per cent to 0 per cent. So it looked then as if we had only just avoided a triple-dip. Today, the data shows that we were nowhere near a triple-dip: after further revision, that first quarter of 2012 showed 0.6 per cent growth. Two earlier quarters, in late 2010 and late 2011, that were originally reported as negative, were revised and showed growth. Talk about damned lies and statistics. I joked to George that I wanted to sue the Office for National Statistics for giving me stomach ulcers and chronic stress.

I'd seen the policy of deficit reduction criticised on TV, derided at the despatch box, and questioned in international forums from the European Council to the G8. Now here it was: Plan A hadn't resulted in the turmoil everyone had predicted. Indeed, our international allies like the US started to tighten their belts.

Still, though, we needed to shift the dial; and we needed people who could help us do so. While a brilliant economist and a shrewd adviser, Mervyn King had become somewhat passive at the Bank of England, and even quite resistant to what we wanted to do on credit easing and activist monetary policy. His term was due to end in 2013, and we were determined to find a governor who could make a real difference.

Three thousand miles away in an office in Ottawa, Dr Mark Carney was doing something special at the Bank of Canada. Like Mario Draghi, he knew that the job of a central bank's governor was not simply to safeguard the banking system and control inflation. He knew that a central bank had a role in supporting growth and tackling monetary problems in the economy. He was an innovator, and the sort of person who would use the Bank's balance sheet to unblock the financial system. A supporter of 'forward guidance' on keeping borrowing costs low, he also chaired the Financial Stability Board, part of the international architecture that was sorting out the financial system in the aftermath of the crash. He was exactly who Britain needed in the hot seat at Threadneedle Street.

There were so many reasons he couldn't do it, not least moving his young family to the UK, but George simply wouldn't take no for an answer, persevered and we got our man. In July 2013 Mark became the first non-British governor in the Bank's history. In office, he moved fast to introduce forward guidance on interest rates, to unblock the financial system by revamping the Bank's liquidity operations, and to modernise the Bank's internal workings.

Business surveys started picking up during the first half of 2013, and on 25 July the GDP figure for quarter two came out as *plus* 0.6 per cent. In other words, in time for the July target Rupert had envisaged.

Within a few months the IMF estimated that our growth in 2014 would be faster than that of any other G7 country. Employment reached a record high. The top 1 per cent of earners were, for the first time since the beginning of the twentieth century – thanks mainly to the reduced top rate of tax – contributing more income tax to the country's coffers than the bottom 75 per cent.

Olivier Blanchard was 'pleasantly surprised'. Vince was vanquished. Boris kept quiet. George and I were vindicated. And Rupert was right: our economy was, finally, going gangbusters.

32

Love is Love

I must have been in and out of the front door of 10 Downing Street thousands of times, but there is one moment I'll remember more than most. As I was leaving, one of the custodians stopped me and said, 'It's because of you that I'm able to marry my boyfriend this weekend.' It was a reminder that politics has the power to make a difference and change people's lives.

Equal marriage was one of the most contentious, hard-fought and divisive issues during my time as prime minister. We would lose party members; one even came to my surgery and tore up their membership card in front of me. It was an issue that I would worry and even wobble over. But I have absolutely no regrets, and it is one of the things of which I'm proudest.

In 'coming out' for gay marriage, in some ways I surprised myself. As I've said, perhaps I was a slow learner when it came to modernisation. I have a terrible tendency to get lost in the endless detail of policy problems – the man under the car bonnet again – and often fail to see the bigger, emotional picture.

I was on the wrong side of Section 28. I also ended up abstaining on – rather than voting against – IDS's rejection of gay couples' right to adopt. I should have proactively supported that right. Even when it came to civil partnerships in 2004, which I was fully in favour of, I remember my arguments: they were great for ensuring that gay couples weren't discriminated against when it came to important details like hospital visiting rights, inheritance and rights for bereaved partners.

I remember Samantha hitting back quite forcefully. 'You've got the right conclusion, but the wrong argument,' she said. I was the difference-splitting, circle-squaring policy wonk focusing on all the practical details, when there was a far bigger point staring me in the face: people should be able to enter into a legal union with the person they love. End of.

I have always believed in freedom of choice. And I have always been a strong believer in marriage – a rather unfashionable view in some ways. There were all the maxims about its benefits: family is the original welfare state; marriage is a framework of commitment. There were all the facts and figures: parents who are married are twice as likely to be together when their child is sixteen than those who are not married but living together.

And for Sam and me there was more than statistics. There was the strength of the vows we made in 1996. Marriage stops you running at the first sign of trouble, and we were proof of that. The pressure that having a disabled child puts on a couple can cause a marriage to break down. Despite the advantages we had, we had felt the strain during the most difficult times with Ivan. But we got through it.

'Pledging yourself to another means doing something brave and important,' I said in my 2006 conference speech. 'You are publicly saying, "It's not just about 'me, me, me' any more. It's about we: together, the two of us, through thick and thin." That really matters. And by the way, it means something whether you're a man and a woman, a woman and a woman, or a man and another man.'

That was what I felt in my heart. Love is love, commitment is commitment – whoever you are and whatever your sexuality. My commitment to equal marriage was the logical next step – though it would take a few people to convince me.

There were many chats with Sam. For a long time she'd been saying, 'A civil partnership is really a marriage, so why don't you call it a marriage, and then it's properly equal? It's not going to make me feel any less married if two gay people want to get married.'

Many of the people around me – particularly George, Kate, Danny, Nick Boles and Michael Salter, my head of broadcast – continued to push on this. We had taken important steps as a Conservative Party on these issues. The next big thing we could do for gay people – indeed, the missing piece of the equality jigsaw – was to allow them to get married.

Despite my increasing keenness on the policy, by 2010 I hadn't quite fully come around to the idea of gay marriage, and it wasn't in our election manifesto. However, the accompanying 'Contract for Equalities', by the then shadow equalities minister Theresa May, said that we would consider the case for changing the law to allow civil partnerships to be called and classified as marriage.

Then, in coalition, we were working with a party, the Lib Dems, that backed gay marriage. I stood before my party – a party that carried so much baggage on this issue – and announced that gay marriage would be our next big, progressive social policy.

The way to do it, I decided, was not to talk about what a departure it was. It was to talk about how the reform was rooted in our fundamental beliefs: 'Conservatives believe in the ties that bind us; that society is stronger when we make vows to each other and support each other. So I don't support gay marriage despite being a Conservative, I support gay marriage because I am a Conservative.' Applause rung out. That short sentence was a giant leap for our party.

Getting it through Parliament was going to be a team effort. Theresa May, who as home secretary had the Equalities brief, met senior Church leaders to try to explain our position. Her equalities minister, the Lib Dem Lynne Featherstone, did an excellent job leading on what became the biggest consultation in government history.

In 2012, when I moved the Equalities brief out of the Home Office and to the DCMS, Maria Miller had the responsibility for driving the beast of a Bill through Parliament. Officials told her it would take a year to get it through; I asked her to do it in six months.

However, I wasn't expecting either the level or the nature of the opposition we faced. Conservative associations were broadly opposed. There was fierce opposition from much of the parliamentary party, which went far beyond the usual suspects. And the cabinet contained opponents too. I knew from the off that Philip Hammond would be opposed, and Owen Paterson joined him.

My director of strategy Andrew Cooper wrote to me on 20 April about the concerns that were expressed in political cabinet. He explained that 'the assertion that Middle Britain' opposed gay marriage was completely false. 'Middle Britain' was *in favour* of gay marriage. The principle that 'gay couples should have an equal right to get married, not just to have civil partnerships' was supported by more than a 2–1 margin, 65 per cent to 27 per cent.

My parliamentary aide Dessie was, as I've said, a Bible-believing Christian. And he, on theological grounds, was not opposed to gay marriage: 'We use our judgement to deal with situations that were never contemplated in the Bible. That judgement, however, ought always to be informed by the principles taught in the Bible, and summarised by Jesus, as loving God and our neighbour.'

Then there was Patrick McLoughlin. He came into my office one afternoon and told me straight, 'You've got to stick with this, boss. I'm an old-fashioned Tory, and a Catholic to boot, but I'm fed up with our party being on the wrong side of the argument every time. I feel embarrassed that I opposed civil partnerships, and I never want to be in that position again.'

Of course I understood people were uneasy about the change. The liberal bigotry which condemns the character of anyone who dares to hold traditional beliefs is itself illiberal – and growing. So I always made it clear that it would be a free vote.

There were two options for the legislation. The first was a one- or two-clause Bill that made same-sex marriage legal, and would allow the first gay marriages to take place in January 2014, but would have to be augmented with secondary legislation. The second option was a 'carry-over Bill' that (unlike most legislation, which fails if not passed within the year) would be more detailed and would spill over into the following year. The latter was safer, but not risk-free. I opted for the quick option. People had waited long enough to be able to marry each other, and we just needed to get on with it.

I also had to square the biggest circle, which was the Church.

The clergy that I knew personally were in favour. Mark Abrey, my local vicar at St Nicholas's Church in Chadlington, was an enthusiastic supporter. So too was the priest at my children's school in Kensington, Father Gillean Craig. I'd see this eccentric, High Church figure when I dropped Nancy and Elwen off, standing at the school gate in his long wool coat and a beret, and I attended the school services he took. 'Good for you on gay marriage,' he said after one. 'We're not all opposed to it or sitting on the fence.'

Except most were. And I was determined to try to neutralise this hostility, producing four safeguards that were known as the 'quadruple lock'.

One: a religious marriage ceremony for a same-sex couple would only be possible if the religious organisation carrying it out had 'opted in' (most didn't, but the Quakers, Unitarians and Reform Jews did). Two: no religious organisation could be forced to marry same-sex couples. Three:

it would not be unlawful discrimination for a religious organisation or representative to refuse to marry a same-sex couple. Four: it was not a common-law legal duty on the clergy of the Church of England and the Church in Wales to marry same-sex parishioners (as it was for hetero-sexual couples).

When I look back now, I think the quadruple lock was excessive. Stipulating that an entire Church must agree to gay marriage if just one wedding is to take place in one church seems over the top.

I was disappointed that Churches took the strong line they did, especially when I knew some of their leading members privately supported it, but kept the 'party line' in public. Justin Welby (who became Archbishop of Canterbury in 2013) made positive noises, but publicly he had to oppose it.

I gave a speech in the garden of Downing Street in 2012 for the annual LGBT reception, and was quite clear about my frustration. 'I run an institution – the Conservative Party – which for many, many years got itself on the wrong side of this argument. It locked people out who were naturally Conservative from supporting it, and so I think I can make that point to the Church, gently. Of course this is a very, very complicated and difficult issue for all the different Churches, but I passionately believe that all institutions need to wake up to the case for equality, and the Church shouldn't be locking out people who are gay, or are bisexual, or are transgender from being full members of that Church, because many people with deeply held Christian views are also gay. And just as the Conservative Party, as an institution, made a mistake in locking people out, so I think the Churches can be in danger of doing the same thing.'

With the legislation prepared and the Church circle sort-of squared, the next matter was passing it into law. Labour and the Liberal Democrats would ensure that the votes were delivered, so there was little danger of it not going through Parliament. But getting gay marriage through the Conservative Party – and I wanted their support – would prove to be like the proverbial camel and eye of a needle.

The Second Reading took place on 5 February 2013, and passed by 400 to 175. Ultimately, 136 Conservative MPs opposed the Bill. Some of them I knew were lost causes. But others shocked me.

Meanwhile, amendments to the Bill were tabled in the Commons and the Lords to equalise civil partnerships and introduce humanist marriages – which would have delayed it – and to hold a referendum – which could have wrecked it.

Despite that, the legislation passed on 21 May 2013, by 366 votes to 161. Overall, 134 Conservative MPs opposed the Bill, accompanied by fifteen Labour MPs, eight Democratic Unionists, four Lib Dems and one independent.

There were parties outside Parliament – MPs inside the Chamber could hear the cheering and singing. But I was at the European Council that day, and found myself being enthusiastically hugged by the bow-tied Belgian PM Elio Di Rupo. The world's second openly gay head of government was embracing the first centre-right politician to have legalised same-sex marriage.

Since then, over 20,000 same-sex couples have been married, and over 10,000 more have converted their civil partnerships into marriages. For many, this meant marriage later in life. Think of all of those years: couples who either had to hide their relationship, or settle for a civil partnership, finally able to declare their love for one another as freely and legally as anyone else.

Even UKIP, which made such a play of opposing gay marriage, eventually said they would not reverse it. Twelve countries around the world have followed suit and legalised it, and more will follow. When I came out for gay marriage in 2011, Barack Obama was still opposed to it. As we legislated in 2012, centre-right leaders in France and Germany were still voting against it.

In 2015 Britain was voted by the Lesbian, Gay, Bisexual, Trans and Inter-sex Association as the best place in Europe to be gay.

As Michael Salter – now Michael Salter-Church, having married his boyfriend Rob – put it, it is about more than the legislation, or even marriage: it is about the message it sends out. In the year I was born, homosexuality was still illegal in Britain. But today, all people growing up in this country know that they are equal in the eyes of the law and society, whoever they are, and whoever they love.

There was something else in this area that we needed to put right.

Alan Turing was the brilliant mathematician whose work cracking Nazi codes at Bletchley Park is thought to have helped to shorten the Second World War by two years. Yet in 1952 he was convicted of gross indecency, and was chemically castrated. He committed suicide shortly afterwards. But it was society that should have been ashamed. After the wartime secrets of that quiet manor house in Buckinghamshire came to light in the 1970s, Turing's genius was finally recognised, and his treatment condemned. I wholly supported those campaigning to absolve him, and it was fitting that 2013, the year in which gay marriage was legalised, ended with a posthumous royal pardon for Turing, granted at the government's request. As promised, after the 2015 election, all those convicted of such offences would be pardoned, and thousands of terrible wrongs put right.

An unhappy by-product of the gay-marriage triumph was that it continued to foment unrest in the parliamentary party. Rebellion seemed to follow rebellion. Despite that, in 2013 we were starting to turn a corner when it came to party management – partly because we had an election in two years and people actually wanted to win, partly because we were seeing a growing economy, and partly because we had got better at managing things, helped hugely by Jo Richards and Laurence Mann in No. 10.

At the same time, the sensible forces in the parliamentary party were beginning to fight back, with the 301 Group and the 2020 Group winning a series of battles. In the 1922 Committee elections in 2012, one serial rebel, Peter Bone, was unseated and another, Stewart Jackson, was defeated. Philip Hollobone was also voted off the Backbench Business Committee. This prevented an unrepresentative minority from having a disproportionate dominance over party affairs and parliamentary business.

In my seven years so far as party leader, there hadn't been much talk of leadership challenges. That is quite unusual for a prime minister – think of all the attempted coups against Gordon Brown and John Major; think of Thatcher and Michael Heseltine. It was also surprising given the fact that we were in coalition, and had a Parliament prone to rebellion.

But in 2013, chatter started. The threat – I was slightly surprised to learn – came from Adam Afriyie, MP for Windsor. He had been holding dinners for potential supporters, I was informed in regular notes from Sam Gyimah, my new PPS (Dessie was now a Treasury whip). Looking at the names, I branded them 'the ultimate B team'. Adam, meanwhile, had real potential as an MP, but was making a bit of a fool of himself.

He made his move by attempting to amend the Bill on the EU referendum to force a nationwide vote before the end of October 2014. Almost all backbenchers agreed that this was simply not a sensible suggestion, and did not leave enough time for renegotiation. The amendment went down in flames, 249 to 15. Even Jacob Rees-Mogg condemned the stunt as unhelpful.

But Adam wasn't the only suspected challenger.

Theresa May popped up in March 2013 to deliver a 'wide-ranging' speech at a ConservativeHome event. Alarm bells ring when someone steps outside their brief. But when they start to wax lyrical about their brand of conservatism, that's when the leadership-pitch klaxon goes off.

Theresa's intervention, however, was something I dismissed at the time. I put it down to her notoriously ambitious and aggressive spads (they were frequently at the heart of rows, and ran the Home Office like a fortress). Michael Gove, on the other hand, was red-faced furious. Even though he himself wasn't one to confine himself to his own brief, he chastised Theresa for her actions during a heated political cabinet meeting. Everyone else stared at the table.

Theresa is sometimes seen as an iron lady, and she can certainly be tough, but in person she is very reserved and rather nervy. She had also been nothing but loyal to me. A couple of days later I had breakfast with her in Downing Street. I thought she was uneasy and embarrassed about the whole thing. I told her I was sorry about what Michael had said, and that I didn't put him up to it. She replied, 'I've always been loyal, and I would never stand against you.'

I didn't feel threatened by Adam or Theresa. In politics it's the anonymous plotters you have to watch out for, and I could never quite shake that feeling of vulnerability.

Despite all that, as 2013 turned into 2014, I was beginning to get my mojo back. Party management was working better. Cabinet camaraderie was much improved. The Policy Unit in Downing Street was firing on all cylinders: I had brought in some really bright people with great ideas, and given the task of leading it, and gearing up for the manifesto, to Jo Johnson, MP for Orpington and brother of Boris.

I also had a new PPS. Sam Gyimah was great, but he was more of a policy person. I needed a Hoover of gossip, who lived and breathed Westminster – with legs so hollow that he or she could spend hours drinking on the Terrace and eating in the tea rooms.

In Gavin Williamson I found that person. He was likeable, fun and different, with the face of a twenty-three-year-old researcher and the mind of a wizened whip. He would try to lighten my mood with stories or quotes about what MPs had been getting up to. 'Which Tory minister has had carnal relations with a Labour MP?' he once asked me as he wandered into my office.

Tony Blair once said that as time passes, a prime minister becomes more capable but less powerful. I'm not sure I felt the same. More and more I felt that I was reaching the top of my game. I had the makings of an A-team around me, I was getting the hang of things, and gay marriage had shown me how you can dramatically change society and lives through sheer persistence and belief. But what about internationally? When action was needed once again to prevent murder, would I be able to take my party with me?

33

A Slow-Moving Tragedy

Just as nobody predicted the Arab Spring, none of us guessed that it would be Syria where that spring would turn to such a bleak, harsh and apparently ceaseless winter.

The rise of social media, the growth of young populations in the Middle East, the spread of Islamist extremism, Sunni/Shia sectarianism and Western anti-interventionism – so many of the trends I witnessed during my time in British politics collided in this eastern Mediterranean country, which would become the site of the bloodiest conflict of the twenty-first century so far.

You can measure the conflict by its scale: nearly 500,000 people dead, twelve million – over half of the population – displaced. You can measure it by its brutality: hospitals barrel-bombed, civilians used as human shields, children gassed to death, whole towns bulldozed and bombed into submission. You can also measure it by the consequences it had for other countries. It almost overwhelmed Lebanon, Turkey and Jordan, whose vast refugee camps I visited. It contributed directly to the rise of terrorism, with the sadism of President Bashar al-Assad helping to transform the ailing al-Qaeda in Iraq into the Islamic State of Iraq and al-Sham – ISIS – the most prolific terrorist group in the world. And it drove more migrants to seek safety in the West, causing angry political conflict in Europe and undoubtedly contributing to the Brexit vote.

Before the uprisings, Syria looked like a relatively stable country. Compared with many others in the region, it was a relatively safe place to live if you were a Christian, Druze, Kurd or other minority group. The country's leader was a softly spoken, suit-wearing, British-educated doctor who seemed committed to a more Western future, and to bringing his country in from the cold.

When William Hague visited him in Damascus in January 2011, just as the Arab Spring had broken out, Assad told him confidently that his regime would avoid the fate of the dictators in Egypt and Libya. Apart from anything else, he said, Syria was united by an ideology of resistance to Israel, and it had been Egypt's foolish attempts to make peace with Israel that had opened the door to revolution.

But appearances were deceptive, and his confidence was delusional. Sure enough, demonstrations began in Syria in March 2011. People protested for the same reasons as their Middle Eastern neighbours: the corruption of their rulers, their desire for jobs and a voice. They believed that there was an alternative to this police state that didn't tolerate dissent, and that this was the time to seize it.

Initially the protests were peaceful. A Foreign Office assessment sent to me at the time rated Yemen and Egypt as 'red' in terms of likelihood of instability, whereas Syria was rated 'medium to low'.

But then Assad began to show the sort of dictator he really was. The regime deployed army and paramilitary units, killing scores of protesters. Within months a multi-factional opposition had formed, as demonstrators armed themselves and resisted his rule.

At this time the Foreign Office view was that Assad could not survive beyond the end of 2012.

Yet far from folding, Assad carried on playing, deploying every card he had. He exploited ethnic divisions between the majority Sunnis and the powerful minority Shias, of which his Alawite sect was a part. Some people who initially rose up against the dictator fell into line because they feared a takeover by the mostly Sunni rebels.

Unlike in Libya, where the opposition had been quick to express their unity, in Syria the opposition was more divided. Over time, a 'national coalition' which we would support would form, but it was deeply divided between various factions.

There was the Free Syrian Army (FSA), which was formed of hundreds of groups, largely supporting a democratic, secular future for Syria.

There were many legitimate Muslim Brotherhood-style Islamic groups, which would accept a Lebanese-style settlement with guaranteed rights and protections for minorities like the Alawites.

But there were also jihadists, like the Al-Nusra Front and later ISIS, who saw Islamist ideology as Syria's future. That was a gift to Assad. He could say that, rather than being a genuine opposition, the revolutionaries were extremists – indeed, he would actually aid those jihadists, buying their oil, releasing them from jail, sparing them his firepower. In other words, there was poison in the antidote. This was one of the factors that condemned Syria to destruction.

Just as the Syrian people fell into sectarian camps, so too the world lined up on the side of either the Sunnis or the Shias, with most of the Gulf countries backing the predominantly Sunni rebels, and Iran and Russia supporting their old ally Assad with equipment and personnel.

These helped to create the conditions for a prolonged conflict. But the most important factor in Syria's plunge into the abyss, the primary reason for the conflict's length and viciousness, was the way the regime chose to fight: depopulating and terrorising rebel-held areas, indiscriminately targeting them with heavy weapons, and cutting off aid supplies. This created a refugee crisis, while also enabling the regime to force Syrians either into territory it controlled, or out of Syria entirely. That denied the rebels the population from which they could draw their manpower. Syria would be destroyed, and Syrians killed and dispossessed by the hundreds of thousands, simply to enable Assad's survival.

Then there was the fact that Syria was known to have large stockpiles of chemical weapons. From mid-2012 I received reports that Assad was testing chemical delivery systems, and that concerned me. It wasn't far-fetched to think that a man who was inflicting such brutality on his people would resort to such tactics.

Of course, there were voices asking us why we wanted to see the back of the Assad regime when we couldn't possibly know what would follow. But when you measure the possibility of a bad outcome were he to go, versus the certainty of continued dreadful brutality and a descent into deepening civil war and Islamist extremism were he to stay, I believed the choice was straightforward. In what sane world could this brute be part of the solution? Who was going to put down their weapons if the political compromise on offer included Assad? And there was a clear British national interest: continued conflict without transition to a new settlement would fuel extremism and terrorism, while destabilising the regime and driving ever-larger flows of refugees.

I was determined that Britain would be in the vanguard, because without our leadership I feared stasis. Back in August 2011 I made a statement, together with Merkel and Sarkozy, calling for Assad to stand down. In August 2012 we agreed to give 'non-lethal assistance' to the rebels, such as body armour, medicines and communications equipment. We had expelled Syrian diplomats from London. We set up a 'contact group' of countries to coordinate sanctions against the regime and assistance for the opposition.

As fighting raged in cities like Homs and Aleppo, I became convinced that the only way we could help to remove Assad and achieve at least the chance of positive change was if two things happened. We needed the diplomatic activity and an international alliance to try to secure his removal via pressure from above, and we needed to help generate the pressure from below by supporting the opposition. The ideals of the former could never be achieved without the latter.

Supporting the opposition never meant Western boots on the ground in significant numbers. An Iraq-style invasion was politically unsellable at home and abroad, and wouldn't have been right in this conflict anyway. But I knew that for the opposition to have any chance against the regime we had to go beyond this 'non-lethal assistance'. They would need our 'lethal assistance' – weapons, training, and back-up from the air. That became my mission: to shift the dial towards this approach.

Legally, we had to build the case that the UK could legitimately provide lethal assistance.

Militarily, we had to get the generals out of their Iraq and Afghanistan mindset, towards a more indirect approach based on training, equipping and mentoring irregular Syrian forces.

Politically, we couldn't continue to palm off all responsibility onto a UN mediator hamstrung by formal diplomacy. We needed to get our hands dirty, with covert deal-making between key opposition figures and Alawites, even with the Assads themselves, so they had an acceptable exit strategy.

Domestically, we had to keep counter-terrorism at the heart of our strategy. Our aim was to get the successor regime and the 'good' opposition to combine forces and destroy the plots at source. The NSC endorsed this approach in early November 2012.

Diplomatically, we had to build the case with other key players. I visited the UAE and Saudi Arabia that month to move them towards supporting the rebels. William had already been to the United Nations and helped to put the case for a strong resolution, which Assad's friend Russia, together with China, unsurprisingly blocked.

The UK then took the assertive step of announcing official recognition of the National Coalition of Syrian Revolutionary and Opposition Forces, supported by the FSA, as the 'sole legitimate representative' of the people of Syria. America followed suit a couple of weeks later.

Once again, all this hinged on US support. Syria was on a different scale to Libya; to have America just 'leading from behind' was not an option. Only it could coordinate a big training and mentoring operation while providing the intelligence, surveillance and air power that would give the opposition the decisive edge on the battlefield. As soon as he'd won the presidential election of November 2012, I told Obama, 'Syria is the foreign policy issue which will define your second term,' willing him to go further and faster now he had four years ahead. He was cautious, but knew we had to look afresh at all options.

In January 2013, Hugh Powell gathered representatives of relevant agencies from the US, France, Turkey, Jordan, Saudi Arabia, the UAE and Qatar to try to forge a serious commitment to integrate mentoring teams with opposition forces inside Syria. The objective was to force the regime's leadership to agree a ceasefire and transitional governance arrangements which would preserve much of the basic state structure. Assad would go, but not immediately.

A mini regional headquarters was agreed. We set up a collective process for monitoring and assessing which opposition fighting groups were sufficiently reliable.

But we needed lethal force, for example the use of special forces and equipment. And we needed a plan when it came to chemical weapons (we had already asked the military at the start of the year to identify potential targets for airstrikes).

At the time, I described the situation as 'right direction, insufficient force'. It just didn't feel as if our allies shared our urgency, compulsion and investment. 'I think the history books will write that Britain was active and on the right side and making the right arguments,' I recorded. 'But it is frustrating that it's not going the whole way.'

Yet if I looked closer, the frustrations weren't solely coming from overseas. Many of them were much closer to home.

For a start, I was pretty frustrated with my cabinet. While George Osborne, Michael Gove and Oliver Letwin strongly supported a more muscular foreign policy, and Nick Clegg was of a similar mind to me, around the cabinet table there was something of an alliance between Iraq War-obsessed Lib Dems and non-interventionist Conservatives who saw few key British interests at stake. Some senior cabinet ministers were sceptical about us becoming further involved.

Our military and security services were, on this issue, a huge source of frustration. From late 2012 I proposed that a small group should meet to discuss Syria, including cabinet members, officials and military top brass, as well as Jon Day (head of the Joint Intelligence Committee) and John Sawers ('C', the head of the secret intelligence service).

David Richards and Sawers offered their cold calculus: that what we had in our toolkit could not guarantee the outcome that we wanted. 'Maybe it's just a wicked problem that cannot be solved,' was the gist. I made the point that just because we didn't have the ability to achieve change on our own, that wasn't a decisive argument for holding back altogether. Indeed, we could only galvanise the necessary international action if we were prepared to take steps ourselves.

'We might make things worse, Prime Minister,' I was told.

'How can we make things worse than a dictator murdering his people?' I would say, exasperated.

A further frustration was that I still wasn't getting the accurate information I needed. Sometimes I felt that I picked up more from odd members of the Syrian diaspora I'd meet than I did from our diplomats.

Nor was I getting the options I needed from officials. National security adviser Kim Darroch wrote to me that there was some reluctance in the Whitehall machinery to put pen to paper with ideas. I wrote back: 'This has to stop – just tell people to caveat what they say!' – pleading for the frankest advice and the boldest ideas, even if civil servants had to hedge their language.

Kim wrote again about the scepticism and the tendency to seek out legal obstacles. As I scribbled on a later strategy document: '. . . of course (HISTORIANS note) we need legal advice. But we should consider what might work and then ask lawyers – not the other way round.' If the military's view was 'all or nothing', the FCO's seemed to be 'almost nothing'.

I understood their nervousness. The aftermath of the invasion in Iraq had been horrendous. There was an inquiry under way. But I also believed it was misguided. Civil servants could only be held to account for the things they advised. It was different for me. Prime ministers are not just responsible for what they do, but for what they don't do. I could see what was going to happen if we didn't act. If Assad won, or clung on, it would mean more towns devastated, lives destroyed, families displaced – and more terrorism here in Britain. There was no doubt: this dictatorial, corrupt, brutal, bloodthirsty leader slaughtering Sunni Muslims while the West stood to one side would be the greatest piece of propaganda for extremism.

I have always believed that just because you can't do *everything* doesn't mean you can't do *anything*. Action might not offer a high probability of success, but inaction pretty much guaranteed an outcome at least as bad. Couldn't the military see that? In March 2013 I even had to overrule a reluctant MoD and order that chemical-weapons-sampling and protective equipment be sent to the FSA. The threat of such weapons played on my mind constantly.

Perhaps my biggest frustration, however, lay with the resident of 1600 Pennsylvania Avenue.

Obama had admitted back in August 2012 that if Assad used chemical weapons, it 'would change my calculus'. That was good. We had a red line, and I trusted 100 per cent that if it were crossed, he would act. But in the meantime, indecisiveness reigned supreme over American policy on Syria. The 'train, equip and mentor' programme that had kicked off by late January 2013 was no more than the *start* of assisting real pressure from below.

As the world dragged its feet, Assad dug in. In November 2012 I went to the Zaatari refugee camp in Jordan to see the effects of his brutality. Rows and rows of white tents stretched far into the desert. Children playing football in the dust stopped and stared. People were desperate to tell me about being bombed and shot out of their houses. What hit me most was the evident permanence of the place – the tarmac roads, proper street signs, souk-like shops. And the size – it was already the fourth-biggest city in Jordan.

By May 2013, I confessed that I thought Assad was beginning to win: 'All the things we and the Americans should have done nine months ago – from proper integrated command centres for the opposition leaders to train and help them – it's sort of beginning to happen now, but it's happening nine months later than it should have done. I don't think it's going to work, because Assad thinks he's doing too well and the peace process is too convoluted.' What's more, I recorded my concern about chemical weapons, which the regime was reportedly using on a small scale.

William Hague returned from the Geneva Conference the following month, with no agreement reached. The opposition wouldn't countenance Assad staying in the long term, and the Alawites in the regime wouldn't risk trying to dump him until it was clear that he was losing.

I was willing to try anything, so I threw a curveball. I could see that Russia, which supported the regime, could prove a big frustration to our efforts. So in the spring of 2013 I attempted some shuttle diplomacy to get the Russians and Americans closer together on the idea of a transition deal built around a new leader from within, or approved by, the regime.

From the landline in my Downing Street office I spoke to Putin. 'Vladimir,' I said, 'I'd like to come and talk to you about Syria – purely and simply about Syria. You're chairing the G20 and I'm chairing the G8, and it's a perfect excuse for us to meet without everyone asking why we're meeting.' He said it was a great idea, and asked if I'd like to come to Sochi to see the preparations for the Winter Olympics.

I was alone in my office, and wanted to get the official view on such a visit. My door was shut, so I picked up the first thing on my desk, a teaspoon, and threw it at the door, hoping it would summon someone. Luckily John Casson appeared and said he thought I should go. From then on, it was known as 'the spoon conversation'.

But before I could get there, something unhelpful happened. We told the US what we had planned, and that caused John Kerry to rush to see Putin in Moscow a couple of days before us. He pitched the same 'Let's make common cause to build a Syrian government that can defeat terror' proposal, but without pinning Putin down on specifics. The danger was that he'd leave a slightly conflicted picture.

Still determined to seal some sort of deal, I flew to the Black Sea city of Sochi on 10 May. This subtropical resort has its own microclimate – it's all palm trees and sandy beaches – and yet it was, incredibly, where Russia would be hosting the Winter Olympics the following year.

It was an extraordinary meeting at Putin's villa overlooking the water. He started off with a stack of cue cards in front of him. We did half an hour on the bilateral relationship – he got about a third of the way through the card tower – referencing everything from British visa policy to the Russian ban on British beef and lamb, which I had been lobbying hard to have lifted. He joked that I seemed willing to trade British cows for Russian businessmen.

Then he pushed the pile to one side. 'Look,' he said, fixing me with his blue-eyed, shark-like stare, 'I don't want to have the normal argument where I say you've got it wrong about Syria and Libya, and you say, no *you've* got it wrong about Syria and Libya, and we have an argument. I want to talk very frankly about how we try and put a sensible plan together.'

Great. That was exactly why I was there. His preoccupation – whether it was genuine or just a posture to discredit the opposition – was with the risks of supporting terrorism. He said that the logic of my position seemed to be that we'd have the Al Nusra Front at the peace deal. 'No,' I hit back. 'I don't want the Al Nusra Front at the peace deal. I want us to unite with a credible unity government of Syria and destroy them.' He smiled. 'Destroy them? Ah, *now* you're talking my language.'

He claimed that his support for Syria was not predicated on supporting Assad personally. I said I understood that a transition from Assad needed credible Alawite and Sunni leaders – including those from within the regime – to work together. We agreed that the UN and Geneva peace process wasn't working. I proposed a P4 approach (the UK, France, the US and Russia). I said the UK would accept a compromise – a ceasefire before any promises of Assad's departure – if Russia accepted that he could not be part of a transitional deal. My vision was for an international conference in Moscow that ended the war with (presumably) the Moscow Agreement. Who cared who got the credit, if it meant getting the right outcome?

Our three hours together turned out to be the most substantive meeting between a Russian leader and a UK leader since Thatcher and Gorbachev. It was followed by a twelve-course lunch, complete with life-sized pineapples made of ice, containing golden spoons laden with caviar, and a caramel chocolate dessert made to look like Big Ben.

After that, Putin was eager to show how he had been preparing for the Winter Olympics. He suddenly appeared wearing dark sunglasses, ushered me into a black Mercedes SUV – just the two of us and a translator – and drove us at high speed to a helipad. We jumped into a helicopter and flew over the Olympic site. He proudly pointed out where they had been storing up snow over the winter, under huge blue tarpaulins. Looking down on the icy scene below, I could feel our relationship thawing. I seemed finally to be getting through to him.

I worked hard to build on this. When Putin visited Britain for the G8 that summer I invited him to Downing Street to honour the Royal Navy veterans who crewed the Arctic convoys supplying our Soviet allies during the Second World War, undertaking what Churchill described as 'the worst journey in the world'.

They had been given the same medal, the *Atlantic* Star, as those who escorted convoys between Britain and America. The Arctic convoy veterans felt that they had been ignored, not least because of the simple fact that the Atlantic and the Arctic are different oceans. I was sympathetic to their case, and helped to secure them a separate medal. It was a joy to bring this campaign to an end, and to have these brave men who had helped defeat Nazi Germany at No. 10.

So here they now were, naval berets on their white hair, their blazers heavy with decorations, being presented with *Arctic* Stars by me, and Ushakov Medals by the Russian president. The whole exercise seemed to soften him. He said in the press conference, 'A lot brings us together in our history, and I hope that we will have a brilliant future lying ahead.'

The Putin–Cameron bond had reached new levels. He was even cracking jokes, though always with a sinister edge. After the ceremony he said to me, 'David, I know you think that I have horns and a tail and don't really believe in democracy . . .' A smile crept in. 'And you know, you wouldn't be entirely wrong.'

It was an alliance that was not to last.

And it was not just Russia. I would need to work almost as hard to ensure that the EU would not frustrate our Syrian strategy.

The arms embargo, imposed in May 2011, damaged our cause, because it prevented the rebels from getting the weapons they needed. It had awful echoes of Bosnia, so at the NSC I pushed for a consensus to end the embargo, though ultimately that would need EU approval. Ed Llewellyn and John Casson discovered when chatting to diplomats that the EU was about to roll over the embargo for another year, blocking weapons from all sides. We had to work quickly to get this altered to a one-month rollover. That gave us time to look for a better alternative formulation.

In the end, it came about in a strange way. François Hollande's office rang mine at the start of the European Council meeting on 14 and 15 March 2013 and asked if I'd talk to him about the Syrian arms embargo. I walked into his room in Brussels anticipating negotiations. There was already a television camera there. I sat down, and he said in English that we really must scrap this arms embargo. I think he wanted to show that he could lead European policy, having been outmanoeuvred by Merkel and me during the previous Council's budget negotiations.

I was delighted to concur, and once again I didn't care if he wanted to make it look like his idea. The EU allowed the embargo to lapse – though further agreements would have to be reached before the Syrian rebels could be supplied with arms. However, some EU countries remained deeply sceptical about helping the rebels. Some were convinced that whatever followed Assad would mean more persecution of Christians; others that it was the Iraq War all over again, and we should stay out of it.

As so often happens in politics, the right result had an unhelpful by-product. Removing the embargo sparked the suspicions of Conservative backbenchers that there was a conspiracy afoot that would take us back to some Iraq-style intervention. As a result, an Early Day Motion was tabled in Parliament on 5 June calling for a vote on any decision to arm the Syrian rebels. I wish I could have dismissed it as the usual suspects causing trouble. But it was more serious than that. Signed by forty-nine MPs, it reflected a large proportion of public opinion which was wary of starting down a slippery slope towards another full-scale engagement.

From now on, the element that was most likely to frustrate British action in Syria wasn't the EU, the US or Russia. It wasn't my cabinet, the security services or the military. It was public opinion, which was played back to me by politicians. If there was a vote in Parliament on taking action – and a vote would be likely – I wasn't convinced that we would get it through. A year earlier, Andrew Cooper's polling had revealed that 46 per cent of the British public would support military action against Syria with a UN resolution, and only 14 per cent without one. Since Russia made a UN resolution impossible, I reflected that only a single, big atrocity committed by Assad would change people's minds and enable me to change my position.

During the G8 summit I hosted in Northern Ireland in June 2013, Obama and I visited an integrated school in Enniskillen, where Catholic and Protestant children learned side by side. It was all poster paint and high-fives, but during the car journey back to the summit, conversation turned to Syria.

'Look,' I said to him as he gazed out of the window at a town that itself had been so wounded by conflict, 'I do think that the continued use of chemical weapons further strengthens the argument that we have to do more.' I added, however, that the state of opinion in Britain on intervention was such that unless there was a major atrocity, we might only be able to help back up America or help with logistics and refuelling when it came to airstrikes.

Obama restated that a big chemical-weapons attack was the only thing that would prompt American action. 'But I want to tell you, David, I'll be very clear that it won't be about regime change. It won't be because I change my Syria strategy, it won't be an excuse to get in there big-time and topple the regime.'

We left the discussion there, and did not instruct our teams to work out in detail what would happen next. Looking back, this was our big mistake. If we had agreed then and there the response we would make to a chemical-weapons attack – and had gone on to specify the resources we would use, the targets we would strike – events could have taken a different and, in my view, far better course.

It was clear that, despite our alignment on so many issues, despite our genuine friendship, on Syria our perspectives were very different. All leaders are steered by history. As I've said, I was deeply affected by the West's failure in the Balkans and in Rwanda. That weighed as heavily on me as the mistakes of Iraq. But for Obama, Iraq was definitive.

As well as my concerns about the US approach to the military track, I had concerns about its approach to the diplomatic track. Because of the scale of the Syrian people's suffering, because of the awfulness of the terrorism being created, I felt that we all should have been prepared to do more to give Russia the status it wanted. I was happy to do that. I never felt Obama was. I still wonder now whether, if Russia had been treated as a diplomatic equal, it could have led to a peaceful outcome.

I know how carefully Obama analysed things, and I am sure his head was telling him that Putin probably wouldn't alter his approach, even with a more direct presidential relationship. But I think he was wrong not to act faster and try harder. 'I think the person who has been absent from it all is Obama. The Americans have not done enough, quickly enough, to back the right people in Syria, and I think Obama has rather missed the boat. It's a slow-moving tragedy,' was how I put it on tape.

The basic difference between Obama and me was that I did not accept that it was inevitable that *any* intervention would put us on a slippery slope to tens of thousands of our troops on the ground. A more limited

Libya-style approach might not have been enough to force the regime to cut a deal, but it certainly would have offered a better chance of doing so than knowingly doing too little too late.

By July 2013 I had a broader assessment: 'The problem with Syria is not excessive intervention leading to disaster; it's lack of intervention which led to Assad being too strong . . . the only rebels with any momentum at the moment are the extremists, and that's because the Western world hasn't done enough to back the good guys.'

Then on Wednesday, 21 August, events took a horrific turn. Hundreds, perhaps thousands, of civilians had been killed in the rebel-held Damascus suburb of Ghouta. It was the long-dreaded large chemical attack. Surely this was the big event that would shock the world and shake us into action. I watched the horrific footage – infants fitting and foaming at the mouth, parents hysterical with fear and grief. The rows of lifeless children in the makeshift mortuaries reminded me of when we lost Ivan.

Poison gas is heavier than air, and sinks down into the shelters where civilians, including children, are hiding from the fighting above. It is totally indiscriminate, and is the most awful way to die. After the horrors it famously wrought during the First World War, the use of poison gas was banned by the Geneva Protocol in 1925. The only comparable incident in modern times was Saddam Hussein's attack on Halabja in 1988, in which 5,000 people were killed.

There was no doubt that Assad was responsible, or about the sort of gas that was used: sarin. Twenty times more deadly than cyanide. Stockpiled by the regime for years. It was carried on rockets that only the regime could access. These were the instruments of evil – and they were the actions of a tyrant emboldened and ruthless.

That was it. Red line crossed. Not only Obama's red line for action but, I thought, a threshold in Britain's collective consciousness – a long-held, visceral aversion to the notorious chlorine, phosgene and mustard gas that had left our soldiers guttering, choking and drowning on the battlefields a century earlier.

I could see two paths opening up before us.

One: there is a chemical-weapons attack, the next day a hundred American Cruise missiles and ten British ones destroy Syrian chemical-weapons command and control sites. I then recall Parliament and say a red line was set, they crossed it, our response was the right thing to do.

Two: you go to the UN, you go to the inspectors, you go to Parliament, you gather all the evidence. You do it more slowly.

I put in a call to Obama to discuss what course of action to take. For four days I waited for him to call back. Four days. A large-scale chemical attack happened midweek, and the leaders of two of the world's biggest military powers and the world's (supposedly) strongest two-nation alliance didn't speak until the weekend.

When he did finally call, Obama said he was considering a brief surgical 'punish and deter' attack, and would like Britain to be part of it. Indeed, he actually said that only Britain really had the capabilities to make a difference, like submarine-launched Cruise missiles. We might be taking action within thirty-six hours, he said. Was I with him? I said yes.

This was almost in the realm of option one – acting immediately and reporting to Parliament afterwards, as we had done with Libya. John Casson rang the MoD and asked them to turn around the submarine we had near Gibraltar. But too much time had elapsed for such a simple approach.

So I decided to write Obama a follow-up letter at this point. First, because of the delay, we were on a timetable where we were going to have to do more things. Second, though Clegg was supportive, he was saying his support rested on making an attempt to secure UN approval. I wrote that I was ready in principle for the UK to take part, but some measures needed to be taken if we were to prevail in the court of public opinion. These were: clarity about the UN inspectors (at that stage the Assad regime had prevented the inspectors' investigations), an attempt to achieve a condemnatory resolution in the UNSC, and a strong public declaration that the US and its allies could not stand aside in the face of a chemical-weapons attack.

The heads of the intelligence agencies, the military chiefs of staff, and those highest up in the national security adviser's team were all of one mind: this was a complete contrast to Iraq, and very targeted, very surgical strikes – a short, sharp shock that would put Assad's chemical weapons beyond use – were doable. Even the Lib Dems supported action, with Nick Clegg confirming that, in spite of his insistence on action at the UN, he was in fact very keen.

I thought Parliament would be supportive too. I knew there would be a desire for UN involvement, which had become, in the national

A surprise visit to British troops in Afghanistan, 6 December 2010.

The first meeting with a foreign leader after becoming prime minister:
taking Afghan president Karzai around Chequers, 16 May 2010.

In Londonderry the public cheers the findings of the Saville Inquiry into the events of Bloody Sunday in 1972, 15 June 2010.

Signing off on the terms of the Scotland referendum with first minister of Scotland Alex Salmond, 15 October 2012.

Speaking to the crowd in Benghazi, Libya, with French president Nicolas Sarkozy and National Transitional Council head Mustafa Abdul Jalil, 15 September 2011.

Local residents in Benghazi bring placards out onto the street.

Queen Elizabeth II attends a cabinet meeting at No. 10 as part of the celebrations of her Diamond Jubilee, the first monarch to do so since the eighteenth century.

The Olympic torch visits Downing Street, 26 July 2012.

A daughter of Downing Street: Florence helping me during the Alternative Vote referendum campaign of 2011.

Liz introduces Larry the cat to Obama at No. 10, 25 May 2011.

The 'treadmill bilateral': a discreet meeting with Obama during the Camp David G8, May 2012.

World leaders at the Lough Erne G8, 18 June 2013 (left to right: José Manuel Barroso, Shinzo Abe, Angela Merkel, David Cameron, Vladimir Putin [obscured], Barack Obama, François Hollande, Stephen Harper, Enrico Letta, Herman Van Rompuy).

Olympics judo with Vladimir Putin, 2 August 2012.

Meeting Angela Merkel at Chequers, 9 October 2015.

The G7 Summit in Bavaria, 7 June 2015 (left to right: Matteo Renzi, Shinzo Abe, Jean-Claude Juncker, François Hollande, Stephen Harper, Donald Tusk, Angela Merkel, Barack Obama, David Cameron).

6 a.m., working through my red box in the flat above 11 Downing Street,
2 November 2011.

The 2015 election campaign, holding up a copy of Labour Treasury secretary Liam
Byrne's infamous 2010 note: 'Dear Chief Secretary, I'm afraid there is no money.'

Making the closing arguments of the general election, May 2015.

Election day fears: reading my draft resignation speech, 7 May 2015 (left to right: Craig Oliver, Clare Foges, Liz Sugg, Ed Llewellyn, Kate Fall, David Cameron).

Visiting Gravesend for the Sikh festival of Vaisakhi, 18 April 2015.

Election night elation: watching the results come in at the Windrush Leisure Centre in Witney, 8 May 2015.

Clapped back in to No. 10, 8 May 2015 (left to right: Samantha Cameron, David Cameron, Chris Martin, Jeremy Heywood).

Rowing (not rowing) with Angela Merkel, Swedish prime minister Fredrik Reinfeldt and Dutch prime minister Mark Rutte at Harpsund, Sweden, 10 June 2015.

Boyko Borisov's
border fence,
4 December 2015.

Playing Wembley: British Indians flock to see
Prime Minister Modi, 13 November 2015.

Calling in at a local pub, The Plough at
Cadsden, with President Xi Jinping of China,
22 October 2015.

A meeting with Donald Tusk and Jean-Claude Juncker during the 18–19 February 2016 European Council, negotiating a new settlement for Britain in the EU.

Inspecting the final text of the renegotiated deal with adviser for Europe and global issues Tom Scholar and Britain's permanent representative to the EU, Ivan Rogers.

Addressing Birmingham students and Remain supporters during a final EU referendum campaign speech, 22 June 2016.

The moment we knew it was over: watching the referendum results come in in the early hours of 24 June 2016.

Nancy, Elwen and Florence collaborate on a letter to the incoming tenants of No. 10, 13 July 2016.

Preparing for my final PMQs as prime minister, 13 July 2016.

A trip to the Reach Academy, Feltham, 12 July 2016, one of the 500 Free Schools given the green light during my premiership.

With the children, moments before leaving No. 10 for the last time, 13 July 2016.

Visiting a lab funded by Alzheimer's Research UK as president of the charity.

consciousness, the test that needed to be passed for people to be satisfied that this wasn't another Iraq. This was complicated by the fact that Russia would just veto a resolution. Obama saw no point in it, but I knew the political importance of at least trying.

I did try speaking to Putin, but any goodwill had dissipated, and the conversations became fractious once more. 'You must accept that these chemical-weapon attacks are from the regime,' I said. 'No, no, no – it's only the opposition that would benefit from this sort of publicity,' he said. It got rather heated, although he ended the call by saying something like 'I hug you' in Russian. I was never sure whether that was a hug in a make-up sense or a mafia sense.

By Monday, 26 August, US Navy warships were in position to attack Syria. But the timetable was slipping still further. Obama had suddenly announced that he would wait for UN weapons inspectors to return from Damascus. But they were detained, and we could hardly launch airstrikes while they were still there. Another card cannily played by Assad.

Because of the delay, we were now well into the realm of option two. We would have to consider recalling Parliament – and whether to hold a vote if we did.

There is no constitutional requirement for a prime minister to ask Parliament before committing British forces to military action. In 2003 we voted on Iraq before the invasion, but in 2011 we voted on Libya only after the first strikes had already taken place. However, I feared that even if I didn't take the initiative I could effectively be forced into it. While the speaker does not have the right to recall Parliament, I knew that he was quite capable of saying that it needed to happen – putting me in an almost impossible position if I said no.

In addition, I felt confident about the case, and thus about the parliamentary position. I believed I had answers to all possible questions; certainties for most doubts. Would it endanger civilians? No, it was targeted on military facilities. Would our troops be at grave risk? No, there would be no boots on the ground. Would it work? Yes, we could take out targets, and it would raise the price on using chemical weapons again. Was it legal? Yes, you're permitted under international law to intervene to protect civilians by deterring the use of chemical weapons.

So during a meeting on Tuesday, 27 August we took the decisions over whether to recall Parliament, hold a debate and have a vote on a substantive motion backing military action, or to make a statement or hold a debate without a vote.

William Hague said yes to a recall of Parliament and a vote. He'd always argued that if there was time Parliament should be recalled, and that this was about the credibility of our word.

Craig Oliver said we had to have a vote if we were recalling Parliament. Otherwise what would we be calling people back for? A conversation?

Hugh Powell later told me how much he regretted not speaking up. For him, the politicians were taking a very big decision about Parliament very quickly, without discussing the risks or alternatives.

As for me, I said I thought it would be difficult, but – and this was the moment when I got it totally and utterly wrong – I also thought the outrage about the chemical weapons would be widely shared, and would stir in everyone what it stirred in me: sheer revulsion and a desire to stop it happening again. Not for the last time, the rational optimist in me got it wrong. But it was done. MPs would be summoned.

So far the biggest obstacles had been trying to convince Putin of the merits of a diplomatic solution, and to push Obama into creating pressure on the ground. It had been a tale of frustrations with cabinet, the military, the security services, Whitehall and the US. But now that a parliamentary vote on military action was on the horizon, there would be another major player to convince: Ed Miliband.

Ed was a far better leader of the opposition than I would ever have admitted at the time. He knew just how much baggage his party was still carrying from Iraq; he had won the leadership after condemning Blair's actions. And he knew something I had also learned during five years as leader of the opposition: you never just hand a blank cheque to your opponent. But I don't think I'd ever have behaved as he did over Syria and the use of chemical weapons.

The day we announced that Parliament would be recalled he came into Downing Street with the shadow foreign secretary, Douglas Alexander, and his chief of staff, Tim Livesey. He posed four questions. Shouldn't we wait for the UN inspectors' full report? What was the legal authority for action? What were the military objectives? How could we avoid escalation or making something worse happen?

The answer to each was clear. A full UN resolution authorising force was a non-starter because we knew Russia would veto it. The legal advice was that our proposed action was lawful, and we would publish a summary. The mission was to deter and degrade, and it was not aimed to have effects over and above that.

He called me that evening to say that the classified intelligence report we had given him proved the case for deterrence. It was indisputable. But then he called again, to say he was uncertain about two points – the timing of the report from the inspectors, and having a proper UN resolution. That was his new red line. I conceded.

The following day I saw him again. I had managed to deliver on both those things – our representative at the UN, Mark Lyall Grant, had circulated a proposed resolution to the Security Council, which the Russians had made very clear they would reject.

Now I was giving Miliband the draft motion he wanted. Still, he said he'd have to take it away and look at it. I said we needed an answer by 4 p.m., the deadline for tabling it.

He didn't ring back until 5.15 p.m., when he said, 'We' – it was very much *we*; you could tell he was under duress from his party – 'We don't think it goes far enough, and can't support it.'

I was astounded, and furious. 'But I've given you what you want – the draft of the proper UN vote and the inspectors' report . . .'

'No, we need a second Commons vote,' he replied.

'A second vote?!' I said. He was pulling red lines out of thin air now. 'For heaven's sake Ed, this is ridiculous. You've never mentioned a second vote before. When do you think you're going to have this second vote?'

'On Sunday. You can reconvene on Sunday,' he murmured.

'This is hopeless. You've got to understand you're putting yourself on the side of Vladimir Putin and Bashar al-Assad by not standing up to this. Ed, we're going to lose this if we don't have Labour's support,' I said.

I'd just been told by the chief whip that we were in deep trouble on our current motion. A third of Tory MPs would vote with us, but another third were very worried about action, and the remaining third were outright opposed to a quick decision to act.

From 5.30 to 6.30 – we'd managed to push the deadline for tabling back and back – I sat with George Osborne, William Hague, George Young, Ed Llewellyn and Nick Clegg deciding what to do. William advised a motion saying that the House would discuss the situation in Syria – anything else risked defeat, and that was too dangerous. I argued for a motion that spelled out the whole thing – go to the UN, listen to the weapons inspectors, have a second vote, everything Miliband wanted.

The chief whip said he thought that was a motion we could carry. So that was the decision.

The tabled motion called for military action 'if necessary' – endorsing the principle of action once weapons inspections were complete. Authorising actual military action would take place only after a second vote.

I went to bed that night feeling relatively confident. I thought that if I were Labour, I'd look at this motion and declare victory. I'd promised a proper UN vote, a report from the weapons inspectors, and a second vote in the House of Commons before action could be taken.

Instead, the first thing I heard on the *Today* programme the next morning was that Labour would oppose the motion. They would put down an amendment with roughly the same requirements as ours, but calling for 'compelling' evidence that the regime was responsible for chemical attacks. No finger of blame pointed at Assad.

I spent the whole afternoon seeing potential Tory rebels. One by one they came into my Commons office. 'Why are we doing this?' 'How can you be sure it was Assad?' 'Won't this just inspire more terrorism?'

We started the day forty short of a majority, and ended it twelve short. All I had left to persuade them with was my speech. Looking back, it was very defensive – every sentence an attempt to exorcise the ghost of Iraq. We weren't talking about invading Syria, I told a packed House. We weren't trying to topple the regime. On this occasion we weren't working with the Syrian opposition. We weren't even voting on immediate action. We were having a vote on whether to have another vote on whether to respond to the large-scale use of chemical weapons – a war crime that Britain had agreed, indeed led the way on, outlawing back in 1925. This was about upholding the rules-based international order. It was the bare minimum from a civilised nation.

Europe had been divided over Iraq; but it was united over responding to the chemical-weapons use in Syria. NATO had been split over Iraq; this time it had made a very clear statement. The Arab League had been opposed to Iraq; but when it came to Syria, it was calling for action.

Ed Miliband's speech was dreadful. I was sure he had wanted to do the right thing initially, but he was no longer willing to do it.

When it came to the vote, it wasn't just the usual suspects who rebelled. It was people like Tracey Crouch, Adam Holloway, Nigel

Mills. People I'd fought alongside at the election, and felt were part of my team. People who were moderate and sensible. David Amess, the independently minded MP for Southend West, said to me, 'I'm not voting about Syria, I'm voting about Iraq. I made a promise to myself I'd never let something like that happen again.' It was official: the Iraq War could now count among its victims the Syrian people.

At 9 p.m. we sat in my Commons office and waited for the division bell to ring. John Casson – the truest colleagues are those as willing to convey bad news as good – told me we were probably going to lose. 'I've got us into a right pickle here,' I said to William. 'I've overreached.'

At 10 p.m. we all trudged through the division lobbies.

Back in my office, we waited for the results.

Labour's motion was defeated by 332 votes to 220.

Then our motion was defeated, 285 to 272. Thirty Conservatives and ten Lib Dems had rebelled. All the Labour MPs in the Chamber had voted against.

It was the first time that a government had lost a vote on a matter of war since 1782, when Parliament effectively voted to end the war against the rebellious American colonies.

My instinct was to concede publicly and generously. That's all you have left when you lose – good grace. I said from the despatch box: 'It is clear to me that the British Parliament, reflecting the views of the British people, does not want to see British military action. I get that, and the government will act accordingly.'

Back in my Commons office, George Osborne, Nick Clegg and I mulled it all over in semi-disbelief. Here we were, not wide-eyed liberal interventionists, but people who believed a line had been crossed by a chemical-weapons attack. We were militarily, legally and morally entitled to respond. But democratically we were not. The rebels, and Miliband, were channelling public opinion. People were told 'legal authorisation to act militarily', but all they saw was dodgy dossiers, WMD and the bungled Iraq campaign.

Our proposed action was merely to launch a limited series of strikes, principally with Cruise missiles. That was what the US, the UK and France ended up doing in April 2018 after a further chemical attack anyway. But in those days of delay it had been ramped up in the minds of the press and MPs to something approaching full-scale war. A copy of that morning's *Sun* still lay to one side of my Commons office. 'Brits Say No to War in Syria' said the front page, with a poll showing that people objected to action by a staggering ratio of two to one.

For the Commons mess, I blamed myself. As prime minister you're not just taking part in international action, you're also managing a party and a Parliament, and in the end I miscalculated.

The following morning I listened to *Today*. Russia welcomed Britain's vote.

Later I spoke to Obama. 'Hey bro, you've had a tough week,' he began. I was emotional. I said how sorry I felt that we couldn't work together. America and our other allies would be out there, in the skies, deterring this monster from any further such attacks. Yet Britain, the world's fourth-biggest military power, time-honoured defender of liberty, decades-long partner of America, would be hanging in the shadows.

Obama couldn't have been more generous about it. He said the special relationship was just that: special. This was a bump in the road, no more. Nothing would happen for forty-eight hours, he said, which I interpreted as meaning an attack on either Saturday or Sunday night. That was a huge consolation: knowing, at least, that America would be doing something.

On the Saturday I had a call from my private office. Obama had made a statement, but it wasn't to announce the number of missiles launched or targets hit. He'd taken a decision the day before – a now famous 'walk in the Rose Garden' – to postpone, and effectively cancel, military action. Apparently the House of Commons' vote 'weighed on the president', as did the opinion polls. Instead, the US had done a deal with Russia for Syria to give up its chemical weapons. As one commentator put it, 'Sometimes a lizard has to lose its tail to survive.' Assad grew back his lizard's tail, and would carry out over a hundred chemical attacks on varying scales in the years that followed.

I was shocked. A British PM had lost a vote, and an American president had lost his nerve. Yet Obama presented this as a decisive, strong-minded move. He even said in 2016 that the decision of which he was most proud was his suspension of strikes against Assad in August 2013. I can't agree. Frankly, his handling of Assad is still the thing I regret most about his entire presidency. If we'd done more to help the Syrian opposition in the beginning, we might have rid the country of this murderous tyrant – and the later war against the ISIS brutes might not have been necessary. With the benefit of hindsight it still looks the same. The risks of action really were proved to be outweighed by the risks of inaction.

From 2014 onwards, as ISIS gained ground, it was the extremists in Syria, and over the border in Iraq, who became the global focus. No one can credibly claim to have foreseen in detail the ISIS breakout in Iraq and capture of Mosul, which I'll describe later. But that extremists would grow in strength if the Syrian civil war continued is precisely what I had been arguing from the very beginning. It would require these jihadists to make Syria a more obvious threat to the West – planting bombs here, killing British and American hostages there – for action there to become acceptable to the public and to politicians.

The failure of August 2013 was, in retrospect, the end of a realistic strategy to solve the civil war in order to tackle the extremists effectively. If the West could not summon the political will for airstrikes, it would not find it to force the regime to the negotiating table via irregular forces either. Just as the Iraq War stands as a monument to many of the perils of intervention, to me Syria is a monument to the perils of non-intervention. It was a wicked problem that could have been solved, but for various reasons, it wasn't.

That, however, is not the whole story. There were other ways the West could demonstrate its commitment to a better world, and we certainly did.

It's February 2016, and I'm standing on a stage in London's QEII Centre alongside Angela Merkel and Ban Ki-moon.

Britain was already the second-largest bilateral donor to Syria. And here we were, hosting a pledging conference that would raise more than any in history – over $12 billion – for food rations, relief packages, vaccinations, health consultations and schooling. This was the great humanitarian cause of the decade, and Britain was not found wanting. Indeed, that was British leadership. At that point we were the only major country committing 0.7 per cent of GDP to overseas aid and 2 per cent to defence.

So while it is fashionable to talk about the UK's shrunken role in the world – even more so after we lost that Commons vote in August 2013 – we cannot underplay the importance of our leadership on these other vital things. Aside from the US, no Western country did more on aid, diplomatic endeavours and the defeat of ISIS than the UK. Of that we can be immensely proud.

34

Leading for the Long Term

While the economy was growing, and we could rightly claim that the medicine was working, as 2013 went on I became obsessed by a different economic question. What *sort* of recovery was under way?

The picture painted by the government graphs was clear: rising investment and increasingly strong growth. But were people feeling this great turnaround in their lives?

The long dole queues that characterised the recovery from recession in the 1980s – there were still three million unemployed in 1986 – had been averted. More than that, we were about to hit record employment. Indeed – and this was to become a favourite statistic of mine – from 2012 to the beginning of 2015, the UK economy created more jobs than the rest of the EU put together.

Yet the picture on living standards seemed less rosy. Wages weren't keeping pace with prices. The Institute for Fiscal Studies (IFS) said this was inevitable: it would have been astonishing if earnings hadn't fallen after the deepest crash in modern history. And on top of that, inflation was climbing due to rising oil prices.

So, how to mitigate this as much as possible?

Our answer was to make every effort to protect living standards. We had been able to 'ease the squeeze' by increasing pensions, freezing council tax and fuel duty, and raising the personal allowance (the latter alone was worth £905 a year to a basic-rate taxpayer by 2017). We had also exempted the lowest-paid public-sector workers from the pay freeze, and – in time – would radically increase the minimum wage.

In my view, the measure to look at wasn't wages, it was take-home pay – income *after* tax. But even on this measure we seemed to be struggling to make progress. The statistics available at the time showed that average disposable income was lower in 2013 than it had been before the 2008 crash. On average, since 2007–08 it had fallen by £1,200 in real terms.

This, it appeared, was a jobs recovery rather than a living standards recovery. Obviously I would have preferred both. But if you had to choose between the two, surely it was better to have more people with the security of work, rather than higher wages for some and continuing unemployment for others.

But there was no doubt that this gave Ed Miliband and Labour an opportunity to highlight falling living standards and point to a 'squeezed middle' demographic: too well off for benefits, but too poor to enjoy the security of the rich. It was one they didn't miss, proclaiming a Conservative-induced 'cost-of-living crisis', a phrase which gained traction in late 2013, as GDP started to go gangbusters but families' finances didn't.

On 24 September 2013 I sat in my office watching Miliband's party conference speech. 'He has been prime minister for thirty-nine months, and in thirty-eight of those months wages have risen more slowly than prices. That means your living standards falling year, after year, after year. So in 2015 you'll be asking, "Am I better off now than I was five years ago?"' Then came the big announcement of the conference: a Labour government would freeze domestic energy prices. It was economic nonsense, but political genius – and maddening, too.

Nonsense, because a price cap would reduce competition and discourage investment – which we badly needed more of – in both the electricity and the gas markets, therefore leading to higher, not lower, prices in the long term.

But genius, because rising energy prices were deeply unpopular, and the companies responsible were seen as profiteering. As a policy it had an immediate 'retail impact', because people could see they would be better off.

And it maddened me because I knew that, as energy secretary in the last Labour government, Ed Miliband had introduced a whole series of measures that had the effect of putting prices up.

When Miliband had finished, I pulled on my running kit. If I had a rare half-hour spare, that's what I'd do: leave by the back gate of Downing Street and do a figure-of-eight loop around St James's Park and Green Park.

I spent a lot of time in the park. Taking Florence to play on the swings after school. Early-morning squats and sprints with my personal trainer

Matt Roberts. And these impromptu runs, when I'd weave through an obstacle course of selfie-snapping schoolchildren, ducks, pigeons, squirrels and, supposedly, spooks (it was rumoured to be a hotspot for the secret services). And other things too . . . Once, as we ran past a couple enjoying a barely concealed spliff, one of the police protection officers urged me, 'Breathe deeply, boss – it might help with the pain.'

On that autumn afternoon in 2013 I pondered our big problem. We'd dragged the economy back from the cliff edge, but delivering prosperity for everyone would take time. Meanwhile, Labour was stitching together disparate issues, like rising energy prices and the cost of living, to form a damaging narrative. All with the election just eighteen months away.

The rise of food banks formed another element of that narrative. Charities offering donations to hard-up families had been around for years – a valuable part of the Big Society. But there had been a recent increase due to the financial crash and rising food prices. A food-bank movement had grown up, and local churches and charities were enthusiastically joining in. This was the case around the world. Unlike Labour, we had allowed Jobcentres to refer people to food banks, thereby radically increasing both awareness and demand. That was the reality. But Miliband peddled a politically toxic mythology: victims of Tory austerity queuing for food parcels.

Then there was a welfare reform that was just as nuanced and reasonable, but equally emotive.

Housing benefit – welfare payments to support those renting in either the private or public sector – was now costing us half as much as the schools budget, and if we were going to control public spending while protecting services, cuts were plainly needed.

It wasn't just about savings, though, but fairness. I believe strongly in a welfare state. But something has gone wrong when, for instance, taxpayers are spending hours commuting to London every day to get to work, while funding the central London homes of those who don't work at all.

In reducing that bill, one change would take on particular potency in Labour's narrative about Austerity Britain. The previous Labour government had changed the rules for private-sector tenants, so that they were only paid housing benefit for the rooms they needed to occupy, rather than for every room in the house or flat they rented. On 1 April 2013 we brought in the same change for tenants in council and housing-association homes. If the home was deemed too large for its occupants – in many cases there were spare bedrooms – the housing benefit they received would be reduced by 14 to 25 per cent. Either they'd have to move to a more appropriately sized property – freeing up bigger houses for bigger families, who badly needed the space – or pay the difference.

Our opponents seized upon this, branding it a 'bedroom tax'. Like the 'pasty tax', the name stuck.

And even though it wasn't a tax, and it was equitable, there was a difficulty with it. While private-sector tenants effectively choose the home they live in, council or housing association tenants have little such choice. And although we were building more council homes – more than Labour, in fact – they were still in short supply, and therefore moving home could be hard. As a result, some people ended up receiving less housing benefit without being able to reduce their costs.

But what was the alternative? If we were to bring housing benefit under control, something had to give. At least this proposal had the merit of levelling the playing field between the public and private sectors.

Add the mythology around this so-called (incorrectly called) 'bedroom tax' to food banks, and we were in danger of losing the economic argument just at the moment the economic recovery should have been winning it for us.

The underlying economic truth was that you can't sustainably grow living standards without rising wages – and wage growth had been sluggish for many years, since long before we came to office. The problem had its roots in decades of underinvestment in skills and infrastructure, and low productivity. We were addressing the effects of a lopsided economy – tilted heavily to the south-east and to finance – but the dividends of long-term decision-making would take time to make themselves felt.

That was how I saw it. We had a big argument: 'Fix Britain's long-term problems' – and Miliband had a popular one: 'Cut prices to help families.' I knew we needed both. Yes, we needed to make long-term decisions to build a strong economy, but at the same time, my ministers and policy unit were told to scour the horizon for things we could do immediately to help with the cost of living, and we announced new measures every week.

Freezing MoT test prices. A crackdown on whiplash-injury fraud, which was adding £100 to the typical car insurance bill. And action on energy bills. Unlike Labour, we had solutions that were workable. We could force companies to put customers on the lowest available tariff. We could also reduce some of the charges on bills that were paying for green schemes.

Some of these schemes were extremely expensive. I was particularly furious about one called ECO, which encouraged people to install insulation in their homes and businesses, but also included very costly options that were increasing bills by an average of £90 a year. My anger landed me in difficulty when it was reported that I said in a meeting, 'Get rid of all the green crap from customers' bills.' I don't remember whether I actually used the 'c' word, but I probably did. The problem was that the phrase was interpreted as an abandonment of our environmentalism. The truth was quite different.

When I started off talking about these issues during the husky days, I was accused of a PR stunt – that the snow was for show, and my passionate posture would melt away once I entered government. In the end, the opposite happened. I ensured that the green revolution continued in the background, without major interventions or speeches.

So often people complain that politicians are all talk and no action. On this, I was all action and not enough talk. We proved beyond doubt that you could both cut carbon and cut the deficit, and we quietly became the greenest government Britain had ever seen. Environmentalism was about the long term – and so was my approach to infrastructure. More investment in railways than at any time since the Victorians. A new line under London – the biggest engineering project in Europe – Crossrail. And one of the most controversial projects I was determined to pursue: 'HS2'. Since privatisation in the early 1990s, the number of rail passengers had doubled, yet we hadn't built a line north of London for 120 years, and the West Coast Mainline from Glasgow to London was full to bursting. By the twenty-first century there was a new type of railway on the scene: high-speed. But Britain had only about 0.7 per cent of the world's high-speed track, namely the Eurostar, or 'HS1'. It was absurd that you could get from London to Paris on high-speed rail, but not from London to Manchester.

It was becoming increasingly clear to me that we needed a new west coast line, and my view was that it needed to be high-speed. We announced our support for it while in opposition, and Labour concurred, setting out plans for HS2, from London to Manchester (between which it would halve the journey time) and beyond. Eventually reaching as far as Edinburgh, it would be like the new spine of a country standing tall in the world.

But while other countries sped ahead with giant infrastructure projects and long-term ambitions, Britain was blighted by a nimbyist aversion to doing anything radical or big or expensive. This was setting us back in the global race. So the debate about HS2 was about more than productivity, or the cost of living, or even our electoral prospects. It was about the will of the West. Did we still have the stomach to do the big things? I believed we did.

For this reason, one battle I regret not confronting was our long-term airport needs. The debate over Britain's airport capacity had been going on for decades. Heathrow was now the busiest two-runway airport in the world, and a third runway was the favourite option for fixing this problem. And yet . . . we decided to push it into the next Parliament.

Another issue lingering on successive governments' to-do lists was the fate of Royal Mail. Turning inefficient state-run institutions into modern and effective service provision is instinctively Conservative. However, when its privatisation was proposed to Thatcher, even for her that was a step too far.

Yet Royal Mail found itself in a highly competitive market, desperate for the freedom to invest, expand and innovate. Stuck in the public sector, it had lost £300 million in 2010 alone, and attempts to reform it, while not getting it back into profit, had set it on a path where it was possible to attract private-sector capital. Germany had shown it could be done successfully with its mail service. And so Royal Mail was sold off in 2013.

For me, the move had a wider symbolism. Modern, compassionate conservatism was seen by some as timid, lacking in radicalism, and just a bit too left-wing for many in the party. This would infuriate me. I would sometimes reel off a list of all the things I did that Thatcher never dared to do, like increasing tuition fees in universities, reforming public-sector pensions, allowing private operators to run state schools, capping welfare, vetoing a European treaty, leaving the European People's Party and privatising Royal Mail.

By the end of 2013 it seemed voters were undecided when it came to the Conservatives' long-term offering versus Labour's instant fixes. Labour remained ahead in voting-intention polls, but people trusted us more with the economy, and preferred me in charge to Ed Miliband.

But I sensed we had the edge. We might have the historical baggage of being cold, uncaring Conservatives, but they had the more recent and more potent baggage of financial profligacy. The polls also showed that the 'bedroom tax' wasn't particularly unpopular. I was given a paper by the Department of Work and Pensions in 2013 which showed a plurality of people supported the policy both in principle and after a more detailed briefing of what it involved.

This was brought home to me during the most extraordinary conversation, in Glasgow Airport of all places. A woman approached me. 'There's something I want to say to you,' she said. 'It's about the bedroom tax.' I thought, 'Oh God, here we go . . .' But she said she was a housing officer, and was totally in favour of it. 'Don't change your tack,' she said as she left with a smile.

There was another policy people were – and still are – desperate to talk to me about: our flagship housing scheme, 'Help to Buy'.

The increasing inability of young people to buy a home of their own was, I believed, a massive political and economic failure. A lack of house-building was now coupled with a crash that had dried up credit, and this in turn led to a huge drop in home ownership – halving for eighteen-to-twenty-five-year-olds in two decades. The average age of a first-time buyer was now thirty-three.

This wasn't just a big deal for the Conservatives – famously the party of aspiration. It was a big deal for Britain. Enormous security, prosperity and identity are derived from bricks and mortar. The opportunity to buy a home was available to me when I was young because I had parental help, and because houses were then so much cheaper. It should be there for others who work hard too, but aren't as fortunate. And the urgency went beyond that: what support would people feel for a global economic system in which they had no stake? What good is capitalism to a generation without any capital?

Of course there were numerous reasons for the decline in home-ownership. The recession had hit families' finances hard. Many small and medium-sized house-builders had gone under. The planning system had become hopelessly bureaucratic. And fundamentally, Britain had not been building enough houses for years.

We were addressing all these things, for example replacing 1,300 pages of planning guidance with just fifty. But the housing market was still flat on its back. And while there were plenty of potential buyers who could afford the monthly mortgage payments, they were simply unable to get a mortgage.

The problem was that the builders wouldn't build unless the buyers could buy, and the buyers couldn't buy unless the lenders were lending. And the lenders weren't lending. Mortgages that had been available for most of my lifetime, that loaned up to 90 or even 95 per cent of the value of the property, simply weren't available. This couldn't be fixed by public spending or changing the tax system. It was a monetary-policy problem. So George and Rupert worked up the idea of Help to Buy – a deliberate reference to Thatcher's Right to Buy – in order to get the market moving. With Lloyds Bank's Antonio Horta-Osorio helping behind the scenes, George launched it in the 2013 Budget.

The first part, the equity loan, would offer a buyer 20 per cent of the value of a new-build home, interest-free, for five years. The second part – the really radical element, never seen before in Britain – was a mortgage guarantee scheme. Since the inability to raise a large deposit – which nervy lenders were now requiring after the crash – was the biggest problem for buyers, we were saying the lenders could go back to offering 95 per cent mortgages, and government would take on the risk.

The criticisms were many. 'It will only help people in London.' 'It's for people buying huge houses.' 'It will just benefit those who already own homes.' 'It won't fix the supply problem – if anything it will drive up demand and make the whole thing worse.'

I had many meetings in the Terracotta Room in No. 10 over six years, but the gathering on 11 November 2013 stands out. The room was full of the first three dozen people who had bought their homes through Help to Buy. Mostly in their twenties and thirties, they stood nervously around the fringes of the room, clutching cups of tea or coffee. Almost every story I was told that day was the same. The deposits they had been asked for were too big. Young people with rich parents could buy houses. So could wealthy foreigners. But hard-working people without the 'bank of Mum and Dad' could not. For all those couples, Help to Buy was a life-changer.

Six years on, there are over 210,000 people living in homes of their own thanks to the scheme. Four-fifths of them are first-time buyers. Ninety-five per cent are outside London. The average cost of their homes is well below the national average. And, crucially, half of those homes are new-builds – contributing to a 41 per cent increase in private house-building since the scheme was launched.

I am always cautious about advocating state interference in free markets, but occasionally a nudge from a long-termist government is needed if you are to help people to fulfil their dreams. And that is what politics should be about.

Sadly, politics is also about whatever nature throws at you. And as 2013 became 2014, nature threw everything it had got at Britain. Coastal storms hit East Anglia, and floods inundated Kent. Rainfall turned the Somerset Levels into lakes, and lashed Wales and the south coast. Then, in February, the cliffs at Dawlish in Devon collapsed, destroying the mainline railway track and cutting Cornwall off from the rest of the country.

I knew by now the dangers of a tardy response to disasters, so I visited several of the places affected. I hadn't yet made it to the Somerset Levels. This low-lying manmade environment was dredged and irrigated centuries ago. But not enough had been done to protect it. Some experts genuinely thought we should let the floods run their course. I was outraged – you can't say that about someone's home, someone's farm. If it's a place where people live, then it needs to be protected.

Sadly, the Environment Agency had failed on that score. It seemed to worry more these days about newts and butterflies than homes and livelihoods. And while the number of homes and businesses that were flooded was just a tenth of that in the floods of 2007, the papers portrayed us as fiddling while Britain was submerged.

To make up for our slow start, we had to go overboard to compensate. I visited Somerset several times, and chaired a series of COBR meetings. I saw the local MP for Bridgwater and West Somerset, Ian Liddell-Grainger, on a weekly basis, and promised the people living there that we would dredge all the waterways – I'd do it myself if I had to.

Something else required quick action that winter too. The coastal route which runs close to the sea, hugging the red cliffs of Dawlish on its way to Cornwall, is rightly famous, but giant waves had swept away a section of the sea wall at this pretty Devon seaside town. In February, I walked along the line that remained, and saw the destruction: the track dangling in the wind like a rope bridge, a defining image of the storm.

I regarded this as a real test for Britain. Were we going to put on a replacement bus service for the next two years, or were we going to get on and fix the track in two months? Thanks to Network Rail's engineers working day and night in their fluorescent orange jackets, there was indeed 'action this day'. And by April I was on the first train back into Dawlish, alongside elated townsfolk and the 'Orange Army' that had saved the day.

Nature's crises don't always come at you like bolts of lightning. An outbreak of bovine TB had led to tens of thousands of otherwise healthy cattle being slaughtered, at huge cost to the taxpayer. The disease was carried, in part, by badgers; it was clear that the TB 'hot spots' contained large numbers of infected badgers (which, as well as infecting cattle, were dying horrible and protracted deaths). No government wanted to grip the problem. And while it was true that cattle movement restrictions and vaccinations would help, there wasn't a country in the world that had successfully dealt with a similar situation that hadn't used some form of culling.

Of course it would be unpopular. People dressed as bloodied badgers became a common sight at protests. 'What about the badgers?' they'd cry. Well, what about the cattle? The lives of the cattle – let alone the livelihoods of the farmers and rural communities – didn't come into it for them. It was another example of the absolutism and illogicality of the animal-rights lobby: opposed to culling, full stop, even though far more animals would be saved than humanely killed, and it would re-establish an ecological system with healthy cattle and healthy badgers.

This was one of those absolutely thankless tasks in politics, when doing the right thing is not very pleasant and not at all popular. Culling had worked in Ireland, and by 2018 we were beginning to see mounting evidence of reduced infection rates in areas where culling had taken place in the UK, with reason to hope that we would finally get on top of the disease over the next decade.

And by 2018 we had learned something else, too. For just as the economic-growth statistics had proved wildly pessimistic (as we saw in Chapter 31), so the cost-of-living figures were wrong too. We know now that living standards actually rose more quickly in the five years after 2011 – the supposed austerity years – than in the five years after 2002, Labour's supposed boom years.

These were also, supposedly, years when the country was led by 'the party of the rich'. Yet during my time in office it was the lowest earners who saw the biggest percentage of weekly wage increases, followed by middle earners. Top earners saw no increase. It was, according to the IFS, the biggest compression of income inequality since the 1970s.

Even without such knowledge, I became more and more content with our answers on the cost of living. But at the same time I was becoming increasingly preoccupied with hardships overseas that we in Britain could only imagine.

35

A Distinctive Foreign Policy

I came of age during Live Aid – quite literally. That great global gig took place on the same boiling-hot July day in 1985 as my brother Alex and I held our joint eighteenth and twenty-first birthday party.

I was a massive fan of Bob Geldof's Boomtown Rats and just about everyone else who was performing – David Bowie, Queen, Elton John, U2 – and had followed Michael Buerk's eye-opening broadcasts from the famine in Ethiopia that had inspired Geldof to organise this sixteen-hour transatlantic concert. As the famously blunt Irishman pleaded, cajoled and swore on air – what is popularly remembered as 'Give us your focking money,' though he never quite put it like that – the duty of rich countries to help poor ones became part of a global consensus, and part of my political creed.

It was one of the joys of becoming prime minister that I could push this agenda, though I would have to wait until the economy started to grow before I had the authority abroad and the political space at home to do so properly.

By that point, in 2013, I also felt as though I'd found my feet on the world stage. I was thinking less about *how* to do things – the summits, assemblies, councils and incoming state visits – and more about *what* I wanted to do to shape the global agenda. Take some risks. Go to places British prime ministers hadn't set foot for years. Confront issues we had overlooked. Forge new relationships. Break old Foreign Office habits. It was a shift from faithfully following the agenda to trying to set it. And it was this shift that I believe enabled me to carve out a distinctive foreign policy.

International development was central to that.

I never saw my support for aid as being at odds with my conservatism. Quite apart from the moral argument about alleviating the most extreme poverty, there was clear national self-interest. We were helping to insure ourselves against the mass migration, terrorism, pandemics and environmental destruction caused by global poverty. We were also gaining 'soft power' by creating the trading partners and allies of the future, and by wielding our influence around the globe.

Over my lifetime, the combination of free markets, smart intervention from states, and people's devotion to the common good had had a spectacular global impact: every single day for twenty-five years, more than 120,000 people had been lifted out of extreme poverty.

The United Nations set a target in the 1970s that each country should spend 0.7 per cent of its national income on overseas development. Britain had committed to this, but had never hit it – only Sweden, Norway, Luxembourg, Denmark and the Netherlands had done so. And then at the Gleneagles G8 in 2005 these promises were reaffirmed by the world's largest developed economies, with the UK pledging to achieve the target from 2013.

My view was simple: we shouldn't make these promises and then break them. Britain should aspire to lead the world in aid and development. 0.7 per cent of GNI equates to less than two pence in every pound of tax paid.

I knew it wouldn't be an easy sell. We weren't rock stars urging fans to feed the world. We were politicians (at a time of anti-politics sentiment) asking for more money (at a time of austerity) to be spent overseas (at a time when the tide was beginning to turn on internationalism). We were also up against a right-wing press determined to reflect those public doubts and make the case against aid – not simply with individual reports rightly exposing stolen or wasted aid, but a sustained campaign against the very principle.

I never wavered. 0.7 per cent was something we could and should do, even in tough times. You couldn't equate the austerity we faced at home with the grinding poverty suffered overseas. I knew that Britain – one of the most generous countries in the world – got that.

But aid also had to change. It could no longer just be about spending lots of money. We had to root out waste. DFID had been rightly respected for its technical expertise. But under Labour it began to resemble a giant NGO, with approaches that were often quite disconnected from the government's other foreign-policy objectives. We needed the hardest of taskmasters to overhaul a system that too often lacked transparency, focus and flexibility, and that was Andrew Mitchell's task when he became secretary of state at DFID in 2010.

We had spent five years in opposition planning how we would combine a growing aid budget with real reform of the aid sector. So when it came to countries that could afford to combat poverty themselves, we ended traditional aid for some (like Russia) and radically reduced it for others (like India). We delivered transparency, publishing full details of every UK aid project online. We established an independent watchdog to monitor spending. We redirected money from poorly performing agencies to the most effective ones, such as the Global Alliance for Vaccines and Immunisations and the Global Fund to Fight AIDS, Tuberculosis and Malaria. And we linked all this to commercial diplomacy, carving out a chunk of the aid programme to help create openings for British business, called the Prosperity Fund.

As well as pushing through these important reforms, I wanted to drive a deeper change in our approach to extreme poverty. For a long time we had been treating the symptoms – disease and hunger – and rightly so. But what about the causes, the deep, knotted roots that tethered people to penury for life?

We have grown up with the idea that 'geography is destiny'. And it's true that some countries are more prone to, say, earthquakes than others. But people's fate increasingly rests on whether they live in a state that has a functioning government and institutions or not. It's often countries with the most resources whose people are the poorest. Geography isn't destiny; governance is.

Look at neighbouring countries with similar starting points and radically different outcomes. Why is Botswana a success story and Zimbabwe a byword for failure? Why is Colombia climbing out of conflict and neighbouring Venezuela in utter chaos? Why are the fortunes of the two Koreas so radically different? The answer is governance and leadership.

Indeed, look back to the situation in Ethiopia that galvanised us in the 1980s. Droughts and food shortages do not make starvation inevitable. Famine is politically inflicted. In 2017, for example, Ethiopia had a drought without a famine, and South Sudan had a famine without a drought.

At a moment when the Western world was losing faith in politics, reflected in the rise of 'anti-establishment' populist parties, the developing world was crying out for strong governance. After all, it is vital in allowing us to live peacefully and happily. If your government is corrupt, you'll have to pay bribes. If there is limited transparency, the wealth and natural resources you should benefit from can be stolen or squandered. If there is no rule of law, you cannot be certain that contracts will be enforced, or that what you have is truly yours. If there is no accurate register of land and property ownership, you cannot borrow against your most valuable asset. If you don't have tax-collecting authorities, you won't have any public services. And if you're denied free and fair elections, you'll never be able to change anything. The rule of law, property rights, decent government, fair elections – these are the things I called the 'golden thread' of truly sustainable development and of real democracy.

I duly increased the proportion of the aid budget spent on countries lacking these things – what we called 'fragile states' – from a third to a half. But in an interconnected world, a prime minister cannot tackle any of the really big challenges – poverty, terrorism, climate change, mass migration – without collective political will and action.

That means round-the-clock diplomacy, using every forum to which you belong and every summit you attend. It means creating new institutions to drive your vision, as I did by establishing the Open Government Partnership in 2011, by which wealthy countries help fragile countries on the path to greater accountability and success. It means marshalling all of the best arguments you can, and sometimes giving your fellow leaders a good shove.

Not that it was always an appealing prospect. When Andrew Mitchell first suggested to me that I chair the UN panel on writing the new global goals for development, my first reaction was 'absolute waste of time'.

Every country on earth is represented in that building in the middle of New York, in what is the only truly global organisation in the world. But it is also, perhaps unsurprisingly, a mess of bureaucracy and frequent paralysis. You're always happy to see it, but equally glad to leave it.

Andrew persisted. In 2000 the Millennium Development Goals had set targets for developing countries and for the direction of aid, from eradicating extreme poverty and hunger to achieving universal primary education. They were not universally met, but significant progress was made in almost all of them. As they were due to expire in 2015, it was unclear what to put in their place.

With the chance to bring clarity, practicality and British interests to the global agenda, I accepted Andrew's challenge, and buried myself in the detail. My co-chairs were the widely respected president of Liberia, Ellen Johnson Sirleaf, and Indonesian president Susilo Bambang Yudhoyono, whose party piece was playing guitar and singing a song he had written about climate change. One of the meetings took place in Liberia, a country which demonstrated what this was all about. Ravaged by war, pillaged by despots, ignored by the world, it was 25 per cent poorer than it was forty years ago. It was so dark when I arrived at night that I asked the driver when we would reach the capital, Monrovia. 'We're here,' he answered, as stray dogs wandered down the dirt tracks.

The Millennium Development Goals had focused people's minds on tackling the most basic outrages of extreme poverty, but there was a sense that the solutions had been handed down from rich countries to poor ones, and that there was no real mention of the environment and climate change. Correcting this was essential, as were three other clear aims I had in mind.

One: the fight against extreme poverty should be the priority. Some campaigners wanted to make tackling inequality the main focus, in part because it would apply equally to wealthy nations as well as desperately poor ones. I thought this was politically correct nonsense. Tackling inequality is an important challenge, but it is nothing compared with the moral outrage of over 750 million people living in abject poverty, without access to clean water, secure shelter or enough to eat. Lose the focus on absolute poverty and we would lose the whole moral force of the campaign for a better world.

Two: there needed to be a far greater focus on gender equality. Quite apart from being morally right, empowering women is the fastest way for an underdeveloped country to develop more rapidly. I wanted the goals to cover girls' education, equal pay and the right for women to own and inherit property. But I also wanted them to cover the things which, sometimes due to cultural sensitivity, had been brushed under the carpet. Every year over ten million girls were forced into marriage. Nearly four million were cut through the horrific process of female genital mutilation (FGM). These things weren't just happening in far-off countries, but in our own. It was time they were ended once and for all.

Three: that governance, corruption and the rule of law – ignored in the original goals – should be put front and centre. This was the aim to which I attached the most importance, because the others depended on it.

The final working meeting was at the UN's headquarters, where we received the near-final draft papers. Yet there was no punch, no crunch, no sign of the simple set of measurable goals we so needed. I had an idea: why didn't we, Britain, just write our own draft of the goals? People might be so blown away by this newfangled notion of clarity that they'd simply read the proposals, agree with them and approve them. Thankfully I had long-time Conservative development expert Richard Parr with me. We sat up late into the night and whittled and worked a long laundry list into twelve key goals, combined with a clear introduction.

I tabled the paper the next day, and chaired the meeting aggressively, giving each committee member no more than ten minutes to talk. One by one, they signed up to the list. Those twelve goals eventually became the seventeen 'Sustainable Development Goals', including that of which I am most proud, SDG 16, which promises access to justice and stronger institutions. The 'golden thread' – or at least most of it – is now something global leaders are united in striving for.

Despite better goals and a more targeted aid budget, it was still a struggle to sell the massive commitment we were making on development. So I decided to highlight its benefits with annual conferences – key 'moments' – to promote our work and seek funding. It was essentially 'Give us your focking money' for politicians – and it worked. At the Vaccines and Immunisation event I held in 2011 with Bill Gates, we arm-twisted enough money out of leaders and foreign ministers to help prevent four million deaths from pneumonia and diarrhoea.

I used this 'convening power' to get people talking about the more controversial subjects. For our 2012 'moment' I seized on the issue of family planning. In Sierra Leone, for example, two-fifths of girls have their first child between the ages of twelve and fourteen. This was both a *cause* of poverty and a *consequence* of poverty – and, frankly, horrifying. But for decades family planning had been something of a taboo subject in the international community.

I wanted to break this taboo and help get basic family planning services to the millions of women and girls who desperately needed them. And I wanted to go further, by calling out practices such as female genital mutilation. I announced that Britain would toughen its laws on FGM, and lead the way in stamping it out in countries like Sierra Leone,

Guinea, Djibouti and Somalia, where over 80 per cent of girls are subjected to the barbaric practice.

'It really should only be discussed by African women,' some officials warned me as I planned this. Ridiculous. It was exactly this sort of cultural relativism that had let us overlook the disgusting crime for fear of offending anyone.

Then there was the Global Summit to End Sexual Violence in Conflict – the largest gathering in history to tackle the stigma and culture of impunity that enabled rape and sexual violence to be used as a weapon of war. William Hague made this harrowing issue his big focus, driven by his tenacious spad Arminka Helic, herself a refugee of the Bosnian War.

As for 0.7 per cent, the opportunity arose for us to enshrine it in law – something we had promised in our manifesto – when the Lib Dem former Scotland secretary Michael Moore won the chance to take a Private Members' Bill through Parliament. Though legislation wasn't my priority, I was still happy to see 0.7 go onto the statute books, committing all future governments to it.

Looking back now at what this aid drive achieved during my time in office makes me even more of a stubborn supporter. Polio, which used to paralyse or kill hundreds of thousands of children each year, is now, with British support, on the brink of being wiped out. Deaths from malaria were cut by 29 per cent between 2010 and 2016 – that's over half a million lives saved.

Thanks largely to 0.7 per cent, British scientists have invented new technologies, from drought-resistant crops to new vaccines and smarter water pumps. Sixty million people have received clean water and sanitation. Eleven million more children around the world go to school, and millions more women and girls have access to family planning. Six million people can claim more secure land and property rights. Twenty-four million mothers, babies and children have been given the nutrition for their bodies and minds to grow to their full potential. Thousands of bright young Brits have volunteered overseas through the International Citizen Service scheme we introduced. As for the starvation that shocked us all at the time of Live Aid, world hunger has nearly halved.

This was about people, not just statistics. In 2011 I visited a vaccination clinic in Nigeria. Amid the heat and humidity, under flickering lights, were rows of women holding their babies. A nurse explained to me that if it wasn't for British aid, many of those babies would be dead.

The aid budget allowed us to act quickly after a typhoon hit the Philippines in 2013, an earthquake struck Nepal in 2015, and, as I'll explain, a deadly virus swept across west Africa. 'Britain stands ready' is a phrase politicians often use, but I knew how much of that readiness came from 0.7 per cent.

The irony is that the people who complain of our waning influence in the world are often the same ones who complain about our development budget. They don't seem to appreciate that our aid commitment gives us soft power unlike anyone else. Britain is there, in every disaster zone. Food packages and tents stamped with the Union flag and 'UK AID' – another Andrew innovation – can be found all over the world. Our aid workers, peacekeepers, medics and volunteers are there on every continent. We don't just stand ready: we stand tall.

One fragile state that became a preoccupation for me was Somalia. The country was a source of terrorism, a driver of mass migration, and the global headquarters of piracy, one of the biggest issues on my desk in 2010, after scores of hijackings and kidnappings, including that of a British couple. It was also, tragically, a land of famine, with an estimated 260,000 people dying of starvation during my first two years in office.

I wanted us to use the lessons learned in Afghanistan to help rebuild this broken, and largely forgotten, country more remotely and smartly. We drove that via the NSC and two international London Somalia Conferences, in February 2012 and May 2013. A key aspect of our approach was separating the unacceptable al-Qaeda part of the al-Shabab movement from deeply conservative and tribal elements that could potentially form part of a future political settlement in the country. It was also about pledging money to stop people starving, funding prisons to lock up the pirates, and helping the Ethiopians and Kenyans defeat al-Shabab.

The results were that in 2013 not one ship was successfully hijacked. Al-Shabab had been pushed back. Somalia had a government. William opened the first Western embassy there since the 1990s. The country is part of the global consciousness – and there is now a consensus to fix it.

Fragile states would remain on the agenda when our turn came to host the G8 Summit in June 2013. For it to be a purposeful, productive and memorable summit, I knew the location was important. I chose Lough Erne in County Fermanagh. I wanted to show Northern Ireland off to the world.

The theme I chose would be important too. It had to tie together the needs of the rich world *and* the developing world, and it had to be achievable.

From the earliest days of party leadership I had been talking about what was going wrong with the global system we lived by. In 2006 I gave a speech entitled 'The Challenges of Globalisation'. In 2007 I said: 'We must create fairness alongside wealth, or we will lose the argument for a dynamic global economy, and lose the benefits it brings.' In 2008, the year the crash happened, I argued that 'We shouldn't replace the free market – we should repair it.'

The Occupy Movement, which camped outside St Paul's Cathedral from the end of 2011 into 2012, represented one solution to the state of affairs: to end capitalism. For me, the solution was to *fix* capitalism. After all, free markets were providing the answers to so many of the ills we faced, from helping people escape extreme poverty to developing the innovations needed to halt climate change.

Now, as G8 host, I had the chance to galvanise the globe about the specifics of how we would fix this global system. I homed in on three things: advancing trade, ensuring tax compliance, and promoting greater transparency. My 'three Ts'.

World trade still hadn't recovered from the crash. As the economic balance of power shifted eastwards, we rich Western nations wouldn't keep growing unless we signed new trade deals and succeeded in new markets. For the countries of the developing world, prosperity would never be achieved by aid alone – they needed the opportunity to trade. And that was being stifled by outdated rules and vested interests.

Aggressive tax avoidance, and its criminal sibling outright tax evasion, were also problems for both rich and developing countries. More and more schemes – not illegal, but morally questionable – were coming to light, depriving people of the revenues they needed and decoupling wealth and fairness. As with most things, the solution had to be international. If one country cracked down on corruption or poor tax practices, its perpetrators would just go somewhere they could get away with it. We had to close off every avenue.

Linked to all this was transparency. Without it you can't see who owns what, who pays what, or where money looted from poor or corrupt countries is hidden. You couldn't separate out the rich world and the developing world on this issue, because so often the stolen wealth was being hidden, for example, in the London property market. That's why I was so keen to legislate for a register of who owned what in Britain – the beneficial owners, the real flesh-and-blood people, not the shell companies they hide behind.

The feel of the summit was important to its success, so I was interested in every last detail, even down to the furniture. The last time a prime minister paid so much attention to a table was probably when Harold Macmillan had the long, narrow cabinet table remodelled to a coffin shape, supposedly so he could see into the eyes of all his ministers. I wanted the G8 table to be small enough to make sure there was no room for additional attendees – translators, private secretaries, even foreign secretaries. We ended up with one beautifully crafted from Irish elm, that remains in Downing Street's Thatcher Room today.

The sun shone on Northern Ireland for those two days. Despite being held in a province that had suffered so greatly from violence, it was the most peaceful G8 ever – and two of the most satisfying days of my time as prime minister.

Even though Syria dominated, there was room to make significant progress on my three Ts.

On trade, the first discussion we had was about the Transatlantic Trade and Investment Partnership (TTIP) between the EU and the USA, which could break down many of the barriers to exports and give an unprecedented investment boost to the British economy. The big announcement was that talks on this, the biggest free-trade deal in history, would begin in July, although it would eventually be crushed under the weight of anti-globalisation, and a fair bit of 'fake news'.

On tax, we agreed that authorities across the world should automatically share information to fight the scourges of tax evasion and aggressive tax avoidance. Multinational companies would have to provide more detailed financial reports about their global activities, which would make it more difficult for them to pay tax only in low-tax countries. This greater openness could help developing countries get the oil and mineral money that was rightfully theirs, and I was pleased when several African leaders we had invited to the event praised the agenda and said that Britain had the credibility to champion it because we had kept our aid promises.

Transparency was always going to be the most difficult one. Leaders tend to be in favour of it in principle and then, armed with endless briefings about why it would not be possible or desirable in their own countries, turn against it in practice. And the spotlight was on Britain for supposed hypocrisy: launching an open-ownership register for ourselves, but exempting our overseas territories like Bermuda, Jersey and the Cayman Islands, whose economies – it was alleged – were built around low taxes and secrecy over ownership.

This was unfair: those territories' practices and procedures were far from the worst in the world. But in order to close off that line of attack I had managed to get ten British dependencies with large financial sectors to agree a convention on information-swapping. We actually invited their leaders over to London on the Saturday before the G8, treated them to a lovely day at Trooping the Colour, and brought them to the Cabinet Room to sign up to rules on tax transparency and mutual recognition that no one had thought possible. Eventually they all committed to drawing up registers of beneficial ownership, although most stopped short of making them open for public view.

(This was genuine progress. It is vital, in the first instance, that tax authorities in different countries are able to see this information. And it immediately put our overseas territories and crown dependencies above many countries in the world, and also many US states. Beneficial ownership registers that are open to the public remain the gold standard, because they allow civil-society organisations and journalists to help with the work of seeing who owns what. In the long run this is essential, not least because tax authorities in some of the most corrupt countries are unlikely to seek information about political or business leaders and what they might own in other countries. In 2018 the UK Parliament took the next step, insisting, via an amendment to a government Bill, on open registers in all the territories for which the UK is ultimately responsible. The backbench MP moving the amendment was Andrew Mitchell.)

The summit ended with a communiqué which, for once, was short, clear and understandable, and is still being followed to this day. As the leaders prepared to depart, Obama asked my team for drinks with his team in his lodgings by the side of the lake. He wanted to congratulate us on a job well done, and knew that the people who really deserved the praise were those who had been working around the clock to make things happen. It was during this friendly exchange that we worked out that Liz Sugg, in many ways the architect of the summit, did the job of fourteen of his staff.

The narrative today is that world leaders were caught out by populism and the anti-globalisation sentiment that fuelled it. As the 2013 G8 shows, that is simply not true. We were confronting the issues head-on. We were being constructive rather than destructive. We were repairing rather than replacing, reinstituting the link between wealth and fairness, and preserving a global system upon which the future of all countries depends. The subsequent success of populist candidates proves how drastically such work is still needed.

The concept of 'action now' had yet to reach another international gathering, the biennial Commonwealth Heads of Government Meeting (CHOGM).

The Commonwealth brings together north and south, east and west. With such diversity and scale – it covers a third of the world's population, its biggest and smallest nations – it has had a unique impact on the world, from sanctions against Apartheid to strides on combatting climate change.

As with many multilateral bodies, it is dysfunctional. But my apprehension about the November 2013 CHOGM was caused by something other than the organisation's usual inertia. One of the fifty-three member states, Sri Lanka, had endured a twenty-five-year civil war, with both the government and the Tamil insurgents accused of torture, rape, extrajudicial killings and disappearances. The war was over, but only after extraordinary bloodshed.

Almost all insurgencies require a political solution, from the IRA in Northern Ireland to the FARC in Colombia, but the end of this war was so brutal that the insurgency was almost entirely snuffed out. Sri Lanka's leader at the time of the final military offensive against the Tamils was Mahinda Rajapaksa. And he was to be our CHOGM host.

When the Indian and Canadian prime ministers announced that they would not attend in protest, there was pressure on me to boycott the event as well, not least from Ed Miliband, who had found another bandwagon to jump on. Buckingham Palace also played rather a cunning game, not confirming whether the Queen would be attending or not. Though Prince Charles was increasingly representing the Queen anyway, I felt her courtiers wanted to see what I would do before they made their decision.

They needn't have worried. Despite the political heat, I thought it was right to go. I should represent the UK at a meeting of the Commonwealth wherever it was being held, and Prince Charles deserved the PM's support. Besides, what would be a more effective challenge to Rajapaksa and his denial of war crimes: staying away and saying nothing, or going there and holding him to account?

I rather began to embrace the idea. If Somalia had been an example of how to use aid, defence and foreign policy to effect change, going to Colombo would be a way of using these conferences to our advantage, and pressurising a nation to do the right thing.

I wanted to visit the north of Sri Lanka, so my first destination was Jaffna, the Tamil city that had been the scene of fierce fighting during the civil war. I was the first foreign leader to visit since 1948, and the day I spent there will live long in my memory. I saw what was left of Jaffna Library, whose priceless manuscripts had been destroyed by fire as government forces tried to eradicate Tamil history.

I visited the Tamil newspaper's offices and met the editor, who had lived in the building for the past three years because he feared for his life. There was a charred printing press that had been shot and burned by regime hoodlums, and the walls were covered with bullet holes where journalists had been murdered.

I went to a refugee camp, whose existence the regime denied. I'll never forget the crowds of women, holding up photos of young men, desperate to tell us their stories. We all had letters thrust towards us about these sons, husbands, fathers and brothers who had surrendered to the military and not been seen since. What had happened to them? Could we help find them?

I flew back to the capital for my bilateral – or rather showdown – with Rajapaksa. His brother, the defence minister, who had supposedly issued orders for the terrible events at the end of the war, reached out to shake my hand. I kept my hands by my side, and sat down.

The truth is that the Tamil Tigers had done some dreadful things, and I did have some sympathy with the case the government put. But I had to make a judgement. Was this a leader who was trying to reconcile his country and have inclusive political institutions that worked for everyone? Or was he a South Asian sectionalist who wanted to secure funding for his own people alone? I thought he was the latter, and I essentially told him so. As we walked out, William Hague said it was the worst-tempered foreign meeting he'd ever been in.

I delayed my return to Britain because I was told that Rajapaksa was asking for another meeting with me. My tough diplomacy had not resulted in 'I'm never speaking to you again'; it had resulted in 'OK, actually I do want to try.' In our one-on-one he pleaded with me. We're doing the right thing, he said. It was going to take time. We mustn't criticise them so much. Didn't I realise that if he did what we asked, if he held a proper inquiry into war crimes, the Sinhalese people would murder him?

Ultimately, as predicted, the Sri Lankan government would offer no concessions and the CHOGM communiqué would make no criticisms. Instead we ended up, as ever, with weak-willed, head-in-the-sand consensus, thanking Sri Lanka for its 'warm hospitality' and praising Rajapaksa for his 'able stewardship' of the event.

The following year, after a good deal of British lobbying, the UN Human Rights Council voted to open an inquiry into war crimes committed during the Sri Lankan civil war. Two years later, Rajapaksa unexpectedly lost office. I was delighted to welcome his successor, President Maithripala Sirisena, to No. 10 in March 2015. For a while, Sirisena looked like the answer. But in 2018 he appointed Rajapaksa to his government, and hopes for that beautiful country were once again looking strained.

I wasn't unused to having my hopes dashed when it came to foreign affairs. Indeed, 2013 brought further disappointment – as well as atonement, frustration and sadness – in my dealings with other countries.

The disappointment came from Burma. I had visited the long-time military dictatorship a year earlier, just after it had taken its first steps towards democracy by holding by-elections. No UK PM had visited since independence in 1948. I met the pro-democracy campaigner Aung San Suu Kyi, who would soon run for the presidency, and reflected on what an amazing story hers could be: from fifteen years of house arrest to transforming her country into a real democracy. However, by the time she came to visit London in October 2013, all eyes were on her country's Rohingya Muslims, who were being driven out of their homes by Buddhist Rakhines. There were stories of rape, murder and ethnic cleansing. The world is watching, I told her. Her reply was telling: 'They are not really Burmese. They are Bangladeshis.'

She became de facto leader in 2015, and the violence against the Rohingya went on.

The atonement related to India. For a long time, friends and colleagues in the British Indian community had encouraged me to go to the Golden Temple in Amritsar. This holiest of Sikh sites had been the scene of a massacre in 1919, when British Indian Army soldiers fired upon a peaceful public meeting, killing hundreds of people.

No serving prime minister had ever been to Amritsar, let alone expressed regret for what happened. I wanted to change both those things, and would do so after the trade mission – the largest in UK history – I would lead to India in February 2013. Ahead of my visit there was an internal row about whether I should say 'sorry'. But ultimately, I felt that expressing regret for what I described in the memorial's book of condolence as 'a deeply shameful event in British history' was appropriate. I knew what it meant to British Sikhs that their prime minister had made that gesture, and I'm glad I did so.

The frustration came from China. Every year, the Dalai Lama visited Britain and asked for a prime-ministerial meeting. Given that China didn't recognise Tibet's independence (and neither did the UK), and that until recently he ran an alternative government in exile, such a meeting would alienate the very people we were trying to develop a relationship with.

But politics aside, this elderly monk who preached peace and kindness was the leader of a religion. So I said I would meet him not as a political leader but as a religious leader, at St Paul's Cathedral, where he was seeing the Archbishop of Canterbury. The Foreign Office said the Chinese would pretend to be cross about this for a couple of months, but it would blow over.

Instead, our ambassador in Beijing was summoned for a dressing-down, and the Chinese government released a statement condemning our action. Ministerial visits were cancelled in both directions. The freeze lasted eighteen months, and only came to an end when George Osborne was invited to visit in October 2013. Boris was also in China that week. He got all the headlines – Boris on a ferry, Boris on a high-speed train, Boris on the drums. But the real action was George quietly brokering things in the background, laying the ground for my trip the following month and discussing a mammoth Chinese investment in the UK nuclear power sector.

I was robust about how I approached the UK–China relationship. We would be critical partners. As I put it to the BBC's Nick Robinson on our arrival: 'I simply don't accept that there is a choice when you come to China that you *either* talk about growth, investment and jobs, *or* you talk about human rights. You do both. And that's what I'm doing on this trip.'

My year of foreign affairs ended with sadness.

I had first met Nelson Mandela when I visited Johannesburg in 2006. In December 2013 I received the news that he had died, aged ninety-five. I led the tributes in the Commons soon after, praising the struggles of a man whose fortitude and determination changed his country and changed the world. A memorial to his life was held in Johannesburg the following week. Tens of thousands of people gathered in the Soccer City stadium – a sea of world leaders past and present, South Africans, banners, face paint, singing, cheering, electric speeches. The booing during the speech of President Jacob Zuma, the country's current leader and the apogee of corrupt leadership, demonstrated once again how desperately we needed to stamp out this scourge.

By the end of the year I had been to twenty different countries, hosted two incoming state visits, held over a hundred bilaterals with other world leaders and attended fifteen international gatherings. Although I still had the same aims, I felt the 2013 me handled things differently to the 2010 me – I was becoming a better leader. But as we entered 2014, all the focus would be on my next big challenge: the general election.

36

The Long Road to 2015

'You campaign in poetry. You govern in prose.' It was the former New York governor Mario Cuomo who famously captured how flowery pre-election promises can turn to dust as the routine business of government is ground out.

In some ways we turned this political dictum on its head. In office we took on causes that were wide-ranging, far-reaching, sometimes even era-defining. Being the first Conservative-led administration anywhere in the world to allow gay people to marry the person they love. Galvanising the globe in the fight against dementia. Keeping our aid promise to the very poorest. Doing our bit to preserve our planet by, albeit rather quietly, becoming one of the greenest governments ever.

There was still a lot to be done, but as we embarked on the long road to 7 May 2015, I knew we'd have to stop talking about many of the things I cared about if we were to be in with a chance of winning. To move from a poetic policy platform to a more prosaic narrative.

Winning was something many commentators had already declared impossible. As the radical and reform-minded Conservative commentator Fraser Nelson put it in 2012: 'By now, it will be clear even to David Cameron that he is on course to lose the next general election.'

The lack of boundary reform had turned a hurdle into a high jump. Plus, the precedents weren't great. While a number of prime ministers had won a second term, since the war only Clement Attlee had held on to power after serving a full five-year term. Even then, he'd only squeaked home, and he had started with a majority. To win, I'd need to get a better result than at the last election. No one had done that after completing a full term since Lord Palmerston in 1865.

I was determined not to make the mistakes of our mixed-messaged, too-many-cooks 2010 campaign. We needed a plan, we needed discipline, and we needed the right people – specifically the right person – leading it all.

I've talked about my admiration for Lynton Crosby. Although the 2005 election campaign was a failure, I bought his key defence – that he was brought in too late: 'You can't fatten the pig on market day,' as he put it. It's true that I didn't like the tone of some of that campaign, but I've never had a problem working with people I don't always agree with, and I could override Lynton when I thought he was wrong.

By the end of 2013 he had moved into CCHQ. I'd see him several times a week in my office, with my team or alone. People like to imagine a brash Aussie stomping in, but in fact Lynton was polite, and even rather deferential. He was also very sensitive to criticism, and I had to convince him several times not to resign over some mishap or other.

But he did know how to enforce focus and discipline. And for that, he had an array of dictums. 'Don't be Connie the commentator' was how he would urge me to stop merely responding to the events of the day. The gnomic 'You need to bell the cat' took on any number of meanings as the campaign progressed. 'Get the barnacles off the boat' was his famous call to streamline our message and ditch side-issues. It became something of a motto.

Lynton's list of things we shouldn't focus on was long. But what about the things we *should* focus on?

Lynton put it frankly: you've been taking tough decisions, working through a programme, all with a clear end in sight, but you simply haven't explained it like that.

So this new device had to be presented as a *plan*, one that linked our record so far with what we had yet to do. It also had to be *economic*. After all, the economy was and is voters' biggest concern. And it had to be for the *long term*. Because after thirteen years of government that effectively said 'Spend for today, to hell with tomorrow,' we had an administration casting its sights further ahead, beyond the Parliament, our political careers and even our lifetimes.

This is how the slightly clunky, but unmistakably clear and simple 'Long Term Economic Plan' (LTEP) was born. We were campaigning in prose.

I could recite its points in my sleep. Reducing the deficit so we could deal with our debts, safeguard our economy for the long term and keep mortgage rates low. Cutting income taxes and freezing fuel duty to help hard-working people be more financially secure. Creating more jobs by backing small business and enterprise with better infrastructure and lower jobs taxes. Capping welfare and reducing immigration so our economy would deliver for people who want to work hard and play by the rules. Providing the best schools and skills for young people so the next generation can succeed in the global race.

When I launched it on 1 January 2014 the starting gun was fired on an election campaign, with over a year and four months to go.

The messaging was crucial. A governing party that has the economy on its side – as we did with the revised GDP figures in 2013 and with real wages rising by mid-2014 – has a big advantage. But it doesn't deliver automatic victory. You need, in a disciplined way, to make your economic effort the central question in the campaign.

When we alighted on the long-term plan that's what we did. Labour, by contrast, hopped from message to message – tax avoidance, Rupert Murdoch, the NHS. They would win skirmishes, but once their 'squeezed middle' rhetoric stopped working for them and the economy started working for us, they never found a consistent issue to fight us on.

The irony was that our parliamentarians, who found it almost impossible to accept lectures from me, were more than happy to do so from an Australian political pro. They were happy to get on board with his message discipline – so much so that the phrase 'Long Term Economic Plan' was uttered 826 times in Parliament in just eighteen months. The prospect of the ballot box concentrates minds.

Just as we needed to focus our messaging, we also needed to focus our resources. This brought with it an awkward question: to what extent would we target Lib Dem seats?

Stephen Gilbert had already drawn up the so-called 40:40 strategy – forty marginal seats we needed to win, forty we needed to hold. Since 2012 ministers had been filing through these seats, and campaign managers were gradually being appointed in each of them. Now Stephen made a pitch to amend the list of seats. He said the Lib Dems had scuppered boundary reform, and as a result we could only win a majority if we won more of their seats. We shouldn't just target a *few* Lib Dem seats, we should go for *all* of them. Especially as they were where my personal appeal was strongest.

Lynton didn't need much persuading. But George and I were less sure. If the strategy failed, and the Lib Dems remained intact, there would be no possibility of a renewed coalition. But if we didn't go for the Torbays,

Twickenhams, Baths and Cheadles, we would probably lose to Labour anyway. Plus, if we had the courage of our convictions, we shouldn't be campaigning with compromise in mind. The decision was to do it.

It was ambitious. Some of the majorities we needed to overturn were huge. But CCHQ was ready – the coffers well stocked, the office well-staffed – all because Andrew had been determined not to run it down in the so-called 'off years' between general elections.

There was, however, one party agenda we failed dismally on: membership. During Francis Maude's time as chairman we took a step backwards when it was proposed that we scrap national membership and return solely to membership via local associations. I foolishly agreed. In this case, devolution didn't work. Some constituency associations with low membership rolls enjoyed running a cosy cartel that picked the candidates for council and parliamentary elections and went to each other's soirées, without bothering to recruit new people. There were even some local parties that, when you called and asked to join (and we did this as part of a 'mystery shopper' exercise), told you their membership was 'full'. We also had no idea how many members there were, and were effectively unable to launch a national mass-membership drive.

Those who argue that what matters is not the number of members, but the number of *active* members, are right to a degree. But a political party needs mass membership for a pool of talent to call on: your candidates, your councillors, your activists, your association chairs, your donors. The bigger the membership, the more representative your party will be (though Ed Miliband showed what a disaster hurried reforms can be when his eye-catching £3 membership drive left Labour open to a far-left takeover).

Although we lacked paid-up members, we did have the activists, whose numbers remained pretty solid throughout the coalition. We also had young supporters, who Grant Shapps was busily convening into a 'Team2015' deployable force, ready for the election. We had more councillors than all the other major parties put together – and again, this held, even though, for most parties in office, seats on local authorities usually plummet. (In the 1980s and 90s, the last time Conservatives were in government, our local government base was almost wiped out; this time we lost fewer than seven hundred councillors and ten councils while I was in power, still leaving us with nearly 9,000 councillors and 191 councils.)

With Lynton full-time, a plan in place and our MPs in line, I felt a surge of confidence. I thought we could win. As I put it at the end of January 2014: 'We're three or four points behind and sixteen months to go with a recovering economy and an opposition leader that people haven't warmed to. I feel pretty bullish.'

There was another reason for that confidence. Politicians often say they don't look at polls. That's a lie. There were two polling questions I paid particular attention to: the leaders' personal ratings, and the parties' ratings on 'Who do you trust to run the economy?' By now the Conservatives had a ten-point lead over Labour on the economy. And the polls on leadership had me at around 40 per cent approval rating and Miliband in the mid-twenties.

That didn't mean we were going to win the election by that margin. Research showed that I was more popular than the Conservative brand, while the Labour brand was more popular than Ed Miliband. However, it still rather put the onus on me, personally, to get out there in those 40:40s with the candidates to explain what the LTEP meant to ordinary people.

Most Thursdays, a lot of Mondays and several two- or three-day stretches were put aside for these so-called regional tour days. They became part of the rhythm of my week.

I loved them. I loved getting out of the No. 10 bunker and having the time to think. I loved the high-fives and selfies in noisy school playgrounds. I loved the banter on the factory floors. I loved the pride in people's eyes as they walked me around their lab, explaining the intricacies of their research. I loved it when the latest couple to have benefited from Help to Buy showed off their new home.

If you followed politics only on Twitter or the *Today* programme, you would assume that everyone is permanently furious. But life isn't like that. Most of the time people are pleased to see you. They tell you how they are feeling and how you're doing. They have things they want to mention to you or raise with you – not shout at you. That's more revealing than any poll or briefing or focus group, and it was a big part of those tour days.

Rude remarks were rare, and as a result I tended to remember them. They often happened in the places you would have least expected. For instance, in a café in Ely, the heart of Conservative Cambridgeshire, I ordered a full English breakfast, and one diner piped up, 'I hope you choke on that.' The staff and customers spent the rest of my breakfast apologising and reassuring me that Ely really wasn't like that.

My schedule would never have been possible without helicopters. I had used them since opposition – it was the only way, short of tele-portation, that I could meet the demands of a modern party leader or prime minister.

I was never very happy in them, though. After recounting one exper-ience to Samantha – I'd just flown somewhere in a helicopter with one pilot, one single rotor, one engine, and a crack in the glass, while it was raining – she made me promise only ever to travel with two pilots and twin engines.

We still had some hairy moments. Once, in opposition, the helicopter I had borrowed was trying to land in a muddy field in West Yorkshire before a nearby party event. Some enthusiastic supporters had laid out a small carpet so we wouldn't get covered in mud as we clambered out. But as we neared the ground the carpet was sucked up into the air, and was seconds away from being caught in the rotors, with potentially disas-trous effects. My children sing an infuriating song about 'dumb ways to die'. Frankly, this would have taken the biscuit.

Then there was the visit to Humberside, where I grudgingly went to support David Davis in his by-election. The constituencies of party leaders and prime ministers attract the single-issue/joke/crank candidates, and between general elections they have by-elections to keep them busy. So DD found himself up against Mad Cow Girl of the Monster Raving Loony Party, a representative of the Militant Elvis Party, and the football presenter turned conspiracy theorist David Icke. As we were about to leave, we found Icke standing next to our helicopter and telling us quite matter-of-factly that it was going to crash. Given that this was a man who believes the world is run by reptiles and who predicted in 1997 that the world was about to end, I didn't think his crystal ball was that reliable. But we did take our seats rather gingerly, and held on a bit tighter than normal . . .

Regional tours were run and delivered by Liz, the indefatigable Martha Varney and the scarily efficient Events and Visits team of civil servants they led. But there were many more people involved. As well as civil servants from the communications team, I was often joined by one or more political press people. There was Craig. There was Alan 'Senders' Sendorek, my fun and dynamic long-time head of political press. There was often Graeme Wilson, the hugely popular, softly spoken Ulsterman who became my press secretary after being the *Sun* deputy political editor. Sometimes there was Giles Kenningham, head of communications at CCHQ, a wheeler-dealer type whose net-practice questions were so aggressive I'd sometimes say, 'Steady on, Giles' – but I'd always be grateful once I got to the interview itself.

Back at No. 10, it was Ameet Gill and Robin Gordon-Farleigh who coordinated all the business of government, via the famous No. 10 grid. They would decide the theme of the day, pumping the Policy Unit for a constant stream of LTEP-related announcements, or Research and Information for LTEP-related milestones.

Indeed, Meg Powell-Chandler and the Research and Information officials worked late into the night cutting the official stats in every direction, so I could turn up anywhere in Britain knowing how many jobs had been created there, how many more children were in good schools, how many Help to Buy homes had been bought, and so on.

Michael Salter set up my broadcast interviews before the tours. There was a way of talking to a dozen or so radio stations called the 'GNS [global news service] round'. You sit with headphones on in front of a microphone, and move seamlessly from BBC Radio Berkshire to BBC Radio Humberside without ever leaving your chair. The biggest risk is geographical failure – woe betide you if you confuse Tyneside with Teesside (as I did once, and was never allowed to forget it).

Clare Foges and Jess Cunniffe would craft speeches for every event I spoke at, and draft the local newspaper op-eds that would coincide with those visits.

So much for 'Cameron's chumocracy' that I read so much about in the papers. The vast majority of my political team had either come to me as researchers in opposition, or arrived over the years via job interviews, CCHQ referrals, or because I'd poached them from think-tanks, newspapers and broadcasters.

Ed and Kate may have been close friends, but they had only become so because we'd worked together over the years. They were both more than qualified for their jobs, and I resented any implication that they weren't there on merit.

So much, too, for the 'old boys' network' I was also accused of running. By 2015, 44 per cent of my special advisers were women. That was not the result of positive discrimination, just an open attitude towards picking the best woman (or man) for the job.

The dictum 'All politics is local' influenced our very deliberate strategy to target local and regional newspapers, radio and TV stations. We calculated that when it came to the media we could bypass the Westminster village and go straight to the actual villages, towns, cities and counties that could help us secure that elusive victory.

Why was that important? Because the national media had become so jaundiced and cynical about politics that it was almost impossible to get them to report what you were doing. Every policy story would turn into a 'process' story (over which Lynton would eff and blind), explaining not what we were trying to do for the country, but what it meant for me or my political strategy – or, frequently, whether it was an attempt to distract from a 'scandal'. Anything that would sell papers in an age when people were increasingly reluctant to buy them.

Local papers were of course going through their own declining sales and frequent closures. But at least they seemed genuinely interested in what you had to say. People mistake that for being soft. It's not: it's because their readers actually care about whether their local road is going to be widened or there will be a new power station in their area. These things make a difference to their lives – much more so than who is up or down in the jungle of Westminster politics. Unsurprisingly, opinion surveys showed that people trusted local newspapers more than twice as much as national ones.

The regional strategy was backed up by the policies of what was becoming a truly localist government. This had been there from the start, and was boosted by the report I commissioned from Michael Heseltine, who had helped regenerate Liverpool in the 1980s and was now calling for a large pot of funding for local growth. It duly came, in the form of the Single Local Growth Fund, and was distributed via Local Growth Deals. There were City Deals, where Westminster funded local authorities to drive economic growth in the way they saw fit. This was a lot of jargon for what was essentially better transport, skills, housing and other infrastructure, as dictated not by central government but by local communities.

George had initially been less of a decentraliser than me. But the experience of government converted him. As chancellor he saw not so much the damage London's dominance did to the rest of the country as he did the missed potential of the north – and, ever the politician, the potential for the Conservative Party to make electoral strides in those parts we had failed to reach for so long.

By the start of 2015 we found ourselves visiting every region together, and launching a LTEP tailored to each area, with specific, tangible promises on what we would do if re-elected.

The biggest focus needed to be in the north of England. The unemployment gap between south and north was stubbornly high at 5 per cent. Some of the poorest regions in western Europe were in this part of our country. George's beloved 'Northern Powerhouse' initiative would become the highest-profile of this devolution revolution. The key point was that the great northern cities of Leeds, Liverpool, Sheffield and Manchester, though individually not a match for London, could be so collectively. Connect them with better transport, play to their strengths in science and industry, boost their cultural assets with government grants, give them the clout that comes with electing a city mayor, and we could help transform the neglected north.

Labour was furious at seeing the blueprint for the north drawn up by a Conservative government with support from Labour council leaders. It felt great politically, and more than that, it felt right.

Another example of the LTEP in practice was our Spring Budget. Keeping interest rates persistently low had meant people could afford mortgages, and helped enable the economy to recover. But those trying to live off modest savings found it difficult. They'd worked hard, put money aside, and were getting a lousy deal.

The existing policy towards private-sector pensions didn't help. In return for tax relief on money you put into your pot, there were restrictions on what you could do with it when you took it out, in particular the requirement to purchase an annuity which would give you an income for the rest of your life. The justification was that if a pensioner blew all their cash instead of buying an annuity, they'd have to rely on the public purse to support them.

But because we'd introduced a higher basic state pension ('the single-tier pension') there would be far less means-testing in future, and therefore far less risk of additional burdens on taxpayers if pensioners ran out of savings. We had always disliked the condescending, nanny-state insistence on making people buy an annuity, but now we had both the means and the rationale to carry the day. George announced from the despatch box what had been kept a secret between us and a select few: that from now on people could do what they wanted with their pension savings – draw it down over time, buy an annuity, or take it as a lump sum. It was cheered to the rafters.

While you can remove barnacles and try to concentrate on your main message, you can't entirely control events. Another expenses scandal was around the corner, but this time the undoing would come about not so much because of the offence itself, or the lack of an apology, but the manner of that apology.

Back in 2012, the *Daily Telegraph* published a story about Maria Miller. Maria had taken the decision to designate her Basingstoke constituency home as her main address, and then claimed £90,000 in allowances on a Wimbledon address that she had purchased years earlier, and shared with her husband, children and parents. This was the controversy: was taxpayers' money being used to house her parents, rather than just her direct family? The subsequent investigation found that she had not deliberately abused the expenses system, but she should pay back £5,800 she had inadvertently over-claimed. It recommended too that she apologise for her uncooperative and combative approach to the inquiry.

'I know you've been through a miserable time,' I told her. Her family, including her elderly parents, had been hounded by the press. She hadn't done what she'd originally been accused of. 'But whatever apology you're going to make, make it in spades. You can't say sorry enough times.' Instead, she ended up giving a thirty-two-second apology in the House of Commons. This was a fatal error. She was encouraged to go, not pushed, and resigned the following day.

It was a real shame to lose Maria. In 2013 she'd been flying high – one of my star ministers who made a real difference as minister for Disability, helped to deliver gay marriage and got us home and dry on Leveson. By 2014 she was back on the backbenches. What a difference a year makes – and in a couple of years I'd know how that felt.

There was an upside to the Maria problem, however. It gave me the chance to bring Sajid Javid, the financial secretary to the Treasury, into the cabinet, and to move Nicky Morgan up another rung of the Treasury hierarchy. The son of a bus driver from Pakistan, Sajid became a vice president of Chase Manhattan bank aged just twenty-five, before charging through the Conservative ranks to win the constituency of

Bromsgrove in 2010. Nicky, a successful solicitor and now the common-sense MP for Loughborough, was as modest as she was impressive.

I looked at people like this and saw the modern, compassionate Conservative Party I had always wanted to build. They'd come of age as junior ministers and were ready for the next step – into cabinet or senior ministeral roles. Along with Matt Hancock, Liz Truss, Stephen Crabb, Anna Soubry, Tina Stowell and Esther McVey, this was my dream team – the line-up I wanted to present to the electorate in less than a year's time. Which meant another reshuffle – and my biggest roll of the ministerial dice yet.

You can spend as long as you like thinking who you want to promote and where to. But reshuffles always come down to the question of how many people you are prepared to fire. It was time to harden my heart.

Gentleman George Young, who had gallantly stepped in after 'pleb-gate', was only ever a temporary appointment as chief whip.

Ken Clarke, now aged seventy-four, had told friends he was willing to start winding down.

Francis Maude, Eric Pickles and David Willetts had each had four years doing jobs they loved, and the changes they had made – civil service reform, local government reform and university reform – were secure. It might be possible to ask them to stand aside after a good innings.

IDS had done less of a good job in Welfare, and perhaps I could finally move him.

Then there was Owen Paterson. He had been a great Northern Ireland secretary, but his time in the Environment job hadn't worked out as well as we'd hoped. Owen seemed to have a tin ear when it came to showing understanding of a controversy, and a more sensitive communicator could surely do better at defusing rural rows.

However, the biggest change was forced upon me. Before the reshuffle William Hague came to see me and said he wanted to stand down at the 2015 election, after twenty-five years as an MP. George tried everything to keep him on the team, even offering to swap jobs with him if he would stay on. But William was determined to go. He would gladly become leader of the House of Commons, he said, understanding that he would look like a lame-duck foreign secretary if he stayed on and everyone knew he was leaving.

By this time I had one overriding test for reshuffles. When your private secretary says to you on the following day that 'X, the new health secretary, is coming to see you,' do you think, 'Yes, great, that's the person I want running the health service,' or do you think, 'Why is this person coming into my office?'

So one question that reshuffle posed was how I felt when I heard the words, 'The education secretary is coming to see you,' and saw Michael Gove bounding in. In many ways I felt energised. Michael was the reforming minister of his generation. Just days before the reshuffle he would be hosting a huge Education Reform Summit at Lancaster House at which ministers from around the world testified to the triumphs of our school freedom agenda. Not only was he one of my best friends, his wife Sarah Vine, a journalist on *The Times*, got on brilliantly with Sam. Our children were at the same school, they had lived a few streets away from us in North Kensington, and we still met up for dinner and they would often come to stay at Chequers.

So why did I move this zealous and charming eccentric from his job as education secretary?

The main reason was that I believed Michael was becoming a threat to our reforms. In the beginning I was glad he took our fight to the failure-preserving, status-quo-defending teaching unions. But we were now four years in, and despite all the successes he was struggling to take even previously reform-minded teachers with him. He had become so unpopular that there wasn't just a danger to the policy, but to the politics.

His enthusiasm and energy expressed themselves in long sermons in cabinet about everything from Islamist extremism to economics. He frequently and unhelpfully ranged well beyond his brief, for example in an article he wrote about the portrayal of First World War generals in the sitcom *Blackadder*. Here was Connie the commentator, and her name was Michael Gove.

In March 2014 he spluttered his way onto rocky ground when, triggered by Jo Johnson's recent promotion to head of the Policy Unit, he declared the number of Old Etonians at No. 10 'preposterous' in an interview with the *Financial Times*. This was unfair, because Jo had a first-class brain and was a convinced Tory moderniser.

With Oliver Letwin and me in the cabinet, there were indeed a large number of people who had come from one school. But that was the world we'd inherited – literally. It was the world we were trying to change with our school reforms, so that one day No. 10 would be stuffed with people from free schools and academies across the country. But it wouldn't happen overnight, and I couldn't bar good people from government just because of the school they went to. I always thought conservatism was about where you were going, not where you came from. Class war was never our thing.

'Just get on with running the Education Department, and stop commenting on everything else under the sun,' I told Michael over the phone. But he couldn't help himself. That summer a letter about an Islamist fundamentalist plot to take over a group of Birmingham schools was leaked. The letter itself was a fraud, but there was little doubt that there was a concerted effort to indoctrinate children with hard-line Islamist values. Michael rightly ordered an investigation into this so-called Trojan horse plot, but then 'sources close to' him (aka him) accused the Home Office of failing to 'drain the swamp' of extremists. The Home Office hit back with a genuine leaked letter, criticising the academies programme and Michael himself.

I couldn't have this mudslinging. We were supposed to be one team, and would soon be going into an election. The leaker in this case was Theresa May's spad Fiona Cunningham. While Michael's briefing was wrong, it was the lesser of the two offences. The Home Office letter was calculated, petulant and frankly unsophisticated. It was accompanied by an accusation that 'Lord knows what more they have overlooked on the subject of the protection of kids in state schools.' I was clear: Michael had to apologise; the spad had to go.

Speaking of bad spads, Dominic Cummings was still in the business of bilious briefing to the papers. Having left government in 2013, he could now put his name to his insults. I knew he was still dripping his poison into Michael's ear as well.

I needed Michael in a top job, but I was beginning to think that a different, less high-profile role might be better for all of us. And he had planted the seed in my mind a year earlier. 'If you were ever to give me another job, I'd love to be chief whip,' he told me. His obsession with people and politics, all that passion and antagonism – perhaps they could be used to recharge the Whips' Office.

Two weeks before the reshuffle I got him into the Downing Street flat and put it to him. 'I think you'll be brilliant at it. You've done four years at Education. All the big changes are made. I want you on the inner team.' He went away to think about it, talked to Sarah, and confirmed the next day that he would do it. I then described to him all my plans for the reshuffle, right down to the lowest junior minister. He was all 'Yes, do this . . . Oh, I wouldn't do that . . . Don't sack *that* person – sack *this* person.'

The next day I was in Wantage, Oxfordshire, visiting the Williams Formula One factory when my plan hit the barriers. Michael emailed to say he had changed his mind. What had happened? I smelt Dominic Cummings, and totally flipped.

Reshuffles fall apart if people go back on their word. This was a job Michael had suggested, that he had accepted, that he had started to do. I rang him and said, 'I don't accept your email. You have agreed to do this job. I've told you everything we're planning. I accept your withdrawal of the email, and I expect to speak to you later on today about how we are going to finalise the reshuffle.' I followed this with a text: 'You must real-ise that I divide the world into team players and wankers. You've always been a team player. Please don't become a wanker.'

As I began my cabinet clear-out the next day, Michael's colleagues were about to sort themselves into team players and . . . people who proved more difficult.

Gentleman George was a gent. Ken was content. Eric and Francis I kept on in the end, but David Willetts was gracious about moving on. David Jones, the Welsh secretary, couldn't have been nicer. He said he'd enjoyed serving, it had been great, and he would help in any way he could. He knew Stephen Crabb would be an excellent appointment.

Dominic Grieve, the attorney general, may have seen the end coming. He was a brilliant lawyer, and had huge respect among the legal profes-sion. But there was a genuine problem. The European Court of Human Rights was finding against the UK for refusing to give prisoners the right to vote. The idea that convicted criminals should have that privilege disgusted me. And to my mind, Parliament was sovereign: this was a matter for MPs to decide, not the courts, particularly not foreign ones. Not according to Dominic: he said our international obligations should override our parliamentary sovereignty. Linked to that, he cited the Ministerial Code, which said we couldn't break the law. So he ended up arguing that it was impossible for us not to propose legislation to allow at least some prisoners to vote, and that as ministers we would have to back it. I fundamentally disagreed. The right of prisoners to vote was not part of any international charter we had signed up to. It was the view of a foreign court, and if Parliament wanted to disagree, it should feel able to. As a result of this disagreement I could see how impossible it would be for Dominic and me ever to reach a more general agreement about reforming the ECHR, the Human Rights Act and the European Court of Justice. So he had to go.

Andrew Lansley had every right to be disappointed. He'd given up his cherished job at Health, and now I was asking him to move again because I needed his job as leader of the House of Commons for William.

Tina Stowell would be Lords leader – she'd been great at taking gay marriage through the Upper House. Philip Hammond would be foreign secretary. He was hard-working, extremely able, and loyal, and I thought he would do a good job in pushing an outward-looking foreign policy. I chickened out of giving Anna Soubry Defence, and made her a minister of state in the department instead. Michael Fallon, who had proved to be the safest pair of hands, would be an ideal fit as secretary of state. Liz Truss, I was sure, would excel at DEFRA.

Esther McVey managed to blab herself out of the secretary of state for culture job when she went on TV during the middle of the Maria Miller crisis and failed to properly support her colleague, despite being asked to do so by No. 10. Instead, she would be IDS's number two at Work and Pensions, attending cabinet. Yes: once again, I failed to move IDS.

The most difficult conversation was with Owen Paterson. He said I was making a terrible mistake, and I should think about it overnight. I said I was sorry, but my decision was final. He was furious, and stormed out of my office.

Things were difficult too with Liam Fox. Three years after his departure from cabinet, I was offering him a way back into government with the job of minister of state at the Foreign Office. He'd be on the National Security Council, a senior government figure, knocking at the door of the cabinet – but apparently he was totally insulted by the proposal.

More seriously, Michael Gove was devastated by what happened when his appointment was announced. None of us had properly focused on the fact that the chief whip only attends cabinet, rather than being a full member. To me, that was just semantics: Michael would be at the table for every discussion. The job also involved a pay cut, although I had offered Michael a flat in Admiralty House to help make up the difference, and so he could live closer to the Commons where he would be spending more time. But when it dawned fully on him, and others, that he'd be taking a £36,000 pay cut, it seemed like a demotion.

Sarah Vine was furious. 'A shabby day's work which Cameron will live to regret,' she tweeted the next day, quoting a Max Hastings op-ed which, with the usual moderation the *Daily Mail* applies to things, described Michael's demotion as 'worse than a crime'.

Appearances matter in politics, including to the people alongside you. I had created a strong team, but tensions and unhappiness were on the rise, and the long-term consequences would be very serious indeed.

37

Junckernaut

So our campaign was under way, but at the same time we faced another obstacle: the election of Britain's seventy-three Members of the European Parliament. This mattered in itself, but it would also have an impact on the political atmosphere, our ability to keep the party together and, crucially, our capacity to unite the right behind re-electing the government.

I had seen several European election campaigns up close, from Margaret Thatcher's disastrous 'You'll live on a diet of Brussels' campaign in 1989 to Michael Howard's aggressively anti-Labour strategy in 2004, which had said far too little about Europe. I had tasted success, too, when we topped the poll in 2009.

This time we needed to demonstrate that, with a strong economy and renewed respect abroad, we could bring powers back from Brussels, and ultimately deliver the in/out referendum we had promised. For anyone wanting real change on Europe, surely that was the winning argument. UKIP *couldn't* give voters the choice. Labour and the Lib Dems *wouldn't*. Only the Conservatives could and would.

Yet there were many factors in UKIP's favour. European elections were prime occasions for protest votes, and now the Lib Dems were in government, UKIP was for many voters a natural 'none of the above' party. Policies like gay marriage and HS2 had driven disaffected Tories into its arms. It would also be able to scoop up voters after the demise of the BNP. In addition, UKIP had been increasing its appeal to working-class voters, particularly in the north, where many felt neglected by Labour.

There was also the Nigel Farage factor. I first met him in 2002, when I was a new MP and he was a new MEP, and we were appearing on Radio 4's *Any Questions* together. People often say 'He's the sort of person you'd have a pint with,' to convey how down-to-earth he is. I've never had a drink with him, but we did have a cigarette as we waited for the show, and I can attest to his amiability.

Yet there are many contradictions. A man who preaches anti-politics, but who has himself been a politician for twenty years. A critic of corporate interests and banking who made his money as a commodities trader in the City. A working-class warrior who went to private school. Someone who bemoaned European immigration, but was married to a German, and lambasted an EU gravy train he'd been riding for years.

At heart, I thought he was easy to understand. I know the type very well. A Conservative who thought 'Enoch was right' about Europe and immigration, who admired Margaret Thatcher for her strength in turning the country round, but overlooked her commitment to our membership of the EU and to making a success of a multiracial Britain.

Once Thatcher was gone and European integration started to accelerate, there would always be a danger to the Conservatives of a breakaway party from the right. This was assisted by the failures of globalisation, and a sense that too many people were being left behind either economically or culturally, or both.

What Farage lacked in working-class credentials he made up for in charisma and an instinctive understanding of his audience. He was also willing to show an unpleasant side. His dog whistles – more like foghorns on occasion – on TB or HIV sufferers coming into the country seemed designed to stir up anger rather than to solve a problem.

It was a mistake of mine to refer to UKIP as 'fruitcakes, loonies and closet racists' in 2006 – not because it wasn't true of some of its members, but because saying so alienated those UKIP supporters we were trying to win back. That said, the party remained a haven for the unsavoury. We alighted on a better strategy, of remaining silent about UKIP while feeding the press details of the eccentric and sometimes obnoxious things their candidates and councillors were saying. There was no end of material, including a claim made by one UKIP councillor that it was my fault that the country had been afflicted by floods – God's wrath for passing gay marriage.

Not that any of this changed anything. Because as it turned out, this election wasn't about gay marriage or HS2. It wasn't about the credibility of Nigel Farage or UKIP. It wasn't even really about Europe, at least not all the European issues. It was about immigration and its link to Europe. And on this Farage had the advantage.

One exchange brought this home to me. Ahead of polling day, Nick Clegg had challenged Farage to two live TV debates. While it would give Farage huge profile, I understood why Nick did it. He was fed up with the press painting him as the Tories' poodle, and wanted to demonstrate his passion about a subject on which he was knowledgeable and had a distinctive position.

During the first debate, Nick had what seemed a great idea. UKIP had a difficult relationship with the truth, and he planned to expose it by producing one of its leaflets from the recent Eastleigh by-election.

'It says here that twenty-nine million Romanians and Bulgarians may come to this country,' he said. 'There aren't even twenty-nine million Romanians and Bulgarians living in Romania and Bulgaria. It is simply not true. So let's have this debate, but let's have it based on facts.'

In any other situation it would have been a zinger. But Farage had a trump card. 'I'm not claiming that twenty-nine million people have the right to come to Britain,' he replied. 'I am claiming that 485 million people have the total, unconditional right to come to this country if they want to.'

Farage had hit on a central political vulnerability of our EU membership terms. We had no hard-and-fast control over immigration as long as we were in the European Union. Even if we argued that the current levels of EU migration were acceptable, indeed even if people agreed with that claim, Farage would always have the argument that we had no control.

On Sunday, 25 May 2014, UKIP won twenty-four seats in the European Parliament, eleven more than in 2009. We lost seven, leaving us with nineteen. It was the first time the Conservatives had ever come third in any national election, and the first time since 1906 that a British election was won by a party other than us or Labour. Farage called it an 'earthquake'.

Perhaps 'tremor' would have been more accurate. UKIP hadn't done as well as we'd feared, and we were only one point behind Labour. Just one in three people who could vote did vote – this was hardly a mass movement of the people. The real test would be at the general election, and I was reassured of our chances a few days later when we beat UKIP at a by-election in Newark by a healthy margin.

But there was one thing I didn't question. If we hadn't already promised a referendum on our membership of the EU, that would have been the central issue of the campaign, and we would have been forced to move after the result. The pressure would have been broad. Those who argue that it was not necessary to hold a referendum often miss this point.

I never thought UKIP's win was an anomaly or an irrelevance. Anti-EU sentiment rumbled right across the continent. The borderline-racist Front National won the election in France. In Germany, the anti-EU Alternative für Deutschland took seven seats. Greece's neo-Nazi Golden Dawn won three seats, while the far-left Syriza won the election. The *Economist* summed it up: 'After the years of financial crisis, the biggest danger to the European project is now economic stagnation and, above all, political rejection.'

José Manuel Barroso's decade as president of the European Commission, the most powerful of all the presidential positions in the EU, was coming to an end. It was the perfect moment for someone to grasp these threats, adapt to the changes and modernise the organisation. Instead, we were presented with the most federalist, integrationist, Brussels beltway front-runners: the former Luxembourg prime minister Jean-Claude Juncker, and the European Parliament president Martin Schulz.

How on earth did this happen?

Previously, presidents had been chosen by the European Council, which was made up of all the elected leaders. It worked on a basis of unanimity: every member state had a veto, so eventually the Council would find someone every country was content with. This is how Margaret Thatcher, John Major and Tony Blair were able to block candidates who were seen as antithetical to Britain's interests.

But then the process changed. Most important of all, the requirement for unanimity in the European Council was dropped. And the Lisbon Treaty required the Council to propose a candidate for the European Parliament to vote on. It did not require European political parties to identify 'lead candidates' in advance of European Parliament elections. But that is exactly what happened. Interpreting Lisbon far more widely than ever intended, each main party grouping in the Parliament would now put forward its own '*Spitzenkandidat*', or top candidate, for the role. They would then use the result to determine which of these 'top candidates' came out on top.

It *sounds* democratic, leaving it up to elected MEPs. But it wasn't.

First, the idea of Europe-wide parties with Europe-wide candidates was a fiction. Hardly anyone would go to the polls thinking, 'I am voting for this or that European president'; they would be voting on a whole host of domestic and European issues. No *Spitzenkandidaten* would be on the actual ballot paper.

Second, the process transferred power away from member states, undermining sovereign national governments and parliaments. The Council, formed of the democratic leaders of those states, would be presented with a lead candidate and be encouraged to endorse them. Countries with genuine concerns could be overridden.

So this changed the terms of our membership quite fundamentally. It altered the relationship between national governments and European institutions. It was a big step towards making those institutions accountable to each other, rather than to their member nations.

A speaking note I took into my meeting with Herman Van Rompuy said: 'It's a fundamental change in the balance of power towards an institution [the EU Parliament] for which there is little love and understanding in the UK, and a major disruption of the sort of balance which could keep us in and engaged . . .'

Jean-Claude Juncker is a European integrationist to the core, an architect of Maastricht and monetary union. He had been a keen supporter of the Constitution that morphed into the Lisbon Treaty, and when the French came to vote on it he made no secret of his lack of regard for their democratic will: 'If it's a Yes, we will say "On we go," and if it's a No, we will say "We continue."' For the EU's survival, for Britain's future in it, and for my renegotiation, there could hardly have been a worse candidate.

His election wasn't inevitable. We might have lost our veto, but many national leaders told me privately that they opposed Juncker's appointment, including Angela Merkel. She would be key. We were the leaders of two of the three largest countries in the EU. If anyone could claim a democratic mandate in blocking the appointment of a particular candidate for Commission president, surely it was us.

Our discussions on the matter started at the beginning of the year, during Merkel's visit to the UK. After her speech in Parliament we had lunch, and I invited her up to the flat for a coffee alone in the kitchen to discuss the two most prominent *Spitzenkandidaten* at that point.

Both Germany's Martin Schulz – put forward by the socialist grouping in the European Parliament – and Jean-Claude Juncker – put forward by the EPP – were unacceptable to Britain, I said, particularly Juncker. Schulz, as a German socialist, was a candidate Merkel would do pretty much anything to block. Personally I found him quite charming, with a love of football and huge knowledge of European and British history. But his answer to every problem was more Europe, much more Europe.

So we were in agreement: I needed Merkel to stop Juncker, and she needed me to stop Schulz. She repeated that intention when I visited Hanover the following month, and we sounded out non-*Spitzenkandidaten* like Ireland's Enda Kenny, the Italian prime minister Enrico Letta, and Danish PM Helle Thorning-Schmidt.

When it came to the European Parliament elections, the EPP finished with 29 per cent of the seats, compared with the Socialists' 25 per cent. The EPP claimed that was a mandate for their *Spitzenkandidat*. 'I won the elections,' proclaimed Juncker.

Merkel was unimpressed with such a presumption. 'There will be a fairly broad tableau of names on the table,' she told a press conference. In other words: this is not a done deal, Jean-Claude.

But then it all began to fall apart. Her party and their coalition allies were angry that she had gone against the candidate they had endorsed months earlier. Juncker also had the backing of Germany's most powerful media group, Axel Springer SE, whose papers published editorials claiming Juncker was a democratically elected president who we mere national leaders were trying to thwart. To them, defending this principle was more important than Britain's membership of the EU.

The pressure was too much for Merkel. After the G7 we met at the British ambassador's residence in Brussels on the evening of 4 June 2014, and she was downcast. 'I think I'm going to have to let you down,' she said. It wasn't our relationship that had changed; it was her circumstances. She didn't quite say, 'It's not you, David, it's me,' but it felt like being dumped. She said it had become such a major political issue that even her mother had called her to tell her to vote for Juncker. I'd heard elsewhere that she was having trouble sleeping.

I didn't give up. We sat up drinking until about 1.30 a.m., and she conceded that despite everything, there was still a chance to block Juncker.

The next meeting she and I had was with Mark Rutte and the Swedish prime minister, Fredrik Reinfeldt, at his equivalent of Chequers, Harpsund. We returned to matters of Juncker. Rutte and Reinfeldt said they wanted to work with me to block him, but would only do so if Germany helped them by preventing it ever coming to a vote.

The four of us sat up after dinner, drinking red wine, this time until 2 a.m. I was tired, but didn't dare leave in case they cooked something up without me. We were waiting for those magical words from Merkel – 'It's all right, we'll block this guy together.' She got close, but they never came.

By the next morning her team had got to her, and she said she was going to have to vote for Juncker. Indeed, at the closing press conference she made some not very subtle comments about 'threats' that had been made about UK membership should he get the job. It felt like game over.

If it was impossible to stop the 'Junckernaut', as we called it, why did I insist on a formal vote on his appointment? Well, I couldn't stop the backroom stitch-up, but at least I could force them to do it in the open. I thought it was important for people to see that I had objected, and that Europe had taken this decision against Britain's view. Plus, there was still the faintest flicker of hope that some countries might side with Britain.

When Van Rompuy came for a meeting in Downing Street, I pressed upon him the need to do things properly. He prevaricated, and said he wouldn't guarantee a vote. 'There is no point continuing with this meeting, then,' I said, standing up and showing him the door. We weren't even halfway through our hour-long slot. He looked stunned. I am not sure a prime minister had ever spoken to a president of the European Council like that before, but I needed him to know that the usual behind-the-scenes deal was not an option. As expected, the news flashed around Brussels instantly.

Soon enough, Reinfeldt and Rutte told their respective MPs that they would support Juncker. I rang the Hungarian prime minister, Viktor Orbán, the one person I knew would stand by his word on this *Spitzen* sham. We were hardly political bedfellows – the once liberal, dynamic politician was becoming increasingly authoritarian. But on this, he was right. And he was my only ally.

By chance, the next European Council was to be an unusual one, beginning in the town of Ypres. We would be commemorating a hundred years since our continent fought so bloodily over this small corner of Belgian land. The Germans were desperate to avoid the symbolism of a bust-up on the former battlefield, and tried, unsuccessfully, to push back the vote to a later summit.

As we stood under the Menin Gate and heard the haunting notes of the Last Post, I showed the Greek PM, Antonis Samaras, and Barroso

where my great-great-uncle's name, John Geddes, was carved onto the
arch, one of 55,000 who fell in Flanders but whose bodies were never
found. Merkel and I talked about the World Cup in Brazil.

At a dinner afterwards, I made a speech. 'We have our differences, and
we're going to talk about them tomorrow. But I want you to know that
my plan is to keep us in the European Union, and so far it's working.'
I explained that since the Bloomberg speech, the UK's support for staying in Europe had gone up rather than down. However, I made it as clear
as I could that the reason I opposed Juncker was that his appointment
and its consequences would jeopardise this. I wasn't the only one who
still hoped he could be thwarted. 'Maybe a miracle will happen tomorrow,' Rutte whispered to me.

Yet miracles only seem to work in one direction in Brussels. And
when we gathered around the familiar table in Justus Lipsius the next
day, Orbán and I were voted down twenty-six to two.

How many do I think actually wanted Juncker? I'd put it at about two.
He *was* the wrong candidate, with the wrong views, at precisely the
wrong time. I had always claimed that 'What I said in Britain, I say in
Brussels.' I'd rather lose than lie. And I'd rather try to do what was right.

The newspapers the next day put it plainly. 'Britain Nears EU Exit',
said *The Times*. 'One Step Closer to Quitting Europe', said the *Telegraph*.

In the meantime, I would have to work with Juncker. Our personal
relationship actually wasn't bad. He was convivial, and enjoyed political
gossip – we would share stories about his old sparring partners John
Major and Norman Lamont. He was also amazingly tactile – a big hugger
and kisser – often with a strong aroma of his trademark scent of brandy
and cigarette smoke

His wasn't the only Commission job that was up for grabs, and Merkel
was already asking me which of them I wanted for Britain. 'After
everything that happened, you can have whatever you want,' she said.
I asked for the financial services portfolio, for which I'd earmarked
Andrew Lansley or Jonathan Hill, the leader of the House of Lords.
'What really matters to me is people who will help Britain with our
renegotiation,' I told her.

Over in Brussels, 'Whatever you want' turned into 'I'll do my best.' We
spent weeks fighting to get Jonathan Hill into the job, including an
unprecedented second European Parliament Committee grilling.
I pointed out to Merkel's team, only partly joking, that the time between

a Merkel promise and a Merkel promise being broken was getting shorter – 'the half-life of a Merkel promise', I called it. She was still an ally, still a friend. I just couldn't trust her quite as I had before, and she knew it.

She urged me to meet the head of the EPP, Manfred Weber, because the Parliament committees would vote on the individual candidates before holding a straight yes or no vote on the whole team. He came to Downing Street, and once again I was fascinated how two countries could look at the EU and see something so different. They see it, with this federalist European Parliament, as a fount of democracy, power and accountability, whereas we see it as a useful forum for mainly economic cooperation between democratic nation states.

There were two ways that the curious incident of Juncker shifted my thinking, and I confessed them both on tape on 2 July 2014.

The first was that it reinforced my fear that I might not succeed in my renegotiation. The second was that it wasn't unthinkable for me not to recommend that we stay in the EU.

As I put it, a referendum and the threat of a UK exit meant that 'We could get a decent set of changes that you could market as guarantees for Britain, but there are moments when I think actually what we want and what they want is quite incompatible. So I don't think it's written yet whether we can get the changes that we need.' I went on to say, 'I am closer to saying if we don't get the renegotiation then there is no point in staying in the European Union. Maybe even I would be able to contemplate an exit. But I haven't got there yet. I am wrestling with this. I am finding it very troubling. But I am convinced we have to deal with this issue.'

Either way, the saga of 2014 made renegotiation and referendum more difficult, but more vital than ever.

Yet my political trials in Europe were fast being put into perspective by what was taking place on the furthest reaches of the continent. Who ever said that the Cold War was over?

38

A 'Small Island' in a Small World

It felt familiar – like something from the Second World War. Tanks rolling into towns. Buildings seized. Roadblocks appearing. National symbols torn down and occupiers' flags hoisted high.

Except this wasn't Adolf Hitler in 1939; it was Vladimir Putin in 2014. And when he invaded Ukraine, he didn't just cross a geographical border – he crossed a political line. Part of a sovereign territory had been seized by force of arms. The rulebook that Europe and much of the wider world had tried to live by since the end of the war had been torn in pieces.

Most people in the West don't think much about this 'international rules-based system'. We don't have to. As physics holds together the material world, politics is supposed to help hold together the human world, giving us order, stability and peace. We respect other countries' borders; they respect ours. We trade with them; they trade with us. We sign up to various organisations – the UN, the WTO, NATO and, yes, the EU – in order to cement that cooperation and make our world a smaller place. It's the quid pro quo that's developed since the twin horrors of Hitler and Stalin, and, except in the Balkans, it has largely held in Europe.

So when those rules are broken, it matters to us. Putin's actions rejected politics, asserting instead that the national interest is better served by old-fashioned, land-grabbing isolationism than alliance-building, rule-abiding internationalism.

Just as the West, including our own Foreign Office and National Security Council, had not seen the Arab Spring coming, so it had failed to predict Russia's annexation of Crimea. Yet it wasn't a total shock. I was a student of Cold War history; I knew what Russia was capable of. I knew Putin, too – over the past four years I had probably got to know him better than any of my predecessors had. I knew how much Ukraine meant to him, how he lamented the loss of Yalta and Kiev.

I also knew – I had seen up close – what Putin had done in Georgia. My response to that invasion had been criticised as opportunistic. Yet when I warned that a weak Western response would embolden Russia I was absolutely right. I was also right when I asked, 'Where next? Ukraine?'

I had been concerned about the volatility of the region for some time. That's why I insisted on going to the rather obscure Eastern Partnership Conference in Lithuania in November 2013. I had a pretty bad relationship with Dalia Grybauskaitė, the country's president and the conference host. As a former EU commissioner, she was very much a 'lie down and obey Brussels' type. But where we did agree was on the need for a tougher approach towards Russia, to defend the Baltic states and show solidarity through NATO.

I understood the vulnerability they felt. For decades they had been suppressed by the Soviet Union. After the fall of the Berlin Wall they had chosen to reassert their nationhood as sovereign states, yet the Russian bear still cast a shadow across their eastern borders, fully accepting neither their sovereignty nor the international rules upon which it was based. In fact, the bear was beginning to growl again, flying sorties over NATO airspace and staging war games and missile tests near the border. There were cyberattacks and old-fashioned trade blockages. These included imports of Lithuanian cheese, and for Grybauskaitė this was the last straw.

The nations that made up this Eastern Partnership with the EU – Armenia, Azerbaijan, Belarus, Georgia, Moldova and Ukraine – may not have figured a great deal in my foreign affairs or national security briefings. But while most countries in Europe fell clearly into either the Western or the Russian sphere of influence, these were at the intersection. Which way they turned would help shape global politics.

In 2013 Ukraine was very publicly trying to make its decision. The president, Viktor Yanukovych, had made much of his intention to enter an Association Agreement on greater trade and security cooperation with the European Union. Indeed, he was meant to be signing it at this very Eastern Partnership conference. But he stalled, suspending preparations to join the agreement and renewing talks with Moscow.

I had first met Yanukovych at Davos in 2007, and found him the worst sort of dictatorial, kleptocratic, Soviet-style puppet. At the Eastern Partnership dinner in Vilnius, after plates of cheese (of course) were handed round with drinks, I sat between him and the Armenian president, Serzh Sargsyan. The Armenians had already opted for the Russian customs union, and it looked as if the Ukrainians were about to do the same.

I had got the lowdown from the Azerbaijani president, Ilham Aliyev. I always found this wily old dictator interesting to talk to about his neighbour Iran, Russian affairs and what was happening behind the scenes. 'It looks like Yanukovych wants more money and a better deal from Russia,' he said, 'so he's playing the EU, pretending he wants an Association Agreement.'

If that was what was happening, Yanukovych had profoundly miscalculated. Because his citizens had become very excited about a potential partnership with the EU, and when he walked out of the conference confirming that he *was* pulling out of the deal, protests erupted in the centre of Kiev. Young people braved the biting winter to turn the Maidan Square into the 'EuroMaidan'.

After his forces fired upon the crowds, killing seventy-seven people, Yanukovych was toppled. There were incredible scenes as ordinary Ukrainians stormed his palace, taking selfies in his gilded bathtubs, holding his Fabergé eggs, even wandering around his private zoo. They had uncovered evidence of corruption on a sickening scale.

I couldn't help being excited by the sense of youth, hope and liberation, the palace-storming desperation for democracy and freedom. But, as in the Arab Spring, when hope opens a door, it is often malign forces that walk through.

While the world watched the west of Ukraine, Putin pounced in the south, the Crimean peninsula. On 27 February, Russian special forces seized the parliament and the airport, roads at Sevastopol and Simferopol were blockaded, crossings to the mainland were blocked and the Russian flag replaced the blue and yellow of Ukraine. Invasion – in Europe, right on NATO's doorstep.

Where did that leave me? Hadn't I spent years cultivating a relationship with this man?

It's certainly true that in 2010 my aim had been to form some sort of positive relationship with Russia, as one of the BRICS countries – Brazil, Russia, India, China, South Africa – with immense growth and trade potential.

Our relationship was starting from a pretty low base. Since the state-sponsored poisoning of the former Russian secret-service agent and defector Alexander Litvinenko in London in 2006, we'd put Anglo-Russian relations in a deep freeze, and no UK leader had held a proper bilateral meeting with a Russian leader since 2007.

At the time, Dmitry Medvedev had been keeping Putin's presidential seat warm until Putin could return under the terms of the Russian constitution (he had already served the maximum two consecutive four-year terms as president, and was now serving a term as prime minister). Medvedev and I met at the G8 in Canada, and he invited me to Moscow. He was articulate, friendly and attuned to Western issues and concerns. Our first conversations were about rapidly changing technology and its potential. He was impressed with Tech City in London, and wanted to promote similar developments in Moscow. Obama had told the major powers in Europe that here was someone we could deal with – and we should try to.

Stepping into the Kremlin for the first time is one of those things you never forget. You walk through anteroom after anteroom of palatial grandeur, each one more ornate than the last. The whole thing is more tsarist than Leninist – they want you to know you're entering the heart of a great empire.

My visit culminated in a great banquet at which we had to drink toast after toast of vodka, giving a short speech after each one. With Russia, a popular and safe topic for me to focus on was our partnership in defeating Hitler. By the sixth shot, even a history buff like me was running out of great battles to toast, and losing my ability to articulate them.

From 2012, when Putin resumed the presidency, we enjoyed a series of moments that held out hope for the future, from the Arctic Convoy ceremony to my Sochi visit. I never forgot that the man I was dealing with was a former KGB officer, and an old-fashioned Russian nationalist. But we needed to have a relationship. It was making a difference on trade, and it could have made a difference on Syria. I was willing to try, and to take risks. But the attempt soon hit serious obstacles. The US experience was broadly similar. Despite Hillary Clinton's attempted 'reset' of relations with Russia, America announced sanctions through the Magnitsky Act. Russia was furious about Western intervention in Libya and Syria. We were all furious it had granted asylum to Edward Snowden.

By the time I arrived at the G20 in St Petersburg in September 2013, things had already cooled between me and Putin. The Russian parliament had just approved a law banning so-called 'gay propaganda' – state-sanctioned homophobia, just months after the UK had legalised gay marriage – and we had made our disapproval very clear. A Russian official then reportedly described Britain as 'just a small island' that 'no one pays any attention to'.

I was asked about the comment at my G20 press conference. It had been relayed to me shortly before I was due at the event, and I had scribbled some thoughts down in the car. It was a joy to depart from the official script and let rip.

'Yes, we are a small island – in fact a small group of islands,' I told the reporters. 'But I would challenge anyone to find a country with a prouder history, with a bigger heart, with greater resilience. This is the country that cleared the European continent of fascism, that took slavery off the high seas. We're a country that has invented many of the things that are most worthwhile, everything from the Industrial Revolution, to the television, to the World Wide Web . . .' As I went on to list many achievements – 'our literature, our art, our philosophy, our contribution, including, of course, the world's language' – I could almost hear the *Enigma Variations* in my head.

When I finally met Putin one-on-one at the end of the G20, I raised the issue of gay rights. It was our iciest encounter yet. He said that Russia's problem was a declining population, and he needed men to marry women and have lots of children.

Although our relationship was in decline, we could still speak informally and he still took my calls. Which was fortunate, because I was to call him two days after the invasion (and three more times after it). I didn't need to ask about his motivation. It was clear that Russia feared the Western sphere of influence, in the form of the EU and, above all, NATO. And Putin was driven by a deep desire for Russian nationhood.

Now that Ukraine's puppet president was gone, a weak and divided neighbour was better than an entire and Western-inclined one. So Russia had resorted to causing chaos in the country. That's why, as well as seizing Crimea, it was simultaneously stoking the unrest in the pro-Yanukovych eastern region of Donbass.

The West had given Putin an inch by letting him make his initial moves, and now he was taking a mile – indeed, several thousand square miles. He might not wish to play by the international rules, but he certainly knew them: like Georgia, Ukraine wasn't a member of NATO, so it wouldn't be forced to retaliate.

I tried to persuade him that there was a way of negotiating a peaceful outcome, respecting Russia's interests in Ukraine. 'We are having a referendum to see if Scotland wants to stay in the United Kingdom,' I said. 'You can have a referendum on Crimea's membership of Russia, but it's got to be fair and legal. What you've done is basically subvert the territorial integrity of another nation state.' But he just didn't accept that. It was as if we were sitting at the same chessboard but playing two completely different games.

Putin said the new Ukrainian government was illegitimate. He said fascists and nationalists were attacking Russians in eastern Ukraine, and the separatists had nothing to do with Russia. In fact our intelligence estimated that 60 per cent of these military forces were Russians, paid by Russia.

For Putin, lying was an art form. Long before the term 'fake news' was coined, Russia had mastered the concept, piping the same spurious line over the internet and across the airwaves, including via a slick new twenty-four-hour English-language news channel called RT (previously Russia Today) – a *Pravda* for the twenty-first century. Add that to everything else it was doing – training Russian paramilitaries in eastern Ukraine, bribing and intimidating those in rebel-held areas, supporting pro-Russian breakaway states – and we were firmly in the territory of 'hybrid warfare'.

The only thing that was missing was a fake election. That was duly added five days after my final phone call with Putin, when, in a hastily organised referendum, 96 per cent of the Crimean people apparently voted in favour of joining Russia. It was now unlikely the peninsula would ever be returned to Ukraine.

William sent me a note setting out the magnitude of the situation. 'This is the most serious foreign policy crisis in Europe since the breakup of Yugoslavia,' he wrote. 'To fail to respond would legitimise Russian actions in Crimea and undermine the credibility of the West and the international system. And the world is watching.'

The answer could never be military action. Obama told me over the phone that it was obvious Russia cared more about maintaining its presence in Ukraine than we did, so we shouldn't pretend we were willing to deploy force. Of course he was right. Given the dangers, there would have been no backing for military intervention. I never considered sending our tanks into Ukraine to face Russia's, or positioning a naval task-force in the Black Sea.

We might not be able to push Putin out of Ukraine's territory militarily, but we could do the next best thing. I saw huge value in a strong diplomatic, sanctions-led response – and, in time, I could see the argument for helping to strengthen Ukraine's armed forces. My mantra became 'de-escalate and deter'.

I resolved to lead the European response. On 6 March there was an emergency meeting of the European Council on the Ukraine crisis. Herman Van Rompuy was set to issue a lame set of conclusions, so I called a pre-meeting of the Germans, French, Poles and Italians in my room before the main meeting. I got them to agree a process for now, and for if things got worse.

This put some people's backs up; nobody who is excluded likes a caucus of states. But it did put Britain in the driving seat of the response. Like the sanctions against Iran, it was one of the examples I'd give during the referendum campaign of Britain using the EU to lead global events.

Unfortunately, my resolve to punish Russia for its actions was matched in some cases by my fellow leaders' reluctance. François Hollande reportedly wanted sanctions to be 'financial, targeted and quick' – he didn't want anything to block the sale of two amphibious assault ships to Russia. Germany, where a good part of the political establishment favoured '*Ostpolitik*', or accommodation with Russia, was also being cautious. Italy, partly reliant on Russian gas, had been a traditional ally of Russia, and was nervous about backing robust action.

Most of the international institutions were, however, responding robustly. The G8, led by the UK, suspended Russia, becoming the G7. The UN passed a non-binding resolution dismissing the Crimean referendum. NATO suspended civilian and military cooperation with Russia.

But anyone who still thought this was a distant issue – another country's problem – was about to be proved wrong. In this small world, no war is distant, and no country immune from its belligerents.

On 17 July 2014, 298 people – Dutch, Australian, British, German, Belgian, Malaysian, Indonesian and others – boarded a regular commercial flight from Amsterdam to Kuala Lumpur. That Malaysian Airlines plane crashed to earth in Ukraine. Everyone on board was killed.

It soon became clear that this had been the work of a surface-to-air missile, launched by a Russian-made Buk system and operated by Russian-backed rebels. It was totally tragic, and absolutely despicable. I spoke to Putin soon afterwards. Again, it was like talking to a brick wall. 'Hundreds of people just died,' I said. 'Nothing to do with us,' was his icy reply.

In the overused but often accurate phrase, while the truth was getting its boots on, Putin's big lie made it halfway around the world. Russia's disinformation – that the whole conflict was Ukraine's fault, that Crimea was legitimately part of Russia, that Ukraine was overrun with fascist thugs – seemed to be more effective than information put out by the EU and NATO.

In an effort to move my fellow European leaders I set out my frustrations in an article for the *Sunday Times*: 'For too long there has been a reluctance on the part of too many European countries to face up to the implications of what is happening in eastern Ukraine,' I wrote. 'Sitting around the European Council table on Wednesday evening I saw that reluctance at work again.' And then what I thought was at the heart of the reluctance: 'We sometimes behave as if we need Russia more than Russia needs us.'

A few days later the leaders came around to a new set of sanctions on Russia, including banning some of Putin's cronies from EU countries and stopping them accessing the money and property they held there. We also imposed an arms embargo (the French eventually agreed to forgo the sale of their warships). And then, a week later, Hollande, Merkel, Italy's Matteo Renzi, Obama and I agreed further trade sanctions, including restrictions on Russian banks' access to our financial markets, which would undoubtedly damage the Russian economy.

At last we were on the right track to deter Putin, even though the EU had yet to wield the really big economic sticks by cutting down or off Russian oil and trade. But what about the de-escalation? Here we made a mistake. By late August the Donbass rebels were losing territory to the Ukrainian army. We should have persuaded Kiev not to provoke Russia further. By continuing the offensive, it did just that. They could not beat Putin's forces; we would not come and help them. By the end of the year there were 10,000 Russian soldiers fighting in Ukraine, as well as more than double that number holding Crimea. Ukrainian losses mounted, its troops were pushed back, and by the time negotiations to achieve a political settlement began they were firmly on the back foot.

I met Putin again at the G20 in September 2014, in the rather warmer climes of Brisbane. Yet things between us were as frosty as ever. I thought about all that had happened since his St Petersburg summit. 2014 was meant to be Putin's year. The Winter Olympics; a Grand Prix, the first in Russia since 1915; the G8 – a real chance to come in from the cold, and into a world of cooperation and compromise. But the lure of zero-sum, narrowly defined nationalism had been too strong. Putin even, rather absurdly, stationed four Russian military ships off the coast of Australia during the summit.

Over the following months and years, Russian jets would frequently be detected flying sorties over the Baltic, and even near UK airspace. I would receive notes saying the RAF had had to scramble Typhoons as Russian planes had been spotted off the Aberdeenshire coast.

As well as micro acts of aggression, Russia's serious crimes continued. Undermining the hard-won peace in the Balkans. Facilitating corruption in its sphere of influence. Propping up Assad in Syria. Meddling in the US elections and, to a lesser extent, the EU referendum. Poisoning a former Russian agent and his daughter on the streets of Salisbury – later causing the death of a woman who was totally unconnected to them. More Russian collateral damage. Meanwhile, any semblance of a UK relationship with Russia is firmly, perhaps indefinitely, on ice.

I was sceptical when originally asked whether the UK should host the 2014 NATO summit, wondering if there would be a substantial enough agenda to make the time and expense worthwhile. But after the success of the G8 in June 2013, and the disaster of the Syria vote in August 2013, and Russia's attack on the rules-based international order, I saw that there were many things it could help us deliver – above all demonstrating the unity and resolve of the Western democracies.

So I announced in September that we would be hosting the summit of twenty-eight NATO leaders, just one year on from the G8. It would also include the fifty nations that had taken part in the Afghanistan mission, making it the largest summit to be held in the UK for decades. I decided it would take place in Wales, at the Celtic Manor golf resort near Cardiff.

There were so many crises unfolding around the world that one commentator went so far as to opine that the summit would be 'the most important one since the fall of the Berlin Wall'. But the spotlight would be on NATO's conspicuous absence in the Ukraine crisis up to mid-2014. The country itself might not have been in NATO, but the crisis clearly impacted on NATO interests, because it attacked the idea of self-determination and was a blow against liberal democracy in Europe. One of the aims of the summit was to firmly discourage the Ukrainians from thinking that NATO was about to ride to their rescue. My briefing pack even stated that a key objective of the Ukraine working session was to encourage the country's new president, Petro Poroshenko, to 'face the reality that Ukraine can't win militarily'.

The spotlight would also be on NATO itself. Just as the international rules-based system was being challenged, so were the institutions that upheld it. The alliance had played a vital role in keeping the peace in Europe. You didn't have to go as far, or be as frank, as Churchill's chief military assistant 'Pug' Ismay, who had said that NATO was necessary to 'keep the Americans in, the Russians out and the Germans down', to see its utility. After seventy years without a major conflict in Europe, except of course in the Balkans, it could claim to be one of the most successful military alliances in history. I supported its continued existence as a self-confident defender of liberty.

While the pattern of threats was changing, I believed NATO was capable of adapting, and that it should remain the cornerstone of our collective security. Like almost all of the international organisations I came across, the closer you got the more concerned you became about waste, bureaucracy, political agendas and sclerotic decision-making. But you also saw the utility in being at the table.

So, for the second year in a row I welcomed the world – two hundred presidents, prime ministers, foreign ministers and defence chiefs – this time to a golf course in Wales. We scattered the fairways with tanks, planes and armoured vehicles, all showing off the best of British. A full-sized model Typhoon parked in front of the hotel entrance made it pretty clear that we were there to talk security, not work on our golf handicaps.

We had to erect marquees for extra meeting space. The Americans' helicopters hovered so low that the leaders could barely hear each other speak. As one thundered overhead particularly noisily, I remember Merkel texting me – so often over the years I'd catch her eye across the table after she'd sent a mischievous message – asking if I was trying to remind her of the Second World War.

I had spent time and effort in the preceding months with the NATO secretary general Anders Fogh Rasmussen to hammer out the agenda. We saw eye to eye on how NATO needed to change.

On terror, it had acted in Afghanistan, but this seemed too much of a one-off. We needed to do more to strengthen the defences of countries at risk, so we agreed security-sector assistance to training missions in other countries.

On Europe, the guarantee of collective security was real, but we needed to make it tangible by stationing more troops in more eastern countries. Britain would play its part by pledging to station 3,500 UK troops in eastern Europe. And we would make sure the so-called rapid reaction force was genuinely rapid – deployable within two to five days – with headquarters provided by the UK.

On modernising defence, we agreed that a fifth of national defence budgets would be spent on new equipment, and that, crucially, our armed forces could operate together by meeting NATO common standards.

Although the eastern Europeans were at the stronger end of the spectrum arguing for a more robust response on Ukraine, while the French, Germans and southern Europeans were not quite so strident, by and large there was unanimity, and NATO's purpose was restated. 'No one should doubt NATO's resolve if the security of any of its members were to be threatened,' the communiqué said. We even got every NATO nation to sign an Armed Forces Declaration, which was effectively the Military Covenant we had created in the UK.

And finally there was a joint declaration to spend 2 per cent of GDP on defence. Rather like the commitment of 0.7 per cent to overseas development, this was a target set out by a multilateral body (NATO), agreed by many countries (in 2006), but only met by a few (the US, the UK, Estonia and Greece). As I've said, the UK became the only major economy to meet both commitments. So much for a small island.

While I had led Europe's response to Russia in the initial phases of the invasion, the peace initiatives were largely handled by France and Germany. Along with Russia and Ukraine, they formed the 'Normandy Format' established during the commemoration of the seventieth anniversary of the D-Day landings.

The reason was simple. Putin had arrived at the event suggesting a large-scale meeting, but Obama and I both thought it inappropriate to overshadow an important ceremony. Also, it seemed wrong to discuss the prospect of legitimising even a small part of what Russia had done just as we were remembering the sacrifices so many had made the last time many of Europe's borders had been redrawn by force. I was quite happy for the UK and the US to watch from a distance, keeping up the pressure on Russia as Germany (with France largely as a makeweight) explored potential deals.

Yes, it looked as if we'd been sidelined, but I didn't care how it looked. I believe I had seen Putin's game for what it was, and responded robustly, earlier than other Western leaders. But what mattered was progress, and ultimately peace. And I respected both Germany's expertise in the region and Merkel's doggedness with Putin.

After the failed ceasefire of the Minsk Protocol in 2014, the subsequent plan to bring peace to the region, Minsk II, was signed on 12 February 2015. But as things stand today the conflict is continuing, if relatively 'frozen' in terms of casualties. Ukraine still has no control of its eastern borders. The rebel areas remain under Moscow's de facto control. And there is no sign of the West, the EU in particular, being prepared to use its economic power to the extent required to make the Russian occupation of the Donetsk region unaffordable.

How should we judge this episode? You could say that Putin won. He has Crimea. He has destabilised a significant neighbour, and effectively controls the eastern portion of the country. But compared with what happened in Georgia, the international coalition was stronger and more successful. Putin's ambitions have in fact been thwarted. He launched this attack because he didn't want a Western-oriented Ukraine, and that is what he now has. Strategically he has lost, not gained, ground.

In other words, the bear has bitten a chunk out of Ukraine, but Ukraine retains its vast economic potential. The West needs to back that; Ukraine needs help in eliminating the corruption, the over-powerful oligarchs and the extreme poverty. In 1990, Poland and Ukraine had the same GDP. But since Poland pursued its European path, its GDP has increased fivefold. Ukraine could do the same.

Something else happened at the NATO conference – a crisis that seemingly came from nowhere.

I had only vaguely heard of Ebola, a virus which first appeared in southern Sudan and Zaire in the 1970s. Incredibly contagious, passed on by contact with bodily fluid, it leads to a slow, painful and bloody death. It had remained largely dormant since. But in January 2014 the disease forced itself onto the global agenda after a suspected outbreak in Guinea.

Governments make contingencies for global catastrophes, and when you lead a government you have an insight into this worst-case world of nuclear war, mass terror attacks, cyber warfare, natural disasters and pandemics. Was Ebola the catastrophe we had been dreading? If so, the world wasn't ready. The World Health Organization, one of the most dysfunctional examples of an international institution, did not call a public health emergency until August 2014, five months after the disease had spread from Guinea to neighbouring Liberia. West African countries, unlike those in equatorial Africa, were completely unprepared for such an outbreak.

In the meantime – and this is why international gatherings are important – leaders ended up taking matters into our own hands. 'We'll help sort out Liberia, if you help Sierra Leone' was the gist of Obama's conversation with me at NATO. France would focus on Guinea.

Sierra Leone's healthcare system was overwhelmed. The authorities had to battle with the local cultural practice of the living touching the bodies of the dead at funerals – only to contract the deadly disease themselves. Misinformation about bogus treatments and the danger of going to hospital was spreading like the virus. The number of deaths grew dramatically. Bodies were left in streets. Entire families were wiped out. This was the brutal, bloody, tragic truth of what happens when a 'what if' becomes a reality. If it continued at that rate, I was warned that there could be a million dead.

But as I chaired the first of many COBR meetings on 8 October, I knew that Britain would be able to act fast. Because of the 0.7 per cent aid commitment, we were good to go. So we set up a command centre in the capital, Freetown, from which to coordinate the fight against the disease. We opened laboratories and built treatment facilities. We distributed public health information and supplied cars, ambulances, beds, safety suits and tonnes of aid supplies. Eventually over 3,000 Britons – soldiers, sailors, scientists, doctors, nurses, aid workers, volunteers – flew to Sierra Leone to assist in the effort.

Meanwhile, I embarked on a mission to backslap, cajole, coerce and even embarrass other countries into action. Half the world came to our Defeating Ebola conference in London. At the European Council I pointed out that the Swedish furniture company IKEA had contributed more to fighting the disease than Austria, Greece, Hungary, Ireland, Luxembourg, Poland and Spain. They promptly increased their contributions.

By the end of the year the tide began to turn. New cases of Ebola were declining, and our volunteers slowly began to return, including the nurses Will Pooley, Pauline Cafferkey and Anna Cross, all of whom had contracted Ebola and been critically ill.

One of the most moving moments I ever had in Downing Street was the reception I held for many of these 'Ebola heroes', who we awarded a special medal for services in west Africa. These people had left the comfort and safety of home for a part of the planet where the most notorious disease in a century was raging. They wore their achievements lightly, and bore their experiences stoically. And I thought to myself: *that's* Britain. We are defined by a quiet, practical, compassionate dedication to doing the right thing – by ourselves and by others. That's *our* type of patriotism.

Looking back, it's incredible to think how much 2014's crises – an invasion on the edge of Europe, a virus in west Africa, and, as I'll come to, a conflict in the Middle East – could preoccupy a country in the North Atlantic.

Even if Ebola reached our shores there would be no pandemic here. Russia hadn't annexed East Anglia. But because of globalisation, these things matter. Trade makes us reliant on other countries. Travel and migration imports others' problems. Technology engages everyone in what's going on in the world. What happens on the streets of Islamabad really does play out on the streets of Bradford. The world is smaller than ever.

Yet on one of the most entrenched foreign policy puzzles, I have to be frank: in the six years I was in Downing Street we made no progress. If anything, the Middle East Peace Process went backwards.

I was wholly for a two-state solution: creating a single Palestinian state, linking the West Bank and Gaza. Some in the West had resorted to a two-*faced* solution: telling each side what it wanted to hear and getting nowhere. I wanted to be tougher on both.

I was – I am – a friend of Israel. This tiny country was a haven for Jews after the most horrific event in modern history. It remains a beacon of democracy in a region of dictatorships. I credit the Conservative Friends of Israel (CFI) group, and its charismatic head Stuart Polak, for opening my eyes, and those of many Tory MPs, to the issues.

But Israel did not always do the right thing. I thought we needed to put pressure on it to stop building illegal settlements, and that work was required on when and how we would take a more aggressive position on recognising the state of Palestine. No people should have to live as permanent refugees.

I also thought, however, that we needed to be tougher on the Palestinian leadership, and their association with Hamas, whose goal was to destroy Israel. Many people condemned the Israeli government for firing rockets into Gaza when Hamas fired rockets into their country. I saw no moral equivalence. One was the army of a democracy; the other was a terrorist organisation. We also needed to get the Palestinians to accept that their bargaining position with Israel was going to get worse, not better, over time. There was no point always holding out for the international community somehow delivering in the future what was not close to being offered today.

Yet every inch of progress was thwarted.

Obama was, I believed, the most pro-Arab, pro-Palestinian president in history. But, as ever, his careful analysis – 'They both need to want peace more than we do,' as he put it to me – meant a reluctance to take risks in order to achieve progress. Plus, he was understandably distracted by the Arab Spring, and anything he *did* propose to put pressure on Israel was rejected by Congress.

There was some hope of the Palestinians moving towards compromise, and I found their president, Mahmoud Abbas, who I had first met in 2007, quite open to the idea. But that hope was dashed when his party announced a unity government with Hamas in February 2012.

Could Israel's prime minister Benjamin Netanyahu be a peacemaker? Sometimes the most bullish figures make the best peace partners. He talked a good game, and had even issued a moratorium on settlement-building. But by 2015 he was making it clear that he would not tackle the issue, and as the settlements grew, a two-state solution became less likely.

I visited Israel for the first time as PM in March 2014. I spoke in the Knesset, made a sobering visit to Bethlehem, where I met Abbas, and

– in one of those small-world moments – bumped into someone in Bethlehem who I used to buy coffee and pastries from on the Golborne Road in west London.

But it was its Holocaust memorial and museum, Yad Vashem, that left the deepest impression, and made me determined to ensure Britain had its own national memorial and museum. There is nothing that can convey the horror of what happened to the victims of the Holocaust, or the scale of it, but the memorials and museums in Jerusalem, Berlin and elsewhere to commemorate and educate are as close as we can get.

On Holocaust Memorial Day, 27 January 2014, I held a reception in Downing Street, inviting fifty Holocaust survivors. One man told me how he had escaped from the Warsaw ghetto through a hole in the wall. A woman showed me her diary, in which her grandfather had written: 'Wherever you go make sure you're a good daughter to the country where they take you.'

It came home to me that these people were coming to the end of their lives, and soon we would no longer have anyone to bear witness to the darkest chapter in human history. It fell to us, right now, to work out how we were going to continue telling their stories, and commemorating the six million people who never made it.

I charged Tim Kiddell, who had written my speech for the occasion, with driving forward a cross-party Holocaust Commission. The group – including Michael Gove, Ed Balls and the actor Helena Bonham Carter – recommended that a museum and memorial should be built, and that they should be at the heart of our democracy, next to Parliament. The project is due to open in 2022.

It wasn't long before things were to flare up again between Israel and Palestine. In June 2014, three Israeli teenagers were kidnapped and murdered. Terror attacks had taken place in southern Israel, via underground tunnels built from Gaza. Gaza launched rockets into Israel; Israel responded ferociously.

I made a statement about it in the Commons, because Philip Hammond was on his way to an EU foreign ministers' meeting, and I was already making a statement on Ukraine. I took what I felt was the correct line: 'The crisis was triggered by Hamas raining hundreds of rockets on Israeli cities, indiscriminately targeting civilians in contravention of all humanitarian law and norms.' I said that Israel should exercise restraint and do all it could to avoid civilian casualties, and

I highlighted the heavy loss of life, but stopped short of calling it disproportionate.

Why my reluctance to use this word? Partly because it is very loaded. If something is disproportionate it is illegal, and if it is illegal it's a war crime. And partly because I thought Israel had a right to defend itself from attack. Any prime minister appreciates that it is easy to request that such defence doesn't result in any deaths, but harder to achieve it in practice.

But what was right in theory sounded harsh. The issue duly played out on the streets of the UK, with hundreds of thousands of people marching past Downing Street condemning Israel's behaviour. MPs with large Muslim populations in their constituencies came under particular fire, their postbags full of outrage. Very sadly, anti-Semitic incidents in the UK began to rise.

And then there was Parliament, where the balance of power had shifted markedly and, extraordinarily, the Labour Party, under its first Jewish leader, Ed Miliband, whipped its MPs to call for the immediate recognition of the state of Palestine.

That August I went off with Sam and the children to Portugal, to a house my mother had rented for the holidays. As we arrived I received a text telling me that Sayeeda Warsi was considering resigning over what she believed was the government's one-sided response to the Israeli conflict.

I asked her to speak to me before making any decision. As I was speaking to her on my BlackBerry, my iPhone buzzed in my other hand. It was a tweet from Sky News saying that she had already announced her resignation. Her own tweet shortly followed. After all those years working together, she had told the media before she told me. It was a sad end: from the first female Muslim in cabinet to the first minister to resign from cabinet via Twitter.

Her resignation said something about the power of global issues in domestic politics: that a UK cabinet minister could resign over a distant war in which Britain played no part.

I spent much of the rest of that holiday on the phone. Israel and Gaza, Ukraine and Russia, Syria and Iraq. The world is never at peace, and as a result, neither is a prime minister.

39

Back to Iraq

On 26 September 2014 I stood at the despatch box in Parliament and made the case for why Britain should, for the third time in three decades, fight a war in Iraq.

ISIS was the successor to al-Qaeda and other fundamentalist groups that had taken root in Syria's war-ravaged plains, like weeds rising from untended fields. It had seized the Syrian city of Raqqa in March 2013, and in June 2014 had overrun Iraq's second city, Mosul, as Iraqi soldiers ran away.

I watched on TV as the ISIS leader, Abu Bakr al-Baghdadi, proclaimed from Mosul's central mosque that the city's capture marked the establishment of the 'caliphate'. This was the utopia many extremists dreamed of: an Islamist state, supposedly modelled on what society was like in the Prophet's day.

The fact that ISIS took on the form of a fully fledged state made it different to anything else we'd confronted. It had an army, a police force, an intelligence service, courts, schools, a health service – even a flag and a national anthem. It received vast revenues from taxes, seized oil fields and captured banks. And it would become notorious for melding medieval methods with modern ones: hanging and flogging, tweeting and blogging.

It is argued that Bush and Blair's invasion of Iraq in 2003 led directly to the rise of ISIS. I don't think that is fair. The aftermath of the invasion certainly spread anti-Western feeling, anger and resentment, which the extremists effectively channelled to recruit more disillusioned Sunnis to their cause. They drew strength from veterans of Saddam Hussein's nationalist Ba'ath Party, which had been dismantled after his defeat.

But more potent was the fact that the Sunnis of Iraq had been marginalised, insulted and ignored by their Shia prime minister, Nouri al-Maliki. A Shia–Sunni settlement had been established, and a functioning state and army were emerging. But once the US withdrew in 2010, Iraq reverted to a Shia-dominated, corrupt, sectarian state. The reason 12,000 mostly Shia soldiers ran from fewer than 2,000 ISIS fighters in Mosul was that they had lost faith in their commanders, who stole their pay and sold their equipment. The UK and the US backed the government, all the while trying to get it to change course. But we failed. And that is the extent to which I believe 2003 indirectly led to 2014.

Once again, while we seemed to be highly capable at intercepting plots – I lost count of the number of atrocities MI5, MI6, GCHQ and the police prevented – no one saw ISIS coming.

To be fair, our attention had been on the Khorasan group, a branch of al-Qaeda that was intent on attacking civilian aircraft. The UK had a unique role in disrupting this, but we had to some degree looked away from Iraq. The war had been traumatic for UK forces and policymakers, and once our troops withdrew in 2009, the machine breathed a sigh of relief that the troubled country would now be almost entirely under the auspices of the US.

Quite right, some might say. After all, what had chaos in the Middle East got to do with Britain? My answer is that it had *everything* to do with Britain.

First of all, Britain was a direct target. We knew these people were planning attacks on our streets. It's what al-Baghdadi urged his followers to do during his call to arms. He may have been focused on creating a caliphate in the Sunni belt between Syria and Iraq, but his ISIS was also a franchise operation, which hosted sub-units of 'foreign fighters' focused on punishing the Western democracies which had brought them up.

Fortunately, a potential ISIS-sponsored marauding gun plot was prevented. A man armed with a knife, who was later found to have ISIS propaganda on his phone, made it as far as the London Underground, but was thwarted. Some, however, would slip through the net, for example a gunman who managed to murder four people at a Jewish museum in Brussels.

There was a second way in which the caliphate affected Britain. Our citizens were victims of ISIS brutality in Iraq. When Mosul fell, ISIS held three Britons hostage: aid worker David Haines, journalist John Cantlie, and Alan Henning, a taxi driver who had been volunteering with a charity taking aid convoys to Syria. Ransoms were sought via video messages.

And when those ransoms weren't paid, more videos would emerge. They followed the same format: the victim kneeling in the desert, the orange jumpsuits, the masked killer, the speech, the beheading. I had never seen anything as chilling or disturbing. Yet I felt I ought to watch them. It wasn't because they were addressed to me personally (by, disturbingly, a man with a strong British accent), but because I felt I owed it to the victims' families to understand what they were going through.

The UK doesn't pay ransoms to terrorist kidnappers. It's the right and sensible stand – in theory. But my God it's hard in practice; especially when you see Italian, French, German and Spanish hostages being released because ransoms have been paid. I was convinced that we shouldn't change our policy, though, because that would put more people in mortal danger. But it made me doubly determined to do everything possible to save those that were kidnapped.

Every time a hostage was taken, I would chair a COBR meeting, kick the system into action, discuss the readiness of our military teams, check we had enough drones, go through all the options for a rescue, and consult fellow world leaders.

And here I worked exceptionally closely with Obama. During every phone call over this period we would discuss the status of our hostages. I pushed and pushed for rescue attempts, and Paddy McGuinness sent me a long note on 26 June 2014 saying that all our work pressing Obama was paying off – US special forces were taking up the baton. I immediately wrote back and said we should support it in every way we could.

On 1 July the moment for the rescue attempt arrived. The US launched a huge operation, with 140 soldiers on the ground. They went to the places where we thought the hostages were being held, only to find they'd been moved and separated a few days earlier.

I was overwhelmed by the bravery of those 140 people who were prepared to risk their lives to save others. And I thought constantly about the hostages, who were enduring a living hell and who we hadn't been able to save this time. We never stopped trying to find our people.

So who were these willing executioners? Appallingly, many of them were British citizens. People who had had all the advantages of an upbringing in a safe, tolerant society full of opportunity, yet ended up in this dystopian desert, murdering innocent people. That was another way Britain was caught up in this war. Because as well as urging attacks on the West, al-Baghdadi encouraged Sunni Muslims all over the world to join the caliphate – and they did. By 2014, five hundred people from across the UK were thought to have left the safety of our country to join this death cult.

I studied their profiles in amazement. An NHS doctor from Sheffield who left his family behind. A teenager from Glasgow whose well-off family said she was radicalised online. Three straight-A schoolgirls from east London. And of course, the masked executioner whose voice I had come to know in those videos, a twenty-six-year-old Londoner called Mohammed Emwazi, known to the press as 'Jihadi John', and one of four of the most prominent ISIS killers nicknamed 'the Beatles'.

I was in no doubt: ISIS had to be confronted and defeated. I never thought that would be done with Western ground troops directly trying to pacify or reconstruct these countries. 'Boots on the ground' – when they were predominantly Western boots – fuelled our enemies' narrative that we were foreign occupiers their followers could be incited to fight against.

But the strategy I set out during PMQs on 18 June 2014 included a vital role for our military. It could take various forms – from hostage rescue attempts to delivering aid supplies. It could also include direct intervention: airstrikes. The reasons for involving our military were even more powerful, I thought, than when we went into Afghanistan to eject al-Qaeda. But I knew I'd be hamstrung by what happened with the Syria vote the year before.

I was confident in our capabilities. We had one of the finest military and intelligence machines in the world. We also had most of the world on our side, and in Iraq a friendly country asking for our help, so help we should. But in both Iraq and Syria the picture was muddied by competing interests.

In Syria, our enemies' enemy was certainly not our friend. Doing a deal with Assad to rid his country of ISIS was out of the question. In any event, given that Assad's brutality was driving ISIS recruitment, such a strategy would ultimately be self-defeating.

Nor in either country was our friend's enemy our enemy. Our NATO allies the Turks treated Kurdish separatists as terrorists, but the Kurdish Peshmerga were the finest anti-ISIS fighters, and we worked closely with them.

Before we could consider any intervention, the first priority was domestic security: specifically, ensuring that our anti-terror legislation was up to scratch. Every time the threats had changed, we'd updated our laws, for example to tackle the IRA and then the emerging domestic threat after 9/11. More importantly, this was a different scale of threat, as those responsible were prepared to die in the act of taking as many lives as possible. Indeed, martyrdom was part of their creed. That had never been the case with the IRA.

In opposition I had voted for many of Tony Blair's proposals for dealing with terrorism, but withheld support for those instruments that seemed too blunt. We needed targeted, tough measures that were in tune with the British way of doing things. But now, with the threat of people leaving our shores to wage jihad – and to return to Britain and commit atrocities – it was clear to me that we needed to do more.

We had already brought in powers to stop suspects from travelling to the region, by seizing their passports. We had broadened the law so that people could be prosecuted more easily for terrorist activity committed abroad, and had removed tens of thousands of items of terrorist material from the internet. Now we would go further, giving police the power to remove passports at borders. As well as being able to deprive dual nationals of their UK citizenship if it was merited, and barring foreign nationals, we would now be able to temporarily exclude British nationals from entering the UK. Airlines would have to comply with our no-fly lists, and share their information. I was astonished to find out how lax airport security was in certain countries – people could literally board a plane in Egypt or Tunisia without being searched. We made sure all countries complied with our security checks.

It wasn't just about treating the symptoms of extremism; it was also about defeating the causes. From tougher powers to ban hate preachers, to proscribing some extremist organisations altogether, we would challenge the false utopia being sold to young, impressionable people. I had grown up at a time when communism held similar promise for idealistic youngsters; and the way we defeated that was by exposing its flaws and demonstrating the virtues of freedom, democracy and the rule of law. It was time to do so again. We did much of this through RICU, which really came into its own when the battle was on to prevent more young Britons being sold a lie and joining the jihad.

But it was not yet time for airstrikes. In June 2014 my NSC summing-up included the words 'not planning military action'. We were still pursuing other ways of confronting ISIS. The US had taken the lead with air support for Iraqi and Peshmerga troops, and in mid-June Obama had ordered US forces to assist Iraqi forces, with a military and diplomatic presence (but, pointedly, no airstrikes). Meanwhile we were supporting local forces by giving them the kit and training they needed, and promoting political change in Iraq.

Then, in August, ISIS advanced into Sinjar, a part of northern Iraq populated by a close-knit community of the Yazidi religion. What followed – the murder of Yazidi men, sexual enslavement of women and girls, and conscription of boys – was truly horrific. Thousands of Yazidis were left trapped on Mount Sinjar, running out of food and other essentials. This was a minority group under siege, and chillingly, the word 'genocide' was starting to be used.

I decided that our military would need to be involved in action, and that there was no time for a parliamentary vote. I came back from my holiday to chair COBR on 13 and 14 August. We were already supporting anti-ISIS forces, but I asked for increased military support for the Kurdish Peshmerga. I authorised Hercules transport planes to join the Americans and drop water and medicine to the besieged Yazidis. Tornado fighter jets also undertook surveillance. Eventually, after a tricky operation, the Kurdish Peshmerga broke the siege, and many victims were able to escape.

From then on I could hear the clock ticking. The ISIS advance continued. More hostages were being taken. More foreign fighters were arriving in Mosul and Raqqa. More plots to cause death and destruction in the West were being hatched. And with ISIS just thirty-seven miles from Baghdad, it wasn't beyond possibility that the Iraqi capital could fall. It was becoming increasingly obvious that the only way we could hope to destroy ISIS was through airstrikes. When, then, would we authorise our fighter jets to take to the skies?

Obama was particularly criticised for his delay in authorising airstrikes. But it was smart politics. We agreed that intervening on behalf of the man who had helped to create the situation, the partisan, Sunni-marginalising Iraqi prime minister Nouri al-Maliki, would be a disaster. Before we did anything big, Maliki had to go.

It was only after pressure from the Shia religious authorities that Maliki agreed to step down in August. In September, Haider al-Abadi was appointed as his successor. He had lived in the UK for many years, and still spoke English with a slight south London accent. While he was clearly no angel, I believed I could work with him.

The appointment of someone who was less sectarian and less prone to corruption was key. Ban Ki-moon was spot-on when he said that 'Missiles may kill terrorists, but good governance kills terrorism.' Just as people turned to the Taliban because it appeared to deliver some form of protection from lawlessness, many embraced ISIS because it promised to provide basic services.

On Saturday, 20 September, Iraq's new government formally asked for military assistance (I had made it clear to Obama that we wouldn't be able to support military action until after the Scottish referendum). On the Sunday evening I assembled a meeting at Chequers with key cabinet ministers. Nick Clegg was very helpful, very pro-airstrikes. Michael Gove was anti, arguing that we didn't have a majority, and therefore shouldn't recall Parliament.

The obvious problem was that it made no strategic sense to attack ISIS in Iraq but not Syria. But not being able to act in Syria wasn't a reason not to act in Iraq. At that moment we had to focus on what we *could* do.

We left with the coalition in agreement. We had the legal go-ahead, based on the fact that Iraq was asking for help to defend itself against ISIS. We were acting as part of an international coalition of sixty countries, many of them from the region. We had public support, with YouGov showing 57 per cent for airstrikes and 24 per cent against. Plus, Obama had already launched airstrikes. There was just one more thing required. I couldn't risk chancing rejection in Parliament again; I needed to be confident of Labour's support before a vote. That meant winning round Ed Miliband.

After the breakdown of trust over Syria, I knew I needed to do things differently this time. It would, I calculated, be easier for him to give his support if negotiations were at arm's length, and led by national security advisers. But I did discuss the issue with him briefly on a couple of occasions, culminating in a phone call while I was at the UN General Assembly in New York.

I told him that I planned to ask for the recall of Parliament. From his tone, I felt he was still looking for an excuse not to support military action. He said: 'What guarantee can you give that there will be Arab countries involved in the bombing over Iraq?'

I rolled my eyes. I said that I was fairly confident, because five Arab countries had just bombed Syria. But I couldn't say 100 per cent: there were complications, because the Iraqi government feared that Sunni states bombing parts of Iraq might provoke a Shia backlash. Miliband said that if we couldn't produce a Sunni, Arab state that was willing to support a bombing action, Labour would find it hard to support us.

Could this be it? Another arbitrary red line from Ed that would scupper the whole thing? It was Wednesday. Parliament would vote on Friday. I had an idea. I had spent a lot of time building a relationship with Mohammed bin Zayed, crown prince of Abu Dhabi. I texted him, then called to ask him to confirm that he would support bombing in Iraq. He agreed.

In cabinet that Thursday I set out the facts. ISIS was a mortal threat. Iraq had asked for support. The legal advice was very clear. We had a strategy to degrade and destroy ISIS. Jordan, Bahrain, the UAE, Saudi Arabia and Qatar had all participated in airstrikes in Syria, and were, 'subject to the permission of the Iraqi government . . . prepared to conduct strikes in Iraq'.

Then there was Parliament. With Miliband whipping his MPs to support the motion, the only remaining question was how many MPs, on all sides, would actually vote for it. Just like the ill-fated Syria vote a little over a year earlier, much would hinge on the speech I gave in the Commons.

I set out the level of ISIS brutality, lest anyone was in any doubt about it. I emphasised that if it were left unchecked, we would face a terrorist caliphate on the shores of the Mediterranean and bordering a NATO member, with a determination to attack our country. But I wasn't naïve. I knew that regaining the territory ISIS had taken wouldn't automatically change the hearts and minds it had captured. Even after ISIS had been dealt with, I knew future prime ministers would be standing at the despatch box dealing with Islamist extremism.

That evening, Parliament voted by 524 to 43 in favour of military intervention (the Labour backbencher Jeremy Corbyn was a teller, and gave the result from the 'no' lobby). I was relieved, but I was also angry. Of the forty-three MPs who voted against, six were Conservatives. And seventy abstained, including twenty-one from our own benches.

This was an organisation that murdered gay people, enslaved women, raped children, maimed, beheaded, even crucified people for minor misdemeanours. It was planning bombings and shootings in Britain. It was beheading our aid workers. What would have to happen for these MPs – who, incidentally, tended to be those perennially calling for higher defence spending – to vote for intervention?

Fortunately they weren't able to derail our action. The following Tuesday, Tornados would be in Iraqi skies, taking out ISIS's sources of finance, its weapons stores and, of course, its ringleaders. We would soon be responsible for the second-highest number of airstrikes, after the US. Meanwhile we gave Kurdish Peshmerga, Jordanian and Lebanese forces millions of pounds' worth of medical supplies, machine guns, mortars and body armour. We trained thousands of Iraqi soldiers, police and Peshmerga. We gave humanitarian assistance to areas liberated from ISIS, and helped to rebuild their schools, police stations and electricity generators

Iraq wasn't the only place we would need our military to counter this extremist menace.

Boko Haram in Nigeria was linked to al-Qaeda, and believed Western education and lifestyles were a sin (the meaning behind its name). It too wanted to institute a caliphate, and like ISIS it would use whatever barbaric means it thought necessary.

In early 2014 a group of its fighters entered the government secondary school in the village of Chibok, seizing 276 teenage girls. They were taken to camps deep in the forest. The Christians among them were forced to convert to Islam. Many were sold as slaves, entering the same endless violent nightmare the Yazidi women had suffered.

As a 'Bring Back Our Girls' campaign spread across the world, we embedded a team of military and intelligence experts in Nigeria, and sent spy planes and Tornados with thermal imaging to search for the missing girls. And, amazingly, from the skies above a forest three times the size of Wales, we managed to locate some of them.

But Nigeria's president, Goodluck Jonathan, seemed to be asleep at the wheel. When he eventually made a statement, it was to accuse the campaigners of politicising the tragedy. And absolutely crucially, when we offered to help rescue the girls we had located, he refused.

Yet again, the problem was a weak government and corruption. The Nigerian army was so hollowed out by venal, politically appointed generals that it was incapable of participating in operations with US/UK assistance. The NSC concluded that we had to play the long game, focusing on a much bigger training effort for the Nigerian military and intelligence forces, and trying to promote more energetic leaders from the younger generation. The Archbishop of Canterbury, as an expert on Nigeria, could be particularly useful on this, and I invited him to join our NSC discussion.

Some of the girls have managed to escape over the following four years, and others have been released, but over a hundred are still missing. Once again, the combination of Islamist extremism and bad governance proved fatal.

How did I feel about all this at the end of 2014? The answer is, depressed. ISIS now occupied an area larger than Britain. A similar brand of terrorism was being wrought by Boko Haram in west Africa, by another ISIS affiliate in north Africa, and by al-Shabab ('the youth') in east Africa, while related groups were springing up in the Philippines, Bangladesh, Afghanistan and the Caucasus. There seemed to be no stopping an evil ideology that seduced minds from the badlands of Syria to bedrooms in Birmingham. When I spoke about the challenge publicly I tried to remain measured and resolute. But privately I did ask myself, would we *ever* be able to defeat this thing?

40

Scotland Remains

It was the evening of Friday, 5 September 2014, and I collapsed on the sofa in the flat. I had just flown back from the NATO summit in Wales, and would have to be back at RAF Northolt first thing for my annual visit to Balmoral.

We loved our trips to Scotland, but this year Sam stayed at home with the children (it was to be Florence's first day at primary school the next day). So I flew up to Aberdeen alone with my thoughts. And there was a lot to think about.

In two weeks the voters of Scotland would decide whether to remain in a Union with England, Wales and Northern Ireland, or to end our 307-year history as a United Kingdom.

This wasn't like the referendum on membership of the European Union. It wasn't a question about whether to remain in a relatively recent association with some neighbouring states, the success of which was at least debatable. It was about the break-up of our country. In my view there was no debate about its success. It is the greatest grouping of nations the world has ever known.

I arrived at Birkhall with the recent news that, for the first time, those saying they would vote Yes – to leave the UK – had taken a lead by two percentage points, 51 to 49. Shortly I'd be having an audience with the Queen at Balmoral Castle: she, the woman who had reigned over the United Kingdom for sixty-two years; me, the man who had allowed a vote on its possible disintegration.

Of course, she was completely charming – they all were. But as Prince Philip showed me the barbecue he had designed to roast grouse and sausages over charcoal when we were all up at the hillside bothy, the referendum was clearly on everyone's mind. They gingerly asked questions about it, but knew they shouldn't express too strong an opinion.

That is the reality of a constitutional monarchy within a parliamentary democracy: a prime minister can instigate a sequence of events that could change the make-up of the country; the royal family can't even express a view on it.

And then the next day at breakfast, there it was in cold print. Among the kippers and the kedgeree was the *Sunday Times*, with the headline 'Yes Vote Leads in Scots Poll'. The Queen wasn't there; she usually had breakfast alone. Instead, I was surrounded by some ladies-in-waiting, equerries, and the moderator of the Church of Scotland. I tried to reassure them about 'rogue polls', about the fact that newspapers loved contrary findings, and what the average poll had been during the long campaign, but I was struggling to convince myself, let alone them.

It had already been the longest campaign in British history, at twenty-eight months. It had also been the most cross-party campaign, with the three-way alliance between the Conservatives, Lib Dems and Labour advocating No testing my ability to work with my political opponents more than any prime minister before me.

It was obvious from the start that our parties would take different roles in what became the 'Better Together' campaign.

With the most Scottish MPs in the Westminster Parliament, and as the official opposition in Holyrood, it made sense for Labour to be the public face of it. That meant putting the even-minded, intelligent former chancellor Alistair Darling at the helm.

The Conservatives were the obvious financers, with Ian Taylor, Andrew Fraser and Donald Houston among the big donors. We put one of our best-known MSPs, David McLetchie, on the board of directors (tragically, David died in August 2013 – a huge loss); installed the director of the Scottish Conservatives, Mark McInnes, at the heart of the campaign; and charged Andrew Cooper with conducting the polling that would help to steer the strategy.

At the start of 2014, Andrew identified several groups of voters, ranging from 'mature status quo', who were certain to vote No, to 'Scottish exceptionalists' and 'blue-collar bravehearts', who were equally committed to the Yes camp. The key demographics were where the battle would be won or lost. The definite No voters made up about 40 per cent of the electorate, and the confirmed Yes voters 30 per cent. That left 30 per cent – a million people – who could go either way.

If they split evenly, the outcome would be a comfortable No victory of around 55 per cent to 45 per cent. If Salmond managed to lure more of them than not, we would be on the rockiest ground.

The voters who felt they shared an affinity with England, Northern Ireland and Wales were likely to be voting No anyway. But among the middle million we needed to win over, a sense of common identity with the rest of the UK was very low. Indeed, it was almost as low as that felt by the most ardent nationalists, those 'blue-collar Bravehearts' who would die before voting to remain under the English yoke.

In other words, those who needed persuading weren't going to be convinced by abstract ideas of history, or an emotional appeal to unity or identity, but rather by practicality. The patriotic case for our historic Union should be made, as I'd discussed with Gordon Brown, but our focus had to be on the wages people brought home, the pensions they received now or in the future, the public services they relied on, and the national institutions that kept them safe.

Were the risks really that grave? People often said to me, 'If you thought that Scotland leaving the United Kingdom would be so bad, why did you allow a referendum?' I've explained why I thought holding the referendum was the right thing to do. Denying it would merely be delaying it; and delaying it would ignite a level of grievance that made independence inevitable.

What was my role to be in this vital battle? Jaunting up from London, representing a party that barely existed north of the border, lecturing Scots – there could be no finer fuel for Salmond's righteous indignation. The message to me, therefore, was to put strict limits on my appearances in Scotland. Fortunately, there *was* a front-facing role for me. In order to create the 'permission' to go negative and outline the risks of independence, we would have to set out the positive case for the Union. There I could make an impact. I could make the romantic, pro-Union case: that you could be Scottish *and* British, and that being the latter was no betrayal of the former.

That was what brought me to the Olympic Park velodrome in London in February 2014. Two years previously I'd seen Team GB, led by Scotland's Chris Hoy, enter the record books in this building. Just think of all the things our four nations – the wider Team GB – had done together, and could still do together. I said, 'Let the message ring out from Manchester to Motherwell, from Pembrokeshire to Perth, from

Belfast to Bute, from us to the people of Scotland. Let the message be this: we want you to stay.'

And when it came to Scotland, I didn't stay away completely. I made some carefully planned visits, during which I underpinned passion and patriotism with the practicalities we knew were key to victory, from a BP oil rig to the vast deck of the new HMS *Queen Elizabeth II* aircraft carrier at Rosyth.

I also authorised the use of the government machine. Keeping the UK together was UK government policy; it was right that the government should set out the facts (Scottish independence was the Scottish government's official policy, and the civil servants there would be supporting that too). So we published fifteen papers detailing exactly how Scotland was better off in the UK, from defence to data roaming. We sent a leaflet to every Scottish household called 'What Staying in the UK Means for Scotland'.

Of course, the biggest risks associated with independence were economic. How would trade across the border work? What about English companies based in Scotland, and vice versa? What would happen to the industries subsidised by the UK taxpayer? And, biggest of all, what about the pound in Scots' pockets? For months Salmond had insisted that an independent Scotland would keep sterling. But that wasn't his decision to make. It was the UK government's. And George and I – and the Lib Dems – were completely aligned in our opposition to his plan.

Currency unions are fraught with difficulty – look at how the euro was stretching the EU to breaking point. An independent Scotland could hardly expect taxpayers in a Union it had just voted to leave to back the use of their currency by a new country. 'If Scotland walks away from the UK, it walks away from the pound,' George declared in Edinburgh that February, playing our most controversial card yet.

By the summer, the nationalists were making *their* most controversial move.

I had always thought the health service was a reason for Scotland to stay in the Union: a great pooling of resources to help those in need, whoever they were, wherever they were in the UK. Instead, during his one-on-one BBC debates with Darling that August, Salmond was alleging that the Tories were gutting this great national institution, and the only way to protect it in Scotland was by leaving the UK.

This had an impact. As we began to receive daily polling, the crevasse between No and Yes narrowed to a crack. The Quad decided we should exercise much more influence on the campaign from London. Danny Alexander became the new unofficial head of the No campaign, chairing meetings every day.

The campaign began to attract criticism. The standard attack was that our campaign was too negative – that we were running 'Project Fear', scaremongering about the risks of separation. But we couldn't let this throw us off our strategy. Andrew Cooper said most voters were not convinced that an independent Scotland would be a disaster, he wrote to me in August. Nor were they sure that independence would jeopardise the pound, jobs or pensions. If the frame of reference was love for Scotland, that would push undecided voters towards voting Yes, not No. What brought them to No was the undeniable fact that No was safe and Yes was risky. In other words: keep framing this as a practical choice more than a patriotic one.

This advice was reinforced by opinion at home.

Our long-standing nanny, Gita, had been with us since shortly after Ivan was born in 2002. When she was revising for her citizenship test, she asked me one of the questions: 'What is the purpose of the cabinet in the government of the UK?' and looked rather surprised when I replied, 'I sometimes wonder.' The cliché was true: we loved her like a member of the family, and when she left to have a baby with her lovely husband we missed everything from her company to her cooking.

We had to advertise for a replacement through an agency without saying where the position was. But we found the perfect person. A young Glaswegian go-getter, Michelle Legowski came to us after stints working on cruise liners. She seemed unfazed by our unusual circumstances, and slotted straight in.

She was also a great sounding-board for the referendum, especially because it was obvious that she was an undecided voter, and so one of the 30 per cent of Scots who would decide the outcome. When she handed over to me in the evening I'd ask her which way she was swaying. For her, as for the friends she'd been discussing the subject with on Facebook, it came down to very practical issues: pensions, pay, taxes, ease of travel. Her voting intention changed from day to day. That kept me focused on the Cooper strategy.

But then came the Balmoral weekend, and the heart-stopping

crossover poll. For people to vote No, it wouldn't be enough to contrast the risk with the status quo. We would have to offer something more.

Douglas Alexander, Alistair Darling and Ed Miliband were proposing a cross-party initiative to show that the commitment to giving more powers to Scotland was genuine, and would be delivered quickly after the 2015 general election. This would be a 'circuit breaker' to arrest the impression of Yes momentum – a vision of change without the need for separation. The reply I scribbled to Andrew Dunlop, who set out the proposal, was unequivocal: 'Whatever it takes!'

The truth was that to win the referendum we needed Labour voters. (As it turned out, our faith in Labour's power in Scotland was misplaced. The party didn't have the activists or the support it once did – as was proven the following year, when it was wiped out by the SNP at the general election.) It was no good Alistair Darling making another speech, or Ed Miliband writing an article. A big Labour figure would have to make a big intervention. Where was that big clunking fist when you needed it?

Gordon Brown told me he wanted to make the devolution promise more real by personally setting out a timetable. I agreed to it. And on the day after the *Sunday Times* poll shock, he made a speech in Midlothian announcing that very timetable.

As he did so, I was in Downing Street holding a reception for business leaders. I had spent much of the time since the poll calling them up and urging, 'You know independence would be terrible – so if you think it, you've got to say it.' Many came out in support of a No vote – BP, Shell, Standard Life, Lloyds, Aviva, Prudential. But those who remained on the fence infuriated me. They didn't have to be politically neutral, and I knew that separation would affect their business – so why not say it?

Ed Miliband then had a good idea. He, Nick Clegg and I should all go up to Scotland on Wednesday morning, leaving William Hague and Harriet Harman to deputise for us at PMQs. It would be unprecedented and dramatic.

I wanted my part of the visit to feel different from previous interventions, and Liz had found the perfect place. Scottish Widows was full of bright young staff with a focus on finance. Instead of lecturing them from a podium, I'd be sitting among them taking questions.

On the way there, I told Craig about what I wanted to say. 'I'm thinking of saying you shouldn't just vote for independence because you want to give the effing Tories a kick,' I said. I wanted it to really hit home that this wasn't a short-term, party-political decision. It wasn't about the next five years, but the next century. 'Go for it,' he said. And I did.

Then, that Sunday, one week after the *Sunday Times* headline, the Queen spoke to some of those gathered outside Crathie Kirk, and said that she hoped Scots would 'think very carefully' about the vote. I was delighted.

The final Monday before polling day, after I chaired our usual 8.30 a.m. meeting, I asked Andrew Dunlop quietly, 'We are going to be all right, aren't we?' He calmly replied, 'Prime Minister, not only are we going to be all right, we'll win by at least ten points.' I thought: I don't care about the exact numbers. Let it be 49–51 if it has to be. Just, *please*, let it keep our country together.

Later that morning I flew with Andrew to the Aberdeen Exhibition and Conference Centre for my final public intervention. My new friend and speech collaborator, Gordon Brown, made lots of comments on my script, in particular emphasising that voting No was a patriotically Scottish act.

Two days before the vote, we unveiled our last-ditch offering to Scotland. The three main UK party leaders all signed up to a further wave of devolution, which would devolve income tax and tax-raising powers. It was splashed across the *Daily Record* as 'The Vow'. This was a repackaging of a promise made back in August, the only difference being that we committed to keeping the Barnett formula, which determines the size of Scotland's annual grant from central government. It was successful in clarifying that healthcare spending was a decision for the Scottish government, finally trumping the SNP's NHS card. It also halted the momentum of the Yes campaign in the final days.

On the eve of the poll, Gordon Brown made the final speech on behalf of Better Together. It was a strange feeling for me. I was moved by him. I was rooting for him.

As I walked down the stairs on the morning the Scottish people would be heading to the polls, I saw my Scotland special adviser Ramsay Jones, who was walking to the front door. 'And . . .?' I asked. 'By ten points,' he replied, echoing Andrew's prediction of a few days before.

What do you do on a day when your country hangs in the balance? As Nancy was due to start secondary school the following year, Sam and I went to look around one option, Holland Park in west London. There I was reminded just how transformational independence for schools through academy status could be. Pupils were learning Latin and Greek. There was a sense of discipline and drive. It had echoes of the school I went to. And yet this was a co-educational comprehensive in the middle of London.

I returned to Downing Street, where the Saltire was flying at full mast.

The previous night, in George's study, we had discussed a losing speech, which Ed was working on. 'We'll have to be best of friends, best of neighbours, best of allies, we'll respect you,' was the tone. Then we had a team meeting on the contents of the speech for a No vote, including William Hague, George and Jess Cunniffe, who was drafting it.

Still, I thought constantly about the possibility of a Yes vote. I had said publicly that I wouldn't resign if we lost. I certainly didn't want to make the referendum in Scotland a referendum on me. That would have been politically unwise when we needed so much Labour support. My true feelings were more complicated. The fact that I'd be the one who precipitated the end of our island story would hurt beyond belief. I saw a lot of senior colleagues that week – Philip Hammond, IDS, William. George said, 'For heaven's sake, don't resign.' Theresa May didn't say anything like that. She possessed monarchic levels of neutrality – no one but Her Majesty played their cards closer to their chest.

All this was weighing on me that night, as camp beds were hauled into offices and Downing Street's offices were turned into dorms. At about 10 p.m. we went to the open-plan press office, Gordon Brown's old command centre, to watch the results come in.

As the turnout was revealed – nearly 85 per cent, higher than for any election in UK history – there was a panic that this might confound the opinion polls. But when Clackmannanshire declared at 1.30 a.m., with a solid majority for No, I told Craig I was going to bed, but that he should keep me posted.

I didn't sleep. And when Craig texted a couple of hours later saying it was looking encouraging, I was straight back downstairs. It wasn't the best moment to reappear: Dundee's result came in, and it was one of the few cities to vote Yes. The lead we'd taken was closing. But Mark McInnes at the count reassured us.

Sure enough, after 4 a.m. the No votes kept rolling in. Two of the happiest hours of my life followed. Everything was going to be OK. At 6 a.m. Salmond conceded. I spoke to him on the phone, and a few minutes later

I went out into Downing Street and made my speech: 'The people of Scotland have spoken. It is a clear result. They have kept our country of four nations together. Like millions of other people, I am delighted.'

And truly I was. The final result was 55 per cent to 45 per cent. Our Union was safe. The question was settled.

One thing I'm glad I didn't do was go straight up to Scotland. We had planned that I would do so, but that morning I felt that what had been intended as a gesture of unity might look like crowing, especially because Salmond had just resigned.

I came in for heavy criticism for bringing up the issue of 'English votes for English laws' in my victory speech. Days before, we had taken a decision to speak about fixing the longstanding wrong that Scottish MPs could vote on matters affecting only English constituencies, whereas, post-devolution, English MPs had no such reciprocal right. Few things exercised Conservative backbenchers more than this 'West Lothian question'. I knew that after all the focus on pleasing and placating Scotland, the cry would go up: what about England?

It had always been on my mind – as far back as 2005 I had commissioned Ken Clarke to look at the issue. William thought we had to seize our moment and ensure that new powers for England moved in lockstep with powers for Scotland and Wales. Although George and Michael Gove were not in favour, we discussed the subject in detail during the preparation for the victory speech.

On the morning of victory, I chose my words carefully: 'We have heard the voice of Scotland, and now the millions of voices of England must also be heard. The question of English votes for English laws – the so-called West Lothian question – requires a decisive answer.' The idea that I could have ignored the issue altogether is nonsense, but I now wish I'd left it until party conference. Friday, 19 September 2014 was a day for magnanimity, nothing more.

In between the referendum and party conference, I had to go off to the UN General Assembly in New York. It was there that I did something else I wish I hadn't done. As I walked through the Bloomberg headquarters with my friend Michael Bloomberg, the former mayor of New York, I confided in him about the whirlwind of the past few days. I said that the definition of relief was ringing the Queen and saying it was all right, and that 'she purred down the line'. But I was being filmed, a microphone picked up my words, and they were around the world before you could say 'royal pardon'. I later made a heartfelt apology to the Queen for commenting on our private exchange.

Five years on from the vote, I truly believe that while the referendum was a risk, it was the right risk. I feel even more strongly that the bigger risk would have been leaving the issue of Scottish independence to fester. The case for a referendum would have grown stronger, the bitterness would have become darker, the British government would eventually have been forced to hold it, by which time the case for independence would have been far greater. We have since seen in Catalonia what happens when a government mishandles the will for a vote on self-determination.

I believe I was also right to take the risk on the terms of the referendum (i.e. give in to almost everything, except a straightforward 'Yes' or 'No' question). Nationalists love grievance, but there was nothing about the referendum they could call foul play on. It was their timing, their franchise, their question – and they lost.

I was right to push the respect agenda, including respecting the fact that the Scots had chosen the SNP as their main party. As I write, they've been in power for twelve years, and the shine is beginning to wear off. Nationalist parties are often bad at governing, and they have done a good job of illustrating this in Scotland.

I was also right to risk an 'in/out' vote, rather than allowing 'Devo Max' – which would have given the Scottish government power over everything except foreign affairs and defence – on the ballot paper. Granted, it would have made a Yes vote less likely. And granted, we ended up delivering something like Devo Max anyway, through the Vow. But I am convinced that having this third option on the ballot paper would have been a conveyor belt to independence. A majority for Devo Max would have been seen as a victory for Salmond. He would have added up the votes for independence and the votes for Devo Max, and claimed that this was an overwhelming mandate for separation. That would have made independence merely a matter of time. Instead, it is a matter that all the parties that support the Union can credibly claim is settled.

There is a lesson here for those wondering how you deal with the rise of populism. We could easily have ignored the separatism peddled by the SNP. We could have maligned them and their form of nationalism. Instead, we identified the anti-establishment sentiment early on, we confronted it, and took the necessary risks.

That was good for our country. But it was also good for our party. Because by driving a respect agenda, by holding the referendum, winning it, and completing the devolution settlement, a Conservative revival no one thought possible was born. In 2016 the Scottish Conservatives overtook Labour as the second-biggest party in Holyrood. Then, in 2017, we snatched so many seats from the SNP that we became the second-biggest Scottish party in the UK Parliament. With the phenomenal Ruth Davidson at the helm, I don't doubt that we could become the biggest.

Just as I'd seen off one so-called anti-establishment threat, another came along.

That year's party conference was meant to be a smooth one. We had won the Scottish referendum. The economy was recovering. Labour was crashing on the all-important measures of the economy and leadership, and were only at 34 per cent in the polls to our 32 – i.e. not far enough ahead.

There was one worry, however. Over the summer our MP for Clacton, Douglas Carswell, had defected to UKIP, triggering a forthcoming by-election. The betrayal was no surprise. Carswell was anti-EU, and a serial rebel who styled himself as an intellectual and an outsider. Perversely – and rather unintellectually – he had stood for the Conservatives when we *weren't* offering an EU referendum pledge, and then switched to UKIP when we *were*. For me, that demonstrated the attraction UKIP still held for some colleagues.

Ever since, we had been on defector watch. I despatched Michael Gove to see every suspect, instructing him to get them to promise, preferably in writing, that they were not going to cross the floor. They all reassured him.

On the Saturday before party conference I was enjoying a rare day in Dean before heading up to Birmingham that evening. We went for a family bike ride, the five of us gliding along the tree-hooded roads of west Oxfordshire, followed by a nice lunch at home. I was getting ready for a reinvigorating post-referendum, pre-conference nap when my phone rang. Two messages in the space of the next ten minutes put paid to any hopes of an afternoon's rest or a freewheeling conference.

I sat at the kitchen table and explained the double mess to Samantha. 'There's this one guy, he's called Mark Reckless,' I said. 'He's a Conservative

MP – well, he *was* a Conservative MP. Because now he's resigned and wants to be a UKIP MP. There will have to be a by-election, as there will be for Carswell. And we'll have to beat this one. Not least because he's an absolute idiot. Then there's this second guy. This one's a minister, Brooks Newmark. Really good man. Well, he was. I mean, he's married, he has five kids, and the *Mirror* has just caught him sending pictures of himself in paisley pyjamas – "sexting" – to a girl on social media who doesn't exist. Who is actually a male reporter.'

Sam took in these two very different types of betrayal. She has a habit when being told about any sexual indiscretion by a man to broaden her remarks to include the entire male gender. 'What is it with you men?' she exploded. 'This time, darling, never mind men,' I said. 'What's happening to my sodding conference?!'

I thought about what should happen next. Could Brooks really claim that it was entrapment? Was he just a bit *sad*? I took the view, however, that it was far graver than that. This was a government minister sharing explicit pictures with a stranger, leaving himself, and the UK government, wide open not just to ridicule, but blackmail. The risk he had taken wasn't just sad, it was serious, and he couldn't remain as a minister.

I saved my real disgust for Reckless. I knew he was anti-EU, but at a lunch he had assured Michael Gove that he had no plans to defect. The night before he resigned, he even left a message on Grant Shapps's voicemail about coming campaigning.

In the end, it was not a nap but treachery that reinvigorated me. Reckless wasn't going to wreck my conference. In fact, I was going to turn this into a positive.

That night at the conference hotel I watched the 10 o'clock news with the team. Everyone erupted into laughter when Reckless appeared. But I knew that to the grassroots, this man wasn't so funny. I myself had taken over from a defector in Witney, Shaun Woodward. I knew how grassroots Tories would be feeling.

I started with the National Convention, an annual gathering of all the association chairmen from across the country. I rallied them with my speech: 'You are the people who stuff the envelopes, walk the streets, bang on the doors . . . and for all this time you got a man who sits on the green benches, and *this* is how he has treated you.'

As I went from area reception to area reception, from the south-west to the north-east, I did the same thing. At each one, the Reckless section of my speech got longer, more animated, more enraged. The crowds loved it, and by the end of the three days the diatribe culminated in the call to arms: 'LET'S GO OUT THERE AND KICK HIS FAT ARSE OUT OF PARLIAMENT.'

By the time I got to my keynote speech, another Clare masterpiece, I was totally pumped. Delivering it on the stage of the Birmingham Symphony Hall, I'd never felt more like the leader I wanted to be. It was proper, undiluted, modern, compassionate conservatism. I ranged from a celebration of Scotland to a homage to our troops leaving Afghanistan. I trumpeted the rise of jobs and the fall of the deficit. I poured scorn on zero-hours contracts, modern slavery and Ed Miliband. I outlined how our LTEP would become 'a plan for you', through reforms to housing, schooling, pensions and the NHS. I spoke about my vision for Britain to lead the genomics revolution, to conquer the sort of rare diseases my family had faced.

There were two big tax promises. If we won the election we would raise the threshold at which people paid the 40p rate of tax from just under £42,000 to £50,000, and the tax-free allowance to £12,500. From modest incomes to middle incomes, working people had a home in our party, and I was proud of that.

What about the two issues fuelling UKIP: immigration and the EU? My approach was to show that they could be tackled moderately and intelligently, without UKIP's populism. I said that we needed controlled borders and an immigration system that put the British people first; that we'd succeeded in curbing migration from outside the EU, but needed to curb it from inside the EU. Here I gave a glimpse into the sort of renegotiation I wanted to pursue in Brussels. I wanted to break the system by which employment agencies signed people up from overseas and not in the UK, meaning they could get immediate access to our welfare system and send benefits payments to their families back home. 'Britain, I know you want this sorted. So I will go to Brussels. I will not take no for an answer. And when it comes to free movement – I will get what Britain needs.'

At an earlier point I had thought I might go further on these entwined issues and set out what was required in more detail, including a control over the numbers coming to Britain from the EU. Eventually I persuaded myself that too much on immigration would just take over the conference.

This part of the speech was one of the early indications of how difficult things would later be in my negotiations with the EU, and in the referendum campaign. Had I gone too far? Did I raise expectations of concessions from the EU that it wouldn't ever make? Or had I failed to go far enough? The issue was hardly just going to go away, so maybe it was better to be more specific. Yes, it would raise expectations, but only expectations the British people already had.

For the moment the speech did the trick, providing a strong response to the defections. But come mid-October, the sleepless nights had returned. Carswell had won his by-election in Clacton by a huge majority, and become UKIP's first ever elected MP. Then in November Reckless won again in Rochester. UKIP was looking worryingly like part of the furniture.

What's more, the British public weren't just making it up about immigration. They were genuinely concerned, and they had reason to be concerned. The numbers *were* out of control. You could see it in crowded doctors' waiting rooms, in ever-expanding class sizes. People wondered whether the new housing being built would be for their children, or for immigrants who hadn't arrived yet. 'I'm spending a lot of time thinking about it,' I confided on tape. On one level, I said, I felt depressed about the problem. Had I helped create it? Had the Tory Party lost touch with its grassroots?

On another level, it was clear that events had created a perfect storm for UKIP. The failure of living standards to recover quickly after a very tough recession. Large-scale immigration. The Eurozone being a mess, making Europe look like the source of all our problems rather than of opportunity. A coalition which, inevitably, people on the right didn't like. Nigel Farage's charisma. Globalisation causing tensions, pressures and difficulties. Add gay marriage to all that, and you had the ideal opportunity for an anti-immigration, anti-Europe protest party.

This was a European (and, eventually, a global) phenomenon. The Swedish prime minister Fredrik Reinfeldt and his government lost the election that month because they'd ignored the dissatisfaction with growing migration and tried to actively embrace it. In France, the Front National was taking off because François Hollande had failed to get to grips with migration.

As with Scotland, I wanted to be a centre-right leader who addressed the problem rather than ignoring it. Yet again, I had political pressure – the rise of UKIP – born of a real problem: rising immigration and the EU's obstinacy. A migration crisis was brewing, and it was only set to get worse – just as we were heading into a general election.

41

The Sweetest Victory

Election victors often say, even if they had been surrounded by doubt, that they always knew they'd come through in the end. I can't say that about 2015. At all. Although I thought there was a better than 50 per cent chance of us being either the largest party in a minority government, or in a coalition of two, or even three, parties, I never gave a majority much more than a wistful 'if only' thought. And anyway, when 2015 dawned, I wasn't thinking about the outcome, only the input. It had been a long road to this defining year, and now we were on the home straight. I braced myself for the hardest slog of my life.

I wondered if elections these days were as gruelling as what my predecessors put themselves through. I thought about previous leaders facing sometimes six public meetings a night, talking without amplification to large crowds, with no helicopters or planes to get around in, and no mobile phones or internet to communicate with. In the 1950 election, Clement Attlee was famously driven on a thousand-mile tour of the country by his wife in their family car (before, apparently, she calmly went to the polling station and voted Conservative).

But politicians back then were not followed by twenty-four-hour news, or scrutinised by social media. They did not have a rainbow of parties to contend with, or a tide of anti-politics and apathy to swim against. I was faced with a more presidential campaign in a less deferential age.

Roads are a favourite metaphor of politicians, and would provide the subject of our first general election poster campaign, which I launched in Halifax on 2 January. 'Let's stay on the road to a stronger economy,' it said, depicting a long strip of tarmac stretching through a rural landscape. The first thing I asked before I approved the poster was where the image came from. I was assured that it was an amalgamation of roads from this country. It turned out after I'd unveiled it that it was in fact just one road: in Weimar, Germany.

A bumpy start to a big year – the gibes wrote themselves. But I had so much faith in Lynton that one pothole didn't bother me much. In fact, it probably bothered *him* more. As I've said, he couldn't stand 'process stories', where the press fixated not on what we were saying, but how we were saying it – all the behind-the-scenes stuff that turned politics into reality TV. Which is why he hated TV debates: the coverage was all about how the candidates looked and behaved, rather than the substance of what they said.

I believe debates are part of the fabric of elections now, and am proud of my role in making that the case. Theresa May's refusal to take part in the debates in 2017 damaged her, and I don't believe any prime minister or leader of the opposition will make the same mistake again. However, after my 2010 experience the downsides were obvious to me, so I saw Lynton's point.

First, while there may be a moment of truth during a debate, or an answer that underlines someone's position, ultimately, like much reality TV, they are largely confected, scripted and rehearsed. And they take up so much of your time.

Second, they give an artificial leg-up to the underdog. As the incumbent, it was all risk for me. If I took part in a debate that included UKIP, it would give Nigel Farage an opportunity to play the insurgent, as Nick Clegg had done in 2010. Debating Ed Miliband had similar potential pitfalls. Because the 'winner' in a TV debate is the person who wins the battle of expectations. And because expectations for Miliband were so low, all he had to do was turn up and not fall over, and he'd be deemed victorious.

I was determined that the debates shouldn't suck all the life out of the rest of the campaign, as they had in 2010, so I suggested some of them be held earlier. The broadcasters reluctantly agreed.

The issue, then, was who was going to take part. Here the broadcasters made a big mistake: they proposed the three main parties plus UKIP. You could see their rationale. UKIP had just won the European elections and had two MPs. But UKIP wasn't the only minority party that had enjoyed mainstream success, or that had an MP at Westminster. Caroline Lucas was the Green Party MP for Brighton Pavilion, and the Greens had done well in European elections. I argued that there was no justification for excluding them.

The BBC agreed. But then they made their next move, and their next mistake. Surely, they argued, with nine MPs between them, the SNP and Plaid Cymru should be included. I think they thought this would annoy me, but I said 'Great.' A row then erupted about whether the DUP and the other Northern Irish parties should take part. I just stood back, watching the ruckus I'd created. Keep going, I thought. At this rate I'd be debating Bus Pass Elvis and the Monster Raving Loonies.

Of course, being the incumbent in an election does give you the advantage of already doing the job you're applying for. But it also has its drawbacks, your vulnerability in TV debates being the perfect example. And there is also the fact that you have less time to prepare and to campaign, because you're still doing your job.

In deciding the sort of campaign we would fight, I was influenced by two key meetings.

The first was held in early January 2014 at Chequers, for the key members of cabinet, plus my core team. Craig Oliver presented Lynton's research explaining the public's view of what we had to offer, based on our record and our reputation. As we talked it through, the picture of what we should focus on began to emerge like a brass rubbing: security. Everything we were doing was about giving people security, and that should be our offering at the election. All the different shades of Conservative around the table agreed.

The second was during the G20 summit in Australia at the end of 2014. I spent the evening with my old Commonwealth centre-right coterie: New Zealand's John Key, Canada's Stephen Harper and Australia's Tony Abbott. We always got together when we could, and kept in contact in the meantime. Sometimes you need people from a long way away to give you some perspective on what's happening right in front of you. And what they said about message discipline, over about four bottles of red wine as we looked out over the Brisbane sky, was illuminating. We had to become far more disciplined. The UKIP surge and the issues of immigration and Europe were boulders in our way. We would need to roll them aside so we could get back on the right road. Preferably a British one.

March's Budget, our last of the Parliament – perhaps our last ever – gave us an opportunity to do just that. George didn't leave any doubt about how strong our economy was. Britain was the fastest-growing major economy in the developed world. Our rate of growth, he took great pleasure in noting, was 50 per cent faster than Germany's, three times faster than the Eurozone's, and seven times faster than France's. We could genuinely say we were rebalancing the economy. Investment was now growing faster than consumption. The north was growing faster than the south. And it was all being done with fairness at its heart: the top 1 per cent of earners were paying over a quarter of all Britain's income tax, the highest proportion on record.

Accusations that the jobs boom was being fuelled by low pay were shot down with an increase in the minimum wage. Tighter regulations defanged the loan sharks. The worst types of zero-hours contracts were outlawed. A massive rise in the tax-free allowance meant that people shouldn't have to rely on foodbanks or loans in the first place. The proposals neutralised so many UKIP, Labour and Lib Dem arguments – shooting so many of their foxes, as the saying goes – that the Lib Dems called it 'the fox-shooting Budget'.

The policies kept coming. Cancelling the proposed rise in fuel duty. Tax cuts for savers. A new Help to Buy ISA. These moves were popular, but I would take issue with anyone calling them 'populist'. Populism is when you conjure giveaways out of thin air – offering gains without any pain. But we'd had the pain. These gains were the product of five years' hard work.

While we were putting our Long Term Economic Plan into practice, Labour was waging a very 'values-based' campaign. Policies, such as rent controls, were chosen for the signals they sent: helping ordinary people while simultaneously targeting big businesses and vested interests. But there was a lucky-dip element to these policies. They didn't add up financially. All this played into Labour's reputation for recklessness.

They hadn't done much to counter that perception. In his no-notes 2014 conference speech Miliband had forgotten to mention the deficit, while shadow chancellor Ed Balls had only been able to name – or half name – one Labour business supporter on *Newsnight*. 'Bill . . .' he said. '. . . Somebody?' the presenter Emily Maitlis suggested. I gleefully explained at PMQs that 'Bill Somebody' wasn't a person, it was Labour's policy.

Their line of attack was that Tory cuts would 'return Britain to the 1930s and destroy the NHS'. But our narrative about Labour returning Britain to the borrowing, overspending and debt of the 2000s was more potent. And while they had prepared themselves to fight a campaign based on living standards, living standards were actually going up.

Yet we soon realised that something even more fundamental was putting people off prime minister Miliband than his pie-in-the-sky policies and addiction to spending.

After the Scottish referendum, the SNP dominated the airwaves: aggrieved, antagonistic, and on course for a big win north of the border at the general election. What worried voters was that, just as we had formed an alliance with the Lib Dems to get into power in 2010, Labour might do the same with the SNP in 2015. Our advertising agency M&C Saatchi produced a devastating poster: a giant Alex Salmond with a tiny Ed Miliband poking out of his breast pocket. Labour in the pocket of the SNP.

There was still an internal argument over whether posters were worthwhile or not. Lynton's unambiguous assessment was that they were 'bullshit'. In 2010 we had spent £7 million on posters; in 2015 we would spend just £1 million. But George and I remained adamant that they could create 'moments' in campaigns. And the poster of Miliband in Salmond's pocket did just that.

In 2010 we had spent £150,000 on all online campaigning, whereas in 2015 we spent £1.2 million on Facebook alone. Obama's campaign guru Jim Messina came over to help us. Traditionally, voter targeting had been based on stereotypes and put into practice through direct mail. Messina developed a more sophisticated system that used social media, direct mail, phone calls and face-to-face contact to identify who might vote for us, and that emphasised different messages to different people.

Labour's campaign was trying desperately to make Miliband look prime ministerial – he was permanently attached to a lectern – but people still saw him as weak. He ruled out a formal coalition with the SNP, but left the door open to some type of less formal cooperation, and did not rule out a confidence and supply agreement with the SNP until the final twelve days of the campaign.

All of this helped us to solve a problem Lynton and I had been talking about for years. The one-Member, one-constituency system often means people vote according to whether or not they like their local MP, regardless of his or her party or the state of national politics. More than ever, we realised we needed to drive home to people the national consequences of this purely locally based voting. 'Losing just twenty-three seats could mean Ed Miliband as prime minister' became my refrain. It was no longer the simple warning to UKIP-fanciers, 'Go to bed with

Farage and wake up with Miliband.' It was addressed to everyone who might have been thinking about straying: 'Go to bed with anyone but the Conservatives and wake up with Miliband *and* Salmond.'

So we had two vital things in this campaign: a great message about ourselves, and a great message about our opponents.

But then I delivered a message that wasn't part of the plan.

I had agreed to do an 'at home' broadcast feature with the BBC's James Landale – an opportunity to cover politics, life, the bigger picture, while hopefully showing people the real me. We were filmed watching Elwen play football for the Chadlington under-10s, and did the interview in my kitchen, as we chopped lettuce and Samantha pottered about in the background. Would I stand for a third term, James asked casually. 'No,' I replied without hesitating. I explained that I was standing for a full second term, that I had a lot more to do, but that there comes a time in politics when a fresh pair of eyes is needed. I cited the talents of Theresa May, George Osborne and Boris Johnson. 'Terms are like Shredded Wheat – two are wonderful, but three might just be too many,' I said, referencing the 1980s advert.

I thought I had done the right thing. I had been honest. I had confirmed, despite suggestions to the contrary, that I would serve a full second term. But I was clear that I wouldn't repeat the mistakes of some of my predecessors, whose premierships were cautionary tales of the dangers of clinging on.

However, something that I thought demonstrated staying power was seen as doing the opposite. We were talking about security, our plan, and seeing through that plan. And here was the boss saying something that sounded as if he wasn't going to be there for it.

Just because I didn't want to cling on beyond a decade didn't mean I held the privilege of power at all lightly. I was desperate to be re-elected, and deeply reflective during the final days before the campaign formally started at the end of March. Soon Parliament would be dissolved, and all our MPs would go off to their constituencies. Who would be coming back, I wondered. Almost all of my special advisers would be resigning and trooping over to CCHQ to fight the election. Would they have jobs to return to in six weeks' time? Sam had already started to pack up the cupboards in the flat, and the pile of blue plastic sacks in the hallway grew by the day. Every 'last of this Parliament' might be the last of my career. My last cabinet. My last PMQs.

That last cabinet brought our time in coalition to a close. It wasn't the end people had predicted – a fiery collapse in the early weeks, or a bitter mid-term decoupling. It was the end of a full five-year term, and a fairly harmonious one at that. I had arranged for the Wychwood Brewery in my constituency to produce a special 'Co-ale-ition' for each minister, and Henderson's in Nick Clegg's constituency provided 'Coalition Crunch' crisps.

The last PMQs – my 146th at the despatch box – got us back on the front foot on tax. Miliband started strong. 'On Monday, the prime minister announced his retirement plans,' he said, predictably referencing my faux pas. 'He said that it was because he believed in giving straight answers to straight questions. After five years of Prime Minister's Questions, that was music to my ears. So here is a straight question: will he now rule out a rise in VAT?'

There had been a question mark over this for some time, and as recently as the previous day George hadn't either confirmed or denied it. But he and I had discussed it – in fact we'd come to the same conclusion before doing so – and had decided to rule out putting VAT up again. Cutting spending hadn't been as painful as some had predicted, and although we weren't saying no taxes would go up ever again, we believed that VAT, National Insurance and income tax should not increase.

I revelled in my reply, as Sam and eleven-year-old Nancy looked on from the gallery. 'In forty-three days' time, I plan to arrange the Right Honourable Gentleman's retirement. But he is right: straight questions deserve straight answers – and the answer is *yes*.' I went on: 'This is a country where unemployment is falling; the economy is growing; the deficit is coming down; in our NHS, the operations are going up; there are more good school places for our children; living standards are rising; inflation is at zero; and there are record numbers in work. All of this could be put at risk by Labour. That is the choice in forty-three days' time: competence and a long-term plan that is delivering, instead of the chaos of economic crisis from Labour.'

The classic election campaign offer: Britain is on the right track, don't turn back.

The following week, Parliament was dissolved. This was it. The final stretch. And I would be spending most of it on a giant, sky-blue battle bus. There were planes and helicopters too. To mark one month before polling day, Sam and I flew to Edinburgh to begin a one-day tour of the UK's four nations, ending at the Royal Cornwall Showground for an evening rally.

On the way home we stopped off at Jon's Fish and Chips in Wadebridge with the candidate for North Cornwall, Scott Mann, and his partner Nicki. It had been an incredible day. I'd made a pie and sipped stout in a brewery. I'd even sat on the *Game of Thrones* 'Iron Throne' in Northern Ireland. But the real highlight was sitting on a picnic bench by the River Camel, wooden forks in our hands, getting to know this lovely couple.

Scott had been a postman in this Cornish town for nearly twenty years. Now he was hoping to make his postal patch his political patch. He was down-to-earth, decent, the most modern and compassionate of Conservatives. This is what it's all about, I thought, all the rallies and interviews and stop-offs and drop-ins. It was about getting people like Scott onto our green benches. Changing our party so we could change our country.

The only interruption to these happy days on the road was the first and only TV debate I'd have to do with my fellow party leaders. In the snap polls afterwards I came third behind the SNP's Nicola Sturgeon and Farage. I'd certainly won my own battle of expectations. And now I could get back to the campaign.

I couldn't remember an election when we'd had so many perfect elements in place – the team, strategy, policies, messaging. Trust in the economy was up. Trust in my leadership was up. But the main polls didn't seem to move. Lynton said to me, 'If the election was tomorrow, PM, you'd lose.' In fact, the only change we did see was Labour taking a 3 to 6 per cent lead in the middle of April.

Things felt good on the ground – better than 2010. But the polls just weren't breaking our way. Perhaps our manifesto, which we were launching that week, would help shift the dial. The three policies from it that we decided to announce during the launch were selected to emphasise the real, tangible, life-changing benefits of our economic plan.

Tenants of housing association properties would be able to buy their homes under Right to Buy. (Previously it was mostly council housing tenants who had this right.)

Parents of three- and four-year-olds would receive thirty hours of free childcare a week, doubling the existing allowance set by Labour.

And anyone earning the minimum wage – at this point, 1.4 million people – would never have to pay income tax again, with legislation stating that the tax-free personal allowance should always increase faster than the incomes of the lowest-paid.

What we were striving to deliver, I said in my speech, was 'the good life'. I meant this primarily in an Aristotelian sense, and it was a mantra I often came back to when I was thinking over policy decisions. But it didn't hurt that the papers presented it in a grow-your-own-veg 1970s-sitcom sense. It was exactly the sort of simple, sunny, down-to-earth tone we were going for.

And yet still the polls didn't budge.

Perhaps reaching out to Britain's minority communities would swing things our way. Opportunity for all – all people, from all backgrounds – wasn't something I dreamed up only at election time. My entire leadership had been founded on raising people's chances while widening our appeal as a party.

During five years in office, we had put that into practice. There were the policies dedicated to removing barriers, such as English-language tutoring (a quarter of British Muslim women didn't speak English). And there were the mainstream policies, from start-up loans to Free Schools to Help to Buy, which were designed to lift everyone up, but actually had a disproportionate effect on minorities. We finished the Parliament with more ethnic minority students at university than ever before. People from those backgrounds were four times more likely to receive start-up loans. A third of participants in National Citizen Service were from minorities, compared with 13 per cent of the population.

CCHQ was at the forefront of this effort. We launched 'friends of' groups to engage with Indian, Pakistani, Sri Lankan, Bangladeshi, Turkish and Tamil communities and more. I was a regular at mosques and temples and gurdwaras, and the election was no exception – I even broke my 'no hats' rule to don an orange headscarf for the Vaisakhi parade in Gravesend in April.

The boldest endeavour in this area was something I hadn't done before. Church attendance among the wider population was dwindling, but legions of worshippers who gathered in warehouses and former bingo halls every Sunday morning were making up for the decline. Many people from these communities had already found a home in our party – British Nigerians were prominent in our associations. But I was sure there were many more potential Conservative voters among these congregations.

In one of the most thrilling episodes of the campaign, I spoke at the Festival of Life, the biggest Pentecostal event in Britain. Going out on stage after 11 p.m., in front of the 45,000 worshippers, I did my best to channel my inner pastor: 'When I was a child, I had a very specific image of what a church was. I thought, to be a church, it had to be an old grey building, with a slate roof and a big spire . . . that it had to have pews and a pulpit . . . But I was wrong – and you prove that. You prove that church is people – that church is a family. And it doesn't matter what the roof is made of, because with your energy, your devotion, your love of Jesus Christ, you raise that roof every time!'

I loved that speech, and I was keen to get onto the next one, which would be about my vision for the year 2020, specifically how we were going to increase the proportion of people from ethnic minorities in higher education, apprenticeships, the armed forces and the police. I also set a target of 20 per cent BME Conservative candidates in winnable seats.

The stage was set early one Saturday morning in Croydon, the marginal seat of one of our hardest-working MPs, Gavin Barwell. I was working up to my favourite part of the speech: a section about what makes Britain the proudest multiracial democracy on earth. 'We are a shining example of a country where multiple identities work,' I told the audience. 'A country where you can be Welsh and Hindu and British. Northern Irish and Jewish and British. Where you can wear a kilt and a turban. Where you can wear a hijab covered in poppies. Where you can support Man United, the Windies and Team GB all at the same time.' Then I ad-libbed: 'Of course, I'd rather that you supported West Ham . . .'

Wait.

I support Aston Villa.

You know when you say something, and you think: did I just say that? You want to rewind. But you can't. It's out there – there's nothing you can do. And if you've got an autocue in front of you, you just have to keep on talking.

So often we misspeak, usually when we're trying to do two things at once. We say 'See you at ten' when we mean eleven. We call someone the wrong name, and don't even notice we've done it.

But this was me, an Aston Villa-supporting prime minister, implying that I supported the wrong team. On camera. On a Saturday. Match day. In a country obsessed with football. Two weeks before a general election. The incident made me look phoney, as if I had always pretended to support Aston Villa, but didn't really. The opposite was the case. I hadn't said a huge amount about being a football fan, but I really was one, and I followed Villa more closely than I let on.

Could it be that we'd flown over West Ham's football ground on our way back to London last night? Or that my political press chief Alan Sendorek and I had been chatting about his team, QPR, facing West Ham that day? Perhaps that had something to do with it, but the thing I most remember was that while I had just said 'Windies' it was 'West Indies' that was written on the autocue, so I had 'West' lodged in my head. I was also beyond exhausted. Brain and mouth had completely disconnected.

'How bad was it?' I asked Liz when we were back in the car.

'Really bad,' she said.

As we ruefully discussed this cock-up on the journey back to Chequers, all our dissatisfaction with the campaign tumbled forth. Even though we were one team with one head, it did feel that there were several competing arguments inside it.

Lynton didn't want any new announcements. He wanted me to go around repeating the message ad nauseam. And he was right. But Craig Oliver said if you've got nothing new to say, you won't get on the news – and he was also right. George Osborne wanted everything tightly controlled, because the potential for disaster was so great. He was right too.

The result was lots of scripted speeches at more formal, sit-down events, with a carefully coordinated backdrop, and a total ban on walk-abouts. Of course, you could say that the West Ham ad lib was an argument for being even more controlled in what we were saying and doing. But Liz and I felt the problem was that there was too much new written material with complex policy ideas. The low point was ending up in a shed factory in Bedford talking about tax-credit taper rates. It wasn't me.

Meanwhile, a press narrative was developing that this was the most boring campaign ever. Partly, they were goading us into taking more risks. But I think there was a truth to it too.

With just ten days to go, Liz and I decided things needed to go up several gears. We met Craig and Lynton at Chequers, and agreed a rethink. No lecterns, no autocue, no scripts. More real people, more rallies, and much more energy. We could use the press's narrative about the campaign being boring to put more fire into it. More passion.

That in itself was slightly contrived. I wasn't tearing up the script – I'd been doing rallies off the cuff anyway, and I'd still be using notes. But I wanted to make people believe that I was making a switch, turning up the energy. This was a process story, the kind Lynton hated. But process seeps into the wider consciousness. Voters would see a more energetic PM, partly because I was laying it on thicker, but also because the media would be presenting it that way.

Soon I was launching our Small Business Manifesto, to coincide with a letter in the *Telegraph* in which 5,000 small-business leaders endorsed the Conservatives. It was a hot room packed with young entrepreneurs, and everyone was standing with their phones in the air like at a gig. 'Taking a risk, having a punt, having a go – that pumps me up!' I told them, talking about the genuine thrill I got from seeing people grow their businesses. I don't think the Institute of Chartered Accountants headquarters had ever witnessed such fervour.

This gave me the confidence for the final 'TV moment', which was a special format of *Question Time*, with Miliband and me being quizzed separately by the audience, moderated by David Dimbleby. This time I brought a prop. Liam Byrne's 'there is no money' letter was tucked in my inside pocket, ready to whip out at any suggestion that Labour could possibly be trusted with the economy again.

How did Miliband do? Well, his big error was to say that he didn't think the last Labour government had spent too much money. The audience actually gasped.

In the green room beforehand we had said, 'Yes, but remember: all he has to do is turn up and not fall over . . .' But then it happened. He tripped off the stage as he left. His stumble was so slight you might have missed it, but in the internet age every gaffe becomes a gif, and every split-second slip is out there forever. It was ungenerous of me to take pleasure in it – I was lucky that Liz had pointed out the trip hazard to me beforehand – but I was so knackered, things were so tight, and we were so close to the finishing line that I was looking for wins wherever I could get them.

Once we were into May, we were motoring. Rallies at Asda HQ in Leeds and at Neasden Temple in north London. To Twickenham, where Vince Cable had a 12,000 majority. To Nuneaton, the bellwether seat we'd be fixated on during election night. We were working so hard. Four hours' sleep a night. Non-stop during the day.

One evening we stopped for a team dinner at the Holiday Inn in

Manchester. 'David! David!' we heard a Lancastrian accent shouting. I could see Liz getting ready to rugby-tackle the interloper when we realised it was the comedian Peter Kay. 'David, are you all right, mate?' he asked. 'Do you need a hug?' Actually, I did. He put his arms around me and told me to keep going. It wasn't a political thing. It was one middle-aged bloke who spent his life on the road saying to another that he understood.

Then came Labour's big announcement. And what an announcement it was. You could see how it came about. They looked at their weaknesses with voters: they had no plan, and while people liked their promises, they didn't believe them. So they decided to kill these two birds with one stone: an eight-foot granite stone. The tablet, quickly dubbed 'the EdStone', was inscribed with six platitudinous pledges, such as 'A strong economic foundation' and 'An NHS with the time to care'. One person dubbed it 'the heaviest suicide note in history'. It gave us a laugh, but it also gave us a kick. We were outraged that Miliband said he wanted to put this eyesore in the garden of Downing Street, and spooked by the possibility of it happening in just five days' time. We needed to do everything we could to stop it.

With every last ounce of energy we hit the final stretch: a journey from one end of the country to the other, starting in Cornwall and ending, via Dumfries, in Carlisle. Over thirty-six hours we saw round-the-clock Britain in microcosm, through walkabouts in town squares, rallies at businesses, and visits to a zoo, a nursery, new homes, a farm and a UPS centre at midnight. We finished up with a huge rally in Carlisle. As I stepped off my prime-ministerial battle bus to greet the crowds, I knew that it really could be the last time.

After that, there was nothing more I could say or do. Sam, Liz, Kate and I got on a helicopter. Someone had given me a bottle of wine, and I opened it as we took off. We didn't have any glasses, which meant downing the contents of our water bottles and decanting the wine into them.

After casting my vote at Spelsbury Memorial Hall early the next morning I joined Michael Gove, William Hague, George Osborne, Ed, Kate, Craig, Liz and Ameet Gill at a venue we had borrowed to plan for potential outcomes.

Scenario one was that we'd have just enough MPs to form a coalition. We worked out what to do in that case: react strongly; say that the Conservatives are the largest party and can therefore provide strong and stable government. We had discussed continuing as a coalition with Nick and Danny, and Oliver Letwin was beavering away on a document, but that was the extent of our preparation. The Lib Dems would have to agree to our promise to hold a referendum on the European Union, but I didn't think that would be an issue.

Scenario two was the zone of uncertainty – the combination of us plus the Lib Dems and the DUP adding up to less than 323. This would be very difficult, because although we'd be the largest party, the SNP plus Labour would be able to stop us forming a government.

We decided that in that case it would be wrong for me simply to announce that we would struggle to form a government, and therefore concede defeat – particularly as we would still, by some distance, be the largest party. Events should be allowed to play themselves out. After all, Labour had ruled out formal cooperation with the SNP, and if they were going to go back on their word, it should be done in plain sight.

As the Conservatives would be the largest party, I would say that in the national interest I was prepared to lead a minority government for a year. I would present a Queen's Speech that would attempt to reach across the aisles. We would propose to abolish the 'bedroom tax', introduce the enhancements to Scottish and Welsh devolution, and because all the major parties accepted the spending review for 2015–16, we'd implement those plans and hold an election in March the following year.

Scenario three was coming behind Labour in any form. I was very clear that even if we were only one seat behind them, that was it. I would immediately resign as PM and party leader.

Clare Foges went to the sitting room to work up the speeches for each of the three outcomes. That afternoon we all sat in the garden at Dean as I read them aloud. 'I'm leaving Downing Street for the last time, we wish Ed and Justine the best, they'll find behind that black door very professional people who will do everything . . .' I was touched that they were all in tears – and I was pretty choked up myself.

My mind was still on the polls, or rather one poll, the exit poll, which was still several hours away. I decided to go for a walk through the fields around Dean, and Ed and George joined me. I put something to them. 'We've thought about coalition, we've thought about losing. Let's imagine we get 315 or more, enough not to need the Lib Dems . . .' This was the scenario no one had even mentioned. Ed told me not to be ridiculous.

I was getting updates through the day. Lynton said we could get three hundred. Stephen Gilbert was sticking to 293. Jim Messina was predicting over 316, but no one told me that, because they didn't want to get my hopes up.

In my head, I had been on 293 seats – lose twenty, win ten was my guess. But as the clock ticked down to 10 p.m., my prediction sank. Tories vote early and Labour supporters vote later, so as the day went on and we received reports of a rising turnout, Labour began to do better and better. Craig was getting calls saying that Labour was starting to brief journalists that, constitutionally, it is who can command Parliament's respect, not who is the largest party, that forms a government. I just kept thinking about all the seats we were going to lose. Would we even be in the 280s?

I made pizza for everyone that evening before we took our places in front of the TV in Dean, in almost exactly the same spots where most of us had sat five years before. The bongs of Big Ben were followed by the voice of David Dimbleby revealing the exit poll. 'Ten o'clock. We are saying the Conservatives are the largest party.' Everyone in the room leapt up. I tried to remain calm. Good. We thought this would happen. What mattered now was whether it was true, and if so, the size of our lead.

The predicted numbers flashed up on screen: a 3-D me with 316 seats, and a 3-D Miliband with 239. More cheers. But I was confused. That was just too good. No one had predicted 316. Craig reassured me. He had edited the BBC's election-night coverage before, and knew John Curtice and the modelling he did. If anything, he thought he was underplaying it.

At about half past midnight, Swindon North vindicated that confidence. We'd held on to the key Labour target with a massive, unexpected swing.

At 2 a.m. I became convinced we were on our way to victory when the critical seat of Nuneaton saw an increased Tory majority.

Then there was my own count, taking place at the Windrush Leisure Centre in Witney. In a room full of gym equipment, my constituency team – Tash Whitmill, Rose Rawlins and Julia Spence – and I gathered in front of a small TV that was showing the biggest election upset in recent years.

The SNP were steadily snatching Labour strongholds, including Gordon Brown's old constituency, Kirkcaldy. They unseated all the Scottish Lib Dems but one. What the SNP were doing to Labour in

Scotland, the Conservatives were doing to the Lib Dems in England. Even Vince Cable lost in Twickenham.

As my own results were about to be called, I took my place on the stage alongside the giant Elmo, the Wessex Regionalists and a guy dressed as a sheikh. I was returned as MP for my beloved Witney, with a bigger majority. I would be back with my amazing constituency teams, including the wonderful Caroline Balcon, my longest-serving staff member.

The speech I gave would be important – it would set up what I'd say later at Downing Street. Craig whispered to me: 'One Nation. Start with One Nation.' I told the crowd: 'I want my party, and a government I would hope to lead, to reclaim a mantle that we should never have lost, the mantle of one nation, one United Kingdom. That is how I will govern if I am fortunate enough to form a government in the coming days.'

There was no longer any 'if', however. Sam and I were driven back to London, helicopters above us, police outriders around us. I was crying. It was a sense of release more than anything else. I hadn't realised until now how great the weight of it all had been.

But before we went home to the flat in Downing Street, there was somewhere I needed to go. CCHQ was my political home. It was the organisation I'd worked for as a young graduate, and where I'd first led a team. I'd been in and out of its various locations in my various roles, as a spad, a candidate, an MP, a party leader, a prime minister – and now the first Conservative prime minister in twenty-three years to win a majority.

'I am not an old man,' I told the assembled staff. 'But I remember casting a vote in '87, and that was a great victory. I remember working, just as you've been working, in '92, and that was an amazing victory. And I remember 2010, achieving that dream of getting Labour out and getting the Tories back in, and that was amazing. But I think *this* is the sweetest victory of them all.'

In the end, 11.3 million people voted Conservative, 600,000 more than in 2010. But it was all the other victories within that which made me so proud.

We finally won the south-west, defeating all fifteen Lib Dem MPs in the region. For the first time since 1970, you could walk directly from Land's End to Westminster and never set foot in a non-Conservative constituency.

We won more Welsh seats than we had for thirty years, increasing the number of our MPs from eight to eleven, taking Gower from Labour for the first time ever.

We held on in Scotland, ending up with the same number of MPs there as the Lib Dems and Labour.

We won the support of modern Britain, with over a million votes cast for us by people from ethnic minorities – an estimated one in three people from those communities. In fact, Sikhs and Hindus were more likely to vote Conservative than Labour – up from 11 per cent in 2005 to 49 per cent. Gay voters, for the first time ever, were almost as likely to vote Conservative as Labour.

Our party was more reflective of modern Britain, as well. In 2005 we had just two BME MPs; now we had seventeen. Then, there were just seventeen women MPs; now there were sixty-eight.

The stories of our new MPs were the story of modern Britain. On our benches we now had people like Seema Kennedy, who was four when she and her family were forced to flee revolutionary Iran. Johnny Mercer, who a few years earlier had been a soldier in Afghanistan. And of course Scott Mann, who just days earlier had been delivering the post in Cornwall.

The moderate, reasonable sensibleness of modernisation which so many people had said would bring down the party had won us new support. Some said I had helped UKIP's cause through my support for gay marriage and the green agenda, yet it didn't split the right – even though UKIP had won 13 per cent of the vote. I think it's because that modernisation was underpinned by those small 'c' conservative issues. It was that, plus our economic stewardship and an effective campaign, that finally sealed the deal with the electorate.

42

A Conservative Future?

I bounded up the staircase and into the flat. Inside, Sam was getting the kids ready for school. Blue bags still filled the hallway, ready for our potential move. But now I knew we were staying put, while the rest of British politics moved around us.

I thought I should try and close my eyes for half an hour that Friday morning to fortify myself for what was ahead. Results were still rolling in, so I switched on the radio by my bedside. As I drifted off, I listened, stunned, as the shadow chancellor, my Commons tormentor Ed Balls, lost his Yorkshire seat. Then I woke to hear that Nigel Farage had failed in South Thanet. And – joy of joys – we *had* kicked Mark Reckless out of Parliament.

Then, within the space of one hour, Farage, Nick Clegg and Ed Miliband all resigned as leaders of their parties.

Amid the tumult, however, there remained a reassuring fixed point in British life. Prime ministers come and go, but there is always the monarchy. Being re-elected does not constitutionally compel you to see the sovereign in person, but I wanted to do so because I thought it was right, not least because this would be a different government. So that lunchtime Sam and I drove to Buckingham Palace, our tally of MPs now 325, just one off a now-assured outright majority.

I was taken through to see the Queen, trying desperately to suppress my excitement.

I said to her that five years ago when I came here, I couldn't tell her what sort of government I was going to form, but this time I could tell her that we were going to have a majority Conservative government. I joked that it might be a small majority, but said it was getting bigger all the time.

Though I didn't know it, at that moment the 326th seat, my neighbouring constituency of The Cotswolds, was being declared for the Conservatives. It was the first outright Tory win since John Major's surprise victory in 1992. We would end up with 330 MPs – what I called my 'small but perfectly formed majority'.

I returned to Downing Street and stood before a wall of cameras to make a speech. Five years earlier I had set out a plan to rescue our economy in the wake of the financial crisis. Now my mission was to heal social divisions, partly created by that crisis, partly built up over many decades.

Step back and Britain was richer, safer, greener and fairer than ever before. But zoom in and you could see the cracks running across the country. The prosperity gap between north and south. The opportunity gap between rich and poor. The barriers to true social mobility. The separatist sentiment in Scotland. The failure of some immigrants to integrate in some communities. The discrimination based on gender, race and religion. Governing for one nation was about mending those fissures.

Had the bookies been right, it would have been Ed Miliband stepping through the big black door. EdStones notwithstanding, he'd fought a strong campaign and I told him so that morning over the phone. He'd done what an opposition leader is meant to do: holding us to account, forcing us to move on policies. He was decent and magnanimous in defeat.

That afternoon he would be making his final official engagement as Labour leader when he joined Nick Clegg and me at the Cenotaph for the seventieth anniversary of VE Day. I felt terrible for Nick. The Liberals had been almost totally demolished, and I knew that was the one thing he didn't want. He desperately wanted to leave his party in a reasonable state. Instead, they were down from fifty-seven MPs to just eight.

We had texted each other a bit through the night – good lucks, well dones, and commiserations, particularly over Danny Alexander's ousting and Nick's inevitable resignation. It was the only fly in the ointment for me – that my triumph was my friend's tragedy.

Now the campaign was over, my new obsession was the size of the task ahead. Governments succeed or fail on whether they make a good impression in their first three months. Even though I was exhausted, it would be foot to the floor for the first hundred days.

On the Saturday after the election the congratulatory phone calls were still flooding in. The difference between now and 2010 was that many of these people weren't just world leaders; they had become friends, particularly Enda Kenny, Angela Merkel, Mark Rutte, Matteo Renzi, Stephen Harper and Barack Obama.

Merkel was full of praise. She said how much she'd enjoyed working together, and looked forward to the next five years. I reflected upon how different the Merkel I had come to know was to the Merkel of popular imagination. People perceive her as collaborative and cooperative – which she is – but this is combined with a ruthless streak. There wasn't much talk about the EU that morning, but I knew she'd be tough on what was to come.

Where Europe did arise was with Obama. Having talked to him a lot about the election in the run-up, it was always clear that he'd thought I'd make it back, one way or another. But after the initial 'Way to go's, his tone changed. 'I hope you've got a plan to keep Britain in Europe,' he warned. 'We want you to stay in the EU.'

The European referendum was at the top of my mind. It was a manifesto promise, people had voted for it, and I would deliver it. It would be my biggest task. But it was also one of many policies from the manifesto – and beyond it – that I planned to deliver in this radical second term.

I couldn't do it alone. I needed the support of parliamentarians to turn these policies into law. It was a small majority. Every vote would count. The first Monday morning back in the House of Commons, I went to address the 22 in Committee Room 14, just as I had in 2010. Often the desk-banging in these meetings is fake – organised by the whips so the press outside can hear. This time it was for real.

There was also the matter of a fully Conservative cabinet to deliver those policies across Whitehall. Before anything else, I talked with Michael Gove about his role. He confirmed that he wanted to move to a department, somewhere he could get his teeth into reform again. Overhauling the prison system would be a big part of my agenda, and he leapt at the possibility of being justice secretary. That would depend on the shuffling I could do elsewhere, so I asked him to sit tight until I sorted it.

In the rest of the reshuffle there were aspects of continuity. George Osborne, Theresa May and Philip Hammond all remained in the great offices of state. But there would be one change for George. When I came to reappoint him as chancellor he asked if my offer to make him first secretary still stood – it was something I'd put to him around the time William had said he was standing down. I was happy to confirm that it did.

There was a case for swapping May and Hammond. But as he hadn't been in the job long enough for that to be fair, I thought I'd wait.

There were the obvious moves, a happy necessity of not needing to leave space for the Lib Dems. David Mundell had shadowed the Scottish secretary job throughout my entire time as opposition leader, and had been the number two in the Scotland Office throughout the whole coalition. It was a very long apprenticeship, and I was delighted finally to give him the top job.

There were the sensible rotations. Chris Grayling seemed a good fit for leader of the House of Commons (he hadn't excelled at Justice, but getting rid of him would anger the right). With his strong business career, Sajid Javid slotted straight into the job of business secretary.

There was some innovation. John Whittingdale, who had once worked for Margaret Thatcher as her political secretary in No. 10, was the chairman of the DCMS Select Committee, and would now lead the department. He knew the brief, reassured the right, and would be a highly capable minister.

There were also some curveballs. No one in my team could see why I wanted to promote Dominic Raab to the Ministry of Justice, after his amendment on human rights legislation had provoked a rebellion of eighty-six Tory MPs in 2014. Yet I knew we'd be better off taking advantage of his ability than suffering because of it. He was clear: 'You have put your trust in me and I will not let you down.'

What to do with Boris? He had re-entered Parliament as the MP for Uxbridge, and his term as mayor of London would end in May 2016. I got him into my office to talk about the options. I wanted him firmly inside the tent from the start. But it became clear that he couldn't sit in cabinet while still being mayor, so we came up with the idea that he could belong to the political cabinet, which would continue to meet every week before the main event.

It was time to harvest the crop of talent from the seeds sown over the past decade or more. Amber Rudd, who had excelled as a junior minister at Energy and Climate Change, would lead that department. Royal Naval reservist Penny Mordaunt became minister for the armed forces (the first woman to hold that position) and Anne Milton became deputy chief whip (another first). Both Anna Soubry and Priti Patel became senior ministers who would attend cabinet, taking the tally of women ministers up to a third. We were steadily ending the idea that the Tories weren't open to talent, to women, to minorities, while at the same time slaying the myth that outright positive discrimination was needed to get to the top.

It was the right time for Eric Pickles and Francis Maude to step down, and they did so graciously. It was also the right time for IDS to go. I was getting increasingly frustrated with the costs of Universal Credit and his failure to come up with plans to reform and reduce spending. But what dissuaded me was – again – the potential backlash from the right. As we'll see, that was a big mistake.

The first policy speech I made in the new government was on health. The Lansley reforms had been too much of a solution in search of a problem. Now we would focus on more tangible, understandable changes that would make the health service friendlier and more flexible. Our headline manifesto promise of a 'seven-day NHS' was a big part of this. Of course, in a literal sense the NHS already operated on a seven-day basis: accident and emergency departments and most hospitals were open twenty-four hours a day, seven days a week, for 365 days a year.

But the NHS was not offering a *genuine* seven-day-a-week service. GP appointments and most non-urgent operations were restricted to week-days. Out-of-hours service was patchy and often of poor quality. With GP surgeries often shut at the weekend, A&E departments came under huge pressure. People don't just get ill during office hours, yet the staffing rotas in hospitals often meant that patients on wards had less access to the most highly trained doctors. There was compelling evidence that mortality rates went up at the weekend.

I knew from my own experience with Ivan that patients and their families got a less good service as the week came to an end. This included the all-important 'magic moment' when you are well enough to go home. If you didn't get discharged on a Friday, you would likely have to wait till Monday to get out of hospital, blocking up beds and putting the NHS under ever more pressure. It was a vicious circle.

So we would attempt this latest reform of the NHS: creating a seven-day health service. This would inevitably involve reforming the pay of junior doctors, a term that covers everyone from newly qualified medics to senior registrars, and who make up about half of NHS hospital doctors.

At that point, 'normal working hours', for which doctors received their usual rate of pay, were 7 a.m. to 7 p.m. on weekdays. Anything outside those hours was paid extra. If we were to guarantee proper cover at other times, we would not be able to afford the existing generous over-time rates. A sensible change would be to extend 'normal working hours' to 7 a.m. until 10 p.m., and to include Saturdays from 7 a.m. to 7 p.m. This was hardly outrageous. Plenty of professionals work shifts that extend late into the evening, and Saturdays, and are paid the same rate for those times as for weekdays. There would still be overtime pay for really antisocial hours – nights and Sundays. What's more, doctors had already had a 48 per cent pay rise in the decade before we took office. And, crucially, we would increase their basic pay by 13.5 per cent to compensate for the loss of this overtime.

And yet I knew it wouldn't be easy. As I knew all too well, the doctors' union, the BMA, was a fearsome opponent. I knew the NHS was sur-rounded by wild myths, for example that the Tories wanted to privatise it. I knew that, while we'd drawn neck and neck with Labour on the NHS in opposition, the Conservative Party had baggage on the issue, and was now some way behind.

I charged Jeremy Hunt with delivering the change. I knew he would be unwavering, and when the heat on him increased – which, the follow-ing year, it did – I would stand by him all the way, including during the first strikes by junior doctors in forty years. (The eventual contract would give a lower pay rise, but created a more complex system of compensated overtime.)

The second policy speech I made was on immigration. This was one of the areas where we'd been held back by being in coalition – specifically on delivering our 2010 manifesto target to get the number of net new arrivals under 100,000 a year.

I believe that people coming from other countries to live and work in Britain are vital to our success as a nation. But you can only maintain consent if those arrivals happen at a rate that allows for integration. Indeed, if you simply say 'Let everyone in,' you're actually harming the cause you claim to champion. I wanted to *control* immigration because I am *pro*-immigration. That's why we reiterated the 100,000 target in our 2015 manifesto – and now we could pursue it with renewed vigour.

Net migration fell for the first two years after 2010, as some of the measures to control migration from outside the EU took effect, particularly the cap on economic migration. But from late 2012 it had started a seemingly inexorable rise, first with increases in EU migration, and then even bigger numbers in terms of non-EU migration. The first Bill of the new government would be packed with measures to try to bring this under better control.

Policies to counter extremism that would simply have been vetoed by the Quad were back on the table too, as was a permanent commitment to maintaining defence spending above 2 per cent of GDP.

And yet to truly become one nation we would have to end the financial hardship and blocked opportunity I'd talked about on the steps of No. 10.

There was a particularly perverse set of circumstances for low-paid people that I was keen to address. Over the years, an absurd system had grown up whereby workers on the minimum wage were having their pay taxed by the government and then given back to them – plus more money – in benefits. It was essentially a money-go-round. The result was a welfare budget that was unaffordable, and that gave out the message, in many cases, that work didn't pay. That was both unfair and unsustainable. We needed to transform Britain from a low-wage, high-tax, high-welfare society into a higher-wage, lower-tax, lower-welfare society.

On top of that, we needed to take action on the problems that prevented people getting good jobs and living a good life in the first place. Problems like family breakdown, poor schooling, debt, addiction, crime, a broken care system and a lack of training.

Our proposals contained things the Lib Dems would have jumped at – like tackling mental illness – and things they would've baulked at – like promoting marriage. They tied together everything we were doing, from cutting the deficit to reforming adoption.

To have any chance of implementing this 'life chances' agenda, we would need to lay the financial foundations in our opening Budget. Immediately this raised a major question about the deficit and public-spending reductions. The manifesto had promised £30 billion more savings over the next two years, and that £12 billion of these cuts would come from the welfare budget. We had to decide whether we should stick to this or slow the pace, pushing our already-postponed deadline for a budget surplus back to 2019.

George just wanted to get on with the welfare changes – to do what we said we'd do within the timeline we'd set out. But I believed that, having earned a reputation for fiscal responsibility, nobody was going to think us feeble for pushing it back a short while. The reputation I badly wanted us to secure was for centre-ground, compassionate, sensible One Nation politics. Going too hard and fast on welfare cuts would risk that. When the crunch meeting came, I put the case for making the welfare cuts over the length of the Parliament. George eventually agreed. But even with this scaled-back programme we were still storing up future troubles with IDS.

The next reform to push through was higher pay, starting with an increase in the minimum wage. At a dinner a few months before, the veteran businessman Sir John Hall, former owner of Newcastle United Football Club, put it bluntly in his strong Geordie accent: 'David, the working people of this country have earned a pay rise, but they're not getting it. Why don't you give it to them, man?'

For too long businesses had got away with paying low wages because they could rely on government topping them up with tax credits. We had been wrong to oppose Labour's introduction of the minimum wage in 1999 – claims that it would increase unemployment had turned out to be incorrect, and businesses were actively supporting it. Now, if we were going to break the cycle of the low-pay, high-benefits economy that was helping to push up welfare bills, we needed that minimum wage to increase.

Labour had promised an £8 minimum wage by 2020. Thanks to the difficult decisions of the past five years, we were the ones that could actually deliver it. Indeed, because the economy was strengthening, the labour market was operating effectively and businesses, with lower corporation taxes, were doing well, we could go further. A compulsory £9 an hour living wage for over-twenty-fives by 2020 – starting at £7.20 the following year – would be the centrepiece of the Budget. When George announced it from the despatch box, it was an electric moment for the Conservative Party. We were finally ending that money-go-round, promoting a greater sense of fairness in our society and faith in our economic system – the actions of a true One Nation government.

If 2010–12 was about finding my feet on the world stage and 2013–14 pushing the agendas I cared about, then 2015 onwards could be the period to enhance Britain's place in the world still further.

Yet my first overseas trip, to the G7 in Bavaria, was overshadowed by a huge row with our greatest ally.

George had been both creative and aggressive in looking for ways to make our relationship with China more meaningful. We agreed – against the advice of the Foreign Office and without the agreement of our traditional allies – that Britain should join the new Asian Investment Infrastructure Bank (AIIB) which the Chinese wanted to establish, based in Beijing. The Japanese objected – vigorously – because there was already an Asia Development Bank, and they feared China's rise. The Americans disliked the plan because they felt it would enhance China's power in the region.

Our view was different. We believed that it was unrealistic to argue that China should take full part in the global rules-based order – joining and paying into institutions we had set up after the war – while at the same time saying that it could never add to these institutions with ideas of its own. More to the point, by joining the new institution at the start, we believed that we could help ensure that it would have proper rules and governance. So we became the first major Western nation to endorse the fund.

It was only when I was sitting around the G7 table with Obama, Merkel and Hollande that I realised Obama's officials hadn't told him about our move. He reacted sharply. 'David, I think you're wrong and naïve, and I can't believe you've broken with Western allies in doing this,' he said.

We had been doing this Treasury to Treasury, with our US counterparts fully informed – it wasn't my fault his officials hadn't told him. For the American president to criticise Britain in front of the French and German leaders in this way was unprecedented, and deeply embarrassing.

Events have proven we were right to join the AIIB. Today it is well established, with rules, governance and personnel befitting a genuine multilateral organisation. Its investments in Asian infrastructure are good for the region and the world economy. Germany, France, Italy and others have followed our lead and joined in. And our decision to join early was transformational in building a better relationship with China.

The decision showed how keen we were to look beyond Europe, and the same was true of trade. That is why I chose south-east Asia for a mammoth trade mission at the beginning of the summer break in 2015.

Indonesia, Singapore and Malaysia were return visits. My visit to Vietnam, though, was the first ever by a British prime minister.

I thought the story of our two countries was one to shout about. We had nurtured the relationship through aid, and as that was slowly withdrawn, the flow of trade continued to increase. As Vietnam embraced market economics, it had grown almost exponentially. From being one of the poorest countries to a vibrant middle-income country. From 58 per cent of people in poverty in 1992 to less than 14 per cent in 2010.

As our convoy drove from the airport to the presidential palace, I looked across the paddy fields towards the skyscrapers of Hanoi, remembering how few cars there had been on that road the last time I had been in the country, as a tourist in the 1990s. After inspecting the honour guard in the rain, I sat down to talk with President Truong Tấn Sang. The memories of so many of these types of conversation blend into one another. Bromides about the importance of good relations. Talking points about trade deals, visa processes and the like. But this stands out. Truong started by explaining 'the two principles of Vietnam's governance', which turned out to be 'One: complete adherence to the principles of Marxism/Leninism' and 'Two: complete introduction of a market-based economy'. He said this with a wry smile, and I replied jokingly that he was my kind of communist. As we walked to meet the rest of our delegations he insisted on holding hands, and throughout the visit I would find his hand reaching out for mine at every opportunity.

Back in Britain, I wondered who I'd be facing at the despatch box as Labour started to consider who would replace Ed Miliband. The front-runner was shadow home secretary Yvette Cooper: effective in the Commons, a formidable intellect, experienced as a minister. The trade unions' pick, Andy Burnham, was less quick on his feet, but equally experienced in cabinet and shadow cabinet. The outsider, Liz Kendall, was the self-styled moderniser, in some ways trying to do to Labour what I had done to the Conservatives, post-2005.

Then there was the wildcard, written off by almost everybody: Jeremy Corbyn.

The far-left Islington North MP said he had only thrown his hat into the ring to promote the hard-left faction of the Labour Party to which he belonged, and that, at sixty-six, he wasn't a long-term contender. He only managed to get the requisite number of backers to appear on the ballot paper because some MPs lent him their votes to broaden the range of candidates. Many later said that if they had thought he had any chance of winning they would never have done so.

And no wonder. This was a man who wanted to leave NATO, scrap Trident, nationalise large swathes of British industry, put up taxes and abolish the monarchy. There were recordings of him describing terrorist organisations like Hamas and Hezbollah as 'friends', and memories of him bringing IRA supporters to the Commons just two weeks after the 1984 Brighton bombing.

I had only ever thought of him as an earnest, eccentric leftie – a sort of hangover from the 1980s Dave Spart type of socialist immortalised in the pages of *Private Eye*. Their world view can be summed up as anti-Western and anti-capitalist. They are so against American hegemony that the USA can do no right, and its enemies are courted as friends. Nothing else would appear to explain Corbyn's consistent support for Iran's mullahs or Putin's Russia. The conventional analysis was that 'Red Ed' had leaned too far left for the electorate. Surely, then, the last thing the Labour Party needed was Comrade Corbyn leading it off a cliff. Which is why the bookies put his chances of winning at 200–1.

But then something incredible happened.

Thousands of people started joining Labour, taking advantage of the recently introduced £3 membership tier that allowed new members to vote for the next leader. Greens, anti-austerity protesters and the rest of the loony left flooded the party. A few Tories joined up too, delighted to be able to throw some more lead onto Labour's sinking ship.

On 15 July 2015 I was heading towards Committee Room 14 to give my end-of-term talk to Conservative MPs and peers before the summer recess when I passed Corbyn in the corridor. I had just seen the latest news on the Labour leadership election, and said to him, 'Jeremy, the *New Statesman* poll has just put you right ahead!' He looked bemused.

As we discussed the implications of Labour's leadership contest in our team meetings, we tried not to be complacent. Perhaps Corbyn was tapping into some profound anti-politics sentiment. Maybe this was our Syriza moment, with the birth of a new radical leftist populist movement. Britain clearly wasn't immune to the populism that was sweeping across Europe. And then, he might actually be better than we thought.

But after about ten minutes of this we'd all say, 'Oh, come on, the guy is a disaster for Labour and brilliant for us.' His rise wasn't just a result of the entryism enabled by the new membership rules. In capturing the centre ground, what had we left Labour with? Cooper couldn't argue for a living wage – we were already doing it. Burnham couldn't call for progressive social reforms – we were making them anyway. Kendall couldn't hammer us over discrimination – we were all over the gender pay gap and equal marriage – and there'd be more to come.

Indeed, as well as capturing it, we'd actually shifted that centre ground rightwards. It became a consensus – at least at that time – that governments had overspent and savings needed to be made; so much so that Labour, under Harriet Harman's temporary leadership, abstained on our Welfare Bill (Corbyn was the only leadership contender who voted against it).

I went off for our annual Balmoral weekend with a spring in my step. Not only was the political picture rosy; we were also going earlier than usual, which meant we would simply be staying at the castle rather than attending the Braemar games. The novelty of spending an afternoon watching the caber being tossed had somewhat worn off.

Instead, I had the usual choice: riding, fishing, walking, deer-stalking, shooting. I hadn't picked up a shotgun for almost eight years, so I thought, why not?

That morning it was pouring with rain as I headed off to the moors. I liked the comedy of walking up a hill, a borrowed shotgun in one hand and a BlackBerry in the other. Then the message came through: Corbyn had won, with 59 per cent of the vote – more than all the other candidates put together.

What a lovely weekend it was. Chris Martin had brought his girlfriend Zoe Conway, and we had a wonderful dinner with William, Catherine and some of her friends. There was tea with Charles and Camilla, and the traditional BBQ cooked by Prince Philip on an otherwise deserted moorside.

I had been clear from the start that Corbyn should receive all the courtesies and briefings due to a new leader of the opposition. Doing things properly meant telling him his invitation to join the Privy Council was in the post, and that I would be seeing him at the Battle of Britain commemoration that Tuesday (royal banquets, war memorials – how these events would pain this pacifist republican!).

He kept asking whether I thought he could do things like PMQs differently. 'Do I have to do it all myself? Can I share it out?' he said. I said it was up to him and the speaker – he could do what he liked.

When the time came to leave Balmoral, our helicopter landed on one of the lawns outside the castle and the pilot let little Prince George sit inside until we boarded. Because the doors had been left open, by the time Sam and I took off, the cabin was full of midges. We spent the journey swatting them with rolled-up newspapers, but we were still bitten to pieces. It seemed I already had a blood feud with the future king of England.

I was in my hotel room in New York, attending the UN General Assembly, when Corbyn delivered his first conference speech as party leader. As I watched it on an iPhone, I mulled how his popularity defied reason.

For a start, there was something of Nigel Farage's fraudulence about this 'man of the people' who was actually a privately educated career politician. His run-ins with the press, and his failure to bow properly at the Cenotaph or sing the national anthem, revealed that he was irritable and petty. The 'kinder, gentler politics' he espoused didn't fit with his overly earnest, aggressively leftist tone.

I concluded that he was in fact a conduit. A repository for anti-austerity, anti-establishment sentiment. People were projecting their hopes and ideals onto him, irrespective of his ability. Something of a 'useful idiot', in Lenin's cynical phrase.

I was in my twenties when the end of the Cold War was proclaimed as 'the end of history'. Western liberal democracy and market economics had triumphed and become the global consensus. It was said there would be no new system, no going back. Yet here we were, with a whole generation in their twenties who had never endured 1970s Britain, or seen the tyranny of the Soviet Union, and who felt that capitalism wasn't serving them. Nationalisation seemed appealing. State power sounded more egalitarian than corporate power. Rules appeared fairer than freedom.

George and I talked about this a lot. It seemed we would have to win the old arguments in favour of freedom, markets and enterprise all over again. We would have to convince people of the merits of equality of opportunity, as opposed to equality of outcome. As I was to put it in my conference speech: 'not everyone ending up with the same exam results, the same salary, the same house – but everyone having the same shot at them'. That was the difference. That was conservatism.

Still, these were halcyon days. Labour had elected someone I was sure was unelectable, the economy was growing, I'd got a majority . . . A devil on my shoulder began to whisper: what would be so wrong about a third term? I could do two and a half or three terms, beat Thatcher's record, and not have been driven mad at the end of it . . .

As soon as I felt my head being turned, I checked myself. Don't even go there. Surely ten years as prime minister and fifteen years as party leader would be long enough. Plus, Sam wouldn't have stood for it. She had been amazing, but only because she knew it would come to an end at some point, and we would get some of our old lives back.

Yet it was dawning on me how difficult it must be being the daughter or son of a prime minister, and how it would become even harder as they grew older.

For now, Nancy was very well-adjusted. In fact, she had suddenly become quite interested in politics. I was on the phone one breakfast time when she asked me who I was talking to. I said it was the Chinese premier. 'Well, tell him to free Ai Weiwei and to stop eating endangered species,' she replied, quick as a flash.

Nine-year-old Elwen was also wise beyond his years, though he once came home rather morose, and said, 'I think people only talk to me because I'm the son of the prime minister.' I felt bad that my job made things hard for him, but after a day he was back to his normal happy self.

The youngest member of the family was only just coming to terms with the strangeness of her family situation. Samantha's mother visited the flat one Monday night, and Flo rushed up to tell her, 'Grannie, Grannie – don't tell anyone, but my dad's the prime minister.' We found out that she had been going around to everyone she met telling them her little secret. Yes, two terms would be quite enough for all of us.

Farce interrupted business, however, when a biography titled *Call Me Dave* was published the Monday before our party conference. I had known it was coming, and that it wasn't exactly going to be a fair and faithful account of my life in politics. Over the months my team and I had joked about what it might contain, coming up with more and more elaborate accusations it might make. But even the most creative (or lewd) among us couldn't have dreamed up its most widely reported claim – the one that came to dominate the book's serialisation and publicity – which was that I'd done something disgusting to a dead pig at a university society initiation.

Here, I can reveal the truth about that story. The first I heard of putting private parts in pigs was when Craig told me about the allegation on the morning of 21 September 2015. My first reaction wasn't anger, or embarrassment, or worry about the impact. It was hilarity. I couldn't believe someone could be so stupid as to research and write a book about me, and include a story that was both false and ludicrous. Anyone who chooses a career in politics requires a thick skin and a sense of humour.

Setting an end date for my time in office had put me in an unusual position. I didn't have unlimited time to do what I wanted to do. And I would set out exactly what that was in that year's conference speech, which this time Jess was writing. Clearing the deficit and ending the money-go-round were still vital, but I also wanted to get to the heart of what was causing all those problems, by intervening vigorously and tackling the deep social problems that had been holding Britain back.

Because the biggest, gravest division in our country was not between north and south or male and female, but between the vast majority who could get by and get ahead in life, and those for whom life was a desperate struggle, and for whom hunger, homelessness, abuse and addiction were a reality.

It wasn't just rhetoric. At one point we had ten departments working on the life-chances agenda, including Education on parenting schemes, the Cabinet Office on a social investment fund and the Treasury on ways to alleviate personal debt.

'We are not a one-trick party,' I told the conference hall in Manchester. We weren't just here to fix our broken economy, but to fix our broken society. To do so, we must enter those no-go zones where politicians often don't dare to venture. I spelled out the facts. A teenager sitting their GCSEs was more likely to own a smartphone than to have a dad living with them. Every day, three babies were born in Britain addicted to heroin. Children in care were far more likely to fail at school, turn to prostitution and even commit suicide than anyone else. And Britain had the lowest social mobility in the developed world: the salaries we earn are more linked to what our fathers were paid than in any other major country.

'Over the next five years we will show that the deep problems in our society – they are not inevitable,' I vowed. 'That a childhood in care doesn't have to mean a life of struggle. That a stint in prison doesn't mean you'll get out and do the same thing all over again. That being black, or Asian, or female, or gay doesn't mean you'll be treated differently.

'Nothing is written. And if we're to be the global success story of the twenty-first century, we need to write millions of individual success stories. A Greater Britain, made of greater expectations, where renters become homeowners, employees become employers, a small island becomes an even bigger economy, and where extremism is defeated once and for all.

'A Greater Britain. No more its people dragged down or held back. No more, some children with their noses pressed to the window as they watch the world moving ahead without them. No – a country raising its sights, its people reaching new heights. A Great British take-off, that leaves no one behind. That's our dream, to help you realise your dreams. So let's get out there – and let's make it happen.'

We talked about reaching 'lift-off' in speeches, and I came off stage about 30,000 feet high. As I drank Guinness in the green room I had never felt more content in my own skin, or more excited about where I was taking the country. We were still doing something that most politicians in the last thirty years hadn't done, which was make cuts, but we were doing it in the right way – the fair way. I felt that I had everything on my side: experience and energy, plus my party and my country. I find it hard to write about that now, because it makes me so sad to think about what might have been.

43

Rolling Back the Islamic State

On 26 June 2015 I was sitting at the European Council table when the Maltese prime minister Joseph Muscat whispered to me that there had just been a terror attack in France. 'Oh God, not again,' I said. It was six months since ISIS gunmen had murdered twelve people at the Paris offices of the satirical magazine *Charlie Hebdo*.

Now, a lorry-driver near Lyon had attacked his employer, beheading him and then driving his car into gas cylinders and shouting 'Allahu Akbar' as firefighters overpowered him. I went over to a grim-faced François Hollande, who was getting up to travel straight back to France. I wished him my sympathy and assured him of Britain's solidarity.

Then it began to emerge that there had been another terror attack, this time in Tunisia. It soon became clear that the vast majority of those killed were British.

The scene was the stuff of nightmares. A young Tunisian man wielding an AK47 had wandered along a popular tourist beach near the city of Sousse, shooting holidaymakers indiscriminately before continuing his spree in a nearby hotel. Thirty-eight people were dead, including thirty British citizens.

Philip Hammond chaired COBR while I flew back. I was shocked by the scale of the attack – the biggest loss of British life to terrorism since 7/7. Yet I wasn't shocked by its nature. This was the sort of thing we had known would happen somewhere, at some stage. Tunisia wasn't particularly high on the list, but there wasn't really anywhere that wasn't on the list. Islamist extremist terrorism was so widespread that there had been five major attacks that same day. As well as France and Tunisia, Kuwait, Somalia and Syria also saw shootings, bombings and stabbings on what was dubbed 'Bloody Friday' by the press.

It wasn't a coincidence. Days earlier an ISIS leader had urged supporters to carry out slaughter to mark Ramadan. It was also a year since the establishment of the caliphate, which had by now engulfed a third of Iraq and nearly half of Syria, and spawned satellites around the world.

As I've said, my thinking on this had shifted somewhat. I knew these fanatics were not representative of the religion as a whole. As one bystander famously shouted when an ISIS-inspired knifeman was tackled on the London Underground later that year, 'You ain't no Muslim, bruv.' Quite.

But however evil they were, however wrong they'd got it, they *were* finding something in Islam to justify their actions. Take the caliphate. The Koran and the Hadith both can be, and both have been, interpreted as advocating the idea of a state for Muslims, headed by a Muslim caliph. For some, the idea of a caliphate harks back to the earliest, and in their view, purer and more successful, version of Islam.

Now, you may be a Muslim who completely disagrees that this caliphate should be modelled on the barbaric times of the past. You may be a Muslim who thinks the idea of the caliphate is complete nonsense. But no one can deny that the idea comes from Islam. To defeat the extremism, I firmly believed, we needed to understand that, just as we had to understand its links with gang culture and crime.

We also needed to understand exactly how the death cult was being spread. 'No one joins ISIS from a standing start' is how I often put it. It is less of a leap to go from being a British teenager to an ISIS fighter or an ISIS bride if you have first been exposed to the Islamist ideology that promotes these anti-democratic values. These weren't just being trumpeted on the internet by Islamist preachers. They were being quietly condoned in some communities. This brought me back to my view that we needed to confront non-violent extremism as well as violent extremism.

However, making this argument brings its own difficulties. Many people hear 'Islamism' being attacked and assume it is everything about Islam, and all the people who adhere to it, that are under fire. The language matters here. I always tried to speak about 'Islamist extremism' and 'Islamist extremist violence', rather than just use the label 'Islamist'. Donald Trump doesn't bother with that distinction. Indeed, he goes in the other direction, frequently referring to 'Islamic terrorism', which in my view is extremely unhelpful.

Another difficulty is that there are Muslims with deeply conservative religious views who are neither Islamists nor supporters of violence. Some of the non-violent extremist behaviour we identify as contrary to modern British values – gender segregation, intolerance towards homosexuality, for example – flow from these conservative views. But they are not limited to hard-line Islam. There are orthodox Jews and fundamentalist Christians who take a similar approach on some of these issues.

How do we chart our way through these waters? We should vigorously stand up for equality, and be increasingly impatient towards these practices wherever they occur, particularly when they encroach on the public sphere. Hearing that Muslim parents often refused to shake hands with a female teacher friend of mine shocked me. Such behaviour doesn't make you a terrorist, but it should be robustly challenged in a free and liberal society like ours.

However, what more directly put people on the conveyor belt towards condoning or practising violence was the *political* aspects of the Islamist extremist world view. We needed therefore to focus particularly on arguments such as the claim that the West is engaged in a conspiracy against Muslim people or Muslim countries, or the belief that only those who want an ISIS-style caliphate are true Muslims.

A few days after the Tunisia tragedy, I spoke at the annual Conservative Summer Party, a gathering of our top six hundred donors and supporters. One by one, I took on the sloppy, misguided and dangerous arguments that stood in the way of a robust and thorough response to such terrorism.

It included a call to arms to get behind all those in Muslim communities – the vast majority – who utterly rejected not just violence, but every sentence in the narrative of ISIS. It was about making sure those who condoned, or excused, or ignored parts of this ideology understood precisely what they were giving rise to. I told the audience: 'We've got to show that if you say, "Of course I don't agree with murder – but I agree building a caliphate is a good thing" . . . or, "Yes I condemn terror – but frankly, Muslims and Christians can't really exist together in a successful society" . . . or, "Violence isn't justified, but suicide bombs in Israel are a different matter" . . . If you say these things, think these things, promote these things, you're advocating the narrative of extremism.'

Such clear thinking would inform my decision-making. And some decisions would be bigger than others.

Back in May 2015 I had held a Ministerial Small Joint Group meeting that tasked the relevant authorities to locate and kill ISIS attack planners. Several targets were identified. The strategy had come back to the NSC, and Michael Fallon authorised the specific operations.

On 21 August 2015, Reyaad Khan, a twenty-one-year-old from Cardiff who had appeared in ISIS propaganda and was plotting attacks on British military commemorations, was killed by an RAF drone in Syria. This was a new departure for our country – the first time a British citizen had been killed by the British state in a country with which the UK was not at war. Another British Jihadi, Ruhul Amin, was also killed in the attack.

Our small group had also agreed with the identification of Junaid Hussain as a target. Along with Khan he had been searching online for people willing to carry out attacks in the UK and worldwide. He was killed in a US airstrike on 24 August (we and the Americans were working hand in glove, sharing the burden, and both ready to strike if the conditions were right).

Then in November, Mohammed Emwazi, 'Jihadi John', was killed, also by a US drone. The death of such a prominent and prolific killer was a huge blow to ISIS.

What were the steps that had led us to this point?

I was determined that we should fight back against the terrorists with everything we had, including lethal force. This determination was only strengthened by the horrors of the beheadings that had begun in August 2014. They had to be stopped.

When terrorists operated in countries where we were working directly as allies of their governments, including on the battlefield, this was straightforward.

In Afghanistan, for instance, the international terrorists were also enemy combatants participating in an armed conflict. Our forces could help to target them, and frequently did.

The situation in Pakistan was more complicated, but still manageable. The terrorists were assisting the Taliban against the government in neighbouring Afghanistan, and so were legitimate targets of the US drone programme. So we could support that.

In Yemen, the Americans and Saudi Arabia were also helping a national government in a struggle against a terrorist insurgency. So al-Qaeda operatives were, again, legitimate targets.

But what about places where there was no government, or where we had no allies? I was sure that, even there, we could and should act. I had seen for myself the effectiveness of the US drone programme in Pakistan and Yemen. In both cases it helped to fundamentally weaken the terrorist organisations – principally al-Qaeda – that threatened us all. The US programme had matured over time. Obama was determined that his actions should be seen to be lawful, and shifted the focus from the CIA to military action by uniformed personnel.

Why, I kept asking, weren't we doing more ourselves, particularly now that the ISIS threat was growing? Why should we accept and work with the US programme, but not supplement it with our own? Surely there would be times when our priorities – and the urgency of hitting some targets – would be subtly different from those of our American friends.

I started to make this argument with our military and intelligence experts. Frequently I would drag them from the NSC discussion around the cabinet table on a Tuesday afternoon into my office next door, where we could have a more private and frank exchange. I wanted us to get into a position where we could act with our own sovereign capability, in concert with the United States, but as a partner providing more of the lethal military capabilities as well as the intelligence.

They responded positively, with candour and can-do. We had the necessary drones, and were getting more. We were key to the intelligence, including on the ground, that could help target those who intended to do us harm. But we needed a plan, operating procedures and clear legal advice. Soon we would have all three.

As well as being the first prime minister since Churchill to have a military adviser in Downing Street, I was also the first to hire a No. 10 legal adviser. I had confidence in the attorneys general that I appointed, but I wanted to know what was and what was not legal as we formulated policy at its earliest stages, not just at the end of the process.

So, in 2014, Andy Hood joined us. He had been a senior legal adviser at the FCO, and immediately made himself indispensable. In 2016 he was replaced in No. 10 by Theo Rycroft, from the attorney general's office. The position has now become permanently established, which in my view has strengthened both the PM's effectiveness and the nation's rule of law.

As I said to the House of Commons in September 2015, the legal advice stated that terrorists plotting to attack our country could be targeted on the basis of self-defence. To pass the legal test for legitimate self-defence, first there needed to be clear evidence that anyone we targeted was planning or directing an imminent armed attack against the UK. And second, we needed to show that any actions we took were necessary and proportionate.

We had the first – and given the circumstances in Syria, we had a strong argument that options other than airstrikes were not available. The Syrian government was not willing or able to prevent attacks by terrorists. If we wanted to disrupt the terrorists' plans, we would have to do so from the sky.

The pilots of yesteryear put on flying caps and goggles; today, drone operators sit with a joystick and a map. Their professionalism and dedication are every bit as valuable. I saw when I visited their base in Lincolnshire how they followed suspects for hundreds of hours to build up a pattern of their movements. I heard about the meticulous work to avoid civilian casualties or harm to any operatives we or our allies might have on the ground. They knew that our drones were only ever as good as our intelligence. I met the people who followed Reyaad Khan and those following Emwazi.

But while we were now able to confront ISIS members in Syria that threatened our country, the picture in that country more broadly was deeply depressing. And it was about to get worse.

In the summer of 2015 Assad was still in power, still massacring his people, still driving radicalisation. And we were still powerless to intervene militarily. But with his resources being reduced and his army bled dry, we comforted ourselves with the thought that he wouldn't be able to go on forever.

And then, with no warning, Russia changed the game entirely. On 30 September it intervened directly in the Syrian civil war, putting its aircraft and soldiers on the front line for the first time, in support of Assad's beleaguered government. The Russians' public narrative was that they were leading the fight against terrorism. They even released a map that divided Syria into areas controlled by ISIS, the al-Nusra Front (including al-Qaeda) and Kurds. There was no mention of the Free Syrian Army.

This wasn't surprising. Syria was Putin's main foothold in the Middle East. The US had effectively opted out of the conflict – and he could opt in and win tactical and strategic advantage, while Iran provided its Shia brothers there with militias. It was another example of the price you can pay by failing to act.

And, wielding his phoney map, Putin could do all this under the guise of 'defeating ISIS'. The truth was that 80 per cent of the early Russian airstrikes didn't target ISIS at all, but the armed opposition groups that were weakening Assad.

I got plenty of expert advice that this would be short-lived, and the Russian forces would not be able to sustain their operations because they were still in Ukraine. Obama also believed Russia wouldn't last the course and that we should leave them to it.

But the bear was mightier than we thought. Putin's intervention saved Assad, and by the end of 2017 his forces had recaptured Aleppo, and the regime had quadrupled the area under its control. Putin won some international respect from the action taken to drive ISIS out of parts of central Syria. I didn't doubt that this was one of his motivations for getting involved. Russia, like all of us, was vulnerable to ISIS's poison. At the end of October 2015, a plane from Sharm el-Sheikh to St Petersburg was bombed by ISIS, killing 224 people.

Two weeks later, Paris was to endure its deadliest attack since the Second World War. A hundred and thirty people were murdered by ISIS in cafés, restaurants, bars and at a rock concert at the Bataclan theatre. The attacks were planned in Syria and organised in Belgium. They were carried out by French, Belgian and Iraqi terrorists, some of whom had fought in Syria and made it back to Europe by travelling, as many had feared, among the hordes of refugees, using false Syrian passports.

This tragedy highlighted the nonsense of the fact that we were acting against ISIS in Iraq, but not in Syria. Syria was where the attacks were being planned and the fighters were being trained. It was the base from which they were radicalising more supporters. ISIS didn't respect the border between the two countries – to them it was one caliphate. It also showed that drones weren't enough: we needed manned airpower to go after ISIS forces in general, not just individual targets. Now was our moment, our opportunity, I thought, to change our policy and properly take the fight to ISIS in Syria.

Although the legal advice for drones and airstrikes was the same, for the latter we would need a parliamentary vote. This was the first time since the disastrous vote in 2013 that I felt we might get such a policy through Parliament. The Paris attack had given ISIS renewed relevance. An opinion poll showed that the public supported airstrikes against ISIS in Syria by 59 per cent to 20 per cent. And the UN would soon signal its support. While there would be no Chapter VII resolution specifically authorising force, there would be a Security Council resolution that we should 'take all necessary measures' against ISIS. Surely that would prove enough this time for our parliamentarians?

However, there were still factors working against us. While the Lib Dems could be trying as partners in government, at least being in coalition had given us a large majority; now we only had a majority of twelve, which could easily disappear when you considered the Conservative backbenchers' rebellious tendencies. The SNP's fifty-four MPs could be counted on to oppose almost any military action. Labour, despite being under the sensible interim leader Harriet Harman, was increasingly left-wing with its new intake. And by the time Parliament met again in September, the party was being led by white-flag-waving pacifist Jeremy Corbyn.

That summer, Gavin Williamson, still my PPS, said that the temptation for Labour MPs to damage me would be too great for such a vote to succeed. He wrote: 'I personally believe that with the forces that you will have stacked against you, including a minimum of twenty rebels on our side, if not substantially more, a vote on Syrian intervention is a vote that you would lose.'

On 3 November 2015 the Foreign Affairs Select Committee published a report saying: 'We believe that there should be no extension of British military action into Syria unless there is a coherent international strategy.' The language – my italics – was pointed: 'In the absence of such a strategy, taking action to *meet the desire to do something* is still incoherent.'

The implication frustrated me. We weren't doing something because we desired to do something – but because we *had to* do something. Our enemies were plotting death and destruction in Syria – Raqqa was ISIS's self-declared capital – and we weren't able to take proper action. The strategy to respond to this threat with airstrikes was entirely coherent.

MPs often find it easier to vote against the process rather than the substance of an issue. Confronted with a question of whether to take military action, you hear a barrage go up: 'Insufficient time for debate.' 'No clear sight of the legal basis for action.' 'Failure to respond to the select committee report.' These are often points people can hide behind so they don't have to make up their mind about the question of substance: 'Do you support the bombing of ISIS in Syria or not?'

I was determined not to let these second-order questions get in the way, so I wanted to do everything correctly. I would respond to the report, and put the question of airstrikes to a parliamentary vote.

What we ended up proposing was more restrictive than it could have been. Unlike the Americans and the French, we were limiting ourselves to only striking ISIS, while they could broaden their attacks to include groups affiliated to al-Qaeda.

The chief whip, Mark Harper, did a great job in persuading our MPs and turning Gavin's predicted rebellion of twenty into single figures. Being careful about the process meant that we won over many Lib Dem and Labour Members by delivering detailed briefings.

There was, however, a screw-up.

The night before the vote I was telling ministers we mustn't make any exaggerated claims or use language that might put Labour or Lib Dem MPs off supporting the motion. We had to be measured and matter-of-fact.

Then, before seeing the 1922 Committee and giving my final speech to colleagues about why they should support the vote, I had a meeting with the DUP to try to secure their support. Nigel Dodds, their leader in the House, had his doubts, but said that I could count on his support, not least because he knew Jeremy Corbyn and the Labour front bench would be opposing us. 'I don't want to go into the lobbies with any terrorist sympathisers,' as he put it.

I have a bad habit of getting phrases that I've heard stuck in my head. And when it came to the peroration of my impassioned 1922 speech to my backbenchers, what did I warn them? 'Don't vote alongside Jeremy Corbyn and a bunch of terrorist sympathisers!'

I had been guilty of the very thing I'd told everyone else not to do. It went down a storm in the room, but then of course it leaked. So the following day I sat there on the front bench knowing that my hot-headed speech to the 1922 had ruined what should have been a dignified parliamentary occasion. Labour MP after Labour MP, including moderate and sensible individuals who wanted to back what I was proposing, asked me to withdraw my remarks.

I feared that a withdrawal would drown out everything else in the debate, and would be an even bigger distraction than the original comment. I also suspected that while there was anger across the floor of the House, the votes would still be there. And a large, stubborn part of me thought, hell, Corbyn *was* a terrorist sympathiser. He praised Hamas and Hezbollah. He invited the IRA for tea after they tried to assassinate the prime minister. He dubbed the death of the world's most prolific terrorist, Osama bin Laden, a 'tragedy' (on a TV channel run by the terrorist-sponsoring Iranian government, which paid him for his appearances). So I stuck to my guns and refused to back down. But inwardly I was kicking myself for such an avoidable error.

There was one MP, however, whose speech carried the day. Sitting next to Corbyn but taking advantage of the free vote, the shadow foreign secretary Hilary Benn urged MPs to confront this evil, and with sixty-six of his colleagues sided with the government. It was one of the best speeches I had ever heard in the House of Commons.

Seven of the usual suspects flouted my three-line whip and voted against airstrikes, but 397 MPs to 223 backed airstrikes against ISIS in Syria – a massive majority.

People don't realise how much this meant to the Americans and our allies like France. They wanted us by their side – not just for the solidarity, but because we could really make a difference. We shouldn't subcontract our security to others and rely on their bravery to keep us safe. We had done the right thing. It was a slightly imperfect way of getting there, but we got there in the end.

By the time I left office in July 2016, ISIS in its current form, with its caliphate, was well on the way to defeat. For all the subsequent bluster and boasting of Donald Trump, he was given a war that was well on its way to being won.

There are many regrets you have after you have left office. You see things you put in place coming to fruition – like the virtual elimination of the deficit, the opening of the five-hundredth Free School, the sequencing of the 100,000th genome – and wish they had happened on your watch.

One of my biggest disappointments was not to have been there in July 2017 when Iraqi forces took back Mosul after a long, gruelling battle. Or to see Raqqa seized by brave Kurdish fighters. Or to congratulate Iraq's prime minister Haider al-Abadi when he announced in December 2017 that his security forces had retaken the last of ISIS's territory in Iraq, freeing slaves, ending tyranny, averting further death and destruction, and restoring hope to the country.

That made us safer at home too. The more square miles the terrorists seized, the more recruits they could attract, the more money they could steal and the more attacks they could plot. But as their territory began to shrink, so ISIS propaganda also began to dwindle, and by the time the fighting to liberate Raqqa began it had disappeared from Syria almost entirely.

Britain can take great pride in our part in that. We were slower to join our allies in attacking ISIS in Syria than I would have liked, but what we did was part of a clear plan.

Overseas, that meant avoiding putting Western troops on the ground where possible, taking time to understand the situation, working with and through allies. This was what I called 'smart intervention'. It aimed to avoid the radicalisation and backlash at home that had been unintended consequences of earlier interventions. It may have been our third foray in Iraq in three decades, but this time, according to the MoD, there was no evidence that any civilians had been killed by our strikes.

At home it meant smarter, targeted measures rather than giant gestures and blanket bans. In 2015, seven major plots were foiled in twelve months. By the end of 2016, 250,000 items of terrorist material had been removed from the internet. In 2017, 150 people had their passports removed. Twenty-five groups were proscribed.

The most significant shift – the cornerstone of this liberal Conservative approach – was the recognition that non-violent extremism is a gateway to violent extremism. Under earlier governments, ministers and officials would routinely engage with groups who stopped short of endorsing violence but whose world view was part of the problem. Now we call them out.

Likewise, while in the recent past some practices had been tolerated out of some misplaced cultural sensitivity, forced marriage was now banned and the law against FGM strengthened. And while previous governments had shied away from asking that British values should be actively promoted in our schools, now teachers were actively promoting democracy, the rule of law, individual liberty, and freedom of speech and religion.

A final area where governments of all colours – including the one I led – were guilty of inconsistency was in their approach to the Muslim

Brotherhood (MB). This organisation was founded in 1920s Egypt in opposition to secularisation, Westernisation and nationalism. It had affiliates and supporters all over the world, from supposedly moderate, democratic Islamists like Recep Tayyip Erdoğan's AK Party in Turkey to violent groups like Hamas in Palestine. In between was any number of political groupings and affiliated bodies, including some constituent parts of the Muslim Council of Britain.

The UK government's approach to the MB was deeply confused. Some departments wanted to shun it, others wanted to reach out to its affiliates, and parts of the Foreign Office believed we needed to engage with it, since its affiliated political parties were winning elections in parts of the Middle East, like that of Egypt's Mohamed Morsi in 2012.

Other parts of the FCO, particularly those responsible for relations with the Gulf kingdoms, wanted us to condemn the MB altogether. Some of our closest allies in the region saw it as the principal source of the extremism and violence with which we were dealing. Mohammed bin Zayed, crown prince of UAE, repeatedly told me that 'Al-Qaeda and ISIS are the plants; the Brotherhood is the seed from which they have grown.'

No one seemed able to give me a straight answer about the nature of the organisation we were dealing with. Meanwhile, our approach seemed to be undermining our work at home and angering our friends abroad. So I ordered an investigation in 2014. John Jenkins, our distinguished former ambassador to Saudia Arabia, agreed to carry it out along with the Home Office's Charles Farr.

The report discovered that while the group didn't formally endorse violence, it praised bombers in Israel, its senior figures justified attacks against coalition forces in Iraq and Afghanistan, and it accepted Hamas as part of the organisation. The conclusion was clear and compelling: 'Aspects of Muslim Brotherhood ideology and tactics, in this country and overseas, are contrary to our values and have been contrary to our national interests and our national security.'

I accepted the unequivocal findings of the report, and set in train the necessary actions, for instance continuing to refuse visas to MB members who promoted extremism and intensifying our scrutiny of the group and its affiliates.

By the end of 2015 we had made vital progress in the struggle against Islamist extremism. A stronger legal framework at home. A growing alliance of like-minded countries abroad. Action against ISIS in both

Iraq and Syria. Much clearer thinking about the struggle in which we were engaged.

After another speech I gave on the subject at Mansion House, I received a letter praising my approach from Tony Blair. He said: 'The violence is horrific, but it is only a symptom of the wider challenge.' That challenge, he said, was the interpretation of a religion.

He was right – and because of that, because we are dealing with an ideology, and one that is entrenched, it is a struggle that will last decades and not years. There will be steps forward but also steps backward. We would soon see that starkly. Because even after the so-called caliphate was dismembered, in 2017 Britain was to endure repeated terror attacks. Indeed, it would rank alongside 2005, 1988 and 1972 as one of the deadliest years at the hands of terrorism in our history.

44

Trouble Ahead

There was a chilling familiarity to the photo of the small Syrian boy washed up on a beach in Turkey, which shook the world in early September 2015. It was the way he lay face-down on the shoreline, in his little blue shorts and red top. As a quadriplegic, Ivan would sometimes wriggle and flip onto his front like that. I recognised that helplessness. I also recognised the awfulness of the second picture, of a policeman carrying the body up the beach. It took me back to February 2009, St Mary's Hospital, and the indescribable sorrow of holding a lifeless child in your arms.

Humans are moved by images of suffering partly because we see in the victims the people we love. We know that the only reason we are here and they are there is luck. Sometimes, such pictures are so powerful – napalm-burned children in Vietnam, skeletal prisoners at Belsen – that they not only capture a crisis, they help change the course of history. So it was with the photograph of three-year-old Alan Kurdi.

This wasn't the only image that represented the largest mass movement of people in Europe since the Second World War – a crisis that would prove one of my biggest tests as prime minister and that would contribute to the vote that would end my time in the job.

Flimsy dinghies bowing under the weight of hundreds of people. Derelict trawlers crammed with thousands. These images crept into the global consciousness as more and more boats set off from north Africa for Spain and Italy.

People have long crossed the Mediterranean to reach the safety and opportunities of western Europe. But it was a convergence of the Syrian exodus, poverty and failure in the Balkans, upheaval in the Middle East, war and corruption in Africa and violence in Asia that was driving up numbers in the 2010s.

In 2013 a boat was wrecked off the Italian island of Lampedusa; 360 of those on board died. Italy felt compelled to set up a maritime mission to rescue others and target the people-smugglers, but still migrants kept making the perilous voyage. In 2014, 170,000 travelled across the centre of the Mediterranean to Europe. In 2015 the figure was 150,000. Nearly 4,000 souls drowned that year.

As the dangers of this journey became more apparent, migrants took to travelling by the western Balkans route across Turkey. Photographs of people crossing to a Greek island, carrying everything they could, cramming into lorries in Macedonia or Serbia, piling onto boiling-hot trains in Hungary and Croatia, became familiar. In 2014, 43,000 migrants had travelled this route. In 2015 this increased to 764,000.

There was a further image that captured the crisis, and this was much closer to home. For years people have tried to enter Britain illegally through the Channel Tunnel, and as the mass movement continued, a camp sprang up in Calais for those most determined to get to our country. At its height, 5,000 people were living in the so-called 'Jungle'. The pictures dominated the front pages in Britain throughout the summer of 2015 – and came with warnings that, since ISIS was at its peak, terrorists could be among their ranks.

The situation produced turmoil in Europe. The continent was finally coming through the worst financial crisis since the war, and was now facing the worst post-war migrant crisis too. And those who arrived were only the tip of the iceberg, representing just 3 per cent of displaced Syrians.

The fact that so many wished to come demonstrated the great strength of Europe: a continent that was prosperous and at peace was a beacon for the world's troubled people. Yet at the same time it demonstrated the inherent flaws in the EU. We had a Eurozone financial crisis in part because you can't have sustainable monetary union without fiscal union. And now we were having a migrant crisis, largely because you can't tear down internal borders without a solid external border.

The failure was also politically dangerous for Britain – and for me. EU leaders wouldn't have much time for my renegotiation talks at this time of crisis. Nor would they have much sympathy with our qualms about free movement of EU citizens between EU countries when they were facing an influx of refugees from outside. And with concerns about immigration as high as ever, images of an inundated Europe – and UKIP making hay with that – could prove hugely detrimental to my goal to win the referendum.

The perception that Britain would be overrun was not just exaggerated, but totally misplaced. Yes, some of those crossing the Mediterranean might see our country as their ultimate destination. They knew it was one of the most welcoming, one of the easiest in which to find work, and above all, one where they would find their own kith and kin (almost every nationality has a home in Britain).

But geography makes that final journey difficult: we are an island. Politics makes it even harder: we aren't in the Schengen no-borders zone. It is difficult to measure flows of illegal immigration, but the number applying for asylum is a useful proxy, as that is the way many illegal migrants seek to regularise their status once they have arrived. In 2015, only 3 per cent of Europe's asylum applications were to the UK. At under 40,000, that was nowhere near our 84,000 peak in 2002. So if the crisis demonstrated the flaws of the EU, it also demonstrated the advantages Britain had of being in the EU while having an opt-out from the no-borders zone – our special status.

The British approach was very much about treating the causes rather than just the symptoms. And our big decisions on aid and defence spending enabled us to do just that.

For example, we were able to take some immediate action in the Mediterranean – which I saw as both a moral duty and an imperative for national security. We despatched Royal Navy ships to assist with the rescue efforts. HMS *Bulwark*, HMS *Enterprise* and HMS *Richmond* saved hundreds of lives. We helped tackle the people-smugglers, gathering intelligence, destroying their boats and arresting those behind the trade.

And we were also able to address the serious problem in Calais. On one single day, 150 migrants took advantage of strikes by French ferry workers to try to get through. The total number wasn't huge, but it was important that we signal both to those seeking entry and to people living in Britain that a border is a border. Not to have responded would have been an invitation to thousands more to try the same route.

It wasn't just our island status or our opt-out from Schengen that we were able to use to sort the situation. Our robust deal with France meant our border guards were in Calais, not Dover. These 'juxtaposed border controls', skilfully negotiated by the last Conservative and Labour governments, enabled us to work even closer with the French to toughen the border on their side of the Channel.

We have all read thrillers about terrorists driving into the tunnel and blowing it up, and thought the idea fanciful. But just because you assume the basic elements of security are taken care of doesn't mean they are. I discovered that there wasn't proper CCTV or lighting through the whole length of the tunnel, nor a combined operations room. As so often, I found that the most important thing to do as prime minister is to ask some of the most basic, often even the dumbest, questions. As a result we installed more fencing, floodlights, CCTV and sniffer dogs – and made Eurotunnel increase their security.

But beyond this, I knew that a whole new approach was needed. To deal with the issue of mass migration you need proper borders and a proper policy – one that deals sensitively with refugees, but also reflects public opinion. The EU didn't have either of these. The existing policy was inconsistent, in that it encouraged more migration, with the result that more lives were lost. Moreover, it didn't have public support, being deeply controversial all across Europe.

I wanted to use our platform in the EU to help shape the continent's response, and made our case at numerous European Council meetings on the subject.

First, I argued, we had to recognise the drivers of this mass migration. The main one, of course, was the war in Syria, since that was where the largest number of migrants came from. I was adamant that we had to keep the pressure on Assad. There could be no lasting peace in the country with the tyrant in power. But in the meantime we should keep up our support for the camps in neighbouring countries. And while Britain was more than doing its bit as the second-biggest humanitarian aid donor, others were not. Part of the reason so many refugees were coming to Europe was that they couldn't work in the camps, and their money was running out. What's more, the amount of money per family for food each day had been cut, because the World Food Programme had a funding shortfall of two-fifths (that was because others didn't donate, though Britain did).

If we kept people safe, well-fed, healthy and educated, we could stop them making the perilous journey to Europe – and keep them in the region, close to the homes to which they could, hopefully, one day return.

Second, we had to do more to stop the problem at its source. The fact was that many of the migrants were coming from Africa, the wider Middle

East, the Indian subcontinent and the Balkans. The top five countries of origin for those entering Italy were Nigeria, Eritrea, Sudan, Gambia and Somalia. Many more were escaping broken or fragile states like Iraq, Afghanistan, Pakistan and Iran. In 2014 the top three countries of origin for refugees arriving in the EU were Syria, Afghanistan and Kosovo.

That meant continuing our humanitarian aid for refugees from ISIS and our assistance for reconstruction efforts when ISIS was chased away. It meant supporting the Afghan government's efforts to bring a measure of stability to the country. It also meant trying to develop compacts with African countries that said we would increase development aid if they took responsibility, implemented proper border controls and accepted people back. It meant ensuring that no country was overlooked in our quest to mend state fragility.

Third, we had to break the thriving business model of the people-smugglers. Desperate migrants were paying more than £3,000 each to board unsafe vessels. Often they were willing to do so because of what they saw on social media, which was that the majority of people made it, Europe wouldn't turn them away, and once they had reached Europe, they could continue onwards to the country of their choice. We had to do something difficult: be willing to turn the boats back.

In European Council meetings I frequently cited the example of Australia. A few years earlier they had ended an influx of immigrants by taking a hard line. They towed back boats; they made it clear that people arriving by boat wouldn't be able to settle in Australia. In the words of Tony Abbott, stop the boats and you stop the deaths.

I also cited the Canary Islands, which had had huge problems with the level of migration from west Africa a decade earlier. The Spanish government invested heavily in maritime patrols and radar systems, did deals with the countries from which the migrants were coming, and – even better than turning boats away – stopped them leaving in the first place.

Fourth, we had to recognise that the problem wasn't just people arriving on Europe's periphery; it was them continuing onwards after that.

The EU's Dublin Regulation requires governments to process asylum applications in the first EU country the seekers reach. The purpose of asylum is to find safety; and once you're in Europe, you're safe. But people who arrived in Italy, Greece or Hungary didn't want to stay there. Most wanted to push on to Austria, Germany, Sweden or Britain, where the hope of jobs and benefits was greater.

And the governments of the arrival countries were encouraging this. The Italians and Greeks were exploiting the right Schengen gave them to receive migrants and palm them straight off onto their neighbours. If they were obliged to keep the people who came in, I thought, surely they would do far more to help control the situation. So that was part of my proposed solution to this chaos: either Schengen had to be suspended, or Dublin had to be enforced. In both cases there had to be proper processes to register people wherever they arrived.

Fifth, enforcing Dublin was only part of a direct policy to strengthen Europe's external borders. It wasn't an impossible task. I frequently backed up the Bulgarian prime minister Boyko Borisov as, at EU Council after EU Council, he made the basic point that Bulgaria had both a sea and a land border with Turkey, but by properly protecting its borders, and through hard work and sensible bilateral deals, it had managed to keep illegal migration to a minimum. Why did Greece consistently fail to do the same thing?

I repeatedly pushed these five arguments with our European allies. The problem was, they looked at the images – the boats, the trains, the Jungle – but didn't see the bigger picture. They didn't see the need to stop the problem at its source, and instead focused almost entirely on what to do with those who had already arrived. As a result, the crisis built, and by the time of the European Council in June 2015, relations were the worst-tempered I'd ever seen.

Interestingly, however, this bad temper wasn't aimed at us. Arrival countries like Italy and Greece were angry that they were being inundated. The places the immigrants were being sent on to, like Germany and France, were furious that these countries were flouting Dublin. All these western European countries were angry with eastern European countries, which didn't want to share solidarity and take migrants when talk of a quota began. And all the while I was thanking my stars that Britain wasn't in Schengen, and I could legitimately insist that we would do our bit, including financially, but would play no part in any quota.

The EU got itself off the hook by saying it was going to introduce a relocation programme for 40,000 people. That was a cursory contribution – equivalent to about four weeks of arrivals. But it had to be delivered, and the summit went on into the night. There was arguing and shouting; some people lost it completely. The atmosphere was terrible. At the end of a very long session, I brokered a compromise between the two sides. Protected by not being in Schengen and by the consistency of my views, I drafted the wording for a voluntary scheme and not a mandatory quota. The European foreign affairs representative, Federica Mogherini, declared in front of everyone: 'Last night, David, you saved Europe.' Not a sentiment I heard very often.

More bitterness was to follow in September, when Angela Merkel forced a quota plan for the resettlement of 120,000 migrants through a meeting of European interior ministers. Consensus was impossible, because the eastern European countries would vote against the measure, and effectively veto it, so the EU powers that be decided that instead of the usual requirement for unanimity, the decision would be made by qualified majority, and the doubters simply outvoted. Ultimately it was futile, because the eastern European countries which were now obliged to accept this new plan simply announced that they would ignore it, and so they did.

Things were precarious enough, but a couple of months later a single decision was to change everything. On 21 August 2015, Merkel suspended the Dublin Regulation for Syrians. Migrants no longer had to register for asylum in the country of first arrival. '*Wir schaffen das*,' she told the German people – 'We can do this.' But to all those immigrants who had arrived in Hungary, Italy and Greece, it meant 'Come to Germany.' To those waiting to board boats around the world, it meant 'Go ahead and make that journey to Europe.' Huge numbers of migrants headed towards Germany, and on 5 September Merkel effectively opened her country's borders to the influx. The step-by-step chancellor had taken a giant leap – and I was convinced it would prove disastrous.

Yet I understood why she had done it. I had convinced myself that the most compassionate thing in the long term was to discourage people from making the journey. But the images of those who did, and those who suffered along the way, made me want to do more to help.

I also knew how much Merkel's hinterland influenced her mentality. She grew up in the wake of the Second World War, when shame hung over her country. Brought up in East Germany, she loathed borders and division. How tempting to cast off the associations and mistakes of the past and make a big, generous gesture.

However, I suspected that the biggest factor was her recent appearance on the German equivalent of *Question Time*. A young Palestinian girl had asked what would happen to refugees from her country. Germany couldn't accommodate all Palestinians, Merkel told her. The girl sobbed, Merkel comforted her, and the clips of her looking awkward and heartless went viral. I suspected it was this moment, rather than the memories of East German checkpoints, that inspired her final decision.

It is easy to criticise Merkel's action, but I have every sympathy with how a moment like that might have changed her approach. It was only a week later that the picture of Alan Kurdi was published, and I found myself in a situation where the mood had shifted, and something had to change. The pressure on me was great. A 'Refugees Welcome' march soon snaked past Downing Street – and they had a point. We have a proud history as a sanctuary for the persecuted. From French Huguenots to Holocaust survivors to Ugandan Asians, people have made Britain their home, and made it the most successful multiracial democracy on earth.

I found myself caught between the increasingly polarised politics that have come to characterise our age. Opinions printed in the *Daily Mail* would have you believe that migrants were all men, predominantly from outside Syria, and very often criminals or terrorists. None should be allowed into Britain. Opposing this were arguments that all migrants were Alan Kurdis, and everyone should be allowed to settle here. Two rival petitions exemplified this divide, with nearly half a million people demanding that the UK 'accept more asylum seekers and increase support for refugee migrants in the UK', while the same number signed another demanding that all immigration be stopped until the defeat of ISIS.

The truth was somewhere in between. The migrants were 72 per cent male. The second-highest number were Afghans and the third-highest were Iraqis, fleeing ISIS. A large number of the people we took in weren't Syrian children, but Afghani teenagers. But by and large they were Syrians. Many were families. People wouldn't make the potentially deadly journey unless they were desperate. We couldn't do everything – but we could do something.

Unlike Merkel, I was determined to keep my head. Moderation was key. I wanted us to take people, but to do so in a way that didn't entice more to make the perilous journey. Crucially, I simply didn't see the point of taking people from inside the EU, because they were safe already. If we were going to take refugees, they had to be from the camps.

We were ahead of the game on this. In 2014 we took the decision to establish the first Syrian Vulnerable Persons Scheme. Led by my foreign affairs private secretary Nigel Casey and developed with the UNHCR, it fitted with our belief that while the best option was for the vast majority of refugees to stay in the region, there was a small number of people – orphans, women and children at risk of sexual violence – for whom asylum in the West was right. When the migration crisis hit in mid-2015, and public, political and international pressure to respond rose, we were ready.

The question was how many we would take. We came up with the number 20,000, to which I added the words 'over the rest of this Parliament'. I knew that saying '20,000 now' would invite calls for 30,000 next year, 40,000 the year after, and so on. There was no right number; we just needed to demonstrate warmth and generosity, and at the same time try to solve the fundamental problem.

Meanwhile, all the problems I envisaged from '*Wir schaffen das*' were coming to pass. Within weeks, 13,000 people were arriving in Germany every day. Nearly a million came that year. The orderly registration of arrivals didn't happen, and the German state effectively lost control. At the same time, Italy, Greece and Hungary were struggling to cope with the influx. Eight thousand people arrived on the Aegean islands in a single day, all ready to take up their personal invitation from Angela Merkel.

What happened next was inevitable. From Croatia to Slovenia to Austria, and later in Norway, Sweden, Denmark and Belgium, the borders started to go up across Europe. Even Germany imposed border controls, and the shine came off the welcoming culture – *Willkommenskultur* – when gangs of migrants attacked women among the crowds celebrating New Year's Eve 2015 in Cologne.

You develop strange, shifting alliances in politics, and I found myself working on aspects of this issue with Hungary's Viktor Orbán and Bulgaria's Boyko Borisov – increasingly being called 'the bad boys of the EU'. Orbán's populism was not mine, and I worried about the direction he was going in and the language he was using. He knew that, but we realised there were, within limits, certain things we could achieve together.

Hungary and Bulgaria were on the outer edge of the EU, and therefore on the front line of the migrant crisis. Yet they'd been effectively abandoned by the failed EU border-policing operation, Frontex, which lacked money and personnel. Therefore they'd taken matters into their own hands. Orbán built a fence on Hungary's border with Serbia and later on that with Croatia, and the flow of migrants to his country stopped. As has been mentioned, Bulgaria has both a sea border and a land border with Turkey, but by building a fence, Borisov kept the masses out of Bulgaria.

I visited Bulgaria at the end of the year – the first British PM to go to the country in sixteen years. I had promised some second-hand British Army Land Rovers to police the border, and in the event we were able to send forty. That evening, Borisov and I dined at a restaurant up a Balkan mountain, and were treated to some authentically Bulgarian entertainment – Balkan dancing and singing, followed by shaman figures who walked across red-hot embers, holding icons of favourite saints.

Meanwhile, the initial EU response was a total failure. It had been intended that 'hotspots' would be set up in Greece and Italy, where people could be properly processed, fingerprinted and checked, but despite repeated announcements, almost nothing happened. These hotspots were supposed to be combined with a relocation scheme to spread 120,000 people around the continent, but that also failed. By Christmas 2015, Britain had settled one thousand of our promised 20,000 – which was more than all the countries signed up to the EU's resettlement quota put together. (Indeed, by the end of 2017 Britain had managed to take 10,000 refugees, compared with only 28,000 across the whole of Schengen. Huge credit must go to Richard Harrington, who I made minister for Syrian refugees. He ran an excellent programme to resettle thousands of these vulnerable people, and they are doing well here.)

Britain continued plugging away, trying to prevent people from making the deadly journeys and to keep them close to home. Working with the World Bank and others, we persuaded the Turks and the Jordanians to grant work permits to refugees. We poured massive sums into Lebanon and Jordan, holding an event at Davos with Jordan's Queen Rania in January 2016 to persuade investors to create jobs in the special economic zone set up in her country. I announced that we would send £20 million in UK aid to help fund Lebanon's overstretched school system. Then there was the historic Syria Conference.

This was 0.7 per cent in action. It didn't just keep people safe, secure, and give them some semblance of a future. It made Britain a leading voice in the EU on the response, and showed an alternative – a better alternative – to simply taking more and more people and continuing the cycle.

And then, finally, my appeals to stop the problem upstream were heeded. It turned out to be a game-changer.

On 18 March 2016, the European Council agreed to a deal with Turkey to stop further migration. Every boat that reached the Greek islands would be sent back to Turkey. And for every Syrian returned to Turkey, one would be taken from the camps and resettled in the Schengen countries. In return, the EU would pay Turkey €6 billion, and would speed up plans for visa-free travel for Turks to the EU.

The terms of the deal were easy to parody as a merry-go-round, but it was entirely reasonable. We needed to 'turn back the boats' in order to stop more people coming, and this agreement did just that. The cash payments were merited – after all, Turkey was looking after 2.5 million migrants. Visa-free travel wasn't the same as the right to work or settle. And, most importantly, it only applied to Schengen countries, not the UK. More to the point – the deal worked. The numbers leaving Turkey for Greece collapsed from around 3,000 a day to fewer than one hundred. As well as that, working with African countries to prevent migration across the Sahara, improving policing and cracking down on people-smuggling, eventually yielded real results. In 2015, over 150,000 people crossed the central Mediterranean illegally. In 2018, just 23,000 did so.

Of course, that did not mean the problem was solved. Hundreds of thousands of migrants remained on the move – they still do. And, tragically, they still die. This proves that the refugee system needs to change completely. The thesis put forward by Oxford University economist Sir Paul Collier, who became something of a guru for me on development issues, is, I believe, the correct one. The best way of supporting people is in neighbouring countries, where they can work, remain close to home, and from which they can return when it is safe to do so.

As for Britain, the crisis proved to me once again that being inside the EU, with our special status, was absolutely to our advantage. Just as we had the single market but not the single currency, so we had influence over migration issues in Europe without having open borders. On this issue – of migrants coming to the continent from outside the EU

– leaving would be of no benefit whatsoever. Many would still choose the UK as their eventual destination. Departing the EU wouldn't help us close or strengthen our borders, but it would mean having no say about how the rest of the EU could deal with this issue more effectively.

We were largely insulated from the refugee exodus, as we were able legitimately to argue that we should be exempt from relocation schemes. This enabled us to sort out problems like the security of the English Channel and to push our own agenda with the Turkey deal. If anything, the crisis demonstrated how much control we *did* have over our borders. It also demonstrated how much sway we had within the EU. Indeed, the comprehensive approach I pushed in all those meetings eventually became EU policy, and is now received wisdom.

The European migration crisis wasn't the same as our immigration problem, which was never economic migrants or asylum seekers arriving en masse. We had a small but significant problem of illegal entry – people clinging to trains or climbing into the back of lorries – but this was dwarfed by people from *outside* the EU coming legally, and overstaying on visitor and student visas. And, of course, by people from *inside* the EU coming in large numbers because of jobs and benefits, and because of free movement without transitional controls.

There is a lesson in the fact that those countries that have taken the largest numbers of refugees – Germany, France, Sweden, Italy – have seen the biggest rise of the far right. But the crisis did hit us, and brought a massive boost to those on the right, and particularly to those campaigning against EU membership. People talk about migration and asylum interchangeably, and UKIP deliberately conflated the asylum crisis with the European issue.

There are so many 'should-have-dones' in politics, and this is one of my biggest. I should have done far more to demonstrate that our approach was different to Merkel's. That we were in a completely different position. That our special status of being in the EU but not in Schengen gave us the best possible insulation from the crisis.

I didn't. And that would clear the way for a single, defining image – a poster, released less than two weeks before the day of the EU referendum vote, looking suspiciously similar to Nazi propaganda, with the words 'Breaking Point'. Behind it was a long line of male migrants trudging across the Croatia–Slovenia border. 'The EU has failed us all', said the strapline. The reality: Britain was largely immune from the EU's failures.

The perception: all these people were on their way to Britain. The result: a blow to the campaign to remain in the European Union that might have proved decisive.

As well as making frequent visits to European capitals, I was looking beyond the continent to strengthen our international alliances. In fact, now I had been re-elected with a majority, leaders were keen to come to me.

A Chinese president hadn't been to Britain in a decade. But in October 2015, Xi Jinping made a visit that would mark a high point in our countries' relationship, with greater trade, investment, tourism and cooperation than ever before.

On these visits you're always trying to come up with 'moments' that might embody your message. With the Chinese, every detail has to be carefully negotiated, usually weeks in advance. We wanted to demonstrate that the hard work that had gone into the relationship meant there was a genuine dialogue. They wanted to show that Xi was a man of the people. So I suggested that he should come with me to have fish and chips and a pint at the Plough, near Chequers. To our surprise, the Chinese agreed.

Though there were cameras snapping away and advisers buzzing around us, it was a chance for me to spend some time with Xi. He has a confident and bullish exterior – he sees himself very much as the big leader, in the mould of Mao and Deng, and projects that image – but behind the scenes I found him reflective and thoughtful. And the pub was a hit. In fact, he enjoyed it so much that a group of Chinese investors later bought the Plough, and plans to launch a chain called 'The Prime Minister's Pub' across China.

Hot on Xi's heels came the new Indian prime minister Narendra Modi. There were several 'moments', including the largest-ever gathering of the Indian diaspora in the UK at Wembley Stadium. Before introducing Modi I told the 60,000-strong crowd that I envisaged a British-Indian entering 10 Downing Street as prime minister one day. The roar of approval was incredible. And as Modi and I hugged on stage, I hoped that this small gesture, like clinking glasses with Xi, would be a signal of the open-armed eagerness with which Britain approached the world.

*   *   *

On 25 November a very personal tragedy struck at the heart of No. 10. At the time when he'd normally be by my side, helping me prepare for PMQs, I received the terrible news that Chris Martin had died of cancer. In the Commons that day his seat in the officials' box was left empty, and I paid tribute to the man who was somewhere between a father and a brother to us all. I described him as 'my Bernard', after one of the central characters in *Yes, Prime Minister*. He had been a central character in my premiership – the perfect PPS.

30 November would be a significant day in the life of our planet, as 160 countries assembled in Paris for the UN Climate Change Conference. They would strike a momentous agreement to limit global warming to less than two degrees above pre-industrial levels. After years of failure, from Kyoto to Copenhagen, it was the most significant step we'd ever taken towards saving the earth, and it cemented our status as Britain's greenest-ever government.

December came, and with it an Autumn Statement that was full of the fruits of our fiscal prudence. We had stabilised our finances enough to guarantee real-terms protection for the police – an increase of £900 million by the end of the decade – and to give the NHS half a trillion extra over five years. The basic state pension would rise to £119.30 a week. Following the shadow chancellor John McDonnell's response – reading extracts from Chairman Mao's *Little Red Book* – the two alternatives of British politics were laid out clearly.

I was in a good position at the end of 2015. Our lead on the economy was at record levels. My personal ratings were high. The migration crisis had shown that I could lead in Europe, and that our special place in the EU immunised us from its worst effects.

But at the same time I was worried. Could I really get things done in Europe – a renegotiation of the terms of Britain's membership – while it was so focused on the migration crisis? And could I really persuade the people of Britain that we were protected from illegal immigration when the issue had become so prominent, and when the scare stories about its implications had only just begun?

45

Renegotiation

My biggest task after the election would be the biggest of my life. It came in two parts: a renegotiation and a referendum. Both were necessary, and both were long overdue.

I didn't see them as separate endeavours; they were two sides of the same coin. I could never have undertaken a proper renegotiation without a referendum. I needed that endpoint to focus the minds of the leaders from whom I was seeking change.

I would never have countenanced a referendum without a renegotiation. A choice between 'Leave' and 'Remain', if all that was available was the status quo, was a false choice: we should attempt real reform. There were genuine problems that needed addressing – those we were facing now, those I saw coming down the track. As I repeatedly told European leaders, it was 'my strategy to keep Britain in Europe'.

Putting it off, ignoring the grievances, letting the EU move in a direction we didn't want while dragging us along with it, would have made leaving – what was by now being referred to as 'Brexit' – not just probable, but inevitable. As I've said, I was convinced a referendum would happen in the near future anyway, and most likely under a more Eurosceptic Tory government which might not even offer the option of reform. It could simply be 'status quo or go' – and it would push for the latter.

Those who say that I just needed some deal, any deal, to enable me to put the question of membership to the electorate are totally wrong. I knew just how important getting the right reform would be to getting the right result. The polling was clear: there was only a majority for remaining in Europe *if* longstanding problems were addressed. Indeed, I was acutely aware of the danger of not getting enough from an intransigent EU to satisfy an impatient public. 'I do worry that what is negotiable is not sellable and what is sellable is not negotiable,' I confessed at the time on tape.

What gave me hope, however, was that I had won concessions in the past. I had cut the EU budget. I had vetoed a treaty. I had got us out of bailouts. I had triggered the 'opt-out' on justice and home affairs that clawed back powers which had been passed from Westminster to Brussels. We had asked for *extrawurst*, and we had got *extrawurst*.

My relations with other EU leaders were strong. The renegotiation and referendum had been signalled long in advance. And I had won an election on the basis of carrying them out.

Added to that, the EU itself was at a significant juncture. Its single currency had just survived a near-death experience. Its central vision of tearing down borders had (almost literally) hit a brick wall. Populist parties challenging its existence were on the march. Now, an important member – the second-largest economy, the biggest financial services centre, the co-founder of the single market – was hovering at the exit. Surely the EU would realise that it had to adapt if it was to remain intact.

When asked what I would do if I couldn't get a serious renegotiation under way, I said repeatedly that I 'ruled nothing out'. I meant it. If our partners in the EU and the European Commission had refused to engage at all, even someone as committed to our membership as me would have had to consider walking away, and to recommend leaving. But if I could initiate proper negotiations and address the problems I was most concerned about, then the heart of my 'renegotiation and referendum' strategy was to throw myself into campaigning to remain.

Writing all this now, I completely accept that my strategy failed to achieve the outcome I desired. But at the time – and subsequently – I believed that the risks of doing nothing were greater. As I have said many times, that doesn't mean I have no regrets, or do not believe that things could have been done differently or better in terms of the negotiation, or the campaign, or indeed the timing. But it is why I don't regret the central decision to adopt a strategy of a renegotiation and referendum.

My focus wasn't just the all-important new settlement. The *process* of holding this historic vote mattered too.

And on the first Monday back at my desk after the general election, I needed to get on and start deciding some of these things. There were more questions than the central matter of when the referendum would be held. Determining the franchise, setting the question, laying down the rules that would determine everything from spending limits to purdah periods – all these things could have an impact.

On all the process questions I faced, I was torn between two considerations.

On the one hand, as this was such an important question, with such huge consequences, there were strong arguments for taking special measures, such as expanding the franchise, setting a threshold and allowing the government to campaign all the way to the end.

On the other hand, special measures such as these would look like – and to a large extent would be – an establishment stitch-up, tipping the balance in favour of remaining.

Heeding the second argument had served me well in the Scottish referendum, and was uppermost in my mind. But looking back, I think I gave it too much weight.

The first question we faced was over a proposal to extend the franchise to sixteen- and seventeen-year-olds, as had happened in Scotland. This could have helped the Remain case. The Lords tabled an amendment for it on the EU Referendum Bill. George was in favour.

But I resisted. I had never supported lowering the voting age before. There has to be a cut-off, and the age at which you can get married without your parents' consent, drink alcohol and work full-time seems like a sensible age to start voting. And I believe that what applies to general elections should apply to referendums too.

I also thought that changing our stance would look completely opportunistic. Plus, Conservative MPs were something like 9–1 against. Any amendment would therefore have had to be carried not just with the help of Labour votes, but with the overwhelming support of opposition MPs. This was not the way to get sceptical MPs on board for supporting my overall approach.

Then there was the quandary over how to phrase the question. I wanted it to be 'Should the United Kingdom remain a member of the European Union – Yes/No?' with 'Yes' being 'stay in the EU on the basis of the new deal'. This mirrored the 1975 formulation, and was what our draft Bill originally proposed. But the law said that the Electoral Commission had to comment, and there was strong pressure to defer to its view.

The Commission recommended that the question be 'Should the United Kingdom remain a member of the European Union or leave the

European Union?' There was little alternative, I felt, but to comply. Had I insisted, and then been criticised by the Commission for not listening to its advice, we would have suffered as a result.

The loss of the positive word 'Yes' from the ballot paper was bad news. And 'Leave' sounded dynamic in contrast to 'Remain'.

We also had to determine whether all UK citizens living overseas should get a vote (currently that right expired fifteen years after moving abroad), and whether EU citizens in the UK should (they could only vote in local elections). Again, both would probably have helped Remain.

The arguments were far from straightforward. Labour – including many who would later criticise me for not doing everything I could to win – would, I believed, have opposed any attempt to extend the franchise to more people living overseas. They feared its implications for other elections, and could have held it up in the Lords. So, again, I decided not to pursue it (and besides, while enfranchising those living in Europe might have got more Remain votes, enfranchising those in Australia, Canada and the USA might have brought more Leave votes).

The biggest call was the question of whether we could have imposed a threshold for any decision to leave the EU. In other words, saying that such a decision needed not just 50 per cent of those who voted, but a certain percentage of the electorate as a whole (some suggested 40 per cent). This mechanism was used in the devolution referendums at the end of the 1970s (Scotland voted 52 per cent to 48 per cent for devolution, but failed to pass the threshold, and the issue was left to fester for another two decades).

An even tougher alternative would have been to say that Leave had to win a majority in each of the four nations of the UK. I thought this suggestion was dangerous: requiring separate majorities would encourage separatism. We were one United Kingdom, and should either stay together or leave together.

I was more tempted by the threshold argument. Having devoured books on British political history, I knew that thresholds were brought about in the 1970s through backbench pressure. I thought there was a good chance this would happen again. Then it would be Parliament deciding, not me trying to fix the result.

Ultimately, the question of a threshold was discussed in the Lords, but it didn't go any further. There was no serious effort in the Commons, and there are many people who raise the issue now who did not at the time.

The reason I didn't make a move on it is that requiring a threshold for one side would have been the ultimate way of strengthening the argument that the establishment was trying to stitch up the result in advance. And what good would it have served if Leave had won a majority, but then been judged to have lost? In exactly the same way as it is delusional to believe we could have gone on forever without a vote on membership, it is delusional to believe that we could have prevented a further referendum by putting our hand on the scale in this one.

Instead I saved my firepower for trying to make sure that the government could operate effectively as close to referendum day as possible. This not only made campaigning sense; it was also because government business with the EU would have to continue during the referendum – and because Remain was the position of the government. Therefore in the draft Bill we did not include a period of purdah. This gave Labour, the SNP (both of which have conveniently forgotten that they did this) and Conservative rebels something to fight, and they pushed through a vote on purdah restrictions. It meant that in those crucial six weeks before the referendum, my brigade of Whitehall staff would be reduced to a platoon of political advisers.

Next to the threshold question, the process decision with the biggest impact would be timing. The latest we could hold the referendum was December 2017. This was written into law. But I always had my eye on 2016. French and German elections would take place in 2017, and it would be far harder to get special concessions for Britain when the two other biggest countries in the EU (and the two other big net contributors) were facing their electorates.

Also, our economy was strong now – but who could be certain about two years hence? The migration crisis had subsided, but what if the summer of makeshift boats on the Mediterranean were to repeat itself? Yes, going relatively quickly would give me less time to negotiate. But I saw greater risks on the horizon. What's more, I didn't think that protracted negotiations with the EU would produce a better outcome – and they might even cause the public to question the wisdom of being a member of such an organisation.

Therefore, at the first major meeting about our EU plans, less than a week after the election, I asked officials to work up 'the most ambitious possible timetable' for the renegotiation and referendum.

I saw Lynton Crosby for dinner on 21 July (we met informally every month), and he agreed that next year was as good a time as any. George was in favour of going early. I finalised the early option over curry with George, though we agreed it could be pushed back if we didn't get a good deal.

I would start the negotiation process at the June 2015 European Council, aiming to agree a deal by the December Council (though it could slide to the next one, in February 2016). The referendum could then take place on the earliest date officials had earmarked: 23 June 2016.

Of course the renegotiation wouldn't be based on the wider-scale treaty change that many had forecast. I had, I admit, thought this wider change likely to happen, because it was so obviously necessary. But I underestimated the desire of the EU's leaders to avoid changes they would have to consult their own voters about. The entire request for change was now something the UK was bringing about unilaterally. As George put it, I planned to start a fight in a room all by myself.

The absence of a general treaty renegotiation and Europe-wide reform wasn't necessarily a disaster. Indeed, it might even have made life easier in some respects. During previous treaty negotiations UK governments had used up enormous amounts of energy batting away proposals for even more integration and political union. The same thing would have happened again. So we were able to get to the point more quickly. And anyway, Europe-wide reform had never been particularly sellable to the British people. I don't think they ever believed that the EU was moving our way – nor do I think they particularly *wanted* it to move our way. They were never going to embrace it like continental Europeans did. What they wanted to know was simply that we could live in our own version of Europe with an annual bill we could stomach, immigration levels we could handle, and rules and regulations we could see the sense in.

So a big part of the ambition was, as I recorded at the time, 'to try and correct the things that British people don't like about Europe' – in particular, the moves towards political union, the excessive regulations and the lack of control over immigration.

But it was also about correcting the things people didn't necessarily dislike, but which were damaging our interests. That, to me, was essential. If you only focus on the things that rankle with voters, you're little better than a populist. Leadership is answering both those qualms *and* the big issues you see coming down the track.

People weren't sitting in pubs bemoaning the dangers of the Eurozone's discrimination against the pound. 'They have their currency, we have ours. That's not going to change, so what's the problem?' was the common attitude. The problem was that the Eurozone was starting to caucus and make rules about the single market and financial services that profoundly affected the UK. We needed safeguards to stop this from happening.

At heart, the ambition of my plan was simple: to shift the EU from its long-held view that all its members were travelling towards the same destination, but at different speeds, and that Britain, the reluctant European, would get there in the end. I wanted to get it across, once and for all, that the UK was not only travelling at a different speed, but that we had a different destination in mind altogether. 'Yes' to the trade and cooperation, but 'no' – indeed 'never' – to political union, currency union, military union or immigration union.

I will forever be criticised by Eurosceptics for not seeking more profound changes. I stubbornly maintain that the sort of changes they wanted were completely undeliverable. More to the point, I continue to believe that different speeds and different destinations *was* profound.

However, the fact that there wasn't a more general renegotiation of the treaties under way did create one other big practical problem for me. In the absence of treaty change, how would we get guarantees that the changes we sought would be legally binding?

It was imperfect, but there was a solution. Officials proudly paraded what they called the 'Danish Model'. It would be agreed that the next time the treaties were opened up, our reforms would be adopted formally. In the meantime they would be fully implemented, and backed by legally binding instruments lodged as a new treaty at the United Nations. That was what happened in Denmark after it rejected the Maastricht Treaty in a referendum in 1992.

I looked at it this way. The federalists had spent years inserting language into one treaty to make the next integrationist treaty inevitable. We were trying to do the same, but for opposite political ends. And also to set a precedent – that reform could be achieved, and further reform pursued.

A less easy problem to solve – indeed, one of the biggest problems when negotiating with the EU – is that once your proposals are distributed to twenty-seven other member states, they are almost certain to be leaked and published. So, when you know you are likely to get less than you initially ask for in a negotiation, how do you avoid the humiliation of that happening in plain sight?

I decided to divide the new terms I was negotiating into four distinct 'baskets'. Within each basket was a bundle of different proposals, some of which I might get more on, some I might get less on. I would try to avoid ever setting out in one place everything that was in each basket. But ultimately I had to get something meaningful within each. MPs and the press would endlessly complain, 'You haven't set out everything you're aiming for.' But to do so would, I am convinced, have been counter-productive. What's more, the basket idea was – though more broadly this is something I failed on – crucial to expectation-management at home.

The first of the baskets I named 'sovereignty', and it focused on bringing powers back to Britain.

There was one very powerful way we could achieve that. The founding Treaty of Rome proclaimed in 1957 that the then European Economic Community was 'to lay the foundations of an ever-closer union among the peoples of Europe . . .'

The phrase 'ever-closer union' had both a symbolic and a real effect.

The symbolism went right back to the complaint made by Eurosceptics about the way the decision to join what was then the European Economic Community was sold to the country by Ted Heath in the 1970s. I had lost count of the times people had said to me, 'I thought we were joining a common market, not a political union.' Yet that phrase, 'ever-closer union', had always been there. I thought that specifically opting the UK out of it would demonstrate that we were serious about different destinations, not just different speeds.

And it wasn't just symbolism. The European Court of Justice would sometimes quote 'ever-closer union' directly in its rulings. More prevalent, and pernicious, was when it *invoked the principle* – often referring obliquely to the 'spirit of the treaties' – in these rulings. It meant that this one small phrase became a key part of the 'ratchet' effect, where powers seemed to flow only one way, towards Brussels. As it was put to me, 'The objective of "ever-closer union" went deep to the heart of the European institutions.' If the UK was specifically carved out of ever-closer union, it would surely help protect us from EU mission-creep brought about by the ECJ.

But that still left the European Parliament, which had been considerably strengthened by the Lisbon Treaty. It was becoming, as the Juncker appointment illustrated, an increasingly strong force for federal rather than national governance of Europe. It wasn't possible to roll back Lisbon and its provisions now it had been ratified and implemented. But we could reform Europe's legislative system, in order to inject a stronger element of national control by introducing a system where national parliaments could, by working together, reject its measures.

What I proposed was a 'red card' system, whereby the Commons and its equivalents could come together to block specific European laws which were not in their national interests. Getting this new measure adopted would require every member state to agree – and the European Parliament itself to accept the change – but I knew there was some support for it from other leaders.

I also said we needed to fully implement the EU's commitment to 'subsidiarity'. This was the declaration in Maastricht that the EU should only act where individual countries' actions would be insufficient – in other words, national where possible, supranational as a last resort. This was one of many things already in the treaties which could make our membership more comfortable, but which simply needed to be heeded.

Then came the second basket: 'competitiveness'.

For many in Britain, particularly on the Conservative side, the whole point of joining the EU had been to make us more competitive and economically successful. That's why the promise of a Common Market was attractive in the 1970s, and why a Conservative prime minister, Margaret Thatcher, had helped to found the single market in the 1980s.

Scrapping tariffs and trade restrictions, agreeing common standards and using the leverage of the EU to open overseas markets for British goods were all seen as positives. Scrapping what were called non-tariff barriers was possibly more important. It prevented other nations from finding spurious reasons not to import our products, and smoothed free trade between our countries.

However, the harmonisation of rules between different countries that the single market required did have some negative consequences. Businesses started to complain about excessive and often costly regulation. The European Commission became overzealous, endlessly seeking out new areas in which to regulate. Brussels issued directives on everything from what hairdressers wore to which lightbulbs we used. Open Europe estimated that EU red tape cost the UK economy £33.3 billion a year.

Again, it was time to rebalance, and we had already made a start. Legislative proposals fell by 80 per cent after we established an alliance of countries fed up with EU overreach and ambushed the Commission before the EU Council in October 2013. More regulations were set to be repealed than ever before. Now we needed to go further, by setting specific targets for deregulation and creating a mechanism to sharply reduce EU legislation.

I didn't just want to mitigate such burdens; I wanted to maximise the benefits of the single market. Our businesses could already trade easily in goods, but the same was not true for services. Theoretically, someone trained as a doctor, engineer or accountant in one country could practise in any of the twenty-seven others. In reality, this was far from being the case. With 80 per cent of our economy made up of services, there would be huge benefits for Britain from completing the single market and giving these professionals unhindered access to a market of five hundred million people. Once again, the advantages were there, they were in reach, we just had to grasp them.

This also applied to new technologies. The single market for digital was far from completed, and British firms stood to benefit if it was. We had just abolished mobile-phone roaming charges and extortionate credit-card fees across the EU. But that had taken years. The next big hurdles included removing barriers to EU-wide e-commerce, and being able to stream TV or music subscription services when in other EU countries. These may seem like small beer, but they were important to consumers, crucial to business, and made membership more conspicuously worthwhile. They also proved I wasn't advancing purely British interests. Completing the single market would make the EU better for everyone.

The third basket was 'fairness' – specifically fairness towards EU countries outside the Eurozone. This was the issue that had led to my showdown in 2011 and the treaty veto – and that George believed was the most important to address. Attempts were being made to move financial services denominated in euros from the City of London to Paris or Frankfurt. And even though we were outside the Eurozone, repeated efforts had been made to drag us in to bailing out Eurozone countries or standing behind Eurozone banks. As late as July 2015 the European Commission had proposed another EU-wide bailout for Greece, despite having promised that it would never happen again. After quite heated confrontations with EU partners and the Commission we were able to protect British taxpayers, but it underlined the need to have legal protections.

Our opt-out from the single currency, negotiated at Maastricht by Norman Lamont, was still a trusty shield, but in places there were now holes in it. The problem was similar to the issue of ever-closer union. The words 'the euro as the currency of the European Union' were written into the European treaties, and could be used – or alluded to – to silence anyone who raised concerns relating to pounds, krone or zlotys.

Consequently, we wanted recognition of existing circumstances, protection of the single market, and assurances that the Eurozone countries could not discriminate against the non-euro countries. To have the words 'the euro is *not* the currency of all EU countries' written into EU treaties would be not just symbolic but operationally significant. The pound could no longer be treated as a second-class currency. There would be no prospect of the UK joining the euro or bearing the costs of another Eurozone crisis. And there would be no common financial destination for the whole EU.

The final basket was immigration. This was the most difficult and politically significant issue. And it was the one I will always look back on and think: could I have done more?

The sense that we hadn't got overall national control of immigration was concrete, easily grasped, and a totally legitimate concern. It wasn't just about the numbers of EU migrants that were coming to Britain, it was about *who* was coming. There were concerns about people arriving and staying under freedom of movement who quite clearly shouldn't have been – criminals, couples in sham marriages, spouses from outside the EU. This was partly a result of mission-creep by the European courts. A series of judgements had helped turn free movement into an almost unrestricted right. They paved the way for anyone who married an EU citizen to acquire EU citizenship, whether they met British immigration requirements or not. That essentially opened freedom of movement to the world, and spawned countless fake marriages.

Theresa May set out meticulously what needed to be done, such as tougher re-entry bans on criminals and the power to deport them if they were already here. Together, we secured agreement from the EU that, for the first time, non-EU nationals would have to meet the immigration requirements of the first country they lived in. Those who wanted to live in Britain had to pass our English-language requirement and our minimum-income requirement, which prevented them being a burden on the taxpayer.

The main immigration issue, though, was the vast numbers of people who were arriving legitimately. And as I have explained elsewhere, immigration offers many positive advantages, economically and culturally, but that only happens when you have a balanced and controlled programme.

After the surge in the numbers coming from the accession countries in 2004 – a million people, rather than the 5–13,000 the Labour government had predicted – arrivals had fallen broadly into balance with the combined number of British people choosing to live abroad and EU citizens returning home due to the recession. But that changed while we were in office. By 2012 there was a sudden uptick in the migration graphs, and by 2015 the net number of arrivals from the EU hit a record 184,000. That is what transformed immigration from an issue that wasn't even mentioned in my Bloomberg speech in January 2013 to an issue that helped propel UKIP to victory at the 2014 European elections.

It was easy to see why so many people were arriving. The continent had been plunged into recession by the crash. Britain was seen as a beacon of jobs, opportunities and stability.

But the imbalance wasn't just down to economic events. It was also down to how freedom of movement had evolved. When the principle was established in 1992, there were just twelve members of the EU, which had broadly similar economies. Now there were twenty-eight members, with wildly differing economies, thanks to the accession of those formerly communist nations in eastern Europe (and then to the way the crash tore through southern European economies). It was inevitable that migration to richer, more stable countries would rise.

It was also – and people forget this – originally meant to be the 'freedom of movement *for workers*'. That was written into Maastricht. Yet two things had gone wrong with this.

The first problem was that the principle was being expanded. It was no longer just people with a job who were coming to the UK under freedom-of-movement rules. Forty per cent of EU migrants were arriving here without a job offer at all. Of course many intended to work, and would work, but some did not.

The second problem was inherent within the principle. The definition within the EU treaties of a worker was someone who worked more than a certain number of hours a week. There could be no differentiation between a worker from another EU country and a UK worker, and successive treaties had expanded these rights over decades.

Combine these two factors with Britain's non-contributory benefits system – under which both workers and non-workers can claim immediately, and as much as anyone else – and you can see why numbers swelled so dramatically, and why action was needed to bring them back in balance.

I did not want to end freedom of movement. I didn't want to stop our citizens' right to live, work, study and retire in other countries, as 1.3 million of them were doing in 2015. My whole political ethos was about giving opportunity to all our people, and I didn't want to do anything that closed doors or put up barriers for the next generation of Britons.

I also didn't want to stop EU workers from coming to this country. We depended on them in our schools, universities, hospitals, businesses – across our entire society. We talk a great deal about the contribution of, say, the Windrush generation who came here in the 1940, 50s and 60s, or the Ugandan Asians in the 70s – and rightly so. But sometimes we forget those like the Balkan refugees who arrived in the 1990s, the hard-working eastern Europeans who since the 2000s have proved vital to our national story, and indeed all Europeans who have made Britain their home, before and since the freedom of movement rules came into place.

And anyway, there was no deal on earth that would have enabled us to stay in the EU and not sign up to the 'four freedoms' – the free movement of goods, services, capital and people. The EU is a quid pro quo: we get full access to the single market only if we accept its conditions.

So I wanted to *reform* free movement. With change, I believed that numbers could become broadly balanced once again. We had shown outside the EU that we could reduce immigration. Numbers coming from the rest of the world were down by a quarter since 2010, almost at late-1990s levels. That was thanks largely to our crackdown on bogus colleges, limits on economic migrants, tougher rules on bringing in family members, and the introduction of exit checks at borders.

Of course we couldn't apply all these things to EU migration. But the existing rules allowed for extra steps to be taken (removing welfare claimants more rapidly, for instance) – and we should take them. There was also the potential for returning rules about free movement to a form more like that which their original authors had intended.

I saw changes in all these baskets as being absolutely essential to the future of the entire EU, not merely to Britain's membership. After all, the euro had some fundamental flaws. You can't have monetary union without fiscal union – as became clear when the financial crash struck, and the euro struggled. Likewise, Schengen had fundamental flaws. You can't have open internal borders without strong external borders – as the migration crisis exposed when it brought open borders under irresistible pressure. And now it was plain to see that freedom of movement had flaws of its own. Unqualified, unrestricted migration within such a large, diverse institution was unsustainable, particularly in such extraordinary times. With the impact of the financial crash, EU enlargement, mass migration and European courts' judgements, it was failing too. Fraying support in the UK was just the beginning of the unravelling.

I truly thought we could find a solution. For a start, we could redefine what it meant to be a 'worker'. Someone coming here under freedom-of-movement rules would have to qualify for benefits through contributions, rather than having an automatic right from day one.

But to make a real difference to numbers, the best option seemed to be designing a mechanism for a cap on numbers, enforced by an emergency brake. This meant that if numbers were too high, they could be brought under control long enough for the underlying drivers that had caused the rise to be addressed.

Assume net EU immigration into the UK reached 100,000 in year x. The emergency brake would then allow you to set a cap at, say, 50,000 in years x plus 1 and x plus 2. Anyone who wanted to come would be able to until the 50,000 number was reached, after which they would have to join a queue. It would be policed by the number of National Insurance numbers allocated to workers. No NI number, no work.

At the outset I thought the advantage of this scheme was clear: it went to the heart of people's concerns about the need to control overall numbers. It was, however, doomed. Business would dislike the restrictions. It would be very bureaucratic. Most significant of all, the British system of issuing National Insurance numbers was simply not fit for purpose. We couldn't seem to align the number of NI numbers with the immigration figures, and did not register EU workers separately from anybody else.

What's more, the promised popularity of the cap might well not have materialised. When you asked people the number at which migration should be capped, the majority wanted it as low as possible. Even '50,000 plus', a modest figure which many thought unrealistically low, was only supported by 15 per cent of people.

And in any case, when we discussed the scheme with officials they said it would have been considered illegal under the treaties – discrimination on the basis of nationality – and therefore impossible to negotiate. Merkel had told me the same. Their argument was that, in a rights-based system, if you meet all the conditions you retain that right whether you are the first migrant or the 50,001st.

There was, however, an alternative: restrictions on welfare for EU migrants.

At first that may sound like a rather roundabout way of reducing numbers. But if you looked at what a draw our welfare system was, you'd begin to see what a difference it could make. I hadn't quite realised the extent of it until my Europe spad Mats Persson set out the figures for me. Of the £25 billion we spent per year on in-work benefits for workers on low incomes, around £2.5 billion went to migrants from the EU and the wider European Economic Area. That money wasn't going to people who had been working hard in Britain for years and paying their taxes. Two in five recent EU migrants were claiming benefits in the UK. Some of them who were working were managing to pay less than £750 in taxes while claiming nearly £14,000 in benefits. Others were claiming child benefit and sending it to children back home, who had never set foot in the UK. And an average family in which one or more members was an EU migrant was claiming almost £6,000 per year in tax credits.

It was easy to see how this had come about. Britain is unusual in that it has a non-contributory benefits system, where you can take out without having put in. And that is a factor that undermined all the arguments about the reciprocity of freedom of movement – 'They have rights in this country that we have in their countries' – because Britons living in EU countries couldn't claim anywhere near what EU citizens living in Britain could, or claim it so quickly.

I suspected that action on this welfare question would actually be more popular than the cap, since I knew how well people responded to our arguments on welfare. I would receive a box note later with figures backing this up. Private polling showed that three-quarters of voters supported a brake being placed on payouts to newcomers. Over 60 per cent thought it would reduce immigration.

I put the policy at the heart of a speech I made on immigration in late 2014, saying we would restore the 'workers' element to freedom of movement. Right now, EU migrants could claim up to £600 a month in benefits. We had already limited that claim to three months. But in the future, I said, we would end it altogether. If a newcomer hadn't found a job within six months, they would have to leave. EU citizens would have to be working for four years before they qualified for the main benefits. Plus we would bring an end to EU workers sending child benefit home.

Officials reminded me that, as with the cap, treating migrants differently to UK citizens would be seen by the EU as a direct contravention of free movement. But I wasn't prepared to let it go. I thought it would be effective, it was fair (because it took account of the different nature of the British welfare system), and it would support rather than undermine the basic idea behind freedom of movement. It was also the approach the British people had voted for: one of our manifesto pledges was to 'Control migration from the European Union, by reforming welfare rules'. And it was a big part of helping us reach the annual migration target – below 100,000 – that was also in the manifesto.

Freedom of movement had worked before. I was convinced that, with these changes, it could work again.

My next task was to convince twenty-seven fellow EU leaders, and the body of the EU, that they should accept the contents of my four baskets – and do so unanimously.

The period between May 2015 and February 2016 felt as if it was spent largely on board a series of ageing RAF planes. Various combinations of the key EU renegotiating team – Liz, Ed, Tom Scholar, Ivan Rogers, Nigel Casey, Daniel Korski, Mats Persson, Craig Oliver, Helen Bower – joined me, travelling from capital to capital, conference to conference and summit to summit. It would become the biggest diplomatic tour in recent history.

I turned up at leaders' pet projects and obscure events. I visited non-EU countries' conferences just to grab a word with their EU attendees. I went to places no British prime minister had ever visited. I hosted the biggest players – Merkel, Hollande, Juncker and European Council president Donald Tusk – at Chequers. I ate my way around the continent. Indeed, over one forty-eight-hour period I had lunch in Rotterdam, dinner in Paris, breakfast in Warsaw and lunch in Berlin – and when

I got home to Oxfordshire for dinner, Manfred Weber, leader of the EPP, was there to meet me.

Wherever I was, my objectives were the same. Get them to agree to post-dated treaty change – the Danish Model. Explain the four baskets, and that I needed the main elements in each. Insist that I was serious about securing a better settlement in order to keep Britain in the EU. Warn of the risks of under-delivering and seeing the UK exit the EU altogether. Above all, get them to see the British perspective. Explain the prize: Britain secure in Europe, and Europe stronger with Britain.

The most important of the conversations started, of course, with Angela Merkel. From pow-wows at the margins of meetings to strolls through the Buckinghamshire countryside, she made it clear that she wanted to help. But it was also clear that she was distracted by the migration crisis. She welcomed the idea of post-dated treaty change, and was gradually softening on ever-closer union. She understood what we wanted on the euro, and while she fundamentally opposed it, she believed that compromise was possible.

She was more sympathetic than most towards my proposed changes to welfare, because Germany had been similarly affected by benefit tourism. But she was adamant that we had to find a way that was non-discriminatory. I assured her that we would make every effort, but impressed upon her: 'If you force the British people to choose between a measure of control over who comes into the country, and staying in the EU, if you give them that binary choice, they will vote to leave the EU.'

The second-most important conversations were with François Hollande. He was amenable to my proposals for treaty change early on, and was open to the other proposals – except one: my proposal to allow non-Eurozone countries to challenge Eurozone decisions. It would remain a point of contention right until the eleventh hour.

Then there was Matteo Renzi: moderniser, reformer, friend. While I was centre-right and he was centre-left, we often seemed to be on the same side. And he wanted to help. But even he would not accept all Britain's proposals. It was the assault on ever-closer union that was hard for the Italians to swallow. Anything that was in the Treaty of Rome was important to an Italian prime minister, he explained.

When it came to the leaders of the 'Visegrad Group' (Poland, the Czech Republic, Slovakia and Hungary), all but Hungary were wholly, implacably, point-blank opposed to my welfare proposals (unsurprising, since 900,000 Poles lived in Britain, many on in-work benefits). The leaders of the Baltic states took a similar view. Yet all the meetings with them were cordial and constructive. The leaders of these countries appreciated the UK's support for their independence from the Soviet Union and their membership of NATO. They wanted to help.

The Latvian prime minister, Laimdota Straujuma, even confided the unease she herself had about the scale of movement that had taken place. One-fifth of her country's people were now living and working in the EU, with over a quarter of those emigrants going to the UK. There was a danger that with so many twenty- to fifty-year-olds emigrating, Latvia's population was hollowing out. This was echoed by one of Poland's senior ministers, who told Ed later that Poland's leadership would be delighted with any solution that discouraged their people from emigrating. One in forty Poles lived in the UK; some places in Poland had become complete ghost towns. But politically they could not say this, and on the welfare proposals their reaction was the same: they couldn't support them. Dalia Grybauskaitė of Lithuania was, as ever, particularly scathing about my requests.

If anything, I seemed to make more headway with two of the men whose appointments I had so opposed a year earlier: the presidents of the European Parliament and the European Commission.

I invited Martin Schulz to London, and made an enormous fuss of him. I took him to St Paul's Cathedral for the two-hundredth anniversary of the Battle of Waterloo, at which our countries had been on the same side: I pointed out that nearly half of Wellington's army spoke German. He liked that.

Jean-Claude Juncker enjoyed his visit to Chequers. The Commission president is treated like a head of state or government when he or she visits other countries. But British governments have always found it hard to see the European Commission as an equal partner, rather than the servant of the European Council and the nation states. A Chequers invitation helped to make an impression Juncker would appreciate.

For all that their leaders did to help, the Commission and the Parliament didn't like my proposals. They even took exception to the proposals for greater competitiveness, believing they trespassed on their own competences. Some of the resistance I was coming up against was political: leaders were benefiting from playing hardball with the UK, particularly if they were on the other side of the left–right political divide.

Much of it was theological. Belgium's Charles Michel opposed everything we were doing – he had a completely unbending view on Europe. For him, ever-closer union was exactly what the EU was about – and everyone had to sign up to it. It was the same with the Luxembourger Xavier Bettel. He referred to me as 'brother' because of the similarity in our looks. We shared dieting tips and exchanged presents. But for him, my renegotiation was heresy.

And a great deal of the opposition was institutional. If I was finding the leaders hard work, their officials were even worse. To them, I was a dangerous heretic stamping on their sacred texts. Often I would report back to Ivan or Tom after a *tête-à-tête* with a fellow prime minister, only to be told that their officials refused to accept what had been agreed. Merkel and Hollande, surrounded by officials deeply imbued with Brussels orthodoxies, were the worst for this.

Time and again I found there was a fundamental misunderstanding between us about the issue of immigration. As Merkel put it to me: 'You have low unemployment, a booming economy, you're growing faster than most of Europe, there is no social crisis. And you are pulling in highly qualified labour, cheaply. Explain to me what the problem is.'

It was a real insight into how differently we saw things. Merkel and others just didn't see free movement as immigration. As far as they were concerned, if you're from inside the EU, you're a worker. If you're from outside the EU, you're a migrant. Indeed, we were constantly told by the central and eastern Europeans that we were not allowed to call them 'EU migrants'. They were 'EU citizens'. In their eyes, 'migrants' were refugees from Syria and Iraq.

I had to explain that that was not how we saw it in Britain. I would start off with my pride in the multiracial democracy we've built, and explain that increasingly when it comes to immigration, race, ethnicity, religion and nationality figure less and less. For us, it was much more about numbers and pressure. Therefore, people saw free movement and immigration as essentially the same thing. Some of the libertarian Leave campaigners would go further, objecting to free movement precisely because it favoured (mainly white) EU migrants over (mainly black and Asian) non-EU migrants.

But quite aside from differences on individual items in the negotiating package, the thing I was picking up from most leaders was that they simply did not think Britain would leave. For them, the referendum was a ruse to get more out of the renegotiation. 'They don't believe we'd leave the EU over this,' one EU adviser wrote to me. And they really didn't. No matter how much I said it.

There were, however, leaders who completely got it. As ever, the northern Europeans, closest to the UK politically, were the most understanding. Lars Løkke Rasmussen and Mark Rutte, the sensible prime ministers of Denmark and the Netherlands, wanted the renegotiation to keep the UK inside the EU and spur much-needed reforms. They understood – and shared – some of our concerns.

The one leader who took a different view from his geographical and political peers was Viktor Orbán. On my visit to Hungary he ushered me out onto the balcony of his grand office in Budapest to speak privately. As we looked across the Danube at the grand buildings of the once-imperial city, he joked, 'I like coming out here; it makes me think I could be running a really big European country.' His view was that the EU desperately needed reform, and he hoped we got everything we were asking for. Indeed, he wished we had asked for even more. But he was a lone voice – and frankly support from Orbán would often turn other countries from passive observers to hostile opponents.

As I retell the story of my European tour, it does read as though I was facing permanent, implacable opposition. Blocs even seemed to be forming, objecting to each basket. And yet I *was* making progress. As I sat down with individual leaders, talked them through my plans, explained how reasonable they were, emphasising how important the proposals – and this renegotiation – were to the future of the EU, I could feel that they were softening.

In addition, the smaller countries' objections counted for less than those of larger nations. It was Merkel and Hollande that this rested on, not Grybauskaitė and Bettel. And with them particularly I was making headway. Slowly but surely, a deal was coming together. Crucially, we were working with the Commission on how to make my benefit restrictions work within a framework of EU law. There would be an emergency mechanism for countries that faced large migratory pressures to restrict in- and out-of-work benefits for new EU arrivals. I would need to argue that this mechanism would be immediately available to the UK, and would last for a long period of time.

As 2015 drew to a close, I reflected on where we were. Since I was spending so much time in the air, an aviation metaphor felt apt. The plane had taken off in June, with the start of the negotiation. It had reached altitude during the visits. Now I could see the landing lights, the runway . . .

But I couldn't touch down yet. The conditions weren't right. In particular, there was no support for the four years of benefits restrictions. I told the Council that December we would come back in February and try for a deal then.

The start of the final push began with a dinner for Donald Tusk and his team in Downing Street in January 2016.

People assume the EU held all the cards: we needed a lot, they needed nothing. But they did need something. They wanted a success after the botched response to the migration crisis, while I had until the end of 2017 to get my deal. I explained to Tusk: 'If you give me what I've asked for – and they're not crazy things – I will take this deal and sell it like mad in the referendum. But if you try and half-change me, there's no need for me to sign up. I've got plenty of time.'

I made sure he understood the implications were even greater than us just going into the referendum campaign with no deal. My own willingness to support our continued membership was in the balance. 'If the February Council denies me these things, and the whole thing breaks up, we are going to have to have a look at our relationship with this organisation,' I told him. 'You've got to understand that this idea that we must go cap in hand to get a mechanism agreed is ridiculous. We're the third-biggest contributor. We are a proud country, and if we are treated like supplicants, then we'll bugger off.' As a Pole, proud of his nationality, I think he got it.

The draft deal that was circulated at the start of February was nearly there. I called Tusk to say I hoped we could minimise any changes from the existing text, partly because every change would be written up at home as a weakening of the agreement or as a defeat.

My mission was no longer to forge a deal, it was now to stop my draft deal being chipped down by fellow EU leaders. But by the time I arrived at Justus Lipsius on the Thursday afternoon for twenty-one hours of non-stop negotiations, two key proposals were already in jeopardy. The protections on financial regulations remained strongly opposed by France and Luxembourg. The four years for benefit restrictions was left blank. And the time period the welfare brake should last (agreed in principle as seven years) was being disputed by the Visegrad countries, as was the end to child benefit being sent outside the UK.

That afternoon I urged the leaders to support the document to the last detail. If they did so I could take it to the British people, win my referendum, and come back to the European Council with a mandate to be a full and active participant.

We then began a working dinner – all about the migration crisis. It descended into a meeting about a meeting, with a whole hour spent discussing which day of the week we should reconvene. This went on until 2.30 a.m. . . .

Tusk was just getting started on the nitty-gritty of the British deal. Dozens of meetings took place between sherpas, lawyers, ambassadors and aides. Tusk and Juncker saw each country or grouping in turn, saying, 'Don't be ridiculous; you are objecting to *that*, but you can accept *this*.' They would try to grind them down. They would also try to grind us down. In the moments I had spare I would grab five-minute naps on the floor of the British delegation room. My final meeting with Tusk was at 5 a.m. The EU is certainly no observer of its own Working Time Directive.

The following day, a 'British breakfast' was scheduled for us to finalise the text of the deal. But we spent so much time haggling with the French, Poles and Czechs that the breakfast became a lunch, then a tea. I wondered if we'd ever eat.

I tried to break the deadlock through Tusk that evening. 'Look,' I told him, 'you've played this very well, but now we've got to put on the table something that is good enough for me to hold a referendum with. If it's less than I want, I'll say forget this, let's do it later on in the year.'

I was quite prepared to do that. Many people had advised me to go further. Lynton and Ameet wanted me to reject the deal and go for it again later in the year. But I favoured the argument made by those who pointed out that if I walked out, I would have to walk back in at some point – and then what? Goodwill, patience and a reasoned, rational approach had got me this far. I thought they could get me over the line.

And I *was* willing to compromise. I had to be. They had made compromises, and I had to be prepared to do so too. The welfare brake, ever-closer union and protections for the pound remained my red lines. I had everything I wanted on ever-closer union: Britain was carved out of it. And the protections for the pound were secure. It was on welfare that movement was needed. I had won the battle that the welfare brake should be immediately available to the UK, and that it would last for a long period of time (seven years). But still the Visegrad countries did not agree.

Tusk proposed two alterations. The first was indexing child benefit to the migrants' countries' living standards, so they could still send the money home, but it would be far less than now, and therefore far less of a draw. The second was that while the welfare brake would immediately apply in the UK, instead of migrants getting no welfare for four years, it would be phased in over four years. After long and excruciating conversations with the team, it was clear that there was no option: I would have to concede on both points. It would remain a good deal, and giving way would enable me to get it over the line.

Finally, at 8.30 p.m., it started: a 'British dinner'. I told the room how much I appreciated what they were doing. I praised all the good things we'd done together, like raising money for Syrian refugees, opening up markets and imposing sanctions against Iran and Russia. 'I know I often don't talk positively about this place, but I do feel positive, and we will come back more positively if we can sort this out,' I said.

Still, the Luxembourger and the Portuguese wanted to raise objections. The Greek wanted to force through his own changes on the back of ours. The Swede wanted to email the paper to his parliament for approval.

Fortunately, the finalised text was fine by Merkel. One by one, it was approved by all the other member states. Then the speeches started. How great the EU was . . . How it had shown the best of itself . . . What an amazing achievement this was for us all . . .

I had one eye on the clock, and I saw that the 10 o'clock news was just starting at home. I left the room to speak to the press as the leaders erupted into applause – a bit for me, a bit, perhaps, at the prospect of going home.

I stood at a press-conference lectern, two Union Jacks on either side of me, a European flag pushed into the corner, and switched on the arguments I would be making for the next four months. I announced that within the last hour I had negotiated a deal to give the UK 'special status' within the EU: we were 'stronger, safer and better off' in this reformed EU where we now had the 'best of both worlds'. And for that reason I would be campaigning to Remain.

I really believed in what I was doing. For all the frustrations, contradictions and hostilities of the past two days, I had seen yet again how
important it was to be at that table, fighting for Britain.

Objectively, the deal was a significant achievement. It reversed a key
tenet of the Treaty of Rome; it was the first meaningful, permanent repatriation of powers by any leader; it redefined what constituted a European
'worker' in EU law; it established that ours was a national welfare system;
it protected the pound, the City of London and our status inside the single
market but outside the Euro; and it entrenched and expanded our special
status. Ivan Rogers argues that this was the biggest feat of all. Not any
particular reform, but the fact that our special status – which had evolved
unofficially over years of opt-outs, from the euro to Schengen – was
finally embedded, and heading for full enshrinement in the treaty.

We had got a deal passed unanimously by twenty-seven reluctant
member states, and we had done it not at a time of reform, but from a
standing start. We had proved that the EU could change, and that I was
capable of effecting that change.

We had got things people said were impossible – even illegal (I joked
at the time that I'd got more benefit cuts through the European Council
than I'd got through our own House of Lords, which was holding up the
welfare Bill at home).

I had got something in each of my baskets. I said I wanted a red card
– I got a red card. I said I wanted us out of ever-closer union – I got us
out of ever-closer union. I said I'd protect the pound – I'd protected the
pound. And so on.

This broadly matched what many mainstream Conservative MPs had
long called for. I look back over the initial 'wish lists', from the one drawn
up by my private office and the policy unit team – Daniel Korski,
Laurence Todd and John Casson – before the election, to that of the
Fresh Start group of Eurosceptic MPs. The headline elements were all
there. They were the things we put in the manifesto; the things I had run
through on the yellow sofas with Merkel before the election; the things
I outlined in a letter to Tusk in November 2015 as we formally began the
renegotiation.

However, I admit that I made mistakes. Since leaving office I have
thought about the renegotiation and its consequences over and over
again. Reliving and rethinking the decisions, rerunning alternatives and
what-might-have-beens. And, of course, I have spoken to people

I worked with in the UK and the EU about what could have been done differently. Talking to me after the vote, Merkel was clear that there was nothing more the EU was willing to put on the table. She was particularly adamant that the EU would never have granted an emergency brake on migration numbers. I have also spent some time running back over what was agreed with President Juncker's point man on the renegotiation, a Brit called Jonathan Faull.

The argument that there was the possibility of some much bigger deal, some grand renegotiation of the UK's entire EU relationship that would leave us with some sort of semi-detached membership within the EU, was always a fiction. There were, for instance, Eurosceptic Conservative MPs who dreamed of staying as a member while being able to accept or reject individual pieces of European legislation at will. They would say that if we really threatened to leave, anything and everything would be possible.

Those who pushed for this scorched-earth renegotiation, of course, branded the final deal a failure. But on their terms it could only ever have been a failure, because the things they wanted could only ever have been achieved outside the EU. Some sort of 'just visiting/overseas member' status *inside* the EU simply didn't and couldn't exist.

Indeed, we are now discovering that a semi-detached membership *outside* the EU is also something of a chimera. Negotiators have been surprised to find the strength of the EU orthodoxy that you are either in the single market, with its four freedoms, but with no serious say over its rules – like Norway; or you are outside it, and treated as a third country.

My biggest mistake was that I had not been more frank from the beginning about all this, and thus about the realities we were going to face. I had allowed expectations about what could be achieved through a renegotiation to become too high. This was partly because I had set out asking for fundamental reform to the EU as a whole, believing that more general treaty change was coming down the track: 'my vision for a new European Union, fit for the twenty-first century' was how I described it at Bloomberg. Such a general renegotiation might have ended up with only modest measures anyway, but the climbdown from multilateral to unilateral reform sent out a signal that we were revising down our ambitions.

I should have done more – particularly once general treaty change was off the table – to focus people's minds on what was really possible. The temptation in politics to answer people's hopes by referring to some future set of talks and treaties is great, but it is a temptation I should have resisted.

In terms of the substance of what was available for negotiation, I should have held firm and avoided the capitulations on the specific welfare measures. They weren't huge concessions, but they wrecked the simplicity of the offer. 'No welfare for four years' and 'no sending child benefit home' became 'not much welfare' and 'not much child benefit'.

And, perhaps because expectations were in the wrong place, the whole process, instead of making people feel the EU had made some valuable concessions, merely underlined for many observers how obstinate it was, and the power it wielded over us. More than one MP later told me the EU's behaviour during the renegotiation was what compelled them to campaign for leaving.

I also could have done a better job at tying the deal to what people wanted. It shouldn't have been 'I wanted a red card – I got a red card.' It should have been 'You wanted control of your laws – the red card gives you that. You wanted to reject a European superstate and a European army – ending ever-closer union guarantees that.' In particular I think I could have been better at explaining the welfare cap – how it would work, and how it would be a success.

Another criticism is that while semi-detached membership wasn't available, there was more we could have asked for and potentially achieved. I am less persuaded of this. Yes, in my renegotiation there was no end to the Working Time Directive, no repatriation of structural funds, no cap on the EU budget. I didn't even attempt these. But it was important to my strategy to go for what I thought was negotiable and what I thought was most important. Other reforms could come at a later date – and again, I failed to explain that this was just the beginning. Besides, could I really imagine anyone switching from Leave to Remain because we'd repatriated structural funds? No.

As well as the mistakes I made in raising expectations, the two things I ponder most are the measures on welfare/immigration, and the timing of the vote.

As I have explained, there were good reasons for 2016 rather than 2017; but my haste was helpful to the twenty-seven, as they knew I was anxious for an agreement. Since the referendum defeat I have agonised over the question of whether I should have paused the discussions about the offer in February 2016. I could have explained that Britain needed more, and that that would take more time. We could have restarted at the end of the year, and held the referendum in the middle of 2017, rather than 2016.

This links to the other question I return to: should I have prioritised, over everything else, an immigration cap, a limit on numbers? Given how the referendum played out, I think that is what could have made the biggest difference. Whatever the polls said about the problems with setting a number, a cap might have provided more of a feeling of control than the welfare deal, which was basically a deterrent. It would also have been more tangible and easier to understand.

What is unknowable is whether it was negotiable – and whether it would have worked politically. The officials who worked with me and Angela Merkel are clear: it was never going to be on offer. And, as with everything else, even if it had been, there would have been caveats. I suspect that such a brake would have been subject to EU approval, specifically by the Commission. It would have been time-limited. Leave campaigners would have been able to pick holes in the policy from the off, and undermine it completely.

What's more, it soon became clear that we are living in a post-truth age. The government could promise a cap that was agreed with the EU and completely legal – but campaigners could still have rubbished it. They could say 'It's a ruse – the EU will simply block the cap.' Or 'Why trust the word of the government?' Or even that none of the deal was set in stone anyway, and it could be disregarded. (In fact they did just that, as I'll describe in the next chapter.)

I have to accept that the deal was a failure, because it failed its main test: to help convince the British electorate to remain in the EU.

But there is one thing I will maintain. This was not a bad deal. We had negotiated significant reform, and created scope for further reform. The subsequent negotiations to leave the EU have shown just how hard it is to negotiate successfully with twenty-seven states and the body of the EU.

And the things I secured are starting to look better and better. They won't be available to us outside the EU. I think particularly of the welfare cap. Compare that with the deal offering a transitional period for the UK leaving the EU which would have gone on until 2021. Had Britain voted to remain, we would by then have had five years of a welfare cap. That could have saved billions of pounds, and helped restore a sense of fairness to our immigration system.

When I look back on 2016, it is not simply that we lost the referendum that disappoints me so deeply. It is that, because of that, we never got a chance to make this deal a reality. I really believe it could have created a place for us in the EU that we could live with for many years to come. Indeed, we could have thrived. Not only would it have been a springboard for further reform, we would have done something that no other EU country had dared. We would have secured important changes, refreshed our mandate for belonging to the organisation, and could have argued strongly for leading the EU in a new, more practical direction.

But back in February 2016 I wasn't just worried about the deal and the process – I knew that the fight ahead was going to be incredibly tough.

46

Referendum

It was New Year's Eve, and we were spending it at Chequers with several other families, including Michael Gove and Sarah Vine. Sarah and I were sitting by the fire in the Great Hall, chatting about the year ahead. I said I was worried about how the Conservative Party was going to sort itself into those who would back a vote to remain in the EU and those who would advocate leaving.

I turned specifically to what her husband intended to do. As a *Times* journalist Michael had been a strong Eurosceptic, and in the past he had argued that we shouldn't fear life outside the EU. I knew this part of his political make-up ran deep. But I also remembered how in opposition he had been instrumental in persuading me that the Conservative Party had to get over its obsession with Europe. Above all, he was committed to our project of modern, compassionate conservatism. He had been by my side for years – but would he be on my side in the biggest battle yet? Sarah said that he would.

I thought she was right, but George was worried. So as 2016 began, we decided that he would work on Michael and I would work on another waverer: Boris Johnson.

Why, when I had the votes of forty-six million people to worry about, was I so concerned about the leanings of these two men?

The Leave side of the campaign was shaping up as a coalition of disgruntled right-wingers and disaffected spads. The presence of these two front-liners would legitimise the cause and help detoxify the Brexit brand. Boris was the most popular politician in the country. Michael was respected by MPs, well liked in the Conservative Party, and had good connections in the centre-right press. I knew how persuasive their popularity and intellectual heft could be. And the polls backed me up: one suggested that if Boris stayed on board then Remain would lead by 8 per cent, but if he went for Leave that lead fell to 1 per cent. Other polls showed the Remain lead would halve if Boris supported Leave.

After his first conversations with Michael, George said things weren't going that well. So the next time we saw him was together, on the yellow sofas in the Downing Street flat. We talked about all the things we had done together, in opposition and in government. About the potential for even more radical change now that we had a majority. Michael was back in a job he loved, with a mission he cared about: reforming and improving Britain's prisons. All that would be put at risk by a divisive campaign, and even more so if the country voted to leave.

George spoke starkly about how he thought the Leave campaign would play out. It might start by being about the technicalities of British sovereignty, but it would soon slip into nativist arguments about immigration. 'The open, liberal Brexit you start off with will turn you into a sub-Farage,' I said. We'd be throwing out of the window all we'd done – and done together, as a team – to make this a modern, compassionate Conservative Party. George made it clear: if Britain voted to leave the EU, everyone, including me, would be finished.

Michael seemed torn – and really pained by the fact. 'My head is in a strange place,' he said. 'For once, I find it hard to articulate. But if I do decide to opt for Brexit, I'll make one speech. That will be it. I'll play no further part in the campaign.'

I found it hard to believe what was happening. Michael was a close confidant. Part of my inner team. Someone I often turned to for advice. Why hadn't he told me before that this might happen? Of course I understood his strong Euroscepticism, but if he was undecided – and it sounded like a 50–50 call for him – wouldn't his loyalty be the thing that brought him down on one side or the other? Not personal loyalty to me, but loyalty to the team, to the project, and to the future of our party and our country. But if he really was going to do this, back Brexit, then I believed him – really believed him – when he said he'd take a back seat.

The second big question mark hung over another former journalist.

I had a lot of time for Boris. I respected his talent. While I found some of his political antics infuriating, there was a reason for his appeal to the public. He was a good mayor of London. He was a great communicator. At his best he was ambitious for Britain, had big ideas and the energy to drive them through.

When I said I wanted him at the heart of my team, it wasn't just a case of it being better to have him inside the tent than outside. It was primarily that his talent made the tent bigger. We had talked before about what would happen when he finished his second term as mayor. Running a big government department would be a chance to show his serious side, I had told him. It would be good for him. I thought it would be good for the government too.

He and I had had many conversations over the years about the European issue. Boris was famously Eurosceptic, since his days covering Brussels for the *Telegraph*. He had strongly supported campaigns for a referendum in the past, not least over the Lisbon Treaty. He had sometimes made trouble for the leadership over the issue, particularly during his starring appearances at the party conference. But he had never argued to leave the EU.

We would talk on the phone and text regularly. Not just because he was a friend, but because he was the mayor of London, and I wanted him to be able to get hold of me whenever he wanted. News about issues in the capital would be mixed in with views about Europe – and frequent challenges to face him on the tennis court. So I fixed a date at the American ambassador's court, where we would be able to play and talk privately. Boris's style on the court is like the rest of his life: aggressive, wildly unorthodox (he often uses an ancient wooden racquet) and extremely competitive.

After our game we sat in the hut next to the court and talked about what would happen next. I started by saying how similar our political outlooks were. We were both One Nation Conservatives – his articulation of this theme at the party conference in October had been particularly powerful. The speeches made by George, Boris and me were virtually interchangeable, even though we hadn't really talked about them to each other beforehand. We had had our disagreements and tensions, but fundamentally we were part of the same team.

I didn't spend long talking about the results of my negotiation in Brussels. I knew that he thought they were disappointing. But I wanted him to accept that getting out of ever-closer union – something he had often spoken and written about – was real progress. This meant that we were improving on the status quo. He accepted that, but still believed that overall it was a missed opportunity.

My pitch was that he had never argued to leave. The logical position for him was to argue that our renegotiation didn't go far enough, but that we should vote to stay, and fight for more change in the future. I explained that I had done what I could without there being more general treaty change – but that treaty change would eventually come, and the opportunity to go further would be there.

'Let's play this out,' I said. 'Assume Remain wins. I'll bring the government back together and make new appointments. You will be a key part of that.' I told him he would have a 'top five' job. He ruminated on what was in the top five, given that he knew I wouldn't move George for him. 'Defence is a top five job,' I said. I was sure the hint was heavy enough to sink in.

I went on: 'I'm not going to be prime minister forever. At the next leadership election – during this Parliament – it will probably be between you and George. Obviously I'm a huge supporter of George, but it should be a fair competition, and you've got every opportunity to win it. This will give you the best possible chance.'

Our discussion continued by text after we went our separate ways. Boris had become quite fixated on whether we could sort the issue of declaring in legislation that UK law was ultimately supreme over EU law. This was a long-running Eurosceptic campaign in Parliament, and I had hoped that it might be addressed by domestic legislation, building on our removal from ever-closer union.

The proposal was for Parliament to pass specific legislation making our Supreme Court the final arbiter of the application of EU law in the UK. I knew that our Supreme Court judges would be much more likely than the European Court of Justice to interpret the meaning of EU treaties in line with the plain words of the text, rather than bringing in their own ideas about what you could stretch the text to mean if you wanted to promote maximum integration. I also knew that they would pay real attention to the fact that Britain was explicitly no longer part of ever-closer union.

This idea was modelled on the sort of constitutional protection some other EU countries – including Germany and the Czech Republic – already had, under which domestic law takes precedence over supranational law.

An argument against was that it would set up a future clash between Europe and the UK. But such confrontations never actually happen, because the EU backs down. So the device acts as a deterrent.

I saw an opportunity to find a 'win' for Boris on this issue that would give him extra cover for coming down on the side of Remain. Oliver Letwin would get to work on this specific bugbear of the most evangelical Eurosceptics, and embarked on a nightmare round of shuttle diplomacy between Boris (helped by the Eurosceptic QC Martin Howe) and the government's lawyers. But those lawyers were determined to defend the purity of European law, and therefore kept watering down the wording Oliver had agreed with Boris and Howe.

In some ways this episode epitomised a problem at the heart of the UK's relationship with the EU. Instead of pragmatically pushing the boundaries in order to make the EU's legal order more tolerable, our officials were determined to play strictly by the rules.

It soon became clear that while Boris cared about this issue, it was secondary to another concern: what was the best outcome for him? I could almost see his thought process take shape. Whichever senior Tory politician took the lead on the Brexit side – so loaded with images of patriotism, independence and romance – would become the darling of the party. He didn't want to risk allowing someone else with a high profile – Michael Gove in particular – to win that crown.

At the same time, he was certain that the Brexit side would lose. So opting to back it bore little risk of breaking up the government that he wanted to lead one day. It would be a risk-free bet on himself. He was doubling down: making doubly sure he would be the next leader.

I kept saying to him: don't take the course that you fundamentally think is wrong for the country.

To be fair, I could see that his agonising over the decision was genuine. He was torn between his head telling him that leaving would be a mistake, and his heart telling him to lead the romantic case for greater independence. I also accept that by holding a referendum I wasn't simply *giving* everyone, including cabinet ministers, a choice: in or out. The referendum *compelled* them to make that choice.

Boris argued with me that this chance – a renegotiation followed by a referendum – might not come again, and so it had to be 'seized'. When I challenged him that 'seizing meant leaving', and that 'out meant out', he would counter that in those circumstances there could always be a fresh renegotiation, followed by a second referendum. (This was a view he would repeat early on during the campaign, only to be rapidly silenced by his new allies in the Brexit camp.)

That said, the fundamental conclusion I am left with is that he risked an outcome he didn't believe in because it would benefit his political career. But I recognise my shortcomings in failing to persuade him; above all, the fact that my renegotiation hadn't done enough to create the conditions for more Eurosceptics to join my side of the campaign. Indeed, one of the greatest miscalculations I made was that I thought small 'e', small 's' eurosceptics like me – in Parliament, in the press and among the public at large – would, like me, see that staying in and fighting, with new reforms agreed, with all the opt-outs secured, with all the advantages over trade and cooperation, was the right course.

Perhaps that is one of the biggest pitfalls in politics. Thinking that others, particularly those you know well, think like you. Often they don't. In the weeks to come I would repeatedly be surprised by MPs, friends, local party members and councillors who I had never heard express the view that we should leave the EU waxing lyrical about how it was their passion. I don't mean to say that they were all opportunists, more that I had given them the chance to think about the issue afresh, and they had decided to take that position.

In the days before the crucial EU summit, George and I saw all the members of the cabinet, some of them more than once. There were those we were certain would be for Leave (Chris Grayling, Theresa Villiers, Iain Duncan Smith and John Whittingdale) and those we thought could be persuaded.

George did an excellent job with Sajid Javid, who was far more pro-Brexit than I had thought. I failed dismally with Priti Patel, who revealed that she had always wanted to leave, but succeeded with Liz Truss and Jeremy Wright, who had both been wavering. All these conversations were far harder than I had anticipated. The latent Leaver gene in the Tory Party was more dominant than I had foreseen.

I was determined to try with everyone.

My conversation with IDS was interesting. I went back to the Maastricht Treaty, when he had been a leading rebel, pointing out how much of it I had managed to put right with my deal. Ours was no longer a temporary opt-out from a common destination; it was the acceptance of different destinations. 'Isn't that what the Maastricht rebels had wanted?' I asked.

For a moment he seemed engaged. He said how much he had enjoyed focusing on domestic policy in government, on his mission to tackle poverty. That it had been good for him personally to spend less time on the European issue. But, as he put it, 'the cell door has been opened' – and this might be the only chance to escape.

By the time I left for the European Council I could already see that the Tory split would be closer to 50–50 than the 70–30 I had hoped for.

However, I didn't anticipate who would be quickest out of the stalls. As soon as I had sealed the deal in Brussels – indeed, while I was still in the building – it was Michael Gove who was the first on the media condemning it.

Then came a special Saturday cabinet, and it was a historic one. For ten years I had fought to keep the Tory Party united over Europe. This was the moment it would begin to divide again, with friends and colleagues taking opposing sides on an issue of fundamental national importance.

And yet, despite all that, it was civilised, dignified, and quite moving. I explained that we were there to do three things: decide the date of the referendum; agree the deal that I had negotiated; and determine whether the government's position should be supporting a reformed EU.

I then insisted on everybody speaking, and doing so in turn. My principal private secretary Simon Case had dug out the cabinet order of precedence, something we had never bothered with, which combines seniority of post and the length of time as a cabinet minister to determine the rank of everyone present, and therefore the order in which I would call them.

Those who said they would be campaigning to leave said they did so with heavy hearts. Those who would campaign to remain made some particularly strong points, some of which I'd go on to use myself. The first was what a fundamental deal this was in terms of the way Europe worked, and the movement of power to Brussels. Our referendum lock was a mechanism for stopping it from going further. Getting out of ever-closer union was the start of taking it back – and this was a genuine breakthrough. The second was something Patrick McLoughlin said to me later that day: 'I would love to live in Utopia, but I expect the EU would probably be there too.'

Everyone around the table – Leavers included – welcomed and supported the deal: a point I made in the room, and afterwards to the media.

I finished off with the words: 'Although we may find ourselves on different sides of the debate about our membership of the EU, throughout this period and afterwards, we will still need to govern and serve the public who elected us. Therefore, it is essential that, despite differences on this one issue, we remain a united and respectful team, and work together on all issues. There is so much more for us to do in government beyond the question of Europe – and we'll only succeed with our ambitious agenda if we continue to work together.'

Shortly afterwards, though, the extent of the split was brought home to me as I watched Sky News. Michael was joined by IDS, Chris Grayling, Priti Patel, Theresa Villiers and John Whittingdale at the headquarters of Vote Leave, which was later to be designated the official Leave campaign by the Electoral Commission. Lined up like that, I realised this was a parallel cabinet. And it had a leader, Michael Gove, who soon after was crowned 'co-convenor' with Labour MP Gisela Stuart. From then onwards, every time I turned on the radio or TV Michael seemed to be there, blasting the deal one minute, saying that EU membership was dangerous the next.

Boris was still to declare. And there was still hope: he admitted to the press that he was 'veering all over the place like a broken shopping trolley'. I was texting him furiously: if you're not sure, do what is right. We agreed to meet in No. 10 on the Saturday. He said that the work on asserting parliamentary sovereignty while in the EU had run into the sand, and concluded it was 'like sucking and blowing at the same time'. But his position was still in play. He was writing two opinion pieces – one in favour of leaving the EU and one in favour of remaining – to help him get his thinking straight. He was texting me throughout these vacillations, and seemed to change his mind at least twice.

But by Sunday afternoon I knew I'd lost him. I was at home in Dean when he texted me. He said Brexit would be crushed 'like the toad beneath the harrow', but that he couldn't look himself in the mirror if he campaigned to remain. 'It's not about you, it's about doing the right thing,' I replied. But it was too late. Nine minutes later he was on TV telling the nation that he had come out for out. I watched from the same chair I'd sat in for the 2015 exit poll. I knew what a serious blow this would be. He was the only leading politician whose favourability rating was higher than mine – crucially on the soft voters we were trying to attract.

So not only was my civil service battalion becoming a platoon; two of my key generals had defected in the first few days of battle.

There was no time to dwell. Over the next four months I would make fifty visits, eleven major set-piece speeches and scores of stump speeches at rallies and fundraising events. I'd write for almost every national newspaper and for local newspapers across the UK. I'd argue for Remain in television debates on Sky News, BBC, ITV, Buzzfeed and Facebook.

Making big arguments about the future of the country was energising. But I would begin to feel a constant sense of paralysis, unable to do what I needed to do, thwarted at every turn, swimming in a quicksand of my own making. It takes a lot to get me down, but the sensation of my feet being nailed to the floor when I needed to be advancing was the worst feeling in my political career.

We made some big mistakes in the campaign – I won't deny it. Nor will I play the blame game: this was a referendum of my making, and a campaign of my choosing. I think about it every day, and turn it all over in my head. As there ended up being only 600,000 votes in it, it's not far-fetched to say that if I had done something differently the result could have been different. It's impossible to say for certain.

What I can say is that throughout it felt as if a sort of cloud was hovering over us. Small to start with, but growing ever larger and darker. Gradually, every killer argument was being drowned out, and every advantage slowly sunk. Every trait of this age of populism – the prominence of social media, the emergence of fake news, anti-establishment sentiment, growing unease with globalisation, frustration over the level of immigration – appeared to conspire against our cause. It wasn't that Leave was besting us in every battle. It was that the physics of politics seemed to have changed. The upper hand became the losing hand, and the higher ground – which we felt we had captured – was surrounded.

At the very start of the campaign we felt like winners-in-waiting. The Remain campaign, named Stronger In, would be headed by Labour (predominantly New Labour) figures such as Peter Mandelson and Jack Straw's son Will, because we knew victory would rest on Labour votes. But it would be steered by Conservatives. In Downing Street we reinstated our Sunday evening meetings. Ameet ran the grid, Liz organised the visits, Craig did the press – each of them no longer just for me and the government, but for all the cross-party constituent parts of the campaign.

Initially, our opponents were in disarray. While Remain engendered cross-party consensus in the form of Stronger In, Leave spawned battles between the rival groups, Vote Leave, Leave.EU and Grassroots Out, that had vied for official designation. They may have had Gove and Boris, but they were also a cauldron of toxicity, including figures like Nigel Farage, Dominic Cummings and the businessman Arron Banks. There was something of the night about them that would, we hoped, put off many voters.

My principal task was to deliver the right messages. Everywhere I went, I explained that Britain was stronger, safer and better off in the EU. That my renegotiation had fixed many of the things that were wrong with our place in the organisation. That we now had the best of both worlds – a special status that gave us access to the largest trade bloc in the world, with opt-outs from the biggest burdens, like the euro, open borders, justice and home affairs powers, and the like. That leaving would be a leap in the dark.

But I went beyond slogans and broad-brush arguments. In Ahoghill, Northern Ireland, I explained how the EU gave farmers access to a market of five hundred million consumers. At the O_2 headquarters in Slough I told them how telecoms was helped by common rules and liberalisation of trade. At BAE in Preston, in front of a Typhoon fighter, I set out how EU membership made defence cooperation easier – and the more retail benefits like keeping the costs of flights and holidays down. In Wales, where almost 50,000 jobs relied on agriculture, I warned of the eyewatering tariffs outside the EU.

Anyone who says that Remain was a metropolitan, elitist outfit, and Leave was the voice of ordinary people, couldn't have been more wrong. This was calf-rearing, call-handling, aeroplane-assembling Britain. They were the ones who benefited most from EU membership and would be hit hardest by leaving. And I was the one going out there and speaking directly to them.

Going into March, I was buoyant. 'I feel I've had a very strong week,' I recorded. 'If anybody wants to know "Does he care about this?" Yes! I really am quite enjoying this campaign, because the arguments are very strong and it is actually quite refreshing to go into a workplace and not to talk about Tory cuts and all the rest of it.' Our chances in four months' time? 'I think it will be OK. I think it is going to be bloody hard work and very close, and it could go wrong. So it's not much better than 50–50.'

My tour outlining the economic advantages of EU membership culminated in a speech at the Vauxhall car plant in Ellesmere Port. It was the first of three speeches intended to cover each element of our three-part slogan. This one focused on 'better-off' – how massively the elimination of tariffs in the 1970s (from 32 per cent on salt, 37 on china and 17 on bicycles to zero per cent on everything) benefited Britain. It might have sounded esoteric, but it mattered – and my job was to get that across.

I wanted to focus in detail on automotives – literally the man under the car bonnet. Because this was one of Britain's, and this government's, greatest success stories. Our ability to export to the EU helped massively with that success. Of every hundred cars made in the UK, forty-four were sold to the EU. That wasn't just because of our proximity; it was because cars were 10 per cent cheaper in the tariff-free single market. And the size of that market meant that complex just-in-time supply chains had become established. Hundreds of parts could cross the Channel several times before becoming a UK-made car.

And all this is what appealed to investors. That was another great success of the government: becoming the world's leading destination for investment like this, after America. Along with our strong, low-regulation economy, it was EU membership that helped to attract that money.

I made it clear that it was the Leave side that was out of touch: 'When I hear people argue that, by being in this single market, we are "shackled to a corpse", I say: "You won't find the people in these industries saying that; or the towns whose employment depends on them."'

Again and again, it was Michael Gove who was getting his hands dirty. First, he came out to say my renegotiation was not legally binding, since the European Court of Justice could override it. We had to scramble government lawyers and the attorney general to say, no, an agreement by twenty-eight member states had the legal force of anything written into any treaty. More significant than the falsity of that claim was the fact that it was a direct attack on me and my integrity. I was naïve, perhaps, but the ferocity and mendacity shocked me.

Despite this, it felt very much – at this stage – as if Remain was on top.

There were two distractions from the campaign, however.

The first related to our replacement of the unassessed Disability Living Allowance (DLA) with the Personal Independence Payment (PIP), which would be assessed. The costs had been getting out of control. Some PIP claims were assessed on the need for mobility assistance, and some court judgements had taken the widest and most lenient interpretation of that – even down to whether someone needed to sit down to put their socks on. I sent George to agree a package of changes with IDS that would ensure the benefit reached those who needed it, and bring the cost down. We announced changes in the Budget on 16 March that would see thousands of people lose PIP or receive it at a reduced rate. A backlash from MPs and disability campaigners followed. Instead of standing by and explaining the decision, IDS distanced himself from it. Such was the reaction that we agreed to look at the proposals again – they weren't set in stone.

Two days later I was on my way back from a European Council when Ed called to say Iain had written me a letter saying he was resigning. I called him immediately: 'You don't need to resign. We can sort all this out. If you don't want to go ahead with any of these changes, we won't go ahead with them.' There were two things at play. I had already agreed to reconsider the plans. And at this precarious time in the campaign his part in the government, that crucial left–right balance, was more important.

Over the next few hours there were several calls between us. I said he couldn't resign without coming to see me first. He maintained that he couldn't go on as things were. He did seem to soften, but then it happened – again. I was talking to him, making the case for him to stay, when my phone buzzed with a Sky News alert: 'IDS resigns'. It was the second time I'd found out from the media that a minister had resigned at the same moment that I was on the phone to them. I was furious with IDS, but determined not to let it throw us off course.

The second distraction came on 3 April, when news outlets began reporting details of a huge leak of documents from the Panamanian law firm and corporate services provider Mossack Fonseca, which had helped many with tax avoidance. While this was a huge story, it seemed irrelevant to the campaign until it was revealed that an investment trust set up by my father, Blairmore Holdings, had used Mossack Fonseca, and that I had held some shares in it before becoming prime minister.

I knew that any accusations that this investment vehicle was established to enable my father or me to avoid tax were unfounded. It was a unit trust, registered with the Inland Revenue, whose price was quoted in the *Financial Times*. There are thousands of these funds, and they are not set up to avoid tax, nor do they help UK citizens to avoid tax. They tend to be registered outside the UK in order to encourage non-UK citizens to invest in them, but there was nothing secretive or underhand about this fund, or its use of a foreign law firm for some advice.

That said, the Mossack Fonseca leak was a landmark moment in the fight against corruption. It showed where some extremely wealthy people, including high-profile politicians, had been hiding their money. Leaders as far afield as Pakistan and Iceland would be prosecuted or would resign. The fact that Xi Jinping and Vladimir Putin had been named in the leak, I thought, made my late father's above-board unit trust a non-story. But I felt defensive because he was my dad, and furious about any allegations over his integrity.

My anger made me blind to the obvious point – the British press were going to have a field day, and would not let this lie until every question had been answered and every detail revealed. I was far too slow to act. It took the intervention of my brother Alex to make me see sense. He came to No. 10, told me we needed to rebut every accusation, and sat in one of the offices helping to get all the expert advice together, including from our father's stockbrokers, lawyers and accountants. For once I saw my brother the QC close-up, using his forensic skills and getting the job done. It was impressive. Combined with my answering over an hour of questions in the Commons on the subject – proving that I had nothing to hide and there was nothing to see – it worked. But we had lost a week.

One advantage we had in the referendum campaign was that we were the government, and the government had a position. I was determined that people should have the facts when they made this momentous decision – including the point that the government they had elected wasn't neutral on this. A precedent was set in 1975, when a leaflet had been sent to every household setting out the choice on remaining in or leaving what was then the EC, and making the government's position clear. After several drafts over several months, 'Why the Government Believes that Voting to Remain in the European Union is the Best Decision for the UK' began to drop on twenty-seven million doormats on 11 April.

I would argue that the points it made have aged well. One of its key arguments was how deep we were into the EU, and how hard it would be to extract ourselves from it: 'The government judges it could result in ten years or more of uncertainty as the UK unpicks our relationship with the EU and renegotiates new arrangements with the EU and over fifty other countries around the world.'

Another point was the imbalance in the relationship, and therefore the difficulties of negotiations: 'Some argue that we could strike a good deal quickly with the EU because they want to keep access to our market. But the government's judgement is that it would be much harder than that – less than 8 per cent of EU exports come to the UK while 44 per cent of UK exports go to the EU.'

On 18 April George carried the economic case further, announcing the findings of a study which showed just how much worse off we'd be outside the EU. Due to the lost trade and investment in the event of Brexit, by 2030 Britain's economy would be an estimated 6 per cent smaller than it would have been, working out as a loss to the average household of £4,300 per year.

This led to uproar that we were being too negative and scaremongering. As in the Scottish referendum, our campaign was branded 'Project Fear'. I believed it was Project Clear. The flipside to 'better-off in' was 'worse-off out', and we should leave people in no doubt about what that would look like.

People who saw this as unhelpfully negative fundamentally misunderstood the battlefield we were fighting on. As with the Scottish referendum, there were those almost certain to vote Leave and those almost certain to vote Remain. It was the undecided voters we needed to focus on, and all the evidence pointed to the conclusion that they would be persuaded by arguments of the head, not the heart. Hence our focus on the benefits of a stable economy, opportunities for their children, good jobs and shared security.

Another advantage of our position was the international support for our continued membership. Our friends wanted us to stay. And, impeccably on cue, the leader of the free world arrived in Britain for a visit. He made a very good point, saying that while it wasn't for him to tell the British people how to vote, he could take a guess at how the rest of the world would respond, including on trade deals. George seized on it. 'Why don't you say that?' he asked. After all, who was better placed to say what America would do in the event of Brexit: Boris Johnson or Barack Obama?

That afternoon, in the FCO's Locarno Room, we stood side by side on a podium as we'd done so many times before, at a press conference. In answer to a question about the referendum he said: 'I think it's fair to say that maybe some point down the line, there might be a UK–US trade agreement, but it's not going to happen anytime soon, because our focus is in negotiating with a big bloc – the European Union – to get a trade agreement done, and the UK is going to be in the back of the queue – not because we don't have a special relationship, but because, given the heavy lift on any trade agreement, us having access to a big market with a lot of countries rather than trying to do piecemeal trade agreements, which is hugely inefficient.' I was delighted with the intervention. Obama was basically saying that all the 'It'll be OK' assumptions from Leave were false. It stood to reason that of course a big country like America would focus on big blocs like the EU.

As Remain were rallying support, Leave were having difficulties setting out what life would look like for Britain outside the EU. Every time they cited examples of other countries that had relationships with the organisation, we shot them down. One day it was Canada – but its trade agreement had taken seven years to negotiate so far, wasn't completed yet and didn't cover services. Plus, Canada sent only 10 per cent of its exports to the EU anyway. Then it was Norway – but Norway paid roughly the same into the EU per capita as us, adopted EU legislation without having any say on how it was formulated, and took in nearly double the number of EU migrants per head that we did. Michael Gove suggested that Britain organise a European free-trade zone with countries like Albania, Bosnia Herzegovina and Ukraine. Even the Albanian prime minister described the idea as 'weird'.

At the end of April, with just two months to go, I took stock. Here we were with near-unanimous cross-party support, and the majority of parliamentarians on-side. We had the resources of government and the endorsements of experts to an extent that had never been seen before. We were deploying six years' experience in office, including two referendum wins and a majority at the general election, to great effect. Above all, we had the economic case nailed down, unanswerable, unquestionable. It felt that we'd captured the high ground.

And yet every position of advantage was undermined.

Our status and resources as a government were tainted by the fact that we were the establishment. We had hoped that putting something out with the badge of Her Majesty's Treasury or the backdrop of Downing Street would give it weight. Instead, it was treated with suspicion and derision. That £4,300 figure came in for particular attack. I would defend it. It was based on British GDP being 6 per cent lower than it would have been after fifteen years, which was not an unreasonable estimate, given how much trade we do with Europe. It was a neat way to demonstrate the real, tangible impact on families' finances. As for the argument about it being too specific, if you don't put down a figure, you'll be asked to name one. And as George reminds me to this day, the Treasury medium-term forecast was correct.

Our voices of support, despite all their expertise, were seen as remote, biased and therefore untrustworthy. Michael Gove even declared, in a defining moment of the campaign: 'I think the people in this country have had enough of experts.' It was an appalling thing to say. Yet it spoke to a deeper problem in our politics. We were living – and campaigning – in an age when feelings were prioritised over facts, where 'experts' could be dismissed as vested interests, elites, the establishment. Michael, one of the most learned, empirical people I knew, had suddenly become an ambassador for the post-truth age.

Our enthusiasm for the Remain case could easily be questioned. After all, we had shifted very quickly from 'big, bossy and interfering' negativity to 'stronger, safer and better-off' positivity. We were asking a lot of people to believe such a handbrake turn. And frankly, we hadn't done enough to trumpet the achievements of the EU, such as ending outrageous mobile-data roaming charges, cheaper flights and reciprocal access to other countries' healthcare systems.

This had a longer gestation than my time in office: British politics has had a strongly Eurosceptic undertone ever since the late 1980s. And there can be no doubt that parts of the press helped to feed this. That said, I accept my share of responsibility for not doing enough to balance the narrative. We should have done more – I should have done more – to mix criticisms of the EU with talking about its very real achievements; not least the two longstanding British objectives of creating the single market and enlarging the EU to take in countries that had emerged from decades of state socialism. In my defence, I would make the point that the EU did not make this easy. Yes, some stories about straight bananas and outlawing British sausages were inflated or invented by over-enthusiastic (or over-pressured) British journalists, but others were not.

More to the point, there was a big-picture 'diet of Brussels' narrative that had helped to shape British politics. In the years I spent leading the Conservative Party, treaty followed treaty, with power after power passing from Westminster to Brussels. More political union, fewer vetoes for individual nations, more money for Brussels, proposals for a European army, cash grabs for Eurozone bailouts – these weren't phantoms made up by the *Sun* or the *Telegraph*, they were real.

And the referendum sent the Eurosceptic press into overdrive. Much of it was predictable – some proprietors and editors were strongly Eurosceptic – but the scale of the onslaught was ferocious. They were also still extremely angry about Leveson, and for some this was a chance for revenge. Over six weeks the *Daily Mail* – circulation 1.5 million – ran eighteen immigration-related front-page stories.

I was particularly frustrated by the *Mail*, because it had never previously argued for leaving the EU. I asked its editor Paul Dacre to come for a drink in the Downing Street flat shortly before the referendum, and asked him why it was now doing so. He said, 'We have always been a pretty Eurosceptic paper.' I replied that I was a pretty Eurosceptic prime minister – it didn't mean I automatically had to argue to leave. 'If you are such a strong and long-standing Eurosceptic, why did you back Ken Clarke to be leader of the Conservative Party?' I teased.

The odd thing about my relationship with the *Mail* and Dacre is that in spite of the fact that he had railed against me as a candidate for the Tory leadership, and frequently ranted about things I did – from gay marriage to green energy – there was some mutual respect. He used to tell me that he admired what was clearly a ruthless streak in me. And although I hated some of what he put in his paper, I knew that he had a brilliance at reaching out and talking to middle England.

I ran through all the arguments with him about how leaving would diminish Britain rather than enhance our position in the world. There were moments when I thought I was getting somewhere. He even admitted that he had in the past flirted with the idea of a European army because he worried about our excessive reliance on the US. But I could tell that ultimately none of it would work: he was on a mission for Brexit.

Instead I tried the *Mail*'s owner, Lord Rothermere. The Leveson Inquiry had put our friendly relationship in the freezer, yet we kept in touch. Over a cup of coffee one morning in my office in No. 10, I simply asked what he thought about the Europe issue. Before he answered, I joked, 'I expect you're a bit like me – the EU drives you mad, but you know we have to be around the table.'

'No, my view is much stronger than that,' he replied. I waited for the diatribe against Brussels. But instead he said, 'I think it will be a disaster if we leave. I may even have to relocate some of my businesses to be inside the EU.'

It has been reported that I went on to ask him to sack Paul Dacre. Frankly, I wish I had – and I wish it had happened. I suspect he does too: two years after the referendum he replaced Dacre with the pro-Remain Geordie Greig. The closest I got was saying, 'Well, if that's your view, why on earth have you got someone editing the *Daily Mail* who is determined to drive us out of the EU?' There was a lot of harrumphing about not instructing editors, and we left it at that. The *Mail* had made its choice.

My difficulties with the media didn't end there. I knew we would not have the support of the *Sun* and probably the *Telegraph*, but I was hopeful that there would be some compensating support from left-leaning papers like the *Guardian* and the *Mirror*. That was indeed the case.

Ironically, almost the biggest problem I had was with the BBC. Of course it maintained its impartiality. And of course its reach and influence remained strong. The problem was, I felt, that it lost its way in terms of understanding the difference between balance and impartiality. The result was, for example, the voices of thousands of businesses arguing for Remain being given equal treatment to just two prominent businesses, Dyson and JCB, coming out for Leave. There were thousands of Remain economists, and a tiny number of Brexiteers, yet the BBC pretended they were equal, giving Economists for Free Trade the same weight as Nobel Prize winners arguing for Remain.

There was an opinion poll at the time purdah kicked in, meaning that the government could play no part in the campaign, that showed Leave ahead. Sterling reacted very clearly, falling immediately. The BBC failed to properly draw the link, even though almost every independent analyst did so. As the old saying goes, the job of impartial media is not to report one person saying it's raining with another person saying it isn't – it is to 'open the bloody window and see who is telling the truth'.

Another weakness of our campaign was that it seemed we were relying excessively on technical arguments, whereas the Leave campaign had the emotional arguments. I tried to put this right by speaking about how the EU had helped to entrench peace in Europe: the 'safer' part of our 'stronger, safer, better-off' triptych. So in the bright, domed atrium of the British Museum I spoke of 'the serried rows of white headstones in lovingly-tended Commonwealth war cemeteries [that] stand as silent testament to the price that this country has paid to help restore peace and order in Europe'. I then asked the question: 'Can we be so sure that peace and stability on our continent are assured beyond any shadow of doubt? Is that a risk worth taking?' The press turned it into 'Cameron Predicts World War Three'.

Leave meanwhile concocted their very own answer to 'better-off'. While we were saying membership was worth £4,300 to each family, they were saying that Britain spent £350 million a week on it. On 11 May, they unveiled their liveried battle bus, emblazoned with the words 'We send the EU £350 million a week. Let's fund our NHS instead.'

It wasn't true. As Boris rode the bus around the country, he left the truth at home. We didn't send £350 million a week to the EU. Our contributions were reduced by almost a third through our rebate, and the rebate could only be ended through unanimity. A British prime minister would have to *agree* to give it away. And the Brexiteers were using the *gross* figure for contributions, not the *net* figure, which took into account the EU spending, including on science and agriculture, that was sent back to the UK. That would have reduced the figure to something like £160 million a week. More to the point, the boost to GDP, and thus to UK tax revenues, through frictionless trade and investment far outweighed any contributions we made. And, of course, we were already spending more than ever before on the NHS. Its budget was ring-fenced, so it wasn't affected by our spending on the EU anyway.

So the bus was disingenuous, it was tenuous – but it was also ingenious. The fact that it was inaccurate actually helped the Leave campaign. They wanted the row, because it continually emphasised the fact that we sent money – however much it actually was – to the EU. Post-truth indeed.

The NHS was voters' top priority, and their top concern was immigration. Since the economy was, the press considered, 'done', immigration was the new topic to focus on.

On 20 May, Michael Gove said that EU immigration would mean up to 'five million extra people coming to Britain' by 2030, and pointed to the EU's stated objective of admission for Albania, Macedonia, Montenegro, Serbia and Turkey.

The £350 million bent the truth, but this new assertion stretched it to breaking point. There was no prospect of Turkey joining the EU for decades, if ever. It had merely applied, and was in talks. The rationale for this was that it encouraged the country to meet the criteria for membership – a free press, the rule of law, open markets, things that benefit us all. Maybe one day, when the world looked very different, Turkey might join. In any case, like every other EU member, the UK had a veto over any new country joining.

Yet when the armed forces minister Penny Mordaunt went on *The Andrew Marr Show* the next day, she denied that the UK could veto the accession of Turkey – and when she was challenged, she repeated it. We were no longer in the realms of bending or stretching the truth, but ditching it altogether. Leave were lying.

'At Turkey's current rate of progress, it would probably join the EU in the year 3000,' was my rebuttal. But that was drowned out the following day when Vote Leave launched a poster warning: 'Turkey (population 76 million) is joining the EU', next to a picture of a British passport. I couldn't believe what I was seeing. Michael Gove, the liberal-minded, carefully considered Conservative intellectual, had become a foam-flecked Faragist warning that the entire Turkish population was about to come and live in Britain. As for Boris, who proudly trumpeted his Turkish heritage, and who had advocated Turkey's membership, he was now backing the false claims about its accession.

It didn't take long to work out Leave's obsession with Turkey. Five hundred million Europeans had the right to come to Britain already. Germans, Italians, Poles . . . Why focus on a country that wasn't a member, and wasn't likely to become one? The answer was that it was a Muslim country, which piqued fears about Islamism, mass migration and the transformation of communities. It was blatant. They might as well have said: 'If you want a Muslim for a neighbour, vote Remain.'

I was being urged by Craig, Ameet and others to rule out Turkey ever becoming an EU member while I was PM. But I felt that would be irresponsible – the country was in the EU waiting room for a reason. And by saying a veto was necessary now was tantamount to accepting that it *could* join shortly.

So paralysis had me in its grip. I was caught between being a campaigner and being a prime minister, and I chose the latter. It truly was asymmetric warfare. I made the wrong choice.

While Leave weren't telling the truth on Turkey, they did have a broader truth on their side. Immigration was a problem. The numbers remained stubbornly high. What followed was possibly the worst timing of anything in my premiership. On 26 May, with less than a month to go until polling day, the Office for National Statistics released its new immigration figures. In 2015, net migration had hit an all-time high of 333,000. It made our 100,000 target look ridiculous. And there was apparently a very simple way of stopping it: ending free movement by leaving the EU.

The reality was of course more nuanced. Over half of that immigration was from outside the EU, so it was completely irrelevant to the referendum. There we must do better. But the numbers from inside the EU would come down because of the new welfare restrictions. Plus, these were unusual times: we were creating more jobs than the rest of the EU put together. It was no surprise that people were coming here. In any case, wrecking our economy by leaving the single market wasn't the answer.

That's how I saw it – and said it – as prime minister. But sitting one evening in a BBC Yorkshire studio after an interview, I saw it from a voter's perspective, as the channel played a five-minute package on Slovakian travellers in Rotherham. Local people were saying, 'I'm moving out. These people are taking over our parks and public spaces and the place is a complete mess,' and, crucially, 'We never voted for this, it's not fair.' Their comments were understandable. That was the moment I realised our argument about the economy was quite complicated, while Leave's argument about immigration was very simple. I thought my worst nightmares could come true. I thought, we could lose this thing.

In the wake of the ONS figures, Gove and Johnson wrote me an open letter criticising the tens-of-thousands immigration pledge, and Boris spoke out in an interview about 'the scandal of the promise made by politicians repeatedly that they could cut immigration to the tens of thousands'. It wasn't 'politicians' they were berating; it was me. It wasn't any old pledge; it was part of the manifesto they had just been elected on. The rules of engagement had been abandoned. This was open warfare.

I was surprised to find new-intake MPs like Suella Braverman striding into the spotlight as raging Brexiteers. But it was the behaviour of the employment minister, Priti Patel, that probably shocked me most. On 28 May she wrote an article for the *Telegraph* critical of the 'wealthy' leaders of Remain, who could never know the downsides of immigration. She explicitly criticised the Conservative manifesto (upon which she had been elected) and the cabinet (of which she was a part). She subsequently used every announcement, interview and speech to hammer the government over immigration, even though she was part of that government. I was stuck, though: unable to fire her because that would make her a Brexit martyr, and would simply fuel the psychodrama in the Conservative Party that we were trying to stop.

By the time we got to June, referendum month, Boris and Gove were pledging to introduce an 'Australian-style points-based immigration system' before the next general election if Britain voted Leave. Quite apart from the fact that we already had the equivalent of a points system (clear categories for immigrants, with some channels such as unskilled labour from outside the EU set at zero), our system was tougher than Australia's, and our level of immigration was much lower. But the real point was that this wasn't Conservative policy, it was their policy. They were setting themselves up as an alternative government.

I refused to rise to it, and pulled my punches. Again and again the option came to hammer Boris and Gove. 'These are now your opponents. They're killing you,' George said. 'You've got to destroy their credibility.' Every time I was shown a mocked-up poster like one of Boris in the pocket of Farage (like Ed Miliband and Alex Salmond in 2015), I vetoed it. I was tethered to the responsibility of my roles – passing up the chance to rule out Turkish accession because I was leader of the country, and passing up the chance to savage these ministers because I was leader of the party. Besides, I just didn't think it would work. Returning fire in 'blue on blue' attacks would just make the campaign look like a Conservative spat, and would encourage others to sit it out. Or so I thought.

I wanted others to fire the weapons in our armoury that we couldn't. Our cross-party clout relied mostly on Labour, but they were AWOL. During the entire course of the referendum campaign Corbyn delivered a handful of desultory speeches about Remain, and went on holiday for part of it. He criticised George's 'fear agenda' and proposed a (narrowly averted) visit to Turkey, of all places, to talk up free movement. The Remain campaign cleared whole days for Labour in advance, but often they would do nothing. Most voters could spontaneously name me as the chief advocate of Remain, but they thought of Barack Obama and Mark Carney as Remainers long before Jeremy Corbyn came to mind.

Perhaps Corbyn wanted Remain to lose. First, the EU doesn't fit into his world view. He's anti-capitalist, and it's a trading organisation. He voted to leave in 1975, and had opposed every treaty – the single market, Maastricht, Nice, Lisbon – that had come before Parliament over the years. Second, he probably looked at the situation and thought that if he hung back and let Britain leave, it would destroy the Tories as the financial crash had destroyed Labour.

As the campaigning stepped up, I made a speech on a City roof garden calling out Leave's lies. I even went to a Hare Krishna centre in Hertfordshire to persuade worshippers. I did everything I could, taking every opportunity, deploying every reasonable, rational, persuasive line, speaking from my heart. But it was like one of those bad dreams where you're trying to shout but no sound is coming out.

At points the dream felt surreal. I found myself at the Oval cricket ground, giving a speech next to a blue Mini. The Lib Dems' Tim Farron stood next to a yellow one, Harriet Harman next to a red one, and the Greens' leader Natalie Bennett beside a green Brompton bike. It was the sort of colourful backdrop we would have used in the general election, and a good visual representation of unity. But somehow (note the absence of Corbyn) it just didn't have the impact we'd hoped for.

Likewise, the appearance of John Major and Tony Blair together on the same platform in Northern Ireland should have been a real 'moment'. And it was a powerful way of showing the risks Brexit posed to the peace process in the North and the open border on the island of Ireland. One of the key reasons we gave for Britain remaining in the EU was because leaving might demand a hard border between Northern Ireland and the Republic, potentially inflaming tensions and undoing years of hard-won peace. Those who claim it was not raised at the time conveniently forget this.

So what to do? The paralysis, that feeling of not being able to make ourselves heard, came back to immigration. I had no clear answer. At a crunch meeting in the flat on the evening of Sunday, 12 June with my team and George, we talked it through. Those from Stronger In wanted at least an intervention making our full case on immigration. Others, including many of my spads, wanted more: a new pledge on reducing immigration, similar to the last-ditch Vow in the Scottish referendum. Some of the Labour figures were calling for a new fund for areas hit hardest by immigration. And many were saying: can't we just try again with Merkel and the EU? Can't we tell them Britain is about to leave the

EU if they don't give us more on immigration? We kept coming back to the importance of maintaining message discipline about the economy, because, as Lynton had put it to me, 'All Leave has is immigration. We shouldn't concede that it is the only battle to be fought.'

A few days later I did have a conversation with my friend Mark Rutte, who held the rotating presidency of the European Council. 'Look,' I said, 'we're in danger of losing this. The problem is the lack of a brake on numbers. Is it worth talking to Juncker, talking to Merkel, trying to come up with something that says we will address this issue?'

He was helpful but sceptical, and the more I thought about it, the more doubtful I became. I talked to Tony Blair and John Major about it. They both agreed that it would just raise the profile of the issue without actually solving it.

So by the time I talked to Merkel, I told her I was going to push on with the plan we had. 'But I want you to know,' I emphasised, 'that this is the major problem – and if we lose, *this* is why we're going to lose.' She simply said that it would be wrong to change tack on migration, and that those who had done so in the German elections had lost.

Still, we desperately needed to change the conversation. So we decided to recommit to Plan A, and force the agenda back onto the economy. On 15 June, with just over a week to go, George announced that if the UK voted to leave, he would be forced to carry out an emergency budget in order to plug the £30 billion hole Brexit was expected to gouge out of the UK's finances. He said he would have to raise income tax, petrol and alcohol duties and inheritance tax, and to cut health, defence and education.

That day, fifty-seven Conservative MPs signed a statement saying they would not vote for any emergency budget. Four party grandees – Norman Lamont, Nigel Lawson, Michael Howard and IDS – published a letter in the next morning's *Daily Telegraph* accusing Remain of attempting to frighten the electorate.

Looking back, I accept that this 'punishment budget' did seem over the top. But the realities of Brexit *were* OTT, and people needed to know them. What's more, the atmosphere was febrile. You had to shout to be heard. I respected George for doing so. He threw everything into the emergency budget, even though he knew it could end his career. It was a contrast to others who wouldn't put their necks on the line and full-throatedly back Remain, in case things went the other way. Those who were hedging their bets in this way simply added to my frustrations.

We were simultaneously trying to make the patriotic, positive case, including plans for a rally in Gibraltar, the overseas territory at the tip of Spain which is proudly, democratically British. Before I departed I was confronted with Leave's most despicable attempt yet to drag the focus back to immigration. As I briefly described earlier, Nigel Farage stood alongside a poster entitled 'Breaking Point' that showed thousands of mainly male, mainly adult, mainly darker-skinned migrants filing across green fields. If Vote Leave's big Turkish hint was a dog-whistle, this was Leave.EU's foghorn.

The parallels with Nazi propaganda were being shared online immediately. And yet in these topsy-turvy, post-truth times, all this went in Leave's favour. If you're criticising it, you're still talking about it – that was their rationale throughout the campaign. If the £350 million was disputed, then good – the subject of the cost of the EU was being raised, whether the figure was accurate or not. Suddenly I realised how much of an advantage their side had in being divided and fractious. Different factions could target different audiences. They were guerrillas to our conventional warfare.

But as our plane came into land at the foot of the famous Rock and our phone signals kicked back in, excitement about the rally and rage over the poster were replaced by other emotions. A Labour MP, Jo Cox, had been attacked in the street on the way to her constituency surgery in Birstall, West Yorkshire. Her condition was critical.

Ed, Liz and I remembered meeting her – as Jo Leadbeater – in Darfur back in 2006. She entered Parliament in 2015, and I knew her as a small, punchy person who always sat in the same place in the Commons, and frequently asked me probing questions about the Syrian refugee crisis.

I immediately called Fabian Picardo, the chief minister of Gibraltar, to cancel the rally and send the thousands of people who had gathered in the main square home. Instead we would meet him and some other dignitaries in private.

It didn't take long for what had happened to become clear. Jo had been on her way to her constituency surgery when a man attacked her with a knife and a gun. I then received a message that she had died of her injuries. The first MP to be murdered since Ian Gow was killed by the IRA in 1990.

Everything else, all the arguments and preoccupations of the past few weeks, suddenly seemed small, distant and irrelevant. A woman had lost her life – a husband his wife, two children their mother, two parents their daughter. Parliament had lost a politician with enormous promise. I felt sick. And I felt even sicker when it became apparent the referendum wasn't irrelevant. The attacker – the murderer – had been heard to shout 'Britain first,' or 'This is for Britain.' Jo had been a staunch Remainer.

I was clear in my mind that the deranged actions of one person shouldn't stop democracy. The referendum would still go ahead, but as a mark of respect to Jo, campaigning should cease for a period. I joined Corbyn in Batley for a moving ceremony to remember Jo, then recalled Parliament after checking with her husband that it was what she would have wanted. The whole House wore the white rose of Yorkshire. 'We are far more united and have far more in common with each other than things that divide us' – words she had used in her maiden speech just a year earlier – echoed through the Chamber. I thought to myself how, in this age of populism, they were needed more than ever.

When the campaign resumed, we were still waking up each morning to the views of the latest expert or industry on the merits of Remain. I thought it was one of our greatest advantages that nearly every voice that mattered backed our case. The voice of major industries: cars, planes, trains, food, pharmaceuticals, farming, fashion, film. The voice of business: the CBI. The voice of many workers: the TUC. Our allies around the world: America, India, Japan, Australia, Canada. The multilateral bodies of the world: the IMF, the WTO, the OECD. Thirteen Nobel Prize winners. The head of the NHS. The former heads of MI5 and MI6. The head of the Church of England. Nine out of ten economists. Stephen Hawking, Tim Berners-Lee and Richard Branson – truly great Britons who so many people admire and respect. 'Maybe it's a conspiracy,' I would say. 'Or maybe all these people are right.'

How relevant, then, were the views of a former government adviser who had emigrated to California four years earlier? Probably worth a nib in a newspaper. But because he had been a friend of mine and a controversial figure in government, the press were keen to hear what Steve Hilton had to say on the referendum. Steve Hilton was ready to oblige. He started by saying that if I wasn't PM I'd want to leave the EU – not true. He claimed that I had been told in 2012 that my migration targets were inaccurate, and impossible to achieve – again, not true: I even checked the paperwork, which showed that while he was director of strategy, net migration was falling. It got down to 150,000, not far off the tens-of-thousands ambition.

Steve stuck around. He happily joined Boris on the bus, and did several interviews. The BBC cast him as the latest character in the Tory soap opera, running his intervention as big news just two days before the vote.

The last week would be about using every tool in my armoury. Every supporter summoned. Every argument deployed. Every advantage exploited. I'd spent hours with Bill Knapp ahead of the final TV debate, a special episode of *Question Time*. It was every bit as brutal as we'd expected. At one point I was asked whether I agreed that I was a 'twenty-first-century Neville Chamberlain'. But that was exactly the red rag needed to bring out my bullishness: 'In my office I sit two yards away from the Cabinet Room where Winston Churchill decided in May 1940 to fight on against Hitler – the best and greatest decision perhaps anyone's ever made in our country, right? Now, he didn't want to be alone, he wanted to be fighting with the French and with the Poles and the others, but he didn't quit. He didn't quit on Europe, he didn't quit on European democracy, he didn't quit on European freedom. We want to fight for these things today, and you can't win, you can't fight if you're not in the room. You can't win a football match if you're not on the pitch.' My finest hour of the campaign.

That's the thing. It's the reason I was so sure about Remain. I was prime minister – not justice secretary, the ex-mayor of a city, the leader of a fringe party, or an ambitious spad. I could say with more authority than anyone the importance of the role the EU played in helping me, as leader of the country, keep Britain safer and make it more prosperous.

On the Tuesday I stood at a lectern outside No. 10 and made my last-ditch case. This was the place I'd gone to tell the British people the biggest, most important things as their PM. Now it was where I was imploring them to listen to me on Europe. To believe me when I said that the EU didn't diminish Britain's influence, it amplified it. That there would be no deal on the outside better than the special status we had on the inside. That this was a decision not for now, but for the future – and it was final.

We began Wednesday, the final day of the campaign, on the road together with Harriet Harman. The battle bus became more of a Noah's Ark as passengers joined us at each visit. We went to a farm, and picked up two farmers. After a garden centre visit with Paddy Ashdown we gained two veterans. We picked up a handbag designer who had just started exporting to the EU, someone from Jaguar Land Rover, and the Green Party's Caroline Lucas. A trip to a school with TV historian Dan Snow left us with a schoolboy, a father, a grandparent and a head teacher on board, followed by a midwife and nurses after we briefly called in to Solihull hospital. The day ended with all these people, and many more, at a rally in Birmingham, where we met Gordon Brown. He delivered a powerful speech, an ideal eve-of-poll message.

After the non-stop speeches, visits, interviews, handshakes and conversations, the day ended, as was so often the case, with just me and Liz. As we sped down the motorway, sipping the beer she had pilfered from Birmingham University, I felt that sense of calm that comes from being able to do no more.

We got back to Downing Street at 8.30 p.m., and I had some dinner with Sam. It was one of those long, light June evenings, and we sat out in the garden, talking about our hopes and fears.

On Thursday, polling day, the two of us went to vote at Methodist Central Hall. All eyes were on how many voters would be doing so too, because the result, we calculated, depended on turnout. We reckoned that if it was moderately high – over 55 per cent – Remain would edge it, since our supporters were more numerous and more likely to vote. No one even mentioned how that logic might change if turnout was unusually high. It just wasn't considered.

Meanwhile, there were independent surveys coming out from the City – hedge funds that had done big ring-rounds – broadly calling it for Remain. Lynton told me it was going to be OK. Jim Messina told me it was going to be OK. Andrew Cooper reported a ten-point lead for Remain. Craig was charging around in a Stronger In T-shirt saying we'd won.

My draft victory speech was an 'open, comprehensive and generous offer' to the nation for a way forward: a route map for healing our divided country after a brutal campaign, and an assurance that we would heed the message from millions of Leave voters that, while we were staying in the EU, the status quo was no longer acceptable: 'As far as Britain is concerned, the political project for further integration is over.' I was looking forward to saying that.

After spending time working up a defeat speech, I let my mind drift to what was more likely to happen next. A reshuffle was shaping up in my head. George to the Foreign Office, and Jeremy Hunt or David Gauke to the Treasury. Boris to defence, or possibly a mega-housing and planning ministry. A few Leavers to replace Remainers. Gove to stay put.

Then there was me. I thought how dramatically my political life-expectancy had been reduced by the brutal campaign. I had already confessed to my core team that I'd probably have to go within a couple of years, even if we won. Maybe I could last until the Commonwealth conference in April 2018, or possibly the party conference that year, doing another two years rather than the full Parliament.

That afternoon Sam and I wandered around the home-cum-workplace we'd come to love. The first-floor and second-floor meeting rooms had been turned into men's and women's dormitories lined with camp beds.

By the evening the state rooms were buzzing with people who had worked on the campaign and a few of Sam's friends. As they helped themselves to the buffet of moussaka, there were some very nervous faces. I found myself going around reassuring people, telling fraught-looking spads, 'It's going to be all right.'

In the Pillared Room, a widescreen TV had been wheeled in front of rows of chairs. I sat in front of it, George by my side, others buzzing around. Ten p.m. came, and though there are no exit polls in referendums, there was a YouGov poll predicting a 52–48 victory for Remain. ComRes had it as 53–47 to Remain. Gibraltar's early result of 96 per cent Remain was inevitable, but nonetheless cheering. By 11 p.m. Nigel Farage had conceded.

We decamped to the Thatcher Room, and sat around the table I'd had made for the G8 leaders three years earlier. Nancy was sitting next to me in her pyjamas. I had one eye on the TV and another on a rather clunky old laptop. On it was a spreadsheet from Jim Messina that set out how many votes were needed from each polling district in order to get a 50–50 result or better. If the real results matched up to or exceeded those figures, we'd be on track.

The first results I saw came through from Tash, my agent in Witney, and seemed to be all right. They needed to be 60–40 in Eynsham ward, and they were 60–40. I thought, 'This is going OK.'

But at midnight Newcastle declared for Remain by just 1,900 votes. It was a worrying deviation from the model – but perhaps it was an aberration.

Then, twenty minutes later, Sunderland declared for Leave by a huge margin. I noticed too that West Oxfordshire's other wards were just a little bit worse than they should be. Before I knew it, Leave was in the lead.

At 2 a.m., results poured in from a cross-section of the country – Brentwood, Flintshire, Middlesbrough, Weymouth and Portland, Merthyr Tydfil, Stockton-on-Tees. They all declared for Leave. Scotland and London's vast and inevitable Remain votes weren't enough to offset them, and we were level-pegging. 'Dad,' Nancy said, 'we're losing.'

With the declaration of South Buckinghamshire, which voted Leave by 51–49, I knew it was over. From quiet confidence to quiet acceptance of defeat – all within the space of a few hours.

After a 2.30 a.m. 'council of war' in my office with Ed, Kate, George and Andrew Feldman, I asked them to leave. I wanted to be alone with my thoughts. I lay on the sofa, knowing I wouldn't sleep, but just needing quiet and calm. The TV was on low, and I tuned back in when the electoral sage John Curtice spoke. Leave were now the favourites. Soon enough, David Dimbleby called it. Britain had reversed its 1975 decision to enter the Common Market, and voted to leave the EU.

What had I shared with my closest advisers and friends? I had told them I was certain about what I needed to do next if we were to lose. It was as clear to me as it had been in 2010 when I formed the coalition, and in 2013 when I pledged the EU referendum. I would have to resign. I had thought about this from the moment I called the referendum. I knew it would be the end of my political career. Perhaps in months, rather than days or weeks, but the end nevertheless. But it wasn't until the Friday before polling day, when George came to stay with me at Dean, that the hypothetical suddenly became the potential.

We had talked through the possible outcomes. First, we knew that a Remain victory, irrespective of how large it was, would still bring big political difficulties, and must be followed by magnanimity. We needed to accept – as I did – the strength of the case about an over-mighty Brussels that our opponents had made.

Then we looked at the losing options. George made the argument that there is always a case for staying on. Who knows where the negotiations would go, what new opportunities there might be. I could only steer towards them if I was in the seat of power. My view was different: if we lost I would have no credibility. I would have to leave. Staying on would simply be delaying the acceptance of a political death that had already taken place.

We wondered whether a very narrow loss would be different to a clear loss. Would there be room for a return to Brussels to ask for more elements to our special deal? Maybe. Perhaps there was a 'triple lock' we could put in place – a vote of the parliamentary party and a vote of the House of Commons that I should continue in office, followed by a new public vote on a new deal, if one was possible. This fictional future survived in my mind for little more than twenty-four hours.

Why? Because all the losing scenarios collapsed into the same outcome: I had lost, and I would have to go.

There was no realistic scenario in which the EU would offer more if I went back to it in the event of a narrow result.

There was no scenario in which the British public – and the Conservative Party – would accept anything other than a prime minister committed to carrying out the verdict of the referendum result.

The principal task of the next PM would be to go back to Brussels and secure a deal for the UK's departure. Personal diplomacy and credibility would be essential. For that, a fresh face would probably be better. As Lord Carrington put it in his memoir, about resigning in 1982, 'the government was in for a hard time and my presence would make it not easier but harder'.

Because I was the person who had led the campaign to stay, and had argued passionately for one side of the argument, I would have no credibility in delivering Leave. It wouldn't just have been hard for me to deliver a policy I didn't believe in – I wouldn't have been able to do so. It wasn't in the national interest for me to stay.

I believed at the time that that judgement was right, and I haven't changed my mind since. Indeed, I would argue that events have borne my decision out. Even with all her protestations that 'Brexit means Brexit and we will make a success of it' – words that I would have found all but impossible to say – my successor found the task of maintaining credibility while trying to deliver Brexit very hard. For me it would have been impossible.

I hated walking away from my job having just won an election. I hated the fact that many members of my team of ministers and advisers, who had worked so hard, would lose their jobs too. I hated abandoning our policies and plans. We had a great manifesto we needed to implement. Losing the 'what might have been' was more painful for me than anything.

So why had I promised that I would stay on if we lost? Initially I hadn't believed that losing would mean an immediate announcement of my departure. But more importantly, I knew that I must do all I could to keep my future separate from the issue of the EU in the public debate. If I had admitted that there was *any* chance of my stepping down if Remain lost, I would have jeopardised the referendum entirely. (This is what would happen to Matteo Renzi in Italy just six months later, when a referendum about reform became a referendum on his future because he said very publicly that he would quit if he lost.)

But I accept that resigning meant going back on my word. Just two days earlier I had stood on the steps of Downing Street and announced, 'Brits don't quit,' and now I was doing just that. I knew that some people would be both angry that I left and angry that I had said I wouldn't – Liz was particularly concerned about the reaction. But I was in absolutely no doubt. Whichever way I looked at it, I couldn't make it work for me or for my country. I was 100 per cent convinced that Britain needed a new prime minister.

That is what Craig, Kate and Ed were urging too. It's what Samantha believed. For them, the main point was that the Tory Party, having largely voted for Brexit and now secured it, would not tolerate me remaining as leader. In that respect the only choice was between leaving at the time of my own choosing, and trying to stay on and being hounded out.

I went upstairs and lay down on the bed for a bit, had a shower, and came back downstairs just before 6 a.m. Everything had been planned very calmly. Samantha and I would go out into Downing Street together. She said, 'I just don't think I can go out there – I feel terrible,' and had a stiff gin at ten past eight, just before we walked hand-in-hand out into the daylight and a wall of cameras.

I praised the Remain campaign, and congratulated the winning side. I said the result must be respected, but, seeking to reassure the markets, which had plummeted by 8 per cent, that nothing would change immediately. But ultimately Britain would need to negotiate a new relationship with the EU, from the outside. I turned to my own part in that. In an analogy I'd thought up the weekend before, I said, 'I will do everything I can as prime minister to steady the ship over the coming weeks and months, but I do not think it would be right for me to try to be the captain that steers our country to its next destination.' I finished: 'I love this country' – I nearly lost it on that word – 'and I feel honoured to have served it, and I will do everything I can in future to help this great country succeed.'

Having just held it together, I came back inside No. 10 and gave Sam a kiss. The staff clapped me in, just like a year earlier. And, just like a year earlier, I went upstairs and made us breakfast.

This time the children had already gone off to school, and I hoped they were all right. Nancy and Elwen had been so engaged in the campaign, and so sweet and supportive to me. I knew they knew I was stressed, because they'd been hugging me more than usual. Nancy had been taking my 'Conservative In' campaign badges and giving them to her friends. There had been a contretemps between her and a bigger girl at the school fair, who had asked if she was for 'out' or for 'in', as in Remain. Nancy replied she was for in. The girl said, 'Well, fuck you.' Nancy replied, 'Well, fuck you too.' Sam and I had never heard her say the 'f' word before she recounted this story. We thought it was a bit shocking, but rather extraordinary.

Totally by chance, later on in the morning I gave my speech outside No. 10, Elwen was due to take part in a school project where they would act out the United Nations having a debate on human rights. They'd been rehearsing, with a German girl in his class playing Angela Merkel, an American boy playing Barack Obama, and Elwen playing me. The teachers asked him that morning if he wanted to go ahead, or if it would be too upsetting given what had just happened. 'I want to do it for my dad,' he replied. His performance apparently had the watching parents in tears.

The whole time, I tried to be the one reassuring people. As members of staff cried, I tried to make jokes. As Samantha wept, I poured her another gin. I got stuck into what was ahead as if it were any other day, and I shed no tears. But when I came to watch a recording of Elwen at the pretend United Nations, defiantly declaring that he was Prime Minister Cameron, it all started to sink in. The significance of what had happened. The shock and the sadness. The fact that I had failed: failed to win the referendum, failed in my vital task of trying to keep Britain in the EU on a better footing. It had been right to give the people a choice. I was sure about that – and I still am today. I couldn't have given the campaign any more. But my regrets about what had happened went deep. I knew then that they would never leave me. And they never have.

47

The End

I had hoped that on Friday, 24 June I would be making a speech confirming Britain's new settlement within the European Union, and looking ahead to several more years in office. Instead, Britain was leaving the EU and I was leaving the job I loved.

I gave little thought to my own emotions. Nor did I dwell too much on the reasons behind the result. These things would come later. I knew that the enormity of what had happened, and the consequences for our country, would stay with me for the rest of my life, that I would turn them over in my head again and again.

But that day my focus was simple. I had to carry on being prime minister, and attend to the responsibilities that accompanied that. I was single-minded in my focus on the task ahead: to muster what dignity I could as I set a sensible course for handing over to a successor.

The first time I left the confines of Downing Street after the result was to visit Buckingham Palace. The Queen and I discussed what had happened, and I explained why I had decided it was best for the country that I resign.

I had already spoken to Michael Gove that morning. I was on autopilot, calmly conceding defeat and offering my congratulations. He sounded more shocked than anyone. He was, many speculated, likely to become Britain's next chancellor.

I had also called Nicola Sturgeon. She was gracious but I knew the SNP would try to exploit the result. Scotland (like Northern Ireland) had voted to remain, yet because it was part of the UK, it would leave. We knew that this divergence, and the political pressure it would bring, was one of the dangers of Brexit, and we had warned about it during the campaign. It was essential to act in a way that reduced the risks, and I instructed Oliver Letwin, who was now mapping out an exit plan, to make sure Scotland's voice was properly heard in the negotiation process.

There were phone calls with the other first and deputy first ministers. I spoke to European leaders and to Obama. To each I said the same thing: 'I had a strategy to keep Britain in the EU. I executed the strategy. It didn't work. I'm sorry.' I told them that we would do what we could to keep Britain and Europe close together, and that my resignation might actually help rather than hinder that. I said I was sad to leave office, but even more sad that Britain would be leaving the EU.

I also received a text from Boris: 'Dave, I am so sorry to have been out of touch but couldn't think what to say and now I am absolutely miserable about your decision. You have been a superb PM and leader and the country owes you eternally.' I replied, 'Thanks, watched your speech, thanks for the kind words. Congratulations.' He was now the favourite to become the next prime minister.

That morning the pound slumped to a thirty-one-year low. It was part of the 'short-term shock' that Treasury analyses had predicted, and that George and I had spelled out in the lead-up to polling day. I felt no satisfaction in being right.

Britain's economy was strong, and our institutions were prepared for such volatility. Mark Carney made a statement that morning to calm the markets. His words were backed up by months of contingency planning between the Treasury, the Bank of England and the Financial Conduct Authority. He outlined the various measures that would be taken to keep the markets functioning, and worked closely with George to make sure the banks were adequately capitalised to deal with events. Thanks in part to what they did, by 28 June the FTSE was rising again. The following month it surpassed its 2016 peak.

There was criticism that the government had not planned for the medium- and longer-term fallout of a Leave vote. Where was the contingency plan for if we left? Wasn't its absence an act of negligence?

Well, we had set out the basic alternatives to remaining: a close partnership with the EU, like Norway; a Canada-style trade deal; or falling back on WTO trade rules. I would argue that subsequent difficulties were due in large part to failing to choose speedily between them. And proper contingencies would depend on our future relationship, which Leave had been careful to avoid specifying. If, for instance, we had begun preparing for a breakdown in relations with the EU – an outcome Leave repeatedly claimed wouldn't happen – we would have been attacked for fearmongering by precisely those who later attacked us for failing to plan.

The 'failure to prepare' narrative has its limits. Unpicking forty years of EU agreements around trade, agriculture, industry, energy, employment rights and many other things would take years, even decades. There was always going to be a limited amount that could be done in the run-up to the vote. And there were always doubts in my mind about whether any future government would really accept contingency work done by one that was foursquare behind staying.

The policy of the government was to remain in a reformed EU. I needed the civil service to be focused on that. Contingency plans would have sapped the effort – and if they leaked, would potentially have been disastrous.

Another risk we warned of during the campaign was how, as the UK was one country against twenty-seven, the EU held the stronger negotiating hand. It wasn't in its interest to make it easy for us. In fact, it was in its interest to make it hard – to deter others from following our move. And so it came to pass, as Jean-Claude Juncker sent a fierce letter around the Berlaymont on Tuesday, 28 June saying that the EU would only begin negotiations after Britain had triggered Article 50 – the mechanism by which a country withdraws from the EU. This was the start of an approach in which the EU only allowed talks to progress to the next stage once the previous stage had been agreed.

When it came to the Conservative Party leadership contest, I knew my responsibility. A departing prime minister shouldn't rush out of the door, nor should they cling to the railings. You are there to serve, whether that takes weeks or months, and not to interfere in your successor's appointment.

Party members would vote on the eventual two candidates after our MPs had whittled down the initial list of contenders through a series of votes. The chairman would need time to make sure the data we held on the membership was up to date and comprehensive. I was willing to continue in the interval, for example offering to attend the upcoming G20 in China if needed. I had said that a new leader should be in place by party conference. That gave us three months.

I knew Theresa May would run. I felt she had a good chance, and could be a good leader. I told Gavin Williamson he should go and help her, and was amazed by the speed with which he took over and ran her campaign. I believed that the toughness I had seen (and been on the receiving end of) would serve her well in Brussels.

However, I knew that the wind was in the sails of the 'Brexiteers'. That meant – and as I've said, this is where the polls pointed – Boris as PM and Michael Gove as chancellor. A couple of years earlier this would have concerned me much less than it did now. Michael had been a passionate reformer and a loyal friend at the heart of my team; Boris had been one of my star players, who I wanted on the pitch.

But to me they seemed to be different people now. Boris had backed something he didn't believe in. Michael had backed something he did perhaps believe in, but in the process had broken with his friends and supporters, while taking up positions that were completely against his political identity. Both had then behaved appallingly, attacking their own government, turning a blind eye to their side's unpleasant actions and becoming ambassadors for the expert-trashing, truth-twisting age of populism.

As it awaited its next occupants, Downing Street became an eerie place. Power was fading like a dimming lightbulb. I was still holding lots of meetings, but the number of attendees was dwindling. Civil servants often had somewhere else to be. Political advisers, all of whom would have to leave with me, were going off for interviews or joining leadership campaigns.

Prearranged commitments in my diary kept me busy, but I was beginning to feel like the political equivalent of *The Walking Dead*. I went up to Cleethorpes for the Armed Forces Day, and felt enormous pride watching the Red Arrows slicing through the sky. In the Downing Street garden we hosted a school performance of Shakespeare, marking four hundred years since his death. It was a medley of plays, including *Julius Caesar*. 'Dad, this could be about you,' Elwen said to me as we listened to those brilliant lines about the most famous political assassination of all. Then there was the hundredth anniversary of the Somme, when I stood in the shadow of the Thiepval Memorial reflecting on the heroism and sacrifice of those dark days. I visited the Farnborough Airshow, where I was surrounded by the cutting-edge kit that was born of our defence review.

Each of these events was a celebration of our country's bravery, creativity and might. It was as though fate was reminding me that Britain was great, and that we could weather whatever storms came our way. But the scale of the decision the country had just taken never left me. Just five days after the result I was back in Brussels for what would be my final European Council.

I saw Juncker on my own. He was sad, but also effusive. He kept saying that the British Army was Europe's best. That we had huge power in the world. That he and I were friends, and how sorry he was for me personally. He then said, 'Of course, I've got to say "No big deal for Britain," but I have to say these things to keep the European Parliament happy. I want to try and make this work.' Angela Merkel assured me we wouldn't have got any more in the renegotiation.

The main meeting of the European Council was warm. I gave an explanation as to why I thought we'd lost: 'The tragedy is this. We've had 180,000 net European citizens coming to Britain every year, but if it was 80,000 we wouldn't be leaving. So you wouldn't be losing the biggest defence payer, the third-biggest net contributor, the second-biggest economy, and one of your two permanent members of the UN Security Council. Essentially we're going because we haven't done enough to fix this immigration problem.' I finished, 'I don't know what you are going to do about it in the future, but whether we are inside the EU or outside the EU, we've got to fix this problem, because it is driving apart countries that should be together.'

The next day was my chance to speak to the party faithful (and not so faithful). The Conservative Summer Party was our annual black-tie fundraiser at the Hurlingham Club in west London. But while I was on stage, it felt as if the political spotlight was shifting elsewhere: to the audience, in fact, where the main contenders, Theresa May and Boris Johnson, were enjoying their meals as they waited to battle it out for the top job.

Yet expectations were upended once again the next morning when Michael decided *against* backing Boris – instead putting himself forward for the leadership. This would inevitably split the vote of Eurosceptic MPs, making things much harder for Boris.

It was all rather strange. Michael had publicly said he never wanted to be leader, and had told me privately in the past that he really meant it. But I could see how it had come about. When you work closely with a prime minister, as Michael had, you can start to think, quite naturally, about the aspects of the job, and which of them you'd do differently. And to be fair to Michael, I think the few days he had had in harness with

Boris had made him wonder if Boris as prime minister was such a good idea. But to be more rigorous, how could he possibly have chosen to back Boris without first deciding whether or not he was a suitable candidate for PM?

Back in my office with George, Ed, Kate and Craig, I prepared to watch Boris's leadership-campaign launch speech. Or at least, we thought it would be his launch speech. But there was another shocking turn of events: rather than going ahead and at least challenging Michael, he declared that he was withdrawing altogether.

George sat there beaming at the TV. 'We have taken Boris out. Now on to Port Stanley!' he said, meaning it was Michael's turn to fall next. He thought it was great news, and that the two of them were both dead in the water. I called Lynton, who said Boris had withdrawn because he knew he wouldn't get the necessary support any more.

I couldn't resist texting the former front-runner: 'You should have stuck with me, mate.' His reply was very Boris: 'Blimey, is he [Michael] a bit cracked or something? Great speech last night, everyone watched and thought we'd all gone insane to lose you and people were looking at me as if I was a leper, but you had eleven hard years of party leadership and six superbly as PM, more than I will ever do. Boris.'

I had seen strong qualities in each of the contenders. Boris had the charisma. Theresa had the competence. Michael had the conviction. But so often the winner comes through not because of their own talents but because of their competitors' failings. And for Michael, one shone through: disloyalty. Disloyalty to me, then disloyalty to Boris.

Since Boris and Michael had pressed the self-destruct button, that left just two women standing: Theresa May and the late entry (supposedly let down by Boris over what job she'd be given in his government) Andrea Leadsom.

I had got to know Andrea a little before she had become an MP because of her passion about improving early-years care for children. She had come to my surgery before she was even a candidate to talk to me about it. I had made her a minister, but was surprised to find her challenging for the leadership so quickly.

While all this rumbled on, I continued with my duties as outgoing prime minister.

I delivered the findings of the seven-year Chilcot Inquiry into the Iraq War. I prepared for a trip to Ghana, Kenya, and to Ethiopia in particular, to see how 0.7 per cent had changed people's life chances there. I also wanted to make good on my promise to visit Somalia, where we had done so much to deliver at least a modicum of stability. I attended my last-ever NATO summit in Warsaw (it was the only time I was to use the new RAF plane – an A330 – which had been converted for royal and ministerial use, and for that brief period it was dubbed 'Cam Force One'). This was the first time I had seen Obama in person after the referendum result. The speech he made was a genuine tour de force.

In another political upset, the maverick businessman Donald Trump had managed to secure the nomination as the Republican candidate for the US presidency. This was despite (or even perhaps, depressingly, thanks to) his protectionist, xenophobic, misogynistic interventions. He seemed unlikely to beat the Democratic candidate Hillary Clinton: the polls put him at five points behind. But hadn't our referendum taught us to expect the unexpected? Hadn't the success of the Brexiteers, as well as the rise of far-left and hard-right parties in Europe, shown us that anti-establishment, divisive, populist politics was the new normal?

Obama's speech drew on this. He identified Trump as being like Putin: a new style of leader, to whom the old rules didn't apply. He said the West needed to be clearer about its values, about what we stood for. And he warned us of the slippery slope that leads from criticising, say, gay people or immigrants to the demonising of whole nations, races and religions. Such rhetoric 'might be popular, but we have to fight it', he said. I couldn't have agreed more.

I spoke to my team about having a contingency plan, just in case the leadership contest ended sooner than anticipated. This was fortunate, because yet another contender decided to self-destruct. That weekend Leadsom had said in a *Times* interview that being a mother gave her 'a very real stake' in the future of the country. After a fierce backlash – how dare she use her opponent's childlessness against her? – she withdrew from the contest. I think the error, which was almost certainly clumsy rather than malicious, had persuaded her that she wasn't ready.

Theresa May was the last woman standing – and would therefore be crowned party leader and prime minister.

That Monday morning, all systems were go: I thought that the right time to leave would be after PMQs on Wednesday. In the car back from Farnborough, Ed got Theresa on the phone. I congratulated her. After reaching Downing Street I went outside and told the cameras about the developments. As I walked back to the black door I thought, 'On top of everything else, this bloody door isn't going to open, is it?' I hummed a tune to keep myself calm – which was inevitably picked up by all the microphones.

I felt relieved that I would be passing the responsibility to a safe pair of hands. And I felt pride in the fact that Theresa would be Britain's – and the Conservative Party's – second female prime minister. As I would taunt Corbyn at PMQs two days later, 'Pretty soon it's going to be 2–0.'

But I also felt pangs of sadness that it was all ending. We had thought we had three years; then we had three months. Now we had just three days.

Nancy was on a school trip in France. I called her and explained that she would have to come back early, as the date for my leaving office was set, and an appointment fixed with Her Majesty. She said she wanted to stay. 'Can't you tell the Queen to do it another day?' she asked.

Liz was also in France, at a wedding. She rushed back, and found Sam and me having a drink on the terrace that evening. Sam was very busy with the fashion business she was about to launch. She had been working hard for the past year, learning to sew and to cut patterns, and filling our dining room with dressmakers' dummies and fabrics. I would come up to the flat to find one friend or another standing on the kitchen table in Sam's latest creation as she fiddled with the hem. Life was moving on. There would be a new business, a new prime minister, a new path for the country.

For my final visit as prime minister I decided to travel to Reach Academy in Feltham, west London. The school was based in a former Jobcentre under the Heathrow flightpath, and had been set up just four years before, but was already providing its pupils with an outstanding education (a year later, its first cohort achieved the sixteenth-best GCSE results in England, with 98 per cent achieving passes in English and maths).

When we came to government in 2010, there hadn't been a single one of these Free Schools. I announced another thirty-one, which would take the total to five hundred. Some were even sending more children to Oxbridge than their independent equivalents. Political gravity had been defied. The best schools could exist in the poorest areas, and they could get the best out of their pupils, whatever their background. You didn't *need* private schooling or selection.

Afterwards, I wandered around the offices of No. 10 as the special advisers packed their things into boxes. Any achievements attributed to me – changing our party, communicating our message, winning elections, turning our economy around – belonged also to my closest team. By serving me, they had served Britain. I felt keenly how much these people had given up. All those holidays cut short, invitations turned down, late nights sitting in the blue glow of their computer screens in No. 10. I wanted to make sure everyone had options and plans – and that what they had achieved was acknowledged.

The UK honours system recognises 'people who have made achievements in public life and committed themselves to serving and helping Britain'. Any member of the public can nominate anyone for an honour. The civil service can make nominations, and frequently does. It's then up to the independent Honours Committee to decide who is worthy of an MBE, OBE or CBE. An outgoing prime minister can also nominate people. Past PMs had done so – Wilson, Thatcher and Major. I thought it was perfectly acceptable to take the opportunity. I knew I'd get flak for it, but then, that was exactly what my team had done for me over the years. Sports stars, actors, musicians and business people had reached the top of their game, made a huge contribution, and been recognised for it. Political advisers in Downing Street had reached the top of their game and made a huge contribution – so why shouldn't they be recognised? I don't regret it.

'The last supper', as it was called, was lasagne in the State Dining Room. This was for the people who had been there from the beginning. Consistent supporters like Tony Gallagher and Michael Spencer. The MPs who first supported my leadership bid, like Hugo Swire and Greg Barker. Advisers, some of whom had been with me for over a decade. Family friends who had been particularly supportive over the past decade.

At 1 a.m., after George and Kate had made speeches, I popped downstairs, where I found my private office still typing away at their computers, getting ready for the arrival of their new boss. I teased them that they'd already forgotten about me. 'It's going to be fine tomorrow,' I said – to reassure myself as much as them.

I was nervous for my last PMQs. I had lots of points to make, things to say, stories to tell, and I was worried I wouldn't get them all in. Over the years I had addressed 5,500 questions from the despatch box – how many I'd actually answered I would leave up to the opposition to decide,

I quipped. But after all those appearances, after a record ninety-two hours of statements, I would truly miss it.

And it was a tribute to that House that I ended with: 'I will miss the roar of the crowd and I will miss the barbs from the opposition, but I will be willing you on. When I say "willing you on", I do not just mean willing on the new prime minister at this despatch box, or indeed just willing on the government front bench and defending the manifesto that I helped to put together. I mean willing all of you on, because people come here with huge passion for the issues they care about and with great love for the constituencies that they represent. I will also be willing on this place. Yes, we can be pretty tough, and we test and challenge our leaders – perhaps more than some other countries – but that is something we should be proud of, and we should keep at it. I hope that you will all keep at it, and I shall will you on as you do.' I ended: 'The last thing I would say is that you can achieve a lot of things in politics. You can get a lot of things done. And that in the end, the public service, the national interest, that is what it is all about. Nothing is really impossible if you put your mind to it. After all, as I once said, I was the future once.'

My side – and a few on Labour's benches – stood to applaud. I'd said everything I wanted to say.

I went back to my office in No. 10 for one last time, and sat and waited for the end with Ed, Kate, George, Craig, my press secretary Graeme Wilson and, with a whole chair to himself, Larry the cat. He wouldn't be coming with me. Downing Street was his home.

Finally I walked down the long hallway from the back of the building to the front door. I was clapped out, just as I'd been clapped in. Just before we reached the door Liz stopped me, Sam and the children in the hallway, having learned from Thatcher's departure that tipping a PM straight out into the street after they've said goodbye to the staff is a recipe for tears. It was her final act of logistical and emotional genius, and it gave me just the right amount of time to gather myself.

I stepped into the street and spoke from the lectern. Florence stood coyly with her head poked between her mum and her sister. She had been nonchalantly talking about moving 'back to the old house', even though she'd never actually been there. 'They sometimes kicked the red boxes full of work,' I said, as I paid tribute to the children. 'Florence, you once climbed into one before a foreign trip and said, "Take me with you."' I looked at her and she started beaming. 'Well, no more boxes.'

Then, my last words in office. 'It has been the greatest honour of my life, to serve our country as prime minister over these last six years and to serve as leader of my party for almost eleven years, and as we leave for the last time, my only wish is continued success for this great country that I love so very much.'

With that I turned to my team, who were assembled outside the front of No. 11, and gave them a wave. As they headed to a pub on Trafalgar Square, I'd be at the other end of The Mall, seeing the Queen. After our conversation she invited Sam and the children in. We were worried about them bowing and curtseying properly, but they behaved impecc-ably. I was so proud of them all that day.

The 'old house' wouldn't be ready for us to move back into for some time, as it was rented out, so we ended up staying at my friend Alan Parker's house for a few nights before we found longer-term digs. It was an odd first evening rattling around in a strange place, rooting for the remote control. It was quiet, too. No duty clerks. No Liz. No red boxes. As I tucked Florence in she asked, 'Daddy, when are we going back home?'

There were still some commitments I was determined to keep. When a memorial is unveiled to fallen police officers, it is convention that prime ministers attend. The ceremony for PC Fiona Bone and PC Nicola Hughes, who had been murdered in the line of duty in 2012, was now in Theresa May's diary, but it would be impossible for her to travel up to Tameside in Manchester on her first day as PM. I was more than happy to go.

That evening I went to Chatsworth in Derbyshire for Patrick McLoughlin's party marking thirty years in Parliament. He was amazed I didn't cancel, but I hate to let people down. It was one of those orange July evenings, and after the speeches were over Liz and I sat on the grass by the river waiting for our helicopter, drinking a bottle of wine and reflecting on what an extraordinary time it had been.

Sam was spending the next few days going back to Downing Street to pack our things. She joked that I didn't want to help. I did – I'm very good at bubble wrap and boxes – but the truth is that I didn't want to go back to Downing Street. It didn't feel right.

So Sam was charged with packing up six years' worth of stuff on her own. I roared with laughter when she told me that when she had finished she put on some music, rolled a cigarette (in the stress of it all we had both started smoking again) and had a final dance around the kitchen

– just as the new residents, Theresa and Philip May, walked in. She showed them around and said we hoped it would be as happy a home for them as it had been for us.

I always said I would like to remain as Witney's MP even after my premiership had ended, and five days after I left Downing Street I returned to the House for the debate I had called on the renewal of Trident. It was the first time I'd been on the backbenches since 2003. Sitting there, I realised what a difficult existence being an ex-PM in the House of Commons was going to be. If you are silent, you are seen as brooding. If you speak, you are viewed as undermining your successor. In the modern age, when scrutiny of MPs is constant and they are expected to have, and to voice, an opinion on everything, it's very difficult for a former prime minister to stay on the backbenches. While I always hoped it would be possible, friends persuaded me that my continued presence would be bad for the new PM and miserable for me. I am sure they were right.

So on 12 September 2016 I announced my decision to leave the House. I would be temporarily appointed a 'Crown Steward and Bailiff of the Manor of Northstead' – yet another parliamentary anachronism through which an MP disqualifies him or herself from sitting in Parliament.

But what about prime ministers who vacate their seats – what are they called? In America, a president remains 'President' for life. In Britain, you don't. I think our way is better. In a democracy, political office is temporary and egalitarian. You enter as a civilian, and you return to being a civilian. Going from prime minister to plain old David Cameron overnight was a blunt but important reminder of that.

In many respects, I lament my political career ending so fast. I'd gone from MP to party leader to prime minister to private citizen in the space of just fifteen years. I was always glad I'd got into Parliament young, because I wanted to be there for a long time, and thought I would be. Instead, I was a former prime minister and a retired MP at the age of forty-nine.

In other respects, it wasn't at all brief. I was leader of the opposition for five years before I won the highest office, which meant that I held the job longer than anyone since Neil Kinnock. I was prime minister for twice as long as my predecessor Gordon Brown, and the longest-serving Conservative Party leader since Thatcher, and before her, Churchill. People commiserated with me for only getting to serve one year of the five I had won at the 2015 election, but I always knew I was going to struggle to make the full term after such a bruising campaign. Had it not been for the Brexit result I would probably have ended up being in power for eight years; instead I was there for six.

Six dramatic years for our country and for our world. The aftermath of the financial crash, the Eurozone crisis, withdrawal from Afghanistan, the Arab Spring, the Syrian tragedy, the invasion of Ukraine, the continued rise of China, the proliferation of technology and social media – this period was densely packed with momentous trends and events. I had crammed it fuller still, with wide-ranging reforms and, yes, referendums.

I look at my time in office as a 'before and after'.

I took over the oldest and most successful political party in the world when it was at one of its lowest ebbs. We'd suffered three consecutive election defeats. We had little to say on anything beyond crime, tax cuts and Europe. We still carried the 'rivers of blood', milk-snatching, poll-taxing, Section 28 baggage collected over decades. Just seventeen of our MPs were women, and only two were from ethnic minority backgrounds. Swathes of the population wouldn't go anywhere near us.

After my modernising mission, things did change. We had refined a progressive conservative doctrine that matched centre-right methods with the issues of the day. We had something to say on the things that mattered to modern Britain – indeed, we became the leading voices on tackling everything from poverty to climate change, gay rights to overseas aid. We now had seventeen BME MPs, sixty-eight women MPs, and a cabinet that reflected modern Britain.

Making the Conservative Party electable was the mission I was given when I took on the leadership. When I became prime minister, my central task was to turn our economy around. It wasn't so long ago that we had had the worst deficit in our peacetime history, were spending more on our debts than on our defences, and unemployment was over 8 per cent.

Now that I was leaving the deficit was two-thirds lower and the economy was growing faster than any other in the G7. More new businesses had been started than at any time in our history, and we were doing more trade with the fastest-growing parts of the world. Most importantly, employment hit a record high, and unemployment was at its lowest rate since the mid-1970s – under 5 per cent and falling.

I thought of the 2.5 million more people who were now in work.

Of the million more running their own businesses. Of the four million people who had paid income tax when we came to power who now didn't pay a penny. And of all those finally earning a National Living Wage.

When I first walked through the door of No. 10 as prime minister, it had seemed Britain's problems – from declining educational standards to ballooning welfare spending– were written in indelible ink. It seemed inevitable that poverty was passed down through the generations, and gay people were denied the rights straight people enjoyed. Patients who had waited months for operations were consigned to mixed-sex wards, and superbugs were rife in our hospitals. There was no pro-Union plan to deal with Scotland's slide towards separation. Terrorism was spreading, and global warming seemed unstoppable.

It was not possible to solve everything in just six years. But we did change the narrative. And we'd done that by helping to change life for millions of individuals. There were now 1.4 million more pupils in 'good' or 'outstanding' schools. Over 200,000 young people had completed National Citizen Service. Nearly three million had started apprenticeships. More young people and more disadvantaged young people were going to university. The result was a country that offered more opportunity to more people.

Some talk about 'austerity' and the difficult decisions we had to take on the economy. But what they don't mention enough is that despite all this, inequality and poverty fell. There were 500,000 fewer people in absolute poverty, including 100,000 fewer children. The rich were paying a higher share of tax than ever before. There were 500,000 fewer children living in workless households, the lowest figure on record. Thirty thousand same-sex people tied the knot in the two years after the law was changed. The elderly were better off than at any time in our history. We were determined to tackle the economic crisis in a different way to Margaret Thatcher, and in that we succeeded. The economy was undoubtedly stronger, and Britain was fairer, and more equal.

Hospital infections, mixed-sex wards, year-long waits for operations were off the front pages because they were largely out of our hospitals. The Cancer Drugs Fund was saving lives. Crime was down by over a quarter. Vast screeds of terrorist material were wiped from the internet and key hate preachers kicked out of the country. Our society was safer, and healthier.

The new respect agenda from Westminster towards the devolved governments was firmly established, bolstered by NATO and G8 conferences hosted in Wales and Northern Ireland. The regions became burgeoning powerhouses of manufacturing, tech and the arts. The lingering wrong of Bloody Sunday was addressed, and relations with our nearest neighbour the Republic of Ireland were at an all-time high. One sour note was the collapse of power-sharing in Stormont, which continues as I write.

Britain had gone all but coal-free. We were a renewables capital of the world, and more importantly had made solar and wind power financially viable. We were at the heart of the worldwide consensus on halting global warming. Our country was cleaner and greener.

The black hole in our defence budget had been addressed, difficult decisions made, and investment in new equipment was now on the up. Countless lives had been saved through our aid budget commitment. To give one example: we had paid for the vaccination of seventy-six million children worldwide against diseases like TB, measles, rubella, pneumonia and diarrhoea. The best estimate is that without that investment, 1.4 million of them would have died. I think of the lives those children have gone on to lead, and the chances they have been given. We were the second-biggest donor in the world during the Syrian tragedy, and led the response to most global crises, from Ebola to the migration crisis. It was undeniable: Britain was standing taller in the world.

Our party showed that you could do more with less. We proved in an increasingly polarised age that politics wasn't either/or – you could be pro-defence and pro-aid; pro-family and pro-equality; pro-public services and pro-fiscal prudence. We demonstrated that you could take the difficult decisions *and* win elections – and that a government could achieve a lot in just six years.

We also showed that coalition could work in this country. We took over a hundred Bills to assent, and implemented a huge programme of reforms, including putting our economy back on course.

When I look at this record, there are some key themes that shine through for me. We kept our promises. Politicians are often accused of breaking their word, but we said we'd honour our UN aid commitment and our NATO defence pledge. Others didn't. We did.

We confronted some of the longest-running sores in our politics. Again, there is a perennial complaint that politicians just kick cans down roads. But by sorting out our defences, building a long-term solution for university funding, addressing the Scottish question and voting reform, we grasped nettles that had been neglected for decades.

We also looked to the future. Whether it was building new railway lines, reforming public sector pensions or renewing Trident, we always had the horizon in mind. Politicians are known for quick fixes and easy wins; we made a beeline for the difficult things. Many will come to fruition long after we're gone.

We gave power away – lots of it. 'Empowering' might be Whitehall jargon, but it applies to so much of what we did. We empowered local groups, through the Big Society, to take over pubs and run their own community centres. We empowered teachers, doctors, council chiefs and prison governors, through our public service reforms and our drive towards localism. We empowered cities and an entire Northern Powerhouse, as well as Holyrood, the Senedd and Stormont, which got greater financial and tax-raising authority. Scotland now has one of the most powerful devolved parliaments in the world. And yes, we empowered people, by giving them the ultimate say on the most important questions.

In fact, the decision to hold a referendum on membership of the EU encompasses all these trends. It was both a promise kept and a sore confronted. It was the ultimate display of handing power to the masses, and it had the long term at its heart.

As I've said, I do not regret holding the referendum. But I deeply regret the result, and I still think Brexit is the wrong path for our country. The best deal for Britain was the one we had, the one I renegotiated. Anything else does not give us the best of both worlds that my deal built on. Most worryingly, our departure, whatever form it takes, jeopardises all the things we've achieved, from the strength of our nation's finances to the stability of our United Kingdom.

We *did* warn that that was the price of a Leave vote. I don't buy the argument that people didn't know what they were voting for, or that misinformation propagated during the campaign means that the result should be annulled. Yes, there was fake news. But there was also the biggest distribution of a leaflet in recent British history. There was a ubiquitous campaign led by the most well known and respected voices. The view that people aren't qualified to vote on an issue that is so complex and important is one that I do not share. The act of electing our MPs is complex and important, but we think people are able to make their minds up on that. So they are competent enough to make their mind up on an issue like EU membership.

The simple truth is that, whatever arguments we on the Remain side put forward, immigration had been too high for too long, and people felt that leaving the EU was a way to do something about it, while simultaneously dealing with an organisation that had, in the view of many, become too big, too bossy and too interfering. The irony of that fact isn't lost on me. The test of our success in government would be whether we could rescue our economy. We achieved that. But in doing so we sparked a jobs boom which attracted record migration to the UK. Our successes really do hold the seeds of our defeat.

I resent the accusation that I had somehow dreamed up the idea of a referendum; that it was an election gambit, and only about the Conservative Party.

I set out the case in the Bloomberg speech in January 2013, more than three years before holding the referendum. There was broad support for the idea, which found expression in the result. It is odd to argue that because a majority decided they didn't want to be in the EU, this proves we should have stayed in without asking.

Yes, the Conservative Party has long been deeply split over Europe, but it is an issue that has vexed every political party, and the British people themselves. How else can you explain the fact that every UK-wide party made a commitment to a European referendum at some stage between 2005 and 2015? (Labour fought the 2005 election on a pledge to hold a referendum on the proposed EU constitution. The Lib Dems backed an in/out referendum in 2010. The Greens backed one in their manifesto in 2015. And UKIP – unsurprisingly – backed a referendum at every election during the period in question.)

The fact remains that, ever since 1975, our membership of the EU had been framed in terms of a plebiscite. By the 2010s, people backed the idea of holding a referendum in polls by a ratio of three to one. By voting Conservative in 2015 they voted for a referendum. In the event, nine out of ten MPs in Parliament voted for the Referendum Bill in June 2015.

I don't argue that the referendum pledge had nothing to do with party politics or public pressure; of course the views of MPs and electoral results were considerations. But so was the genuine problem of trying to get the right settlement for the UK. Are critics really saying that public concerns and political pressure should play no part in these things? Surely that's part of what politics and democracy are: listening to people and responding to their grievances.

Again, when looking at the EU debate I hope people will come to see it in the wider context. Britain has always been different, always been carving itself out of things rather than throwing ourselves head-first into them. Remember: Thatcher had got us the rebate. Major (and Brown) had kept us out of the euro. Even the arch-Europhile Blair carved us out of justice and home affairs powers. I had got us out of bailouts, treaties and further powers going to Brussels. But the rip-tide dragging powers from Britain to Europe had become too strong. Being in the single market, but out of the euro, made that current even stronger and more dangerous, particularly as the euro lurched into a long-running crisis. The only way to deal with this was to be radical, to entrench our special status, and to extract ourselves from certain elements – in particular the goal of ever-closer union. We needed to repatriate powers we'd lost, and to do so with the ultimate goal of a referendum. Yet what I achieved evidently wasn't enough to convince people.

However, I maintain that while leaving the EU is not *my* choice, it's a *legitimate* choice for the sixth-biggest economy in the world, and something that, like everything else in our history, we can make the best of. Instead of being *in* the EU but *out* of the bits we really didn't like, such as the single currency, we can be *out* of the EU but *in* the things we need, such as a strong trading relationship and security cooperation. Instead of being reluctant tenants, we must become contented and cooperative neighbours. That is how I see it.

It was a letter from Tony Blair that reassured me as I took office, and in the days after I left it was a missive from his predecessor, John Major, that gave me succour, particularly his words on the wisdom of calling a referendum.

Politics is a brutal business, and there are only a very few who will understand so profoundly how you and Samantha will be feeling today.

Although the last few weeks must have been truly awful, you have carried yourself with a dignity few could have matched. And you have a great deal to be proud of: saving the economy; reforming education and health; meeting our foreign aid obligation for the first time; enabling gay men and women to achieve a legal status they have long deserved; and much more.

You have also won two elections; led the first successful coalition in [peacetime] for 100 years; changed the perception of the Conservative Party; and brought forward far more women into the parliamentary party. I know of no one else in our party who could have achieved all this.

I am sure that the referendum result will be weighing heavily on you, but let me offer some comfort. I believe history will see this in a much wider perspective than everyday scribes. History will acknowledge that a referendum was becoming inevitable – if not now, later – and delay would have made Brexit more likely.

I concur with Sir John on the inevitability of a referendum. And I also believe in the inevitability of a reckoning for the EU. The issues have not gone away. Indeed, I see European leaders fighting the same battles I fought. In choosing the new Commission president, presidents and prime ministers have abandoned the *Spitzenkandidat* system. Meanwhile, new proposals come forward for a European army, and other EU leaders baulk at them. The European Commission demands that Italy revises its budget and threatens sanctions. Once again, the answer in Brussels seems to be 'more federalism, more integration, more Europe'.

I knew at the time how important it was to try to change things – that the problems were so big they demanded a big, bold response. If we didn't, it would only result in people flocking to the extremes.

'Populism' might not have been a common term until the end of my premiership, but as a force it was there all along, playing to people's grievances and pulling them towards parties that would, in my opinion, only inflame matters.

The economic grievances that followed the financial crash, and the cultural grievances that resulted from unprecedented levels of migration, affected most of the Western world. At least Britain was a country with a government that was admitting them, confronting them, and trying to assuage them by creating a fairer economy and a stricter immigration system. Rather than being batted around by these forces, we were standing strong against them, or at least trying to. The EU referendum was the biggest example of that.

And surely that is a vital part of what democratic politics is about. It is about representing your people, responding to their objections – regardless of the risk to your own position.

The prevailing narrative today is that liberal democracy, once proclaimed the winning ideology of the twentieth century, is now in retreat. You can see why this argument has gained traction when so many so-called 'strongman' leaders are arguing – and in some cases demonstrating – that you can succeed by ditching the building blocks of democracy, such as the rule of law, a free press and open economies.

I disagree. Ultimately, the success of a nation still depends on the freedom of its people. The yearning for democracy is stronger than ever. In a world that is richer and more connected than ever before, we have more power than ever before to deliver accountable government. That is why, even in these uncertain times, I remain optimistic. I still think our best days lie ahead of us. Indeed, I am dedicating some of my post-political life to helping countries build real, representative democracies with a robust rule of law. After all, that is the only thing that guarantees prosperity, security and opportunity for the long term.

It is my ambition to write to every future prime minister, just as my predecessors wrote to me. I think about doing so well into my old age, and wonder who will be sitting at that desk and opening the envelope. Maybe someone who was elected as an MP on those memorable election days in May 2010 or 2015. Maybe someone who went to a Free School or did National Citizen Service. Maybe someone who came to this country as a Syrian refugee. Or someone who walks up Downing Street with their partner – husband and husband, or wife and wife – as proudly and nervously as Sam and I did in 2010.

Whoever they are, I will tell them this. That Britain is the greatest country on earth. Our greatness is derived not from our size, but from our people – their decency, their talent, and that special British spirit. There is no need for new ideology or systems, we have the best one here: democracy. We are lucky that this political system enables politicians to act upon what I think motivates most of them: the national interest and public service. And if you listen hard, beyond the sound and the fury, you will hear that this quiet patriotism and belief in democracy, public service and the national interest is what unites people too. Remember that as you pick up the baton and lead. I will be willing you on as you do.

Index